# Impacts of Mobile Use and Experience on Contemporary Society

Xiaoge Xu
*Mobile Studies International, Singapore*

A volume in the Advances in Human and Social
Aspects of Technology (AHSAT) Book Series

Published in the United States of America by
    IGI Global
    Information Science Reference (an imprint of IGI Global)
    701 E. Chocolate Avenue
    Hershey PA, USA 17033
    Tel: 717-533-8845
    Fax: 717-533-8661
    E-mail: cust@igi-global.com
    Web site: http://www.igi-global.com

Library of Congress Cataloging-in-Publication Data

Names: Xu, Xiaoge, editor.
Title: Impacts of mobile use and experience on contemporary society / Xiaoge
  Xu, editor.
Description: Hershey, PA : Information Science Reference, an imprint of IGI
  Global, [2019] | Includes bibliographical references.
Identifiers: LCCN 2018037297| ISBN 9781522578857 (hardcover) | ISBN
  9781522578864 (ebook)
Subjects: LCSH: Internet--Social aspects. | Mobile computing--Psychological
  aspects. | Mobile communication systems--Use studies. | Life cycle, Human.
Classification: LCC HM851 .I454639 2019 | DDC 004.01/9--dc23 LC record available at https://lccn.loc.gov/2018037297

This book is published in the IGI Global book series Advances in Human and Social Aspects of Technology (AHSAT) (ISSN: 2328-1316; eISSN: 2328-1324)

# Advances in Human and Social Aspects of Technology (AHSAT) Book Series

Ashish Dwivedi
The University of Hull, UK

ISSN:2328-1316
EISSN:2328-1324

## MISSION

In recent years, the societal impact of technology has been noted as we become increasingly more connected and are presented with more digital tools and devices. With the popularity of digital devices such as cell phones and tablets, it is crucial to consider the implications of our digital dependence and the presence of technology in our everyday lives.

The **Advances in Human and Social Aspects of Technology (AHSAT) Book Series** seeks to explore the ways in which society and human beings have been affected by technology and how the technological revolution has changed the way we conduct our lives as well as our behavior. The AHSAT book series aims to publish the most cutting-edge research on human behavior and interaction with technology and the ways in which the digital age is changing society.

## COVERAGE

- Philosophy of technology
- Activism and ICTs
- Information ethics
- Technology Dependence
- Computer-Mediated Communication
- Human-Computer Interaction
- ICTs and human empowerment
- Technology Adoption
- Technology and Freedom of Speech
- ICTs and social change

IGI Global is currently accepting manuscripts for publication within this series. To submit a proposal for a volume in this series, please contact our Acquisition Editors at Acquisitions@igi-global.com or visit: http://www.igi-global.com/publish/.

# Titles in this Series

*For a list of additional titles in this series, please visit: www.igi-global.com/book-series*

701 East Chocolate Avenue, Hershey, PA 17033, USA
Tel: 717-533-8845 x100 • Fax: 717-533-8661
E-Mail: cust@igi-global.com • www.igi-global.com

# Editorial Advisory Board

# Table of Contents

### Section 1
### Investigating the Impacts of Mobile Use

**Section 2**
**Investigating the Impacts of Mobile Experience**

# Detailed Table of Contents

## Section 1
## Investigating the Impacts of Mobile Use

*Mobile is used by people of different age groups from the youngest to the oldest in different countries. It is also used in different areas for different purposes. Its reach and penetration are expedited and exponential. Its uses and impacts are multi-dimensional and multi-directional, being constantly examined from different disciplinary perspectives using different methods, generating growing amount of scholarship on how mobile has been used and its impacts on both actors and activities. In this section, we have nine chapters to provide most recent research findings to serve as an updated account of the renewed academic efforts in continuously describing, explaining and predicting the impacts of mobile use on actors and activities in the world. As far as actors are concerned, this section covers adolescents, college students, working professionals, and retirees, involving both males and females in different countries. In terms of activities, this section involves use of mobile for academic studies, personal entertainment, dating, food posting, mobile social media, and use of mobile apps. In terms of approaches, this section highlights interdisciplinary and integrated methods. In terms of countries involved, this section covers Australia, China, Colombia, India, Oman, Pakistan, New Zealand, Turkey, UK, and US.*

# Chapter 1

Equipped with psychological theories of development, the authors of this chapter argue that the impacts of mobile technology should be situated in and understood within their different contexts of use. Besides their advocacy of a situated and integrated approach to understanding the psychological determinants of mobile use and the impact on development, the authors also call for increased interdisciplinary and integrated working to capture the complexity of technology and behaviour and the recognition of individual differences.

# Chapter 2

The chapter explores the role of smart phones and mobile apps in the process of third age formation in Zhengzhou, a second-tier city in China located in central Henan province. The term 'third age' refers to a transition period from active work to retirement. Compared with the previous generation, the demographic approaching retirement in China today is more digitally literate, although this varies accordingly in Zhengzhou, a second-tier city. The use of digital technology offers people a different kind of retirement. This study shows that an increasing number of people around retirement age (55-65) in Zhengzhou are using smart phones and apps to reimagine the possibilities of post-work lifestyles. The research asks if the use of mobile apps is changing peoples' perspectives on traditional responsibilities and peoples' expectations of retirement.

# Chapter 3

As an indispensable part of their lives, mobile has been playing an important role among undergraduate students in influencing their lives and academic studies. Although gender has been proven to be an important factor in distinguishing males and females in using the internet and other digital media, what are the gender differences in mobile use and habits? The authors of this chapter offer an updated answer to that question by surveying a selected sample of undergraduate students in Turkey.

In addressing the question as to whether mobile dating is a fancy way to fall in love, the author of this chapter ignited further interest in examining the issue by reviewing up-to-date studies on mobile dating, examining mobile dating applications and individual users' motivations at different stages of mobile dating. Furthermore, the author also included LGBTQ studies in her investigation. Finally, the author identified various problems related to mobile dating that invite further examination and solutions.

To post food on social media has become a frequent source of fun and joy in life for many mobile users. In investigating such a common scene on Instagram among its young users, the authors of this chapter investigated the relationship between social activity, personal traits like narcissism and shyness, and uses and gratifications from posting food photos on Instagram. Uses of Instagram for posting selfies were also examined for comparison. Results showed that while posting food photos were associated with social activity, posting selfies were associated with shyness. Narcissists were more likely to involve in posting both food photos and selfies. Implications of the results in explaining the generation of visual contents on social media are discussed.

## Chapter 6
*Hafidha S. AlBarashdi, The Research Council, Oman*
*Abdelmajid Bouazza, Sultan Qaboos University, Oman*

Combining a survey and focus groups as a mixed-methods research, the authors of this chapter examined the functions, types, and motivations of smartphone usage and gratifications. Furthermore, the authors also investigated the rates, symptoms, and reasons of smartphone addiction. Still another achievement of the authors was to look into the relationship between smartphone usage, gratifications, and addiction with academic achievements among college students. On top of identifying three levels of addiction, the authors also located distinctive traits of these levels. This chapter provided an interesting example of how mixed-methods research can be employed to investigate mobile use, gratifications, and addiction. It is expected from the editor that this chapter would lead to more comparative studies between or among countries or cultures using by using a mixed-methods research to triangulate and/or complement findings of using different research methods. The inclusion of this chapter in this volume is also meant to invite further studies to investigate the gap between what mobile users want and what they actually get from using mobile as well as its related experience on both the normative and the empirical sides.

## Chapter 7
*Sadia Jamil, University of Queensland, Australia*

Through examining use of mobile in Pakistan's Sindh province, the current chapter presents a unique and interesting case of the socio-economic impacts of mobile use on users' lifestyles. Although there exists an obvious divide between urban and rural areas in terms of impacts of mobile use, the case of Pakistan could serve as an alert to scholars that why mobile use remains limited in narrowing the gap between urban and rural areas against a backdrop of mobile being widely believed to be able to play a big role in narrowing the social and economic gap between urban and rural areas. The author of this chapter found that mobile use was also gender-biased in rural areas, resulting in a gap between males and females as far as social and economic impacts of mobile use on their lifestyles.

## Chapter 8
*Li Zhenhui, Communication University of China, China*
*Dai Sulei, Communication University of China, China*

China is well known for its wide and increasing commercial use of mobile social media for various purposes in different areas, ranging from online shopping to social networking. Such a popular commercial use was insightfully examined in relation to social relationship in the age of mobile internet, which enables people of either weak or strong connections to socialize anywhere anytime, leading to scenarios where mobile social media can be leveraged for profits. In what way can user experiences be guaranteed while platforms' value-added targets be achieved at the same time? In addressing that question, the authors of this chapter examined the commercial use of mobile social media in the context of complicated social networks. It is expected from the editor that further studies are to be carried out to comprehensively and comparatively examine the same topic in different countries or cultures.

### Chapter 9
*Biplab Lohochoudhury, Visva-Bharati, India*

With a bird's-eye-view of the journey of exploring mobile use and its impacts in India, one of the biggest mobile markets in the world, the author of this chapter provides an analytical and insightful review of earlier studies on mobile communication development and utility-driven usage. Although country-specific, this chapter provides a very interesting, insightful, and invaluable model to the world, that is, the model of creator-audience matchmaker. As demonstrated by the two cases in this chapter, the creator-audience matchmaker model can be used as a heuristic tool for understanding mobile app development and success in India.

### Section 2
### Investigating the Impacts of Mobile Experience

*Behind each mobile use lies mobile experience as each use of mobile generates each unique mobile experience, which can be defined as an outcome of actors' mobile use in mobile-related activities. The seven chapters in this section deal with seven different topics with one common theme, that is, mobile experience and its impacts.*

### Chapter 10
*Danielle McKain, Robert Morris University, USA*

As the world is moving towards experience economy, consumers are paying more and more attention to memorable and fun experience beyond a product or service. Learners are the same, especially when learning goes mobile. Mobile learning has been examined in different areas ranging from forms and formats to features and functions. Mobile experience in learning, however, has not yet fully examined. After identifying mechanisms to measure and evaluate mobile learning experience, this chapter reviewed what mobile learning resources could be leveraged to enhance mobile learning experience, followed by recommendations for further studies.

### Chapter 11
*Rong Hu, University of Nottingham – Ningbo, China*
*Xiaoge Xu, University of Nottingham – Ningbo, China*

In the age of experience economy, students are also attaching greater importance to enlightening and entertaining experience in learning. Earlier studies on mobile learning experience have generated increasing amount of knowledge on experience-oriented learning. Through reviewing earlier studies on mobile experience in learning Chinese as a second language and recommending dimensions and directions for further studies, this chapter aims at inviting further studies on how to enhance mobile learning experience in general by applying the six stages of mobile experience, assisted with the 3M (mapping, measuring, and modeling) approach with a focus on the gap between the normative and the empirical dimensions of mobile experience in learning Chinese in the context of Confucius Institutes around the world.

# Chapter 12

After closely examining the experiences of mobile translation in which people engage with translation on mobile platforms in the contexts of healthcare, crowdsourcing, and machine and translator training, the authors have identified a tightly intertwined relationship between mobile translation and machine translation. They have also found that the technological side is more dynamic than the user side in the case of mobile translation and machine translation, which may lead to a gradual reduction of people learning foreign languages and a possible loss of professional translators and language specialists. When it comes to contextual and textual translation, however, human translators currently outperform mobile or machine translators. Although human contribution will be determined by translation scenarios or specific translation tasks, the human-mobile/machine interaction in translation deserves further studies. It is imperative to compare mobile use and experience in human-mobile interaction related to translation in different cultures or countries so as to locate similarities and differences. Furthermore, it is also expected from the editor that further studies should focus on mapping, measuring, and modeling those identified similarities and differences.

# Chapter 13

Mobile has proven to a most successful tool for bottom of pyramid markets as it is the most affordable means to generate utilities for mobile users of lower income strata of pyramid and to provide them with quick and low-cost access to information, government systems, business opportunities, access to education, and health. This chapter presents the landscape of contemporary experience through mobile phones in the field of economic value creation, social value, and health among the bottom of the pyramid in India.

# Chapter 14

Defining mobile experience as a process of usage, affordance, roles, and impacts of mobile phones during wars and conflicts, the author has located similarities and differences in mobile experience in war and conflict reporting among professional journalists, citizen journalists, governments, militaries, rebels, NGOs, activists, and communities. After exploring the changes and trends related to mobile experience in war and conflict reporting, the author also offered specific dimensions and directions for further studies.

Mobile has become a mainstream medium for news consumption on the go. To cater to the growing demand for mobile news, traditional news providers have switched from "mobile too" to "mobile first" strategies. To enhance mobile news communication, it is imperative for mobile news providers to stay abreast of mobile news consumers' changing expectations of mobile news experience in a news app. It is equally imperative to identify the gap between news consumers' expectations and what mobile news experience is embedded in a mobile news app. Using a mobile experience index, the authors of this chapter have located the venue and extent of the gap through conducting a survey of mobile news app users and a comparative analysis of indicators of mobile news experience in selected news apps.

# Chapter 16

After a comprehensive review on mobile tourism experience, the authors have identified the benefits and impacts of mobile use to tourists and their experiences. Besides locating similarities and differences in using mobile for tourism, the authors have confirmed that mobile empowers tourists to get more from their vacations and to have more flexible planning, resulting in satisfaction and accomplishment. This chapter enlightens tourism operators, among other stakeholders, on the opportunities for contextualized mobile advertising, which would attract and convert tourists into potential customers.

# Foreword

As you read these words you are all reading a book, *Impacts of Mobile Use and Experience on Contemporary Society*, but the way you experience this edited and timely book differs. Clearly, our experience of any form of text depends on our previous knowledge and how we interpret and make sense of all the new information, arguments and conclusions that we are all exposed to. However, our differences in experiences do not stop there. Possibly, you read a printed copy while sitting in an armchair, and have disconnected from all of your personable, portable and connected technologies. Alternatively, you have tuned to one such technology to experience this book, like a laptop, a tablet or your smartphone. This choice influences your experience. While we nowadays take mobile technology for granted (Ling, 2012), mobile *phones* and its successor the smartphone have not always been around.

Imagine for a moment what your life would be without your smartphone. Imagine your routines as you wake up in the morning, and the role your mobile device plays in these experiences. Imagine your commute to and from school, university or work, and what your mobile device enables you to experience and do during this time. As we continue, we can imagine and reflect on the diverse roles our mobile devices play in our everyday lives, and for society as a whole. Originally designed for connecting people via calls, mobile phones have morphed into portable computers that are ubiquitously available. The development and global distribution of smartphones, connected to evolving mobile ecosystems of customized applications, have made possible for a multitude of uses and experiences. People have developed frequent habits of mobile use, and these have become an integral part of how we communicate with others (Ling, 2012), to how the news media produce mobile news (Westlund & Quinn, 2018) that we may access and experience to learn about the world around us (Westlund, 2015). The plethora of ways in which we use, and rely on, our mobile devices nowadays are fascinating. Many do not even reflect on their technological dependence of mobile media and communication in their everyday life. This is strikingly different compared to the years preceding the advent of contemporary smartphones with touchscreens. For example, an extensive focus group study with young adults carried out 2006 in Sweden revealed a strong unwillingness to make use of their mobile *phone* for other purposes than mobile communication. They actually appreciated those moments on the go when they could not get connected to the internet or access other news media sources. At the same time, some were interested in significantly expanding their mobile use (Westlund, 2007). This study, published immediately before the launch of the iPhone, was featured in one of the first books gathering social scientists around the topic of mobile media.

Emerging as a subfield within media studies as well as communication, research into mobile media and communication has grown markedly over the last decade. During the early 2000's there were relatively few social scientists researching mobile communication, but with the diffusion of smartphones there has been a growing interest for research among policy makers resulting in public inquiries (Gómez-Barroso et. al. 2010), and also amongst communication and media studies scholars.

Some key milestones throughout the 2010s include, but are not limited to, special issues guest edited by Naomi Baron in the leading journal *New Media & Society* in 2010, and in 2011 by Rich Ling and Heather A. Horst. In 2013 the journal *Mobile Media & Communication* was launched, and the same year Chumskey and Hjorth (2013) published the edited book *Mobile Media Practices, Presence and Politics. The Challenge of Being Seamlessly Mobile.* In the following years more important edited handbooks were published, including *The Routledge Companion to Mobile Media* (Goggin & Hjorth, 2014); *Interdisciplinary Mobile Media and Communications: Social, Political and Economic Implications* (Xu, 2014); *Emerging Perspectives on the Mobile Content Evolution* (Aguado, Feijóo & Martínez, 2015); and *Handbook of Research on Human Social Interaction in the Age of Mobile Devices* (Xu, 2016). In the subsequent years, research into mobile media and communication has continued to develop and grow in significance. As we enter 2019 mobile media and communication has established itself as a natural part of everyday life and in the routines of diverse organizations around the world. Moreover, mobile media and communication continues to have an important position in the scholarly field. For example, this year comes with the publishing of the *Handbook of Mobile Communication,* edited by Ling, Lim, Fortunati and Goggin (2019, forthcoming). As is common with handbooks, it covers a broad range of themes and topics relating to mobile communication around the world, and mostly summarizes and reflects on existing research. Handbooks have a fundamental value for students and scholars who want an entry point into the field, and learn what is currently known about different topics. This book on the other hand, more deliberatively seeks to advance the research frontiers in the more specific area of mobile experiences and uses.

This book edited by Xiaoge Xu marks a welcome contribution to this field of research. The book offers eighteen chapters. This includes the introductory chapter (providing a research framework) and the final chapter (discussing dimension and directions for future research), both authored by the editor. The remaining sixteen chapters divides into two sections: 1.) impacts of mobile use, and 2.) impacts of mobile experience. The first section presents the readers with nine articles that covers a vast terrain of studies into mobile uses, ranging in focus from adolescence, third agers and gender, to behaviors related to mobile dating, food posting via Instagram, as well as other studies in the intersection of mobile and social media. This section also contains articles dwelling into socio-economic impacts, mixed methods, and creator-audience matchmaking, respectively. The second section contains seven articles that focuses on impacts of mobile experience, with chapters exploring such inquires in the salient cases of mobile learning, translation experiences, journalism, and tourism. While covering developments from several corners of the world, like other books, it predominantly contains analyses from Asian countries. This is a most welcome feature of this book, and thus helps expand and deepen the geographical gaze. Ultimately, this book is set to bring further fuel to the continued developments of mobile media and communication as a research area.

Moreover, I encourage researchers to adopt a holistic and sociotechnical approach in their future study of experiences in relation to mobile media and communication, researching different agents involved. Keeping a focus on audiences is important, but also the social actors operating and developing content and communication via mobile platforms, as well as the mobile platforms as such (i.e. technological actants). Scholars can and should consider the overall interplay of actors, actants and audiences in diverse activities (e.g. Lewis & Westlund, 2015) relating to mobile media and communication activities, as has been done in recent review of research into mobile news and mobile journalism (López-García, Silva Rodríguez, Vizoso, Westlund and Canavilhas, 2019). Ideally researchers should engage across disciplines and engage in multiple methods, as suggested by Xu in this book, to get a more comprehensive understanding of how actors and audiences experience the various affordances, services and content that mobile media put on offer.

*Oscar Westlund*
*Oslo Metropolitan University, Norway*

## REFERENCES

Aguado, J. M., Feijóo, C., & Martínez, I. J. (2015). *Emerging perspectives on the mobile content evolution*. IGI Global.

Chumskey, K., & Hjorth, L. (2013). *Mobile media practices, presence and politics. The challenge of being seamlessly mobile*. New York: Routledge.

Goggin, G., & Hjorth, L. (2014). *The Routledge companion to mobile media*. New York: Routledge. doi:10.4324/9780203434833

Gómez-Barroso, J. L., Compañó, R., Feijóo, C., Bacigalupo, M., Westlund, O., Ramos, S., … Concepción García-Jiménez, M. (2010). *Prospects of Mobile Search*. European Commission, JRC, Institute for Prospective Technological Studies (IPTS), EUR 24148 EN 2010.

Lewis, S., & Westlund, O. (2015). Actors, Actants, Audiences, and Activities in Cross-Media News Work. *Digital Journalism*, *3*(1), 19–37. doi:10.1080/21670811.2014.927986

Ling, R. (2012). *Taken for grantedness: The embedding of mobile communication into society*. Cambridge, MA: MIT Press. doi:10.7551/mitpress/8445.001.0001

Ling, R., Lim, S. S., Fortunati, L., & Goggin, G. (Eds.). (2019). Handbook of mobile communication. Oxford, UK: Oxford University Press. (forthcoming)

López-García, X., Silva Rodríguez, A., Vizoso, Á., Westlund, O., & Canavilhas, J. (2019). Periodismo móvil: Revisión sistemática de la producción científica/Mobile journalism: Systematic literature review. *Communicar*, *59*, 2019–2. (forthcoming)

Westlund, O. (2007). The adoption of mobile media by young adults in Sweden. In Mobile Media 2007 (pp. 116-124). The University of Sydney.

Westlund, O. (2015). News Consumption in an Age of Mobile Media: Patterns, People, Place and Participation. *Mobile Media & Communication*, *3*(2), 151–159. doi:10.1177/2050157914563369

Westlund, O., & Quinn, S. (2018). Mobile Journalism and MoJos. In H. Örnebring (Ed.), *Oxford Research Encyclopedia of Communication*. Oxford, UK: Oxford University Press. doi:10.1093/acrefore/9780190228613.013.841

Xu, X. (Ed.). (2014). *Interdisciplinary mobile media and communications: Social, political and economic implications.* Hershey, PA: IGI Global.

Xu, X. (Ed.). (2016). *Handbook of research on human social interaction in the age of mobile devices.* Hershey, PA: IGI Global.

# Preface

"Mobile is reshaping not only who we are but also what we do in this mobile world" (Xu, 2016). I repeated that remark since 2016 on various occasions besides its presence on the homepage of the website of *Mobile Studies International* (http://msi.wiki). It has been proven to be true by abundant evidence around the world. Equipped with, among others, artificial intelligence, mixed reality, big data, cloud computing, and machine learning, mobile is becoming more popular and powerful than ever, reshaping to a greater extent who we are and what we do in this mobile world.

Mobile is reshaping us, including the way we talk, the way we communicate, the way we live, and the way we see our world. Mobile is used more for text chatting, voice recording or live video talks than phone calls. Mobile is the window to the world for most of us as we seek and follow news more on mobile than the print or electronic media. It is reshaping our way of working as mobile is more used as a personal working station, where we email, chat, talk and work anytime anywhere cross time and space barriers of all kinds. Mobile is enabling us to work from home, from café, from bus stations or even from inside cabs.

Use of mobile is bringing about diversified mobile experience to us, enabling us to enjoy fun, enjoyable and memorable memories and senses in mobile parenting, mobile dating, mobile payment, mobile shopping, mobile healthcare, mobile storytelling, mobile advertising, mobile marketing, mobile journalism, mobile music, mobile gaming, mobile filmmaking, mobile e-sports, and mobile entertainment. All these activities enriched by mobile use have left us enticing, entertaining, engaging, empowering, enlightening, and enhancing experience.

As mobile use and mobile experience are closely and tightly interconnected, interdependent and even interwoven to such an extent that it is hard to separate them. Previous studies, however, have largely examined them separately as indicated by the existing literature. To remedy this situation and to invite further studies along this new direction of integrating mobile use and mobile experience in investigating their impacts on actors and activities, the editor of this volume has designed a research framework of a proposed 3M approach (mapping, measuring and modeling) to investigate both mobile use and mobile experience at the normative and empirical levels as well as the gap between the two (Xu, 2018). The proposed framework presented in the first chapter of this volume is designed to serve as a guideline as well as an invitation for further efforts to conduct joint efforts nationally, regionally or globally to continue to investigate the changing impacts of mobile use and experience on this changing world.

To initiate the global efforts, in early January 2018, the editor of this volume invited scholars from around the world to contribute their various studies to this global comparative investigation of the impacts of mobile use and experience on contemporary society. Within less than a year, scholars from Australia, China, Colombia, India, New Zealand, Oman, Pakistan, Portugal, Turkey, UK, and US have submitted their respective chapters after a few rounds of amendments. All the chapters, ranging from Chapter 1 through Chapter 16, are thematically divided into two parts. The first part, consisting of nine chapters (1-9), deals with the impacts of mobile use while the second part, consisting of seven chapters (10-16), handles the impacts of mobile experience.

The investigation of the impacts of mobile use in the first part covers the determinants and impacts of mobile use during adolescence in UK (Chapter 1), the impacts of mobile use on retirees or the Third Agers in a second-tier city in China (Chapter 2), the gender differences in mobile use and habits in Turkey (Chapter 3), the impacts of mobile use on mobile dating (Chapter 4), food posting on Instagram among university students in New Zealand (Chapter 5), use of smartphone, gratifications and addiction among university students in Oman (Chapter 6), socio-economic impacts of mobile use in Pakistan (Chapter 7), the commercial use of mobile social media and social relationship in China (Chapter 8), and use of mobile apps in the framework of creator-audience matchmaking in India (Chapter 9).

The second part involves investigations of mobile experience in mobile learning through reviewing earlier efforts, reviewing existing mobile learning resources, and providing recommendations on how these resources can be leveraged to enhance mobile learning experience in US (Chapter 10), reviewing earlier studies on how mobile learning has been leveraged to enhance mobile experience in Chinese learning as a second language (Chapter 11), the current state and future directions for enhancing mobile translation experience in the world (Chapter 12), mobile experience at the bottom of pyramid in India (Chapter 13), mobile experience in war and conflict reporting (Chapter 14), mobile news experience in the cases of *New York Times* and *The Guardian* (Chapter 15), and smart tourists' experience in the case of smartphones used to enhance leisure travels (Chapter 16).

The concluding chapter by the editor is designed to provide further dimensions and directions for further studies of the impacts of mobile use and experience around the world. Using the recommended 3M approach (mapping, measuring and modelling), coupled by the gap focus (the gap between the normative and empirical) and comparative orientation assisted with mixed methods research, the concluding chapter examined mobile features, mobile journalism, a new approach to explanation and prediction of mobile experience, and how mobile experience can be maximized to secure sustainable development as examples of dimensions and directions for further studies of mobile use and experience.

## REFERENCES

Xu, X. (2016). *Mobile Studies International*. Available at http://msi.wiki

Xu, X. (2018). Comparing mobile experience. In K. Norman & K. Kirakowski (Eds.), *Wiley Handbook of Human-Computer Interaction Set* (Vol. 1, pp. 225–238). Wiley.

# Acknowledgment

Strongly encouraged for more than 11 months by the concerted efforts among 28 contributors from 10 countries, as the editor of this volume, I have finally wrapped up the whole editing process.

First of all, I would like to express my gratitude to the publisher IGI Global for agreeing to consolidate two volumes into one and also for their insightful suggestions about the new title.

My deepest gratitude should be given to all the contributors of this volume for their valuable and insightful contributions.

I would like to thank my family for their tolerance of my absence and less attention for them. I am also deeply grateful to my grandchildren, Peggy Xu and Jonny Xu, for being one of the major sources of my courage, strength, and aspiration.

# Introduction

Mobile has been connecting more and more people around the world. By 2017, 5 billion people are connected to mobile services in our wirelessly wired mobile world, including 3 billion mobile Internet users, and by 2025 the number of unique mobile subscribers will reach 5.9 billion, accounting for 71% of the world population (GSMA, 2018). While most of the developed countries approach saturation, mobile growth will be driven by such developing countries as India, China, Indonesia, Pakistan, Bangladesh, and African and Latin American countries, according to the Mobile Economy 2018 released by GSMA.

Mobile has been widely used worldwide by almost everyone, ranging a small kid (who is still learning to speak) to a granny (who is losing clarity in speaking). Mobile has reached almost every corner of the world, wherever a human being happens to be, within coverage of a mobile communication network. Mobile has been used in almost all human activities, ranging from banking to dating. Such wide and growing use of mobile has generated diversified and unique experiences in different areas such as mobile news consumption, mobile filmmaking, mobile photography, mobile news reporting, mobile translation, mobile healthcare, mobile learning, mobile teaching, and mobile tourism.

Use of mobile has brought about tremendous and enormous impacts on our lives. Instead of going to a bank, we can conduct all kinds of banking activities and transactions on mobile. Instead of paying by cash or debit/credit card, we pay directly from mobile in a café, a restaurant, a shopping mall, a taxi, a grocery store, or a canteen. Instead of shopping in a store or a mall, we buy or sell everything on mobile, ranging from toilet paper to wine. Mobile has also removed space, distance, time, or even language barriers, enabling us to stay connected with people from all over the world. We can date, socialize and make friends with new people on mobile.

Mobile use has also changed the way we work and do business tremendously. To save energy and cost, with the permission from line managers, we can choose to work on mobile anytime anywhere as long as we get our respective jobs done. We communicate and collaborate through social media with colleagues within the same institution or company and beyond. We can also get business or projects done on mobile without physically working together. Our mobile is our office. Our mobile is our world. It has changed our work routines and practices. "Mobile too" used to be a common practice in many companies. But now more and more companies have changed their policies from "mobile too" to "mobile first". It is also becoming increasing urgent for companies to focus on "mobile only" since mobile has become the most popular and powerful medium in the world.

Behind mobile use lies unique mobile experience. Every new use of mobile has brought about new experience as well. When mobile was enabled for texting, we had a choice to text instead of calling so that we would not disturb others at work. When mobile was further enabled to take photos and to record videos, we had a totally new experience. With new mobile technological advances being added to mobile, we are experiencing more diversified and unique experience, enabling us to live, communication, and work pleasantly, productively, and effectively in a more experience-rich fashion.

To investigate mobile use and experience as well as their impacts in this ever-changing mobile world, we need a better framework to investigate the impacts of mobile use and experience. To that end, we would like to propose a focus our examinations on actors and activities involved in mobile use and experience. In the following sections, we begin with our elaboration of our proposed investigation framework, followed by our discussion on how it can be applied in investigations of mobile use and mobile experience. Assisted with a mixed methods research, the framework is situated in the context of experience economy.

## INVESTIGATION FRAMEWORK

Conceptually, "actor" in this chapter is defined as any participant in any interaction with mobile for different purposes. It can refer to a small kid using mobile for playing a game or merely watching a video, movie or a cartoon. It can also refer to a granny using mobile for staying connected with her grandchildren via video calls. And it can also refer to anyone in between along the diversified spectrum of different actors who are involved in one way or another in using mobile for different purposes. By activity, we refer to any areas of human activity of using mobile to engage in different events or processes to achieve different goals. It can refer to such activities as journalism, advertising, marketing, governing, shopping, business, parenting or dating. Both actors and activities constitute the core of mobile use and experience, therefore they can be used to serve as the framework for any investigation of mobile use and experience as well as their impacts.

Within the actors-and-activities framework, an investigation of mobile use by different actors will not only look into demographic factors of different actors but also their motivations, expectations, and perceptions of mobile use. And mobile use is also connected with examination of different activities, in which mobile is used. A comparison is also expected to be carried out to locate similarities and differences in mobile use and experience. Furthermore, we also recommend the 3M approach of mapping, measuring and modelling (Xu, 2018) mobile use and experience to the framework. Also recommended is the gap focus, that is, the gap between the normative or expected mobile use and experience and their empirical ones. The framework is also accompanied by the recommended mixed methods research, which is used to generate triangulated, complementary, accurate and reliable findings as well as consequent robust conclusions.

In this mobile world, mobile devices and services are no longer the ultimate important things in mobile users' mind. What is ultimately important is the experience beyond the product or service a mobile user is exposed to. It is not trendy but current to focus on mobile experience as our world is moving towards experience economy. To investigate mobile use is the first step, a crucial step towards a better understanding of mobile experience. Without knowledge of how mobile is used, it will be hard to examine mobile experience.

## MOBILE USE

By mobile use, we refer to use of mobile devices and technologies by different actors in different activities for different purposes. Mobile is used by actors of different gender, age groups, occupations, social classes, ethnic groups, religions, cultures, and countries. Mobile use can be grouped into the following categories: (a) informational, (b) relational (Lee, Kwak, Campbell, & Ling, 2014), (c) extractive, and (d) immersive (Humphreys, Von Pape & Karnowski, 2013). Mobile use can also be demographic since actors of different demographic features may differ, such as mobile use by healthcare professionals (Ventola, 2014), by students and teachers (Organista-Sandoval & Serrano-Santoyo, 2014), by the handicapped (Brandenburg, Worrall, Rodriguez & Copland, 2013), by adult learners (Hashim, Tan & Rashid, 2015), social uses of mobile phones among university students (Chuma, 2014), mobile devices uses among youth (Stald, Green, Barbowski, Haddon, Mascheroni, Sagvari, & Tzaliki, 2014), teens' mobile Internet use (Lin, Zhang, Jung, & Kim, 2013), use of mobile as status instrument among youths (Abeele, Antheunis & Schouten, 2014), and socio-demographic gaps in mobile use (Lee & Kim, 2014).

Mobile use also differs when it comes to activities, in which mobile is used, such as use of mobile in radio broadcasting (Rosales, 2013), learning (Santos & Ali, 2012; Kafyulilo, 2014), health (Fox & Duggan, 2012), healthcare (Jennings, Ong'ech, Simiyu, Sirengo, & Kassaye, 2013), road safety training (Reychav & Wu, 2014), the meteorological early warning system (Meissen, Faust & Fuchs-Kittowski, 2013), space weather measurements (Pankratius, Lind, Coster, Erickson, & Semeter, 2014), roles, work and life (Battard & Mangematin, 2013), mobile decision support and business intelligence (Power, 2013), use of mobile for proximity marketing and smart Cities (Sneps-Sneppe & Namiot, 2013), use of mobile in social networking (Salehan & Negahban, 2013), mobile assisted language learning (Duman, Orhon, & Gedik, 2015), mobile social games (Wei & Lu, 2014; Chen & Leung, 2015), and mobile news use (Chan, 2015).

As a better way to investigate mobile use by different actors in different activities, we would like to recommend a new way of investigating mobile use according to its purpose. For instance, mobile can be used in the following ways: personal assistant, commerce, shopping, banking, healthcare, government, parenting, journalism, public relations, advertising, branding, social media, film-making, audio-recording, video-recording, music-sharing, instant messenger, education, entertainment, economy, creativity, elections, advocacy, activism, romancing, dating, gaming, and gambling. These purposes can be further grouped into the following types: (a) personal assistant, (b) business, (c) government, (d) journalism, (e) persuasion, (f) entertainment, (g) social media, (h) love, (i) family, (j) production, and (k) education.

To investigate mobile use by different actors in different activities, we would like to suggest a 3M approach (Xu, 2918), that is, to map, to measure and to model mobile use. To map is to locate where mobile use is located. To map mobile use, we can look into the following areas: (a) personal assistant, (b) business, (c) government, (d) journalism, (e) persuasion, (f) entertainment, (g) social media, (h) love, (i) family, (j) production, and (k) education. Furthermore, we will also have to examine mobile use by actors of different gender, age groups, occupations, social classes, ethnic groups, religions, cultures, and countries. To measure mobile use is to gauge its width, depth and frequency. To measure mobile uses is to gauge the frequency, width and depth of using mobile in different areas, ranging from mobile government to mobile dating. For effective and better measurement, mobile use can be conceptualized and operationalized differently according to the different natures and objectives of investigating mobile use. To map and measure mobile use, however, is not the ultimate goal. Ultimately, the goal of mapping and measuring mobile use is to develop a model to describe, explain and predict changes and patterns in mobile use or similarities and differences in a comparative study. To model mobile use, we have to look into the political, economic, cultural and social factors of mobile use in a country. We also have to take into consideration the demographics of mobile users, including age, gender, education, income, sexual orientation, marital status, race, occupation and level of mobile savviness. Furthermore, we also need to examine different motivations, expectations, and perceptions of mobile use as differences in these areas may also shape different use of mobile.

Among those identified factors, which factor or factors play a bigger part in shaping use of mobile? And how does a factor or do factors shape use of mobile? Is there any pattern being formed among any shaping factors? As mobile use can be both normative and empirical, mobile users have different expectations of what mobile use should be and do to them and what expectations of mobile use are actually materialized or practiced can also be different. How similar or different is mobile use on the normative or empirical side? How similar or different is mobile use in terms of the gap between the normative and the empirical? What kinds and ranges of gaps exist between the normative and empirical sides? These are the questions to be addressed in further studies. Answers to these questions and more will generate a model for describing, explaining and predicting changes and patterns mobile use.

## MOBILE EXPERIENCE

Closely related to mobile use is mobile experience as each use of mobile generates each unique mobile experience, which can be redefined as an outcome of actors' use of mobile in mobile-related activities, largely derived from a working definition, which states that mobile experience is both a process and an outcome of a user's interaction with a product, a service, a content or their different combinations (Xu, 2018). Situated in an interactive, personalized, immersive, and mobile context, mobile experience can fall into a wide spectrum ranging from the worst to the best (Xu, 2018). Mobile experience has become increasing important since our world is moving towards to experience economy, where consumers are giving more attention and importance to the process and outcome of interacting with a content, product, service of their different combinations.

Among earlier studies on mobile experience, some were largely culture-oriented as they investigated the distinctive cultural inclinations in user experiences related to mobile phones and mapped cultural models to mobile phone user interface design (e.g. Eune & Lee, 2009). Others largely focused technology-centric. For instance, a technology-oriented study looked into functionality, usability, and experience in a combined fashion (e.g. McNamara & Kirakowski 2005). A user-oriented approach study examined usefulness, ease of use, hedonics, aesthetics, and pleasure/fun (e.g. Mahlke, 2005) while another user-centric study located the challenge of continuous involvement of the user and the need to integrate knowledge into the development process which is increasingly interdisciplinary (Moor, Berte, Marez, Joseph, Deryckere, & Martens. 2010).

As suggested by the names of the above-cited approaches, they have their own focuses on different elements, i.e. culture-based, technology-centric, or user-oriented. And it is obvious that if we focus one element, we tend to omit or ignore another element, which may be important in our understanding or comparison of mobile experience. What is also scarce is the related scholarship on the impacts of mobile experience on different actors and activities since conceptually mobile experience remains at its infancy stage although there have been sporadic related studies.

To fill this gap, similar to our recommended 3M approach to mobile use investigation, we it is also our recommendation to map, measure and model mobile experience. To map mobile experience is to identify mobile experience is located while measuring mobile experience is to gauge the extent to which mobile experience presents itself in different activities. The ultimate objective, however, is to model mobile experience, in other words, to describe, explain and predict changes and patterns in mobile experience or similarities and differences in mobile experience in a comparative study (Xu, 2018). The same 3M approach (mapping, measuring and modeling) can also be used to investigate or compare the impacts of mobile experience on different actors in different activities.

Specifically, for a better and effective investigation, the proposed a 6-dimension model includes: (a) enticement, (b) entertainment, (c) engagement, (d) empowerment, (e) enlightenment, and (f) enhancement. This 6-dimension model is also a 6-stage process and outcome of actors' interactions with products, contents, or services in different activities. To be enticed is the very first essential step for us to enjoy a full-scale mobile experience. Without being enticed, it would be hard for us to start to enjoy other aspects of mobile experience. To be entertained is another important stage, where once we are entertained, we are closer to be engaged, which is another important step, without which we will move away. More important than being enticed, entertained and engaged is to be empowered. This is especially true in the mobile age, where everyone can leverage mobile devices to be a story teller or to contribute to storytelling about an event, issue, community, or a nation. Beyond being enticed, entertained, engaged, empowered is to be enlightened and enhanced, the two final stages of the hierarchy of human desires for better experience. To be enlightened is to obtain a better understanding of what is being consumed so as to be a better consumer while to be enhanced is to enhance our consciousness, skills, and abilities as the final stage of mobile experience (Xu, 2018).

## MIXED METHODS RESEARCH

To effectively map, measure and model mobile use and experience and their impacts, mixed methods research should be employed instead of purely quantitative or qualitative methods. By mixed methods research, we refer to a combination of quantitative and qualitative research methods. As mixed methods research has been gaining acceptance among researchers, it has become a valid alternative research design, offering richer insights into the phenomenon being studied and allowing the capture of information that might be missed by utilizing only one research design (Caruth, 2013).

The purposes of using mixed methods research, according to Venkatesh, Brown, & Bala, (2013), include the following: (a) complementarity, (b) completeness, (c) developmental, (d) expansion, (e) corroboration or confirmation, (f) compensation, and (g) diversity. To obtain mutual viewpoints about similar experiences or associations is what complementarity meant to achieve while completeness refers to ensuring total representation of experiences or associations. By developmental, they refer to new research questions derived from other research methods or hypotheses to be tested, which is derived from other research methods. With expansion, they mean to clarify or elaborate on the knowledge gained from other research methods while corroboration/confirmation are designed to evaluate the trustworthiness of inferences gained from other research methods. To counter the weaknesses of other research method by employing the other research method is to what compensation is meant to obtain. The last purpose of using mixed methods research is to maintain diversity by obtaining opposing viewpoints of the same experiences or associations (Venkatesh et al., 2013).

Although mixed methods research possesses some weaknesses, its strengths outweigh its weaknesses. For instance, mixed methods research has, among others, the following weak points: (a) it can be difficult for either qualitative or quantitative researcher and (b) it can be more time consuming and expensive (Cronholm & Hjalmarsson, 2011). Nevertheless, mixed methods research is stronger than any single method used in the following areas: (a) it can add meaning to numeric data while add precision to narrative data; (b) it can handle a wider range of research questions; (c) it can triangulate the results of different research methods; (d) it can present a more robust conclusion (Cronholm & Hjalmarsson, 2011).

Mixed methods research can be leveraged in mapping, measuring and modeling mobile use and experience. In terms of mapping mobile use and experience, both quantitative methods and qualitative methods are employed to locate where they are located. The same rule can be applied to measuring mobile use and experience. Although modeling focuses on testing hypotheses, qualitative methods may also be used to either triangulate or complement the results of hypothesis testing. In brief, in mapping, measuring and modeling mobile use and experience, once numeric data are generated from quantitative methods, they can be either triangulated or complemented with the narrative data resulted from qualitative methods.

## EXPERIENCE ECONOMY

The proposed framework of actors and activities is situated in the backdrop of experience economy. By experience economy, we refer to a particular economic system, which emphasizes selling experience (Pine & Gilmore, 1998) beyond its carrier, be it a product or a service. In modern societies, consumers expect more beyond a product or a service. Experience has its four realms, that is, entertainment, educational, esthetic and escapist (Pine & Gilmore, 1998). In light of what Boswijk, Thijssen, & Peelen (2006) have observed, experience features the following ten characteristics: 1. There is heightened concentration and focus. All five senses are engaged. 2. One's sense of time is altered. 3. One is touched emotionally. 4. The process is unique for the individual. 5. There is contact with the 'raw stuff', the real thing. 6. One both does and undergoes something. 7. There is an element of playfulness (flow). 8. One has the feeling of being in control of the situation. 9. There is a balance between the challenge and one's own capacities. 10. There is a clear goal (Boswijk, Thijssen, & Peelen, 2006). To satisfy these characteristics of experience is a must, according to Boswijk, Thijssen and Peelen (2006), for the creation and co-creation of meaningful experiences, which begins with focusing on the meaning of human experiences. Experience can also be conceptually and empirically expanded to include being enticed, entertained, engaged, empowered, enlightened, and enhanced (Xu, 2018) when actors are engaging in different activities such as learning, shopping, working, dining, sight-seeing, watching a movie, entertaining, or interacting with a product and/or service.

In experience economy, work is a theatre and every business is a stage (Pine & Gilmore, 1999) while experiences of all kinds can be created (Sundbo & Darmer, 2008) for consumers to be willing to pay for enjoying. The more enjoyable, memorable and fun an experience is, the more willing consumers are to pay. This general trend has become even more apparent in the mobile world, where mobile experience is king although a content, a product or a service remains fundamental important. According to Pine and Gilmore, the passionate advocates of experience economy, our world is already at the stage of an experience economy, "where experiences supplant services as the predominant economic offering in terms of GDP, employment and especially actual value" (Pine & Gilmore, 2013, p.26). With more and more people prefer experience over possessions, more money is spent on experience rather than actual goods or products. Against this general backdrop, it is imperative to investigate mobile use and experience in the context of experience economy.

## CONCLUSION

Mobile use can be investigated as a daily interactive process of using mobile in different activities for different purposes. Both normative and empirical use of mobile can also be examined to locate the gap between the two. Moreover, a comparative study is also highly recommended to locate similarities and differences between or among actors of different gender, age groups, occupations, educational levels, social classes, ethnic groups, religions, cultures and countries. Further comparison can also made to identify similarities and differences in using mobile in different activities, such as journalism, advertising, marketing, entertainment, education, propaganda, storytelling of different kinds, parenting, and dating. The impacts of mobile use on actors and activities can also be examined by using mixed methods research so that the results of using different research methods can be triangulated or complemented to secure more accurate findings.

In a similar manner, mobile experience can also be examined or compared as an outcome of interaction with mobile-related activities by using mixed methods research to triangulate or complement the results of different research methods. It can also be investigated at the normative and the empirical level. The gap between the two should also be examined to identify changes and patterns as well as similarities and differences in mobile experience. As for the proposed six different indicators of mobile experience, they can be further conceptualized and operationalized according to different purposes of investigation in different activities.

As the mobile world keeps changing, similar or different changes have also taken place in different countries. To describe, explain and predict similar or different changes in mobile use and experience as well as their impacts on actors and activities, it is essential and crucial to conduct inter-cultural or international comparative studies since this world is becoming increasingly inter-connected, inter-dependent and inter-influencing. What has changed in mobile use and experience in a country may also spillover in another country and keep rippling to more countries. Advances in mobile use and experience in one country will also shortly spread out to other countries.

The actor-activity framework accompanied by the 3M approach and the gap focus is designed to serve as a guideline for further studies in the hope that it will lead to and guide further studies of mobile use and experience as well as their impacts on actors and activities to continuously contribute to the body of knowledge on mobile use and experience.

## REFERENCES

Battard, N., & Mangematin, V. (2013). Idiosyncratic distances: Impact of mobile technology practices on role segmentation and integration. *Technological Forecasting and Social Change*, *80*(2), 231–242. doi:10.1016/j.techfore.2011.11.007

Boswijk, A., Thijssen, T., & Peelen, E. (2006). *A new perspective on the experience economy. Bilthovenm The Netherlands*. The European Centre for the Experience Economy.

Brandenburg, C., Worrall, L., Rodriguez, A. D., & Copland, D. (2013). Mobile computing technology and aphasia: An integrated review of accessibility and potential uses. *Aphasiology*, *27*(4), 444–461. do i:10.1080/02687038.2013.772293

Caruth, G. D. (2013). Demystifying Mixed Methods Research Design: A Review of the Literature. *Mevlana International Journal of Education, 3*(2), 112–122. doi:10.13054/mije.13.35.3.2

Chan, M. (2015). Examining the influences of news use patterns, motivations, and age cohort on mobile news use: The case of Hong Kong. *Mobile Media & Communication, 3*(2), 179–195. doi:10.1177/2050157914550663

Chen, C., & Leung, L. (2015). Are you addicted to Candy Crush Saga? An exploratory study linking psychological factors to mobile social game addiction. *Telematics and Informatics*.

Chuma, W. (2014). The social meanings of mobile phones among South Africa's 'digital natives': A case study. *Media Culture & Society, 36*(3), 398–408. doi:10.1177/0163443713517482

Cronholm, S., & Hjalmarsson, A. (2011). Experiences from sequential use of mixed methods. *Electronic Journal of Business Research Methods, 9*(2), 87–95.

Duman, G., Orhon, G., & Gedik, N. (2015). Research trends in mobile assisted language learning from 2000 to 2012. *ReCALL, 27*(02), 197–216. doi:10.1017/S0958344014000287

Eune, J., & Lee, K. P. 2009. Analysis on Intercultural Differences through User Experiences of Mobile Phone for Glocalization. Proceedings of International Association of Societies, 1215–26.

Fox, S., & Duggan, M. (2012). *Mobile health 2012.* Washington, DC: Pew Internet & American Life Project.

Hashim, K. F., Tan, F. B., & Rashid, A. (2015). Adult learners' intention to adopt mobile learning: A motivational perspective. *British Journal of Educational Technology, 46*(2), 381–390. doi:10.1111/bjet.12148

Humphreys, L., Von Pape, T., & Karnowski, V. (2013). Evolving mobile media: Uses and conceptualizations of the mobile internet. *Journal of Computer-Mediated Communication, 18*(4), 491–507. doi:10.1111/jcc4.12019

Jennings, L., Ong'ech, J., Simiyu, R., Sirengo, M., & Kassaye, S. (2013). Exploring the use of mobile phone technology for the enhancement of the prevention of mother-to-child transmission of HIV program in Nyanza, Kenya: A qualitative study. *BMC Public Health, 13*(1), 1. doi:10.1186/1471-2458-13-1131 PMID:24308409

Kafyulilo, A. (2014). Access, use and perceptions of teachers and students towards mobile phones as a tool for teaching and learning in Tanzania. *Education and Information Technologies, 19*(1), 115–127. doi:10.100710639-012-9207-y

Lee, H., Kwak, N., Campbell, S. W., & Ling, R. (2014). Mobile communication and political participation in South Korea: Examining the intersections between informational and relational uses. *Computers in Human Behavior, 38*, 85–92. doi:10.1016/j.chb.2014.05.017

Lee, J. H., & Kim, J. (2014). Socio-demographic gaps in mobile use, causes, and consequences: A multi-group analysis of the mobile divide model. *Information Communication and Society, 17*(8), 917–936. doi:10.1080/1369118X.2013.860182

Lin, W. Y., Zhang, X., Jung, J. Y., & Kim, Y. C. (2013). From the wired to wireless generation? Investigating teens' Internet use through the mobile phone. *Telecommunications Policy, 37*(8), 651–661. doi:10.1016/j.telpol.2012.09.008

Mahlke, S. (2005). Understanding Users' Experience of Interaction. *Proceedings of the 2005 Annual Conference on European Association of Cognitive Ergonomics*, 251–4.

McCarthy, J., & Wright, P. C. (2005). Putting 'Felt-Life" at the Centre of Human-Computer Interaction (HCI). *Cognition Technology and Work, 7*(4), 262–271. doi:10.100710111-005-0011-y

McNamara, N., & Kirakowski, J. (2005, July). Defining usability: quality of use or quality of experience? In *Professional Communication Conference, 2005. IPCC 2005. Proceedings. International* (pp. 200-204). IEEE. 10.1109/IPCC.2005.1494178

Meissen, U., Faust, D., & Fuchs-Kittowski, F. (2013). WIND-A meteorological early warning system and its extensions towards mobile services. In EnviroInfo (pp. 612-621). Academic Press.

Moor, K. D., Berte, K., Marez, L. D., Joseph, W., Deryckere, T., & Martens, L. (2010). User-Driven Innovation? Challenges of User Involvement in Future Technology Analysis. *Science & Public Policy, 37*(1), 51–61. doi:10.3152/030234210X484775

Organista-Sandoval, J., & Serrano-Santoyo, A. (2014). Appropriation and educational uses of mobile phones by students and teachers at a public university in Mexico. *Creative Education*.

Pankratius, V., Lind, F., Coster, A., Erickson, P., & Semeter, J. (2014). Mobile crowd sensing in space weather monitoring: The Mahali project. *Communications Magazine, IEEE, 52*(8), 22–28. doi:10.1109/MCOM.2014.6871665

Pine, B. J., & Gilmore, J. H. (1998). Welcome to the experience economy. *Harvard Business Review, 76*, 97–105. PMID:10181589

Pine, B. J., & Gilmore, J. H. (2013). The experience economy: past, present and future. *Handbook on the experience economy*, 21-44.

Power, D. J. (2013). Mobile decision support and business intelligence: An overview. *Journal of Decision Systems, 22*(1), 4–9. doi:10.1080/12460125.2012.760267

Reychav, I., & Wu, D. (2014). Exploring mobile tablet training for road safety: A uses and gratifications perspective. *Computers & Education, 71*, 43–55. doi:10.1016/j.compedu.2013.09.005

Rosales, R. G. (2013). Citizen participation and the uses of mobile technology in radio broadcasting. *Telematics and Informatics, 30*(3), 252–257. doi:10.1016/j.tele.2012.04.006

Salehan, M., & Negahban, A. (2013). Social networking on smartphones: When mobile phones become addictive. *Computers in Human Behavior, 29*(6), 2632–2639. doi:10.1016/j.chb.2013.07.003

Santos, I. M., & Ali, N. (2012). Exploring the uses of mobile phones to support informal learning. *Education and Information Technologies, 17*(2), 187–203. doi:10.100710639-011-9151-2

Sanz-Blas, S., Ruiz-Mafé, C., & Martí-Parreño, J. (2015). Message-driven factors influencing opening and forwarding of mobile advertising messages. *International Journal of Mobile Communications*, *13*(4), 339–357. doi:10.1504/IJMC.2015.070058

Sneps-Sneppe, M., & Namiot, D. (2013). Smart cities software: customized messages for mobile subscribers. In *Wireless Access Flexibility* (pp. 25–36). Springer Berlin Heidelberg. doi:10.1007/978-3-642-39805-6_3

Stald, G. B., Green, L., Barbowski, M., Haddon, L., Mascheroni, G., Sagvari, B., ... Tzaliki, L. (2014). *Online on the mobile: internet use on the smartphone and associated risks among youth in Europe*. Academic Press.

Sundbo, J., & Darmer, P. (Eds.). (2008). *Creating experiences in the experience economy*. Edward Elgar Publishing. doi:10.4337/9781848444003

Venkatesh, V., Brown, S. A., & Bala, H. (2013). Bridging the qualitative-quantitative divide: Guidelines for conducting mixed methods research in information systems. *Management Information Systems Quarterly*, *37*(1), 21–54. doi:10.25300/MISQ/2013/37.1.02

Ventola, C. L. (2014). Mobile devices and apps for health care professionals: Uses and benefits. *P&T*, *39*(5), 356. PMID:24883008

Wei, P. S., & Lu, H. P. (2014). Why do people play mobile social games? An examination of network externalities and of uses and gratifications. *Internet Research*, *24*(3), 313–331. doi:10.1108/IntR-04-2013-0082

Xu, X. (2018). Comparing mobile experience. In K. Norman & K. Kirakowski (Eds.), *Wiley Handbook of Human-Computer Interaction Set* (Vol. 1, pp. 225–238). Wiley.

## ADDITIONAL READING

Abdesslem, F. B., Parris, I., & Henderson, T. N. H. (2010). Mobile experience sampling: Reaching the parts of Facebook other methods cannot reach.

Aggarwal, M., Grover, S., & Basu, D. (2012). Mobile phone use by resident doctors: Tendency to addiction-like behaviour. *German Journal of Psychiatry*, *15*(2), 50–55.

Aguado, J. M., & Martínez, I. J. (2007). The construction of the mobile experience: The role of advertising campaigns in the appropriation of mobile phone technologies. *Continuum*, *21*(2), 137–148. doi:10.1080/10304310701268679

Akkucuk, U., & Turan, C. (2015). Mobile use and online preferences of the millenials: A study in Yalova. *Journal of Internet Banking and Commerce*, *21*(1).

Andersson, T. D. (2007). The tourist in the experience economy. *Scandinavian Journal of Hospitality and Tourism*, *7*(1), 46–58. doi:10.1080/15022250701224035

Ballard, B. (2007). *Designing the mobile user experience*. John Wiley & Sons. doi:10.1002/9780470060575

Baum, T. (2006). Reflections on the nature of skills in the experience economy: Challenging traditional skills models in hospitality. *Journal of Hospitality and Tourism Management*, *13*(2), 124–135. doi:10.1375/jhtm.13.2.124

Beltagui, A., Candi, M., & Riedel, J. C. (2012). Design in the experience economy: Using emotional design for service innovation. In Interdisciplinary Approaches to Product Design, Innovation, & Branding in International Marketing (pp. 111-135). Emerald Group Publishing Limited.

Benedikt, M. (2001). Reality and authenticity in the experience economy. *Architectural Record*, *189*(11), 84–87.

Beranuy, M., Oberst, U., Carbonell, X., & Chamarro, A. (2009). Problematic Internet and mobile phone use and clinical symptoms in college students: The role of emotional intelligence. *Computers in Human Behavior*, *25*(5), 1182–1187. doi:10.1016/j.chb.2009.03.001

Bigné, E., Ruiz, C., & Sanz, S. (2007). Key drivers of mobile commerce adoption. An exploratory study of Spanish mobile users. *Journal of Theoretical and Applied Electronic Commerce Research*, *2*(2).

Bille, T. (2012). The Scandinavian approach to the experience economy–does it make sense? *International Journal of Cultural Policy*, *18*(1), 93–110. doi:10.1080/10286632.2011.561924

Billieux, J. (2012). Problematic use of the mobile phone: A literature review and a pathways model. *Current Psychiatry Reviews*, *8*(4), 299–307. doi:10.2174/157340012803520522

Binkhorst, E., & Den Dekker, T. (2009). Agenda for co-creation tourism experience research. *Journal of Hospitality Marketing & Management*, *18*(2-3), 311–327. doi:10.1080/19368620802594193

Birch, S. (2008). *The political promotion of the experience economy and creative industries*. Samfundslitteratur.

Biswas, A., Donaldson, T., Singh, J., Diamond, S., Gauthier, D., & Longford, M. (2006, June). Assessment of mobile experience engine, the development toolkit for context aware mobile applications. In *Proceedings of the 2006 ACM SIGCHI international conference on Advances in computer entertainment technology* (p. 8). ACM. 10.1145/1178823.1178834

Borowy, M. (2013). Pioneering eSport: The experience economy and the marketing of early 1980s arcade gaming contests. *International Journal of Communication*, *7*, 21.

Boruff, J. T., & Storie, D. (2014). Mobile devices in medicine: A survey of how medical students, residents, and faculty use smartphones and other mobile devices to find information. *Journal of the Medical Library Association: JMLA*, *102*(1), 22–30. doi:10.3163/1536-5050.102.1.006 PMID:24415916

Boswijk, A., Peelen, E., Olthof, S., & Beddow, C. (2012). *Economy of experiences*. Amsterdam: European Centre for the Experience and Transformation Economy.

Boswijk, A., Thijssen, T., & Peelen, E. (2007). *The experience economy: A new perspective*. Pearson Education.

Botha, A., Calteaux, K., Herselman, M., Grover, A. S., & Barnard, E. (2012). Mobile user experience for voice services: A theoretical framework.

Botha, A., Herselman, M., & van Greunen, D. (2010, October). Mobile user experience in a mlearning environment. In *Proceedings of the 2010 Annual Research Conference of the South African Institute of Computer Scientists and Information Technologists* (pp. 29-38). ACM. 10.1145/1899503.1899507

Chang, T. C., & Huang, S. (2014). Urban tourism and the experience economy. *The Wiley Blackwell companion to tourism*, 220-229.

Chang, W. L., Yuan, S. T., & Hsu, C. W. (2010). Creating the experience economy in e-commerce. *Communications of the ACM, 53*(7), 122–127. doi:10.1145/1785414.1785449

Cherubini, M., & Oliver, N. (2009). A refined experience sampling method to capture mobile user experience. *arXiv preprint arXiv:0906.4125.*

Choliz, M. (2010). Mobile phone addiction: A point of issue. *Addiction (Abingdon, England), 105*(2), 373–374. doi:10.1111/j.1360-0443.2009.02854.x PMID:20078493

Chung, N., & Kwon, S. J. (2009). The effects of customers' mobile experience and technical support on the intention to use mobile banking. *Cyberpsychology & Behavior, 12*(5), 539–543. doi:10.1089/cpb.2009.0014 PMID:19772441

Church, K., & Cherubini, M. (2010, September). Evaluating mobile user experience in-the-wild: Prototypes, playgrounds and contextual experience sampling. In *Proceedings of the Workshop on Research in the Large: Using App Stores, Markets and other Wide Distribution Channels in Ubiquitous Computing Research* (pp. 29-32).

Dahlbom, B., & Ljungberg, F. (1998). Mobile informatics. *Scandinavian Journal of Information Systems, 10*, 227–234.

Degen, M., Melhuish, C., & Rose, G. (2017). Producing place atmospheres digitally: Architecture, digital visualisation practices and the experience economy. *Journal of Consumer Culture, 17*(1), 3–24. doi:10.1177/1469540515572238

Delagi, G. (2010, February). Harnessing technology to advance the next-generation mobile user-experience. In Solid-State Circuits Conference Digest of Technical Papers (ISSCC), 2010 IEEE International (pp. 18-24). IEEE. doi:10.1109/ISSCC.2010.5434067

Djamasbi, S., McAuliffe, D., Gomez, W., Kardzhaliyski, G., Liu, W., & Oglesby, F. (2014, June). Designing for success: Creating business value with mobile user experience (UX). In *International conference on HCI in Business* (pp. 299-306). Springer, Cham. 10.1007/978-3-319-07293-7_29

Dodson, L. L., Sterling, S., & Bennett, J. K. (2013, December). Minding the gaps: Cultural, technical and gender-based barriers to mobile use in oral-language Berber communities in Morocco. In *Proceedings of the Sixth International Conference on Information and Communication Technologies and Development: Full Papers-Volume 1* (pp. 79-88). ACM. 10.1145/2516604.2516626

Donner, J. (2005). User-led innovations in mobile use in sub-Saharan Africa. *Vodafone Receiver, 14.*

Donner, J. (2006). The use of mobile phones by microentrepreneurs in Kigali, Rwanda: Changes to social and business networks. *Information Technologies & International Development, 3*(2), pp-3.

Donner, J. (2008). Research approaches to mobile use in the developing world: A review of the literature. *The Information Society, 24*(3), 140–159. doi:10.1080/01972240802019970

Donner, J. (2009). Blurring livelihoods and lives: The social uses of mobile phones and socioeconomic development. *Innovations: Technology, Governance, Globalization, 4*(1), 91–101. doi:10.1162/itgg.2009.4.1.91

Donner, J., & Escobari, M. X. (2010). A review of evidence on mobile use by micro and small enterprises in developing countries. *Journal of International Development, 22*(5), 641–658. doi:10.1002/jid.1717

Dresselhaus, A., & Shrode, F. (2012). Mobile technologies & academics: Do students use mobile technologies in their academic lives and are librarians ready to meet this challenge? *Information Technology and Libraries, 31*(2), 82–101. doi:10.6017/ital.v31i2.2166

Economides, A. A., & Grousopoulou, A. (2008). Use of mobile phones by male and female Greek students. *International Journal of Mobile Communications, 6*(6), 729–749. doi:10.1504/IJMC.2008.019822

Einav, L., Levin, J., Popov, I., & Sundaresan, N. (2014). Growth, adoption, and use of mobile E-commerce. *The American Economic Review, 104*(5), 489–494. doi:10.1257/aer.104.5.489

Ek, R., Larsen, J., Hornskov, S. B., & Mansfeldt, O. K. (2008). A dynamic framework of tourist experiences: Space-time and performances in the experience economy. *Scandinavian Journal of Hospitality and Tourism, 8*(2), 122–140. doi:10.1080/15022250802110091

Eriksson, N. (2012). User experience of trip arrangements: A comparison of mobile device and computer users. *International Journal of E-Services and Mobile Applications, 4*(2), 55–69. doi:10.4018/jesma.2012040104

Falchuk, B., & Loeb, S. (2008, November). Towards" Guardian Angels" and Improved Mobile User Experience. In *Global Telecommunications Conference, 2008. IEEE GLOBECOM 2008. IEEE* (pp. 1-5). IEEE. 10.1109/GLOCOM.2008.ECP.345

Ferreira, H., & Teixeira, A. A. (2013). *'Welcome to the experience economy': assessing the influence of customer experience literature through bibliometric analysis (No. 481)*. Universidade do Porto, Faculdade de Economia do Porto.

Fiore, A. M., Niehm, L., Oh, H., Jeong, M., & Hausafus, C. (2007). Experience economy strategies: Adding value to small rural businesses. *Journal of Extension, 45*(2), 1–13.

Freire-Gibb, L. C. (2011). The rise and fall of the concept of the experience economy in the local economic development of Denmark. *European Planning Studies, 19*(10), 1839–1853. doi:10.1080/09654313.2011.614391

Garcia, A. (2012, October). Trust and mobile media use in schools. []. Taylor & Francis Group.]. *The Educational Forum, 76*(4), 430–433. doi:10.1080/00131725.2012.707566

Garrett, B. M., & Jackson, C. (2006). A mobile clinical e-portfolio for nursing and medical students, using wireless personal digital assistants (PDAs). *Nurse Education Today, 26*(8), 647–654. doi:10.1016/j.nedt.2006.07.020 PMID:17011674

Geven, A., Sefelin, R., & Tscheligi, M. (2006, September). Depth and breadth away from the desktop: the optimal information hierarchy for mobile use. In *Proceedings of the 8th conference on Human-computer interaction with mobile devices and services* (pp. 157-164). ACM. 10.1145/1152215.1152248

Goose, S., Schneider, G., Tanikella, R., Mollenhauer, H., Menard, P., Le Floc'h, Y., & Pillan, P. (2002, January). Toward improving the mobile experience with proxy transcoding and virtual composite devices for a scalable Bluetooth LAN access solution. In *Mobile Data Management, 2002. Proceedings. Third International Conference on* (pp. 169-170). IEEE. 10.1109/MDM.2002.994414

GSMA. (2018). The Mobile Economy 2018. Retrieved from https://www.gsmaintelligence.com/research/?file=061ad2d2417d6ed1ab002da0dbc9ce22&download

Guerreiro, T., Jorge, J., & Gonçalves, D. (2010, August). Identifying the relevant individual attributes for a successful non-visual mobile experience. In *Proceedings of the 28th Annual European Conference on Cognitive Ergonomics* (pp. 27-30). ACM. 10.1145/1962300.1962309

Guex, D., & Crevoisier, O. (2015). 8 A comprehensive socio-economic model of the experience economy. *Spatial Dynamics in the Experience Economy*, 119.

Hagen, P., Robertson, T., Kan, M., & Sadler, K. (2005, November). Emerging research methods for understanding mobile technology use. In *Proceedings of the 17th Australia conference on Computer-Human Interaction: Citizens Online: Considerations for Today and the Future* (pp. 1-10). Computer-Human Interaction Special Interest Group (CHISIG) of Australia.

Hatfield, J., & Murphy, S. (2007). The effects of mobile phone use on pedestrian crossing behaviour at signalised and unsignalised intersections. *Accident; Analysis and Prevention*, *39*(1), 197–205. doi:10.1016/j.aap.2006.07.001 PMID:16919588

Hayes, D., & MacLeod, N. (2007). Packaging places: Designing heritage trails using an experience economy perspective to maximize visitor engagement. *Journal of Vacation Marketing*, *13*(1), 45–58. doi:10.1177/1356766706071205

Irshad, S., & Rambli, D. R. A. (2014, December). User experience evaluation of mobile AR services. In *Proceedings of the 12th International Conference on Advances in Mobile Computing and Multimedia* (pp. 119-126). ACM. 10.1145/2684103.2684135

Ishii, K. (2004). Internet use via mobile phone in Japan. *Telecommunications Policy*, *28*(1), 43–58. doi:10.1016/j.telpol.2003.07.001

James, J., & Versteeg, M. (2007). Mobile phones in Africa: How much do we really know? *Social Indicators Research*, *84*(1), 117–126. doi:10.100711205-006-9079-x PMID:20076776

Järvinen, S., Peltola, J., Lahti, J., & Sachinopoulou, A. (2009, November). Multimedia service creation platform for mobile experience sharing. In *Proceedings of the 8th International Conference on Mobile and Ubiquitous Multimedia* (p. 6). ACM. 10.1145/1658550.1658556

Jun, J. W., & Lee, S. (2007). Mobile media use and its impact on consumer attitudes toward mobile advertising. *International Journal of Mobile Marketing*, *2*(1).

Kalogeraki, S., & Papadaki, M. (2010). The Impact of Mobile Use on Teenagers' Socialization. *The International Journal of Interdisciplinary Social Sciences: Annual Review*, *5*(4), 121–134. doi:10.18848/1833-1882/CGP/v05i04/51690

Kamilakis, M., Gavalas, D., & Zaroliagis, C. (2016, June). Mobile user experience in augmented reality vs. maps interfaces: A case study in public transportation. In *International Conference on Augmented Reality, Virtual Reality and Computer Graphics* (pp. 388-396). Springer, Cham. 10.1007/978-3-319-40621-3_27

Karlson, A. K., Bederson, B. B., & Contreras-Vidal, J. L. (2008). Understanding one-handed use of mobile devices. In *Handbook of research on user interface design and evaluation for mobile technology* (pp. 86–101). IGI Global. doi:10.4018/978-1-59904-871-0.ch006

Kenny, R. F., Van Neste-Kenny, J. M., Burton, P. A., Park, C. L., & Qayyum, A. (2012). Using self-efficacy to assess the readiness of nursing educators and students for mobile learning. *The International Review of Research in Open and Distributed Learning*, *13*(3), 277–296. doi:10.19173/irrodl.v13i3.1221

Kim, B. (2013). The present and future of the library mobile experience. *Library Technology Reports*, *49*(6), 15–28.

Kim, B. (2013). *The library mobile experience: Practices and user expectations*. American Library Association.

Kim, B., & Ball, M. (2010). Mobile use in medicine: Taking a cue from specialized resources and devices. *The Reference Librarian*, *52*(1-2), 57–67. doi:10.1080/02763877.2011.521733

Kjeldskov, J. (2002, September). "Just-in-Place" information for mobile device interfaces. In *International Conference on Mobile Human-Computer Interaction* (pp. 271-275). Springer, Berlin, Heidelberg.

Koehler, N., Vujovic, O., & McMenamin, C. (2013). Healthcare professionals' use of mobile phones and the internet in clinical practice. *Journal of Mobile Technology in Medicine*, *2*(1), 3–13. doi:10.7309/jmtm.76

Kristoffersen, S., & Ljungberg, F. (1999, August). Mobile use of IT. In *the Proceedings of IRIS22, Jyväskylä, Finland*.

Kristoffersen, S., & Ljungberg, F. (1999, September). Designing interaction styles for a mobile use context. In *International Symposium on Handheld and Ubiquitous Computing* (pp. 281-288). Springer, Berlin, Heidelberg. 10.1007/3-540-48157-5_26

Krontiris, I., Langheinrich, M., & Shilton, K. (2014). Trust and privacy in mobile experience sharing: Future challenges and avenues for research. *IEEE Communications Magazine*, *52*(8), 50–55. doi:10.1109/MCOM.2014.6871669

Kroski, E. (2009). How to create a mobile experience. *Library Technology Reports*, *44*(5), 39–42.

Kukulska-Hulme, A. (2009). Will mobile learning change language learning? *ReCALL*, *21*(2), 157–165. doi:10.1017/S0958344009000202

Lee, A. (2010). Exploiting context for mobile user experience. In *Proceedings of the First Workshop on Semantic Models for Adaptive Interactive Systems (SEMAIS)*.

Lee, D., Yi, M. Y., Choi, J., & Lee, H. (2011). Measuring the mobile user experience: Conceptualization and empirical assessment. *SIGHCI 2011 Proceedings. Paper*, *3*, 2011.

Lee, J. K., & Mills, J. E. (2010). Exploring tourist satisfaction with mobile experience technology. *International Management Review*, *6*(1), 91–111.

Li, H., & Hua, X. S. (2010, October). Melog: mobile experience sharing through automatic multimedia blogging. In *Proceedings of the 2010 ACM multimedia workshop on Mobile cloud media computing* (pp. 19-24). ACM. 10.1145/1877953.1877961

Liang, T. P., & Yeh, Y. H. (2011). Effect of use contexts on the continuous use of mobile services: The case of mobile games. *Personal and Ubiquitous Computing*, *15*(2), 187–196. doi:10.100700779-010-0300-1

Lonsway, B. (2013). *Making leisure work: Architecture and the experience economy*. Routledge.

Lorentzen, A. (2012). The development of the periphery in the experience economy. In *Regional development in Northern Europe* (pp. 34–47). Routledge.

Lorentzen, A. (2013). Cities in the experience economy. In *The City in the Experience Economy* (pp. 19–36). Routledge.

Lorentzen, A., & Hansen, C. J. (2009). The role and transformation of the city in the experience economy: Identifying and exploring research challenges.

Lorentzen, A., & Jeannerat, H. (2013). Urban and regional studies in the experience economy: What kind of turn? *European Urban and Regional Studies*, *20*(4), 363–369. doi:10.1177/0969776412470787

Loureiro, S. M. C. (2014). The role of the rural tourism experience economy in place attachment and behavioral intentions. *International Journal of Hospitality Management*, *40*, 1–9. doi:10.1016/j.ijhm.2014.02.010

Lu, J., Liu, C., Yu, C. S., & Wang, K. (2008). Determinants of accepting wireless mobile data services in China. *Information & Management*, *45*(1), 52–64. doi:10.1016/j.im.2007.11.002

Lupton, D. (2014). The commodification of patient opinion: The digital patient experience economy in the age of big data. *Sociology of Health & Illness*, *36*(6), 856–869. doi:10.1111/1467-9566.12109 PMID:24443847

Mallat, N., Rossi, M., Tuunainen, V. K., & Öörni, A. (2009). The impact of use context on mobile services acceptance: The case of mobile ticketing. *Information & Management*, *46*(3), 190–195. doi:10.1016/j.im.2008.11.008

Manthiou, A., Lee, S., Tang, L., & Chiang, L. (2014). The experience economy approach to festival marketing: Vivid memory and attendee loyalty. *Journal of Services Marketing*, *28*(1), 22–35. doi:10.1108/JSM-06-2012-0105

Marshall, T. (2011). Moving the museum outside its walls: An augmented reality mobile experience.

Mehmetoglu, M., & Engen, M. (2011). Pine and Gilmore's concept of experience economy and its dimensions: An empirical examination in tourism. *Journal of Quality Assurance in Hospitality & Tourism*, *12*(4), 237–255. doi:10.1080/1528008X.2011.541847

Mendoza, A. (2013). *Mobile user experience: patterns to make sense of it all*. Newnes.

Menozzi, M., & Hofer, F., N pflin, U., & Krueger, H. (. (2003). Visual performance in augmented reality systems for mobile use. *International Journal of Human-Computer Interaction*, *16*(3), 447–460. doi:10.1207/S15327590IJHC1603_4

Morgan, M., Elbe, J., & de Esteban Curiel, J. (2009). Has the experience economy arrived? The views of destination managers in three visitor-dependent areas. *International Journal of Tourism Research*, *11*(2), 201–216. doi:10.1002/jtr.719

Morrison, A., Mulloni, A., Lemmelä, S., Oulasvirta, A., Jacucci, G., Peltonen, P., ... Regenbrecht, H. (2011). Collaborative use of mobile augmented reality with paper maps. *Computers & Graphics*, *35*(4), 789–799. doi:10.1016/j.cag.2011.04.009

Naftali, M., & Findlater, L. (2014, October). Accessibility in context: understanding the truly mobile experience of smartphone users with motor impairments. In *Proceedings of the 16th international ACM SIGACCESS conference on Computers & accessibility* (pp. 209-216). ACM. 10.1145/2661334.2661372

Nakhimovsky, Y., Eckles, D., & Riegelsberger, J. (2009, April). Mobile user experience research: challenges, methods & tools. In CHI'09 Extended Abstracts on Human Factors in Computing Systems (pp. 4795-4798). ACM. doi:10.1145/1520340.1520743

Nehra, R., Kate, N., Grover, S., Khehra, N., & Basu, D. (2012). Does the Excessive use of Mobile Phones in Young Adults Reflect an Emerging Behavioral Addiction? *Journal of Postgraduate Medicine Education and Research*, *46*(4), 177–182. doi:10.5005/jp-journals-10028-1040

Nordin, N. A., Shin, W. H., Bin Ghauth, K. I., & Bin Mohd Tamrin, M. I. (2007, September). Using service-based content adaptation platform to enhance mobile user experience. In *Proceedings of the 4th international conference on mobile technology, applications, and systems and the 1st international symposium on Computer human interaction in mobile technology* (pp. 552-557). ACM. 10.1145/1378063.1378153

Oh, H., Fiore, A. M., & Jeoung, M. (2007). Measuring experience economy concepts: Tourism applications. *Journal of Travel Research*, *46*(2), 119–132. doi:10.1177/0047287507304039

Park, M., Oh, H., & Park, J. (2010). Measuring the experience economy of film festival participants. *International Journal of Tourism Sciences*, *10*(2), 35–54. doi:10.1080/15980634.2010.11434625

Park, Y. J. (2015). My whole world's in my palm! The second-level divide of teenagers' mobile use and skill. *New Media & Society*, *17*(6), 977–995. doi:10.1177/1461444813520302

Patro, A., Rayanchu, S., Griepentrog, M., Ma, Y., & Banerjee, S. (2013, December). Capturing mobile experience in the wild: a tale of two apps. In *Proceedings of the ninth ACM conference on Emerging networking experiments and technologies* (pp. 199-210). ACM. 10.1145/2535372.2535391

Pérez-Sanagustín, M., Ramirez-Gonzalez, G., Hernández-Leo, D., Muñoz-Organero, M., Santos, P., Blat, J., & Kloos, C. D. (2012). Discovering the campus together: A mobile and computer-based learning experience. *Journal of Network and Computer Applications, 35*(1), 176–188. doi:10.1016/j.jnca.2011.02.011

Pine, B. J., & Gilmore, J. H. (1998). Welcome to the experience economy. *Harvard Business Review, 76*, 97–105. PMID:10181589

Pine, B. J. II, & Gilmore, J. H. (2000). Satisfaction, sacrifice, surprise: Three small steps create one giant leap into the experience economy. *Strategy and Leadership, 28*(1), 18–23. doi:10.1108/10878570010335958

Pine, B. J., & Gilmore, J. H. (2011). *The experience economy*. Harvard Business Press.

Pine, B. J., & Gilmore, J. H. (2014). A leader's guide to innovation in the experience economy. *Strategy and Leadership, 42*(1), 24–29. doi:10.1108/SL-09-2013-0073

Pine, B. J., Pine, J., & Gilmore, J. H. (1999). *The experience economy: work is theatre & every business a stage*. Harvard Business Press.

Po, S., Howard, S., Vetere, F., & Skov, M. B. (2004, September). Heuristic evaluation and mobile usability: Bridging the realism gap. In *International Conference on Mobile Human-Computer Interaction* (pp. 49-60). Springer, Berlin, Heidelberg. 10.1007/978-3-540-28637-0_5

Poulsson, S. H., & Kale, S. H. (2004). The experience economy and commercial experiences. *The Marketing Review, 4*(3), 267–277. doi:10.1362/1469347042223445

Quadri-Felitti, D., & Fiore, A. M. (2012). Experience economy constructs as a framework for understanding wine tourism. *Journal of Vacation Marketing, 18*(1), 3–15. doi:10.1177/1356766711432222

Racadio, R., Rose, E., & Boyd, S. (2012, October). Designing and evaluating the mobile experience through iterative field studies. In *Proceedings of the 30th ACM international conference on Design of communication* (pp. 191-196). ACM. 10.1145/2379057.2379095

Robinson, S., Marsden, G., & Jones, M. (2014). *There's not an app for that: mobile user experience design for life*. Morgan Kaufmann.

Roto, V., & Kaasinen, E. (2008, September). The second international workshop on mobile internet user experience. In *Proceedings of the 10th international conference on Human computer interaction with mobile devices and services* (pp. 571-573). ACM. 10.1145/1409240.1409354

Roussou, M., Katifori, A., Pujol, L., Vayanou, M., & Rennick-Egglestone, S. J. (2013, April). A life of their own: museum visitor personas penetrating the design lifecycle of a mobile experience. In CHI' 13 Extended Abstracts on Human Factors in Computing Systems (pp. 547-552). ACM. doi:10.1145/2468356.2468453

Sarmento, T., & Patrício, L. (2012). Mobile Service Experience-a quantitative study. *AMAServsig2012*.

Schou, S. (2008, January). Context-based service adaptation platform: Improving the user experience towards mobile location services. In *Information Networking, 2008. ICOIN 2008. International Conference on* (pp. 1-5). IEEE.

See-To, E. W., Papagiannidis, S., & Cho, V. (2012). User experience on mobile video appreciation: How to engross users and to enhance their enjoyment in watching mobile video clips. *Technological Forecasting and Social Change, 79*(8), 1484–1494. doi:10.1016/j.techfore.2012.03.005

Seifert, A., & Schelling, H. R. (2015). Mobile seniors: Mobile use of the Internet using smartphones or tablets by Swiss people over 65 years. *Gerontechnology (Valkenswaard), 14*(1). doi:10.4017/gt.2015.14.1.006.00

Seo, Y. (2013). Electronic sports: A new marketing landscape of the experience economy. *Journal of Marketing Management, 29*(13-14), 1542–1560. doi:10.1080/0267257X.2013.822906

Sey, A. (2011). 'We use it different, different': Making sense of trends in mobile phone use in Ghana. *New Media & Society, 13*(3), 375–390. doi:10.1177/1461444810393907

Shah, C., Gokhale, P. A., & Mehta, H. B. (2010). Effect of mobile use on reaction time. *Al Ameen Journal of Medical Sciences, 3*(2), 160–164.

Sidali, K. L., Kastenholz, E., & Bianchi, R. (2015). Food tourism, niche markets and products in rural tourism: Combining the intimacy model and the experience economy as a rural development strategy. *Journal of Sustainable Tourism, 23*(8-9), 1179–1197. doi:10.1080/09669582.2013.836210

Sims, F., Williams, M. A., & Elliot, S. (2007, July). Understanding the Mobile Experience Economy: A key to richer more effective M-Business Technologies, Models and Strategies. In *Management of Mobile Business, 2007. ICMB 2007. International Conference on the* (pp. 12-12). IEEE.

Sköld, D. E. (2010). The other side of enjoyment: Short-circuiting marketing and creativity in the experience economy. *Organization, 17*(3), 363–378. doi:10.1177/1350508410363119

Smidt-Jensen, S., Skytt, C. B., & Winther, L. (2013). The geography of the experience economy in Denmark: Employment change and location dynamics in attendance-based experience industries. In *The City in the Experience Economy* (pp. 37–52). Routledge.

Song, H. J., Lee, C. K., Park, J. A., Hwang, Y. H., & Reisinger, Y. (2015). The influence of tourist experience on perceived value and satisfaction with temple stays: The experience economy theory. *Journal of Travel & Tourism Marketing, 32*(4), 401–415. doi:10.1080/10548408.2014.898606

Sundbo, J. (2009). Innovation in the experience economy: A taxonomy of innovation organisations. *Service Industries Journal, 29*(4), 431–455. doi:10.1080/02642060802283139

Sundbo, J., & SËrensen, F. (Eds.). (2013). *Handbook on the experience economy*. Edward Elgar Publishing.

Sundbo, J., & Darmer, P. (Eds.). (2008). *Creating experiences in the experience economy*. Edward Elgar Publishing. doi:10.4337/9781848444003

Tan, J., Ronkko, K., & Gencel, C. (2013, October). A framework for software usability and user experience measurement in mobile industry. In *Software Measurement and the 2013 Eighth International Conference on Software Process and Product Measurement (IWSM-MENSURA), 2013 Joint Conference of the 23rd International Workshop on* (pp. 156-164). IEEE. 10.1109/IWSM-Mensura.2013.31

Therkildsen, H. P., Hansen, C. J., & Lorentzen, A. (2013). The experience economy and the transformation of urban governance and planning. In *The City in the Experience Economy* (pp. 115–132). Routledge.

Tjong, S., Weber, I., & Sternberg, J. (2003). Mobile, youth culture, shaping telephone use in Australia and Singapore. *ANZCA03 Australian and New Zealand Communication Association*.

Ureta, S. (2013). Mobilising poverty?: Mobile phone use and everyday spatial mobility among low-income families in Santiago, Chile. In *Understanding Creative Users of ICTs* (pp. 105–114). Routledge.

Walsh, S. P., White, K. M., & Young, R. M. (2010). Needing to connect: The effect of self and others on young people's involvement with their mobile phones. *Australian Journal of Psychology*, *62*(4), 194–203. doi:10.1080/00049530903567229

Weilenmann, A. (2001). Negotiating use: Making sense of mobile technology. *Personal and Ubiquitous Computing*, *5*(2), 137–145. doi:10.1007/PL00000015

White, M. P., Eiser, J. R., & Harris, P. R. (2004). Risk perceptions of mobile phone use while driving. *Risk Analysis: An International Journal*, *24*(2), 323–334. doi:10.1111/j.0272-4332.2004.00434.x PMID:15078303

Yan, Q., & Gu, G. (2007, July). A remote study on east-west cultural differences in mobile user experience. In *International Conference on Usability and Internationalization* (pp. 537-545). Springer, Berlin, Heidelberg. 10.1007/978-3-540-73289-1_62

# Section 1
# Investigating the Impacts of Mobile Use

*Mobile is used by people of different age groups from the youngest to the oldest in different countries. It is also used in different areas for different purposes. Its reach and penetration are expedited and exponential. Its uses and impacts are multi-dimensional and multi-directional, being constantly examined from different disciplinary perspectives using different methods, generating growing amount of scholarship on how mobile has been used and its impacts on both actors and activities. In this section, we have nine chapters to provide most recent research findings to serve as an updated account of the renewed academic efforts in continuously describing, explaining and predicting the impacts of mobile use on actors and activities in the world. As far as actors are concerned, this section covers adolescents, college students, working professionals, and retirees, involving both males and females in different countries. In terms of activities, this section involves use of mobile for academic studies, personal entertainment, dating, food posting, mobile social media, and use of mobile apps. In terms of approaches, this section highlights interdisciplinary and integrated methods. In terms of countries involved, this section covers Australia, China, Colombia, India, Oman, Pakistan, New Zealand, Turkey, UK, and US.*

*In their chapter on mobile use during adolescence, Melody Terras and Judith Ramsay recommended that impacts of mobile use should be examined in different contexts as mobile use can be shaped by the interactions of mobile users within different contexts. They also suggested that such examination should also be interdisciplinary and integrated in order to shed light on the complexity of technology and behavior of young mobile users. Their recommendations were framed within psychological theories of development.*

*Mobile is popular not only among young users but also among retirees. How mobile has been used among retirees, especially for those who just retired with plenty of time and money to spend in cities? An answer was provided by Chen Guo, Michael Keane and Katie Ellis in their chapter on impacts of mobile use on third agers, or in other words, retirees, in a tier-two city in China.*

*As an indispensable part of their lives, mobile has been playing an important role among undergraduate students in influencing their lives and academic studies. Although gender has been proven to be an important factor in distinguishing males and females in using the Internet and other digital media, what are the gender differences in mobile use and habits? By surveying a selected sample of undergraduate students in Turkey, Beliz Dönmez located gender differences in mobile use and habits.*

*In the chapter to explore whether mobile dating is a fancy way to fall in love, Lu Sun ignited further interest in examining the issue by reviewing up-to-date studies on mobile dating, examining mobile dating applications and individual users' motivations at different stages of mobile dating, including LGBTQ studies. Finally, the author identified various problems related to mobile dating that invite further examination and solutions.*

*To post food on social media has become a frequent source of fun and joy in life for many mobile users. In investigating such a common scene on Instagram among its young users, through investigating use of mobile to post food on Instagram, Wan Chi Leung and Anan Wan identified six gratifications sought from posting food photos. Furthermore, they also located predictors of personal involvement and frequency in posting food on Instagram.*

*To investigate smartphone usage, gratifications, and addiction by applying mixed research methods of a survey and focus groups, Hafidha S. AlBarashdi, and Abdelmajid Bouazza examined the functions, types, and motivations of smartphone usage and gratifications. Furthermore, the authors also investigated the rates, symptoms and reasons of smartphone addiction. Still another achievement of this authors was to look into the relationship between smartphone usage, gratifications and addiction with academic achievements among college students. On top of identifying three levels of addiction, the authors also located distinctive traits of these levels.*

*Through investigating mobile usage and its socio-economic impact in Pakistan, Sadia Jamil presents a unique and interesting case of the socio-economic impacts of mobile use on users' lifestyle. Although there exists an obvious divide between urban and rural areas in terms of impacts of mobile use, the case of Pakistan could serve as an alert to scholars that why mobile use remains limited in narrowing the gap between urban and rural areas against a backdrop of mobile being widely believed to be able to play a big role in narrowing the social and economic gap between urban and rural areas. Another interesting and imperative case brought about by this chapter is that mobile use could also be very gender biased in rural areas, resulting in a gap between males and females as far as social and economic impacts of mobile use on their lifestyles. Pakistan is not alone when it comes to the divide between urban and rural areas as well as between male and female mobile users in terms of socio-economic impacts of mobile use. This chapter can serve as a stepping stone to further studies of the same topic in other developing or underdeveloped countries.*

*China is well known for its wide and increasing commercial use of mobile social media for various purposes in different areas, ranging from online shopping to social networking. Such a popular commercial use was insightfully examined in relation to social relationship in the age of mobile Internet, which enables people of either weak or strong connections to socialize anywhere anytime, leading to scenarios where mobile social media can be leveraged for profits. In what way can user experiences be guaranteed while platforms' value-added targets be achieved at the same time? In addressing that question, the authors of this chapter examined the commercial use of mobile social media in the context of complicated social networks.*

*Through case studies of mobile apps by using the creator-audience matchmaking model, Biplab Lohochoudhury provided a bird's-eye-view of the journey of exploring mobile use and its impacts in India as well as an analytical and insightful review of earlier studies on mobile communication development and utility-driven usage. Although country-specific, Lohochoudhury's chapter offered a very interesting, insightful, and invaluable elaboration of how the model was applied to secure a better explanation and prediction of use of mobile apps in India.*

# Chapter 1
# Mobile Use During Adolescence:
## Determinants and Impacts

**Melody Terras**
*University of the West of Scotland, UK*

**Judith Ramsay**
*Manchester Metropolitan University, UK*

## ABSTRACT

*Equipped with psychological theories of development, the authors of this chapter argue that the impacts of mobile technology should be situated in and understood within their different contexts of use. Besides their advocacy of a situated and integrated approach to understanding the psychological determinants of mobile use and the impact on development, the authors also call for increased interdisciplinary and integrated working to capture the complexity of technology and behaviour and the recognition of individual differences.*

## INTRODUCTION

Mobile phones are ubiquitous in everyday life and are having a profound impact on the way people of all ages work, learn, play and communicate. Children are the fastest growing population of mobile phone users: recent research in the UK indicates that mobile ownership and use is highest among children aged 12-15 years, with 55% owning their own tablet and 83% owing a smartphone (Ofcom, 2017). Despite the widespread use of smartphones, understanding of the scope and scale of their impact on children's development remains limited and the evidence base concerning the psychological factors that determine their impact and use is currently highly fragmented (Terras & Ramsay, 2016).

Initially research focused on smartphone use by adults, but in recent years there has been increased research on smartphone use by children. In the beginning, research was quite demographic and rather descriptive in nature focusing on documenting levels of ownership and patterns of use. More recently research has begun to explore socio-demographic and psychological factors in order to understand how and why smartphones are influencing, not only personal and societal practices concerning communication and internet use, but also the opportunities and challenges this presents for children's behaviour and

DOI: 10.4018/978-1-5225-7885-7.ch001

development. An influential example of this type of research are the recent large-scale European studies that investigated the factors that influenced children's (9-16 years) use of the internet and smartphones. The results indicate the significant and complex influence of a range contextual factors such as parental and family use and attitudes concerning mobile devices and societal and cultural norms (Livingstone et al., 2015; Masheroni & Olafsson, 2015). Such findings are an important and timely reminder of the crucial influence of context in determining behaviour and highlight that it is imperative to consider the range of contextual factors that influence the use of mobile devices and how this use can impact on development.

Mobile devices, especially smart phones not only offer children increased opportunities for communication, both by speech and text, but the internet access they support provides access to a wide array of information, resources, social media and online spaces where they can learn, socialise, apply and develop their cognitive and social-emotional skills. Smartphones afford children a range of complex digital contexts to operate in and interact with i.e. smartphones provide access to "new contexts of development and being" (Terras & Ramsay, 2016, p. 2) and it is therefore essential to fully understand the nature of these contexts, the impact they may have on development, and the processes through which their influence occurs. We propose that such questions will be most successfully addressed by adopting a psychological perspective: the primary aim of psychology is to identify and understand determinants of behaviour. In particular we draw upon theory and research from developmental psychology as it has a well-established tradition of recognising and mapping the influence of context on development. Therefore, in this chapter we offer a psychologically-informed critical review and synthesis of the existing literature and identify how development is being influenced by smartphone use and the contexts and processes through which this occurs. We discuss how the increasing use of mobile technology and the internet access it supports, is changing the nature of existing developmental contexts and creating new online contexts for psychological development. Two key developmental contexts: parents and peers are considered in-depth and we illustrate how mobile technology, especially smartphones are influencing parenting practices and the growth of parental mediation strategies; peer group influences on mobile use and the impact of mobile use on social behaviour and relationships are also examined. The discussion will focus specifically on adolescence as this age group of children are the heaviest users of smartphones and it is essential to understand the impact smartphone use is having during this formative phase of psychological development. We conclude by summarising the insights that this psychological perspective offers and demonstrate how viewing smartphone use through a psychological lens, drawing on theory, research and methodology, helps to provide a framework within which to situate future research that is more explanatory in nature and helps to shape the future research agenda.

## THE DEVELOPMENTAL IMPACT: SMARTPHONES OFFER NEW CONTEXTS FOR DEVELOPMENT

Adopting a psychological perspective, especially consideration of developmental psychology theory, research and methodologies can offer insight into factors that determine the use and impact of smartphones (Terras & Ramsay, 2016). Psychological theories of development recognise the important influence of context for example, Bronfenbrenner (1979) advocated an ecological approach to understanding development and characterised how the different contexts, or ecosystems, that a child inhabits can influence development either directly or indirectly. He identified contextual influences at a number of levels: *Microsystem* e.g. home, school, peer activities; *Mesosystem* which represents the links and interconnections between the different microsystems, the *Exosystem* that captures less direct influence such as mass media, political and economic influences, and the *Macrosystem* which representations social and cultural values. Such an approach highlights the importance of contextual influences on development and reminds us of the need to explicitly consider the influence of online and technologically mediated contexts. Smartphone use not only occurs directly within traditional *microsystem* contexts such as the home and school, but the internet connectivity that smartphones provide enable children to access an exciting new *virtual micro-system* with unprecedented access to news media and information concerning not only social but also educational, economic and socio-political content that can influence development.

The significant influence of social context and socio-cultural practices on development were also recognised by Vygotsky (1987), and his insights concerning the importance of language and communication, social interaction and the scaffolding of behaviour by adults and more experienced peers has had a profound and enduring impact not only within psychology but also in educational and developmental science contexts (Hedegaard, 2009). Smartphones allow children to access wide ranging opportunities for communication and socialisation either via email, texting or via the increasing number of social media platforms and messaging apps such as You Tube, Facebook, Twitter, Snapchat, Instagram, WhatsApp and online special discussion/blogging forums and multimedia platforms. Internet use is no longer passive, it is highly active and participatory and offers a range of interactive opportunities for cognitive, social and identity development (Terras, Ramsay & Boyle, 2015).

Viewing smartphone use through a psychological lens not only supports the specification of behavior, and theory that captures contextual influences; a psychological perspective also brings methodological rigor with its focus on the scientific method. Such an empirical approach allows examination of not just correlations between behavior and outcomes, but a means to establish causal relationships. Of particular relevance to Smartphone use is the use of longitudinal studies to track how behavior changes over time, an essential element of understanding the use of mobile technology across the lifespan. The question then is to establish the nature and processes that underlie this influence.

## Understanding the Process of Influence: Development With and Through Technology

Yan (2018) argues that although child development has always been influenced by technology, "mobile phone use plays a unique role" (Yan, 2018 p. 6) due to the very high rates of ownership and use, the immense opportunities for personalisation they offer, and their multi-function nature that offers a wide range of technical and developmental affordances that allow children to develop both *with* and *through* technology. *Development with technologies* capture that fact that technology, including smartphones, are part of everyday culture (e.g. part of Bronfenbrenner's *Exosystem*) and can therefore influence development. *Development through technologies* captures the ability of technology to act as a developmental mechanism, in Vygotskian terms it is a tool to support understanding and development as they are used directly by the child. This characterisation is important as it offers a potential explanation of the processes through which smartphones influence development and is very much in keeping with a psychological contextualised view of development by recognising how the wide spread use of smartphones has altered existing contexts and created new contexts for development and being.

Specifically, Yan (2018) identifies two mobile contexts or cultures that influence development by providing opportunities to learn, model and receive scaffolding of behaviour: (1) General Mobile Culture (Goggin, 2012) that embodies the general socio-cultural context which includes attitudes and behaviour concerning smartphone use, and (2) Youth Mobile Culture (Vanden Abeele, 2016) which captures the context of peer influence, especially practices and norms concerning smartphone use, that is so important to adolescent development. Yan (2018) identifies how the context, and the technology used within it, can influence development either indirectly as a moderator, or directly as a mediator. A moderator variable influences the strength of the relationship between variables, a mediating variable explains how or why there is a relation between two variables (Baron & Kenny, 1986). Understanding the nature and process of the influence of smartphone use on development is essential and psychologists have long recognised the importance of specifying the nature of contextual influences on development, specifically whether a variable (an aspect or character of the environment or a person) determines or only influences behaviour i.e. whether it functions as a mediator or a moderator. The key question then is whether smartphones act as mediators or moderators on development?

Smartphones may have an exceptionally strong influence on development as they have the ability to act as both mediators and/or moderators depending on the context. At the *Exosystem* level they "can be considered a developmental moderator that changes developmental processes by adding an extra element" (Yan, 2018, p. 7). At the *Microsystem* level they can function as "a developmental mediator (i.e. tools and signs) that changes developmental processes by adding an internal element" (Yan, 2018, p. 7). The increasing use of smartphones by children means that the developmental process is becoming more influenced by their mediation effects and highlights the very strong and almost unavoidable impact of mobile technology on development (Yan, 2018). Therefore, it is essential develop a detailed understanding of the use of smartphones, contextual influences, and the impact of smartphones on development.

## SMARTPHONES: IMPACT AND USE

The technological sophistication of mobile devices and the online opportunities they afford are developing at a rapid pace and will continue to do so in the future, therefore it essential that we understand their impact on development. Recognition of the impact of technology is not new, and research attention has been paid to the influence of radio, television, computer games and the internet in general. However, the increasing use of mobile devices, especially smartphones, not only offers new opportunities but also challenges to smartphones users and society as a whole. Smartphones are increasing in popularity and fast becoming the mobile device of choice for a number of practical reasons such as decreasing cost, improved *wi-fi* connectivity and faster and more stable internet access, but also for more psychological reasons such as the high degree of personalisation they offer and the wide range of Apps that enable anytime and anywhere 24/7 access to opportunities to socialise, shop, learn and be entertained by the music, games, film and TV etc. of choice.

In recent years smartphone ownership has rapidly increased, with 77% of adults in the US (Pew Research Centre, 2018) and 74% of adults in the UK owning a smartphone (Ofcom, 2018). Smartphones are also becoming the preferred, and in many instances the sole, means of accessing the internet and these trends in ownership and use are also evident in children. Increasing numbers of children now own a smartphone or mobile device and are accessing the internet from a very young age. Recent data indicates that children's use of mobile devices develops and changes across time: initially children use mobile devices such as tablets to watch TV and to access the internet. However, this pattern of use and access changes as children grow older and acquire their own smartphone and begin to access the internet not only to passively view content but to actively use a range of Social Media. For example, in the UK, only 1% of 3-4-year-old children have a smartphone but 21% own a tablet which 71% use to go online, with 53% of children being online for nearly 8 hours a week. At ages 5-7 years 5% have a smartphone, 35% own a tablet which 63% use to go online, with 53% of children being online for an average 9 hours a week and 3% have a social media profile. By 8-11 years of age, 39% have a smartphone, 52% own a tablet which 46% use to go online, with 94% of children being online for an average 13.5 hours a week; 23% have a social media profile and report that the TV or their tablet would be the device they would miss the most. By the time children reach adolescence their use of, and reliance on, smartphones has increased substantially with 83% of 12-15-year-old children owning a smartphone, 55% own a tablet which 49% mostly use to go on line; but now 29% mostly use their smartphone to go online, with 99% of children being online for an average 21 hours a week. 74% of adolescents have a social media profile and report that their mobile phone would be the device they would miss the most (Ofcom, 2018).

Although research on mobile devices is often multidisciplinary, smartphone use by children is receiving increased attention from psychologists due to the recognition of its potential influence on physical, cognitive and psycho-social development. The first study that focused specifically on smartphone use examined the attentional impact of smartphone use on driving (McKnight & McKnight, 1993), and over the years interest has continued to develop: by 2016 there were 2344 published studies examining a wide range of issues, with medical interventions, radiation exposure and texting initially receiving the most attention. More recently, research is examining issues such as cyberbullying, sexting, mental health and mobile addiction (Yan, 2018). This research clearly indicates that children are spending more time online and that smartphone use is having a clear impact on behaviour. We therefore need to understand why smartphones use is so appealing and the contextual influences on their use.

## Psychological Determinants of Use

It is essential to remember that a full understanding of the developmental impact of smartphones, requires appreciation not only of the reasons why the technology is used, but also the opportunities that using this technology provides. Both these factors are considered by the two main approaches that have examined the reasons why people use technology: Uses & Gratifications Theory (Katz, 1959) and the Technology Acceptance Model (Davies, 1986).

Uses & Gratifications Theory attempts to explain why individuals use technology and thereby focuses attention at the level of the individual. As the name suggests, this theory aims to capture how technology is used (uses) and the benefits it brings (gratification). Over time the theory has been refined to provide a more detailed understanding of technology related behaviour with research driven by this approach addressing the underlying psychological motivations for the behaviour- the uses and gratifications sought, and the consequences of the behaviour- the gratifications obtained (Raacke & Bonds-Raacke, 2008). Research indicates that ppeople use mobile phones for a range of reasons including sociability, social support, entertainment and instrumentality (Leung & Wei, 2000). Research has also identified the motivations for text messaging such as accessibility/mobility entertainment, relaxation and escape, information seeking, organisation and co-ordination, socialisation and social support, and revealed some interesting gender differences with women valuing accessibility/mobility, relaxation and escape, and coordination more than men (Grellhesl et al., 2012). As expected, a major factor in the use of social networking sites such as facebook and twitter is to maintain and form new social connections (Raacke & Bonds-Raacke, 2008; Chen, 2011). Uses and Gratification theory has also extended our understanding of interesting differences in the motivations to use different types of social networking sites. For example, people tend to use snapchat to meet more personal needs such as emotional support, advice and to express negative emotions as well as connecting with others (Phua, Jin & Kim, 2017). Smartphones are frequently used to access social media and research using a Uses & Gratifications approach has demonstrated the importance of considering the different uses and gratifications of facebook users and the relationship between emotional and affective attitudes towards facebook use and wellbeing (Dhir & Tsai, 2017) and offered insight into problematic social media use by demonstrating how personality traits predict different motives of use, preferences for different social media platforms and how a combination of these two factors predict problematic social media use (Kircaburun et al (2018).

The Technology Acceptance Model also offers insight into users' behaviour and has proved particularly useful as it provides a framework to help understand the nature and directionality of the relationship between different influences on technology use which is key to understanding the mediating and moderating influence of smartphones on children's development. The Technology Acceptance Model (Davies, 1981) explains why people use or do not use technology and is based upon The Theory of Reasoned Action (Fishbein and Ajzen, 1975) and the Theory of Planned Behaviour (Ajzen, 1985), well established psychological theories that aim to identify the determinants of behaviour. The key components are Behavioral Intentions, Attitude towards the behavior and the Subjective norm: behaviour is determined by the Intention to perform it, and this Intention is determined by a person's attitude toward the behaviour and the subjective norms of the behaviour and Perceived Behavioral Control. The theory of planned behavior predicts that when there are more positive attitudes, supportive subjective norms and high degrees of perceived control, the stronger the behavioral intention, therefore the more likely a behavior is to occur. When applied to the use of smartphones, studies have shown that perceived ease of use and perceived usefulness are key determinants of use. Other relevant psychological influences include social inclusion as an important motivation for use, as well as instrumental use i.e. usefulness and importance in to achieving goals (Park et al., 2013). Interestingly perceived usefulness and perceived ease of use have been shown to be important factors in smartphone dependency i.e. people are more likely to become dependent on smartphones once they appreciate its benefits and ease of use. Here too social inclusion motivations play an important role (Park et al., 2013). Such findings are important reminder that although the technology used to access to such sites (smartphones) and the technologies that underpins these platforms may be the same, the reasons that people choose to use them and the benefits they gain from doing so may be very different.

## CONTEXTS OF USE

The significant influence of the home as a context for development is well recognised and as children grew older, especially when they reach adolescence, peer influence becomes increasingly important; therefore, it is essential to understand how smartphones are used within these contexts and how they influence development by creating new contexts for development and being.

### The Contextual Influence of the Home and Parents

Research demonstrates that the home environment is becoming increasing rich in technology and that parental use and attitudes are strongly influencing their child's use of technology from a young age (Marsh et al., 2015; Stephen et al., 2013), especially the use of smartphones and social media (Terras,

Yousaf & Ramsay, 2016). In this modern age, parents not only have to manage their child's use of technology, but also their own. Parents now have to supervise and monitor both off and online behaviour and develop their parenting styles to meet these needs. Although the vast majority of parents appreciate the benefits of technology, they also wish to minimise disadvantages such as lack of sleep and exercise and reduced opportunities for face-to-face social interactions due to excessive use. Of particular concern are the risks associated with internet access such as exploitation, access to inappropriate adult content, bullying, mental health issues and possible internet and smartphone addiction (Haddon & Vincent, 2015) and parents frequently monitor and limit technology and internet use of their child.

A key concern that parents express about smartphones is not related to the device itself but related to the internet access it allows. Internet access via smartphones is more difficult to monitor than tablet or laptop access due to the small size of smartphones that makes their use less noticeable (Vincent, 2015). Emerging evidence indicates that parental concerns over excessive use are well founded: research indicates a relationship between the amount of time spent using mobile devices and psychological well-being, with lower leisure use associated with lower levels of depression (Krcmer et al, 2013). The nature of the activity may also have an impact: recent evidence demonstrates the negative impact of excessive social media use, with high levels being associated with less happiness, increased depression and sleep disturbance (Lin et al., 2016; Levenson et al, 2016). Given these findings parents rightly have concerns over the impact of smartphones on development and are even more likely to monitor and regulate their use. Regulation of technology, in all its forms, is a newly emerging parental responsibility which is especially challenging given the rapid pace of technological development.

The nature of parent-child relationship is widely accepted to influence child development (Bowlby, 1969), and emerging evidence suggests parenting styles and parental attachment may also influence how children use smartphones, with weaker parental attachment being associated with negative smartphone behaviour such as internet addiction and cyberbullying (Chang et al, 2015), and more nurturing parental styles associated with less addictive use (Bae, 2015). The ways in which parents manage their child's behaviour is traditionally captured by the concept of parenting styles (Baumrind, 1966) and this approach has recently been expanded to include how parents regulate the use of technology using the concept of parental mediation. Livingstone et al. (2011) examined how parents of 9-16-year olds across 25 European countries managed their child's use of mobile devices and the internet and identified five main parental mediation strategies. The first two strategies involve active mediation of internet use and also a separate strategy to address internet safety, with active mediation based on discussion and collaborative use to demonstrate safe practice. Discussion not only appears to support safe use, but evidence also suggests it may help to reduce some of the more negative aspects of internet media use experienced by adolescents such as reduced life satisfaction (Boniel-Nissim et al, 2015) and compulsive internet-related behaviours (Van den Eijnden et al., 2010). Other parental mediations strategies include the imposition of restrictions either through the use of technical filters or simply setting time limits on use and content. A final strategy involves post-hoc checking of internet activities. From these findings it is clear that technology is impacting on the home environment and producing changes in parental practices that in turn influence their child's behaviour and development.

Parental use of technology, especially smartphones has been shown to influence the nature of their children's use, yet many parents do not fully recognise the extent of their influence (Plowman et al., 2012), and thereby underestimate the powerful influence of social learning (Bandura, 1977) concerning the use of technology. Recent evidence suggests that the increasingly use of technology within the home is not only influencing technology related behaviour, but it may also be having a more generalised detrimental impact on children's behaviour generally. McDaniel & Radesky (2018) examined the concept of *technoference* which reflects technology-based interruptions in parent–child interactions in young children (3 years) and identified a predictive relationship between levels of *technoference* and parental reports of externalising and internalising behaviour problems displayed by their children. Although they looked at young children, the presence of this relationship is concerning, and the nature and directionality warrants further study as evidence from older children also indicates a more generalised negative impact on behaviour with parental relationship quality being associated with social functioning in both off and online contexts such as poor school connectedness, more deviant peer relationships and gaming addictions (Zhu et al., 2015).

## THE CONTEXTUAL INFLUENCE OF PEERS

Peer influences play an essential role in development and become increasingly important during the adolescent years (Santrock, 1998). Smartphones offer an increasing array of opportunities for communication and social interaction via a variety of social media platforms. Children and young people are avid users of social media, which is generally accessed via a smartphone, which provides opportunities to interact with their peer group and friends to receive social support and learn social norms (Subrahmanyam et al., 2008). Online and technology mediated interactions with peers are becoming an increasingly important context for development as peer interactions are an important means of not only social skills but also identity development (Zarbatany et al.,1990; Reis and Youniss, 2004). Online contexts offer important new contexts for self-presentation and identity development and recent findings from older adolescents raises important issues for consideration with respect to how younger children may use social media to create and manage their online identity. For example, Johnson & Ranzini (2018) explored how self-presentation motives influence the sharing of entertainment content (e.g., music and films): individuals motivated to conform to peer group norms shared music and film that was in keeping with group norms, while those aiming to represent themselves in the best possible light shared more impressive music and films.

Emerging evidence demonstrates the increasingly role of peer influence via social media not only on the use of social media itself, but also on behavior more generally. Mishra et al (2018) explored online content in the form of electronic word of mouth (eWOM) concerning consumer experiences and the purchasing of goods. Findings indicate that the eWOM behaviour of teenagers is strongly influence by peer norms, especially for boys. The results also indicate interesting potential gender differences in online peer influences, as females tend to place more emphasis on the perceived credibility of the source. Teenager health behaviour is strongly influenced by social norms, especially peer influences and research demonstrates these influences are operating via social media exposure with recent empirical studies illustrating the influence of social media exposure to e-cigarettes influencing e-cigarette beliefs and use (Pokhrel et al, 2018), prodrinking messages and binge drinking (Yang & Zhao, 2018).

Although the posting of photos and videos and other user-generated multi-media content underpins social media content and use, it is important not to underestimate the importance of written verbal communication- mostly commonly in the form of text messages. Text messages were the first and are the most established form of mobile mediated communication, and like technology itself, the use of text messages evolves over adolescence. Insight into their evolving use can be gained by using a longitudinal method. For example, Coyne et al (2018) examined how text messages were used over a 6-year period

(13-18 years). Texting peaked during mid adolescence and four different types of texters were identified: perpetual, decreasers, moderators and increasers. Most participants fell into the perpetual category. Interestingly excessive use of text messages at age 13 years was associated with negative outcomes such as increased levels of depression, anxiety, physical and relational aggression at age 18 (Coyne et al., 2018). These findings remind us of the need to recognize individual differences in technology-related behavior and the importance of considering how behavior changes over time as both these factors exert an important influence on development.

The increased sophistication of cameras on smartphones and the ability to share photos has resulted in the increased growth of picture messaging and the posting of images on social media. Picture messaging also supports sexting behavior. Rice et al (2018) examined a sample of 12-18yrs and found that 17% were both sending and receiving sexts, and 24% only receiving sexts. These findings are comparable but higher than existing data e.g. Klettke et al. (2014) whose literature review on sexting, in samples of children 18yrs and under, reported a lower rate of 10.2% - 18.63%. It is important to note that prevalence rates are difficult to establish due to the different definitions of a sext i.e. whether it includes text and a picture or only a picture. Regardless of the actual figures, available data indicates it is a significant issue and concern is justified especially due to the relation between sexting and risky behavior, being sexually active, exploitation and privacy. The increasing occurrence of sexting behaviour by adolescents needs to be addressed and managed with further research required to determine the reasons that underlie this behavior. Unsurprisingly, peer influence plays a significant role: having a peer who sexts means adolescents were twice as likely to send a sext and 13 times more likely to receive one (Rice et al., 2018). Adolescent smartphone use and online behaviour seems to be as open to peer influence as more traditional non-technology mediated behaviours and it is essential that future research explores the complexity of peer influences in the technology mediated contexts of development that smartphones provide access to.

## SUMMARISING INSIGHTS AND RECOMMENDATIONS FOR FUTURE RESEARCH

Adopting a psychological perspective in terms of theory, research and methodology offers considerable insight into smartphone behaviour now and in the future as it can provide a framework to organise future research and practice. In this section these insights are summarised and their implications for future researched considered in the following section.

### The Utility of a Psychological Perspective

In this paper we have discussed how viewing smartphone use by adolescents though a psychological lens can enrich our understanding of the contextual influences on their use and highlights the importance of considering how technology use in general and smartphone use in particular is creating new online contexts for development and being. The insights offered fall into four main areas:

### The Situated Nature of Development

Psychological theories of development remind us that behaviour does not occur in a vacuum, it is situated in a wider context and we have discussed how different contexts, and the behaviour of people within them, can influence the use of smartphones. The adoption of such a situated and ecological approach helps to identify and characterise factors that influence smartphone use and provide a framework to help understand how smartphones may influence development either as a mediator or moderator, thereby informing the research agenda for the future.

## Similarities and Differences in Behaviour in Online and Offline Contexts

The increasing use of smartphones and its associated impact on development necessitates the need to understand how behaviour and development is being impacted by these technologies, and how online behaviour differs from offline behaviour- what are the similarities and differences and whether behaviour in physical contexts and virtual contexts? Are they subject to the same influences? By way of example, consider the difference between the online and offline concept of a friend, online social networks enable users to see the connections and relationships between people in a way that is not possible offline and users build on these connections and establish connections with their friends' friends (Notley, 2009), in an online context "a friend may be someone, one has never met, and will never meet, face-to-face" (Terras, Ramsay & Boyle, 2015, p. 31). Yet despite these differences, similarities across contexts, remain especially with respect to potential determinants of behaviour. For example, the influence of peer behaviour remains powerful in both on and offline contexts especially in supporting the development of social skills and identity and peer influences have been identified in the use of smartphones for sexting behaviour. The influence of parenting remains strong, although parenting techniques are evolving to meet the demands of the challenges of mobile and Web 2.0 technology, they still remain a major influence on development and behaviour with respect to the use of technology.

## The Fragmented Nature of the Existing Evidence Base

At present the evidence base concerning the use and developmental impact of smartphones is highly fragmented descriptive with research studies often relying on self-report measures which tend to provide inaccurate data on behaviour and atheoretical in nature (Terras & Ramsay, 2016; Legris, Ingham, & Collerette, 2003). The situated approach advocated by psychological theories offers a framework for integration and also provides a means not only to describe behaviour, but also to support consideration of the underlying causes and processes with key developmental mechanisms such as observational and social learning, and scaffolding offering explanations for peer and parental influences on smartphone behaviour. Although we have advocated the benefits of adopting a psychological perspective in this chapter, future research concerning smartphones should be multidisciplinary in order to aid integration and fully capture the complexities of technology, behaviour, how it changes over time, and the contexts in which it used (Shaw et al., 2018). Whatever perspective is taken, it is essential that future research retains a situated approach to capture contextual influence. We only have to look to evolutionary history to be reminded of how we develop and adapt to our environment, be the environment physical or virtual.

## The Importance of Individual Differences

A key element of a psychological approach involves the recognition of individual differences in behaviour and the need to understand motives of use. Behaviour is similar in both offline and online contexts as it is determined by psychosocial variables such as personality, need for social and emotional support, motivation. To date consideration of the influence of a full range of individual differences in determining the use of smartphones has received limited research attention and it is essential that all these factors must be addressed by future research.

## IMPLICATIONS FOR FUTURE RESEARCH

As noted above these psychological insights help identify important areas for future research and these are grouped into two main areas below:

### The Psychological Function of Smartphones: The Importance of Individual Differences

In order to fully understand how smartphones are influencing development, the influence of contexts and how contexts are influenced by individual differences in the socio-emotional profile of the user requires future research to more closely address the psychological function of smartphones. Although we currently have a good understanding of what smartphones are used for e.g. texting, social media, gaming, watching videos etc., greater understanding of the psychological function of smartphones is required to fully appreciate their developmental impact (Terras & Ramsay, 2016). Interesting insights into their role and function comes from recent research exploring perceptions of smartphones and the purpose they serve, and clearly demonstrates how smartphones are perceived as essential by children and young people as they enable them to feel a part of and participate in everyday life. Smartphones serve the purpose of Being across a number of domains Being Mobile enables not only 24/7 internet access but more importantly it allows them to participate in physical and virtual contexts and activities simultaneously; Being Social reflects communication and participation in social media with smartphones regarded as being a necessary tool to support their social life, Being Educated captures the use of smartphones to access and use formal and informal online learning opportunities; Being Entertained reflects access to entertainment such a videos, movies and games, Being Me captures how online activities support the development of identity, self-expression, and socio-emotional well-being; lastly smartphones were viewed as a means to Be safe as although young people recognised the risks associated with online activity these were outweighed by the opportunities provided to make contact with parents and other figures of trust and authority if required (Vincent, 2015).

Consideration of individual differences reminds us that technology is used differently, for different motives, and that some individuals may be more likely to develop problematic behaviour. The increasing concern over the potential for internet, gaming and smartphone addiction has been matched by increased research to understand determinants and identify who is most at risk. Although existing research offers interesting insights e.g. evidence suggests an association between the quality of mother-child attachment and mobile phone dependency (Toda et al., 2008), and emerging evidence demonstrates the influence of individual differences such as extroversion, self-esteem and extent of self-monitoring in problematic smartphone phone use (Bianchi & Phillips, 2005; Kim & Hahn, 2015), much like other research concerning the impact of smartphones it often fails to offer explanations, only associations. Therefore, future research must be more theory driven, use longitudinal designs and exploit the benefits of statistical analysis and modelling techniques to further understand the nature of negative influences. An interesting example of research involving structural equation model using mediation analysis is that of Seo et al (2016) who demonstrated a positive predictive relationship between mobile phone dependency and depression in adolescents. Future research should also be informed by available evidence from slighter older populations, for example research in college students offers insight into the relationship between the amount of smartphone use and psychopathology, with heavier users scoring higher on recognised indicators of psychopathology such as ruminative thinking and poor emotion regulation (Elhai & Contractor, 2018). So the influence of these factors should be explored in adolescent samples to help identify problems early and inform the development of interventions.

In the future, it is important not just to focus on those most at risk, given the widespread use of mobile devices and the internet, it is essential to promote safe practice for all, and future research should examine factors that support the positive use of smartphones and aim to identify factors that help to mitigate against negative impact. A promising start is the work of Robertson et al. (2018) who examined the importance of resilience in preventing internet and online gaming addiction. Resilience is a key concept in mental health and developmental psychology (Rutter, 2000; Luthar, Cicchetti & Becker, 2000), and is defined as "a dynamic process encompassing positive adaptation within the context of adversity" (Luthar, Cicchetti & Becker, 2000; p. 543). A resilience perspective aims to capture individual differences in how people cope, or fail to cope, and helps to identify those most at risk. A resilience perspective not only focuses on risk, it also encourages consideration and identification of protective factors that mitigate against risk and negative outcomes. A resilience approach is well suited to examining smartphone use as it adopts a situated perspective which explicitly considers the social context in which behaviour occurs and how these contextual factors, both external such as the role of parents and peers, and internal such as self-esteem and personality, can influence functioning over time and could serve as an informative framework within which to situate future research examining both the positive and negative impact of smartphone use on development.

## The Importance of Methodological Rigour and Neuroimaging

Methodologically sound research studies are key to advancing understanding of the developmental impact of technology. Much of the existing evidence base is limited by poor research design in terms of sample size and their representativeness, and this is highly detrimental given the importance of individual differences and the complex nature of contexts. The majority of studies are cross-sectional and although they offer insight into differences between the use of mobile technology across different age groups and in different contexts, they do not permit the detailed study of how behaviour changes over time which is necessary to identify the directionality and causality of factors influencing smartphone behaviour. For example, although parental influences on their child's use of technology is well documented, the causal nature- does it moderate or mediate behaviour- remains unclear (Hui-Lien et al, 2016). Longitudinal studies are necessary to track and understand how and why behaviour changes over time and use of longitudinal research is adding insight into smartphone behaviour. For example, we have discussed how the use of a longitudinal approach has developed understanding of how texting behaviour may changing across adolescence and also reminds us of the presence of individual differences in the use of smartphones i.e. not all adolescents use technology in the same way or even in a consistent manner (Coyne et al., 2018).

The increasing use of neuro-imaging and the emergence of the interdisciplinary field of social neuroscience, offers exciting new methods and opportunities to study the impact of mobile technology at not only a behavioural, but also at a neural level. The recent work by Sherman and colleagues (Sherman et al., 2016; Sherman et al 2018) is an excellent example of the benefits this approach offers both now, and in the future. The importance of peer group acceptance is well documented in the developmental literature where it is generally conceptualized as being qualitative in nature as these influences are difficult to quantify. However, the opportunity to explicitly "like" content on social media offers a straight forward way to quantify peer feedback and this aspect has been capitalized on by recent research examining how individuals respond (Sherman et al., 2016; Sherman et al 2018). This research not only quantifies influence, they also attempt to explain it and advance our understanding of the processes involved by using brain imaging techniques. In doing so they have identified how the brain responds to social media images: the rewarding benefits of receiving positive feedback and approval on social media are reflected in neural activation in brain areas sensitive to reward. These insights are further supported by a recent review of the literature on adolescent neural development which illustrates how the adolescent brain is sensitive to the effects of technology, especially peer influence such as acceptance and rejection on social media, and reminds us of the importance of studying individual differences in technology use and how these differences may influence brain development. Hence, future research should explore not only how technology may impact on behavioural and neural development, but also seek to identify which individuals are at highest risk for negative impacts (Crone & Konijn, 2018). The potential benefits of using neuroimaging data can be further maximised by following the advice of Foulkes & Blakemore (2018) who also remind us that the psychosocial development of adolescents is supported by functional and structural brain development; but the insights this approach currently offers is constrained by the fact that most existing neuro imaging studies tend to neglect individual differences in brain development. Therefore, future research should employ longitudinal designs with fully representative and sufficiently large sample sizes to ensure individual differences are fully explored. It is essential that future research explores individual differences in brain development as such differences may underpin variation in technology related behaviour and therefore require in-depth study in the future. It is clear that the increased use of neuroimaging has immense potential to increase our understanding of the impact of technology at a biological level and thereby the possibility of an explanation for the intensive use and powerful influence of social media in the future.

To conclude, in this chapter we have illustrated the advantages of viewing the developmental impact of smartphone use by adolescents through a psychological lens: it has developed our understanding and helped to inform the future research agenda; highlighted the importance of theory and methodological rigour in determining directionality of influence and establishing causality; focused attention on context and motives for behaviour; highlighted the importance influence of parents and peers; emphasised the need to recognise individual differences; demonstrated the insights that can be gained by considering the similarities and differences between online and offline behaviour and their associated influences and determinants; illustrated the benefits of neuroimaging; and reminded us of the importance of supporting good practice and building resilience. Although a psychological perspective is highly informative; it is not the only valuable source of information and we propose that future research needs to become more integrated and interdisciplinary in order to capture the complexity of technology and behaviour and how these develop over time and that future research should embody key psychologically informed tenets by retaining a situated approach, being methodologically rigorous and increase its consideration of individual differences.

# REFERENCES

Ajzen, I. (1985). From intentions to actions: A theory of planned behavior. In J. Kuhl & J. Beckman (Eds.), *Action-control: From cognition to behavior* (pp. 11–39). Heidelberg, Germany: Springer. doi:10.1007/978-3-642-69746-3_2

Bae, S. M. (2015). The relationships between perceived parenting style, learning motivation, friendship satisfaction, and the addictive use of smartphones with elementary school students of South Korea: Using multivariate latent growth modeling. *School Psychology International*, 36(5), 513–531. doi:10.1177/0143034315604017

Bandura, A. (1977). *Social Learning Theory*. Englewood Cliffs, NJ: Prentice Hall.

Baron, R. M., & Kenny, D. A. (1986). The moderator-mediator variable distinction in social psychological research: Conceptual, strategic, and statistical considerations. *Journal of Personality and Social Psychology*, 51(6), 1173–1182. doi:10.1037/0022-3514.51.6.1173 PMID:3806354

Baumrind, D. (1966). Effects of authoritative parental control on child behaviour. *Child Development*, 37(4), 887–907. doi:10.2307/1126611

Bianchi, A., & Phillips, J. G. (2005). Psychological predictors of problem mobile phone use. *Cyber-Psychology and Behavior: The Impact of the Internet. Multimedia and Virtual Reality on Behavior and Society*, 8(1), 39–51.

Boniel-Nissim, M., Tabak, I., Mazur, J., Borraccino, A., Brooks, F., Gommans, R., ... Finne, E. (2015). Supportive communication with parents moderates the negative effects of electronic media use on life satisfaction during adolescence. *International Journal of Public Health*, 60(2), 189–198. doi:10.100700038-014-0636-9 PMID:25549611

Bowlby, J. (1969). *Attachment and loss* (Vol. 1). Attachment.

Bronfenbrenner, U. (1979). *The ecology of human development: experiments by nature and design*. Cambridge, MA: Harvard University Press.

Chang, F. C., Chiu, C. H., Miao, N. F., Chen, P. H., Lee, C. M., Chiang, J. T., & Pan, Y.-C. (2015). The relationship between parental mediation and Internet addiction among adolescents, and the association with cyberbullying and depression. *Comprehensive Psychiatry*, 57, 21–28. doi:10.1016/j.comppsych.2014.11.013 PMID:25487108

Chen, G. M. (2011). Tweet this: A uses and gratifications perspective on how active Twitter use gratifies a need to connect with others. *Computers in Human Behavior*, 27(2), 755–762. doi:10.1016/j.chb.2010.10.023

Coyne, S. M., Padilla-Walker, L. M., & Holmgren, H. G. (2018). A Six-Year Longitudinal Study of Texting Trajectories During Adolescence. *Child Development*, 89(1), 58–65. doi:10.1111/cdev.12823 PMID:28478654

Crone, E. A., & Konijn, E. A. (2018). Media use and brain development during adolescence. *Nature Communications, 9*(1), 588. doi:10.103841467-018-03126-x PMID:29467362

Davis, F. D., Jr. (1986). *A technology acceptance model for empirically testing new end-user information systems: Theory and results* (Doctoral dissertation). Massachusetts Institute of Technology.

Dhir, A., & Tsai, C. C. (2017). Understanding the relationship between intensity and gratifications of Facebook use among adolescents and young adults. *Telematics and Informatics, 34*(4), 350–364. doi:10.1016/j.tele.2016.08.017

Elhai, J. D., & Contractor, A. A. (2018). Examining latent classes of smartphone users: Relations with psychopathology and problematic smartphone use. *Computers in Human Behavior, 82*, 159–166. doi:10.1016/j.chb.2018.01.010

Fishbein, M., & Ajzen, I. (1975). *Belief, attitude, intention, and behavior: An introduction to theory and research*. Reading, MA: Addison-Wesley.

Foulkes, L., & Blakemore, S. J. (2018). Studying individual differences in human adolescent brain development. *Nature Neuroscience*, 1. PMID:29403031

Goggin, G. (2012). *Cell phone culture: Mobile technology in everyday life*. Oxfordshire, UK: Routledge.

Grellhesl, M., & Narissra, M. (2012). Using the Uses and Gratifications Theory to Understand Gratifications Sought through Text Messaging Practices of Male and Female Undergraduate Students. *Computers in Human Behavior, 28*(6), 2175–2181. doi:10.1016/j.chb.2012.06.024

Haddon, L., & Vincent, J. (2015). *UK children's experience of smartphones and tablets: perspectives from children, parents and teachers. LSE*. London: Net Children Go Mobile.

Hedegaard, M. (2009). Children's development form a cultural-historical approach: Children's activity in everyday local settings as foundation for their development. *Mind, Culture, and Activity, 16*(1), 64–82. doi:10.1080/10749030802477374

Hui-Lien, C., Chien, C., & Chao-Hsiu, C. (2016). The moderating effects of parenting styles on the relation between the internet attitudes and internet behaviors of high-school students in Taiwan. *Computers & Education, 94*, 204–214. doi:10.1016/j.compedu.2015.11.017

Johnson, B. K., & Ranzini, G. (2018). Click here to look clever: Self-presentation via selective sharing of music and film on social media. *Computers in Human Behavior, 82*, 148–158. doi:10.1016/j.chb.2018.01.008

Katz, E. (1959). Mass communications research and the study of popular culture: An editorial note on a possible future for this journal. *Departmental Papers (ASC)*, 165.

Kim, J., & Hahn, K. H. Y. (2015). The effects of self-monitoring tendency on young adult consumers' mobile dependency. *Computers in Human Behavior, 50*, 169–176. doi:10.1016/j.chb.2015.04.009

Kircaburun, K., Alhabash, S., Tosuntaş, Ş. B., & Griffiths, M. D. (2018). Uses and Gratifications of Problematic Social Media Use Among University Students: A Simultaneous Examination of the Big Five of Personality Traits, Social Media Platforms, and Social Media Use Motives. *International Journal of Mental Health and Addiction*, 1–23.

Klettke, B., Hallford, D. J., & Mellor, D. J. (2014). Sexting prevalence and correlates: A systematic literature review. *Clinical Psychology Review*, *34*(1), 44–53. doi:10.1016/j.cpr.2013.10.007 PMID:24370714

Krcmar, M., & Cingel, D. P. (2016). Examining two theoretical models predicting American and Dutch parents' mediation of adolescent social media use. *Journal of Family Communication*, *16*(3), 247–262. doi:10.1080/15267431.2016.1181632

Legris, P., Ingham, J., & Collerette, P. (2003). Why do people use information technology? A critical review of the technology acceptance model. *Information & Management*, *40*(3), 191–204. doi:10.1016/S0378-7206(01)00143-4

Leung, L., & Wei, R. (2000). More than just talk on the move: A use-and-gratification study of the cellular phone. *Journalism & Mass Communication Quarterly*, *77*(2), 308–320. doi:10.1177/107769900007700206

Levenson, J. C., Shensa, A., Sidani, J. E., Colditz, J. B., & Primack, B. A. (2016). The association between social media use and sleep disturbance among young adults. *Preventive Medicine*, *85*, 36–41. doi:10.1016/j.ypmed.2016.01.001 PMID:26791323

Livingstone, S., Cagiltay, K., & Ólafsson, K. (2015). EU Kids Online II Dataset: A cross-national study of children's use of the Internet and its associated opportunities and risks. *British Journal of Educational Technology*, *46*(5), 988–992. doi:10.1111/bjet.12317

Livingstone, S., Haddon, L., Görzig, A., & Ólafsson, K. (2011). *Risks and safety on the internet: the perspective of European children: full findings and policy implications from the EU Kids Online survey of 9-16 year olds and their parents in 25 countries*. Academic Press.

Luthar, S. S., Cicchetti, D., & Becker, B. (2000). The construction of resilience: A critical evaluation and guidelines for future work. *Child Development*, *71*(5), 543–562. doi:10.1111/1467-8624.00164 PMID:10953923

Marsh, J., Plowman, L., Yamada-Rice, D., Bishop, J. C., Lahmar, J., & Scott, F. (2015). *Exploring Play and Creativity in Pre-Schoolers' Use of Apps: Final Project Report*. Available at: www.techandplay.org

Mascheroni, G., & Ólafsson, K. (2016). The mobile Internet: Access, use, opportunities and divides among European children. *New Media & Society*, *18*(8), 1657–1679. doi:10.1177/1461444814567986

McDaniel, B. T., & Radesky, J. S. (2018). Technoference: Parent distraction with technology and associations with child behavior problems. *Child Development*, *89*(1), 100–109. doi:10.1111/cdev.12822 PMID:28493400

McKnight, A. J., & McKnight, A. S. (1993). The effect of cellular phone use upon driver attention. *Accident; Analysis and Prevention*, *25*(3), 259–265. doi:10.1016/0001-4575(93)90020-W PMID:8323660

Mishra, A., Maheswarappa, S. S., Maity, M., & Samu, S. (2018). Adolescent's eWOM intentions: An investigation into the roles of peers, the Internet and gender. *Journal of Business Research*, *86*, 394–405. doi:10.1016/j.jbusres.2017.04.005

Notley, T. (2009). Young people, online networks, and social inclusion. *Journal of Computer-Mediated Communication*, *14*(4), 1208–1227. doi:10.1111/j.1083-6101.2009.01487.x

Ofcom. (2017). *Children and Parents: Media Use and Attitudes Report 2017*. Accessed from https://www.ofcom.org.uk/research-and-data/media-literacy-research/childrens/children-parents-2017

Ofcom. (2018). *Adults Media Use and Attitudes Report 2018*. Accessed from https://www.ofcom.org.uk/__data/assets/pdf_file/0011/113222/Adults-Media-Use-and-Attitudes-Report-2018.pdf

Pew Research Centre. (2018). *Mobile Fact Sheet*. Accessed from http://www.pewinternet.org/fact-sheet/mobile/

Phua, J., Jin, S. V., & Kim, J. J. (2017). Uses and gratifications of social networking sites for bridging and bonding social capital: A comparison of Facebook, Twitter, Instagram, and Snapchat. *Computers in Human Behavior*, *72*, 115–122. doi:10.1016/j.chb.2017.02.041

Plowman, L., Stevenson, O., Stephen, C., & McPake, J. (2012). Preschool children's learning with technology at home. *Computers & Education*, *59*(1), 30–37. doi:10.1016/j.compedu.2011.11.014

Pokhrel, P., Fagan, P., Herzog, T. A., Laestadius, L., Buente, W., Kawamoto, C. T., ... Unger, J. B. (2018). Social media e-cigarette exposure and e-cigarette expectancies and use among young adults. *Addictive Behaviors*, *78*, 51–58. doi:10.1016/j.addbeh.2017.10.017 PMID:29127784

Raacke, J., & Bonds-Raacke, J. (2008). MySpace and Facebook: Applying the uses and gratifications theory to exploring friend-networking sites. *Cyberpsychology & Behavior*, *11*(2), 169–174. doi:10.1089/cpb.2007.0056 PMID:18422409

Reis, O., & Youniss, J. (2004). Patterns in identity change and development in relationships with mothers and friends. *Journal of Adolescent Research*, *19*(1), 31–44. doi:10.1177/0743558403258115

Rice, E., Craddock, J., Hemler, M., Rusow, J., Plant, A., Montoya, J., & Kordic, T. (2018). Associations Between Sexting Behaviors and Sexual Behaviors Among Mobile Phone Owning Teens in Los Angeles. *Child Development*, *89*(1), 110–117. doi:10.1111/cdev.12837 PMID:28556896

Robertson, T. W., Yan, Z., & Rapoza, K. A. (2018). Is resilience a protective factor of internet addiction? *Computers in Human Behavior*, *78*, 255–260. doi:10.1016/j.chb.2017.09.027

Rutter, M. (2000). Psychosocial influences: Critiques, findings, and research needs. *Development and Psychopathology*, *12*(3), 375–405. doi:10.1017/S0954579400003072 PMID:11014744

Santrock, J. W. (1998). *Adolescence: Exploring Peer Relations*. New Delhi: Tata McGraw Hill Publishing Company.

Seo, D. G., Park, Y., Kim, M. K., & Park, J. (2016). Mobile phone dependency and its impacts on adolescents' social and academic behaviors. *Computers in Human Behavior, 63*, 282–292. doi:10.1016/j. chb.2016.05.026

Shaw, H., Ellis, D. A., & Ziegler, F. V. (2018). The Technology Integration Model (TIM). Predicting the continued use of technology. *Computers in Human Behavior, 83*, 204–214. doi:10.1016/j.chb.2018.02.001

Sherman, L. E., Greenfield, P. M., Hernandez, L. M., & Dapretto, M. (2018). Peer influence via Instagram: Effects on brain and behavior in adolescence and young adulthood. *Child Development, 89*(1), 37–47. doi:10.1111/cdev.12838 PMID:28612930

Sherman, L. E., Payton, A. A., Hernandez, L. M., Greenfield, P. M., & Dapretto, M. (2016). The power of the like in adolescence: Effects of peer influence on neural and behavioral responses to social media. *Psychological Science, 27*(7), 1027–1035. doi:10.1177/0956797616645673 PMID:27247125

Stephen, C., Stevenson, O., & Adey, C. (2013). Young children engaging with technologies at home: The influence of family context. *Journal of Early Childhood Research, 11*(2), 149–164. doi:10.1177/1476718X12466215

Subrahmanyam, K., Reich, S. M., Waechter, N., & Espinoza, G. (2008). Online and offline social networks: Use of social networking sites by emerging adults. *Journal of Applied Developmental Psychology, 29*(6), 420–433. doi:10.1016/j.appdev.2008.07.003

Terras, M. M., & Ramsay, J. (2016). Family digital literacy practices and children's mobile phone use. *Frontiers in Psychology, 7*. doi:10.3389/fpsyg.2016.01957

Terras, M. M., Ramsay, J., & Boyle, E. A. (2015). Digital media production and identity: Insights from a psychological perspective. *E-Learning and Digital Media, 12*(2), 128–146. doi:10.1177/2042753014568179

Terras, M. M., Yousaf, F., & Ramsay, J. (2016). The relationship between Parent and Child Digital Technology use. *Proceedings of the British Psychological Society Annual Conference.*

van Den Eijnden, R. J., Spijkerman, R., Vermulst, A. A., van Rooij, T. J., & Engels, R. C. (2010). Compulsive Internet use among adolescents: Bidirectional parent–child relationships. *Journal of Abnormal Child Psychology, 38*(1), 77–89. doi:10.100710802-009-9347-8 PMID:19728076

Vanden Abeele, M. M. (2016). Mobile youth culture: A conceptual development. *Mobile Media & Communication, 4*(1), 85–101. doi:10.1177/2050157915601455

Vincent, J. (2015). *Mobile Opportunities: Exploring Positive Mobile Opportunities for European Children, POLIS.* London: The London School of Economics and Political Science.

Vygotsky, L. S. (1978). *Mind in society: The development of higher psychological processes.* Harvard University Press.

Walrave, M., Heirman, W., & Hallam, L. (2014). Under pressure to sext? Applying the theory of planned behaviour to adolescent sexting. *Behaviour & Information Technology, 33*(1), 86–98. doi:10.1080/0144929X.2013.837099

Yang, B., & Zhao, X. (2018). TV, Social Media, and College Students' Binge Drinking Intentions: Moderated Mediation Models. *Journal of Health Communication*, *23*(1), 61–71. doi:10.1080/1081073 0.2017.1411995 PMID:29265924

Zarbatany, L., Hartmann, D. P., & Rankin, D. B. (1990). The psychological functions of preadolescent peer activities. *Child Development*, *61*(4), 1067–1080. doi:10.2307/1130876 PMID:2209178

Zhu, J., Zhang, W., Yu, C., & Bao, Z. (2015). Early adolescent Internet game addiction in context: How parents, school, and peers impact youth. *Computers in Human Behavior*, *50*, 159–168. doi:10.1016/j.chb.2015.03.079

# Chapter 2
# Impacts of Mobile Use on Third Agers in China

**Chen Guo**
*Curtin University, Australia*

**Michael Keane**
*Curtin University, Australia*

**Katie Ellis**
*Curtin University, Australia*

## ABSTRACT

*The chapter explores the role of smart phones and mobile apps in the process of third age formation in Zhengzhou, a second-tier city in China located in central Henan province. The term 'third age' refers to a transition period from active work to retirement. Compared with the previous generation, the demographic approaching retirement in China today is more digitally literate, although this varies accordingly in Zhengzhou, a second-tier city. The use of digital technology offers people a different kind of retirement. This study shows that an increasing number of people around retirement age (55-65) in Zhengzhou are using smart phones and apps to reimagine the possibilities of post-work lifestyles. The research asks if the use of mobile apps is changing peoples' perspectives on traditional responsibilities and peoples' expectations of retirement.*

## INTRODUCTION

The aging society is a global phenomenon. Populations are increasing in most countries and people are living—and working longer. But some countries are aging faster than others; this has economic and social ramifications; there is a need for appropriate policies to manage peoples' expectations as they exit the work force. Nowhere is the aging society 'problem' more acute than the People's Republic of China (PRC), which is suffering from a self-inflicted population time bomb, the One Child Policy. Instituted in 1978 by Deng Xiaoping to curb excessive population growth, the One Child Policy determined that each

DOI: 10.4018/978-1-5225-7885-7.ch002

Chinese family would have no more than 'one child'. The repercussions of this policy, including a gender imbalance favouring males, are now being felt, although the policy has now changed to a 2-child policy.

To understand aging, we must first consider life stages. The approach adopted in this paper is 'ages'. In the literature four ages are noted; the first age is childhood, the second depicts career and mid-life, the third retirement, and the fourth, old age and decline. Our discussion of the 'third age' depicts people's transition from an age of 'independence, maturity, responsibility and working'—i.e. the 'second age'—to one of personal achievement and 'fulfillment after retirement' (Laslett, 1987). However, the third age, conventionally understood as concurrent with retirement, is far less understood than the other ages. In China this third life (st)age has received little academic attention. This chapter fills that void.

Before the One Child Policy was introduced, and even into the 1980s and 1990s, new retirees were regarded as respected elders of the family; seniors, whether they were uncles or aunts, or grandparents, would often be tasked with looking after young children. Congregating in parks and playing mah-jong, many third agers would exchange stories of their lives, and perhaps share photos or letters. In the 2000s, however, a combination of economic reforms and new technologies, together with the changing shape of families, has seen the advent of a different modality of third age, one in which people have more autonomy. The most significant new technology impacting on peoples' autonomy is the mobile phone and many studies have looked at youth culture use of mobile phones. However, it is important to note that people in China who are in their third age today are avid mobile phone users; they share videos and stories; they use cameras to record their travels or to make video movies; they search out apps to enrich their lives. Many still go to the parks and play mah-jong of course but in comparison to the past there are many more lifestyle options.

Our goal is to provide a new understanding of the relationship between third-agers and their mobile phones. The first part of the chapter compares the aging society, namely, a problem to be solved, with the digital society, identifying a spectrum of opportunity and potential. We begin with some baseline data on population change in China, then we identify reasons for increased longevity. We briefly discuss the Chinese government's concerns in regard to managing this changing demographic. With the focus turning to a digital society, the question is: how is this group using technology to be more fulfilled? For instance, does the autonomy gained by access to digital technology take the pressure off the welfare system?

The second section turns to understanding the relationship between the third age and the digital age. This section expands the definition of 'third age'. The term is now used by a number of social scientists investigating population trends (Gilleard & Higgs, 2002, 2008; Gilleard, Higgs, Hyde, Wiggins, & Blane, 2005; Laslett, 1994). The third age is then linked with the concept of 'digital capital' (Park 2017). Digital capital, derived from Bourdieu's notions of economic and social capital, allows us to evaluate the digital literacy of this demographic and with this the potential of fulfilment. The context of digital capital in this paper is the increasing use of mobile phone and apps in China, and the distinctive phenomenon of 'mobile use only' among retirees in China (Deng, Mo, & Liu, 2014; Donner, 2008).

In this section, we ask the following research question: How are third agers using mobile technology to achieve personal fulfilment? The discussion also investigates the role of 'mobile-use only' in retirees' daily life. The next section looks at research findings of mobile use by early retirees in Henan province, located in central China. The final section draws conclusions, including looking at some of the negative aspects of the digital society.

## THE AGING SOCIETY AND THE DIGITAL SOCIETY

The aging population is a global phenomenon and impacts on all societal sectors. According to the World Health Organization (2018), by 2050, the percentage of the global population over 60 will reach 22% globally, nearly double the percentage of 12% experienced in 2015. The aging population is growing rapidly because people are living longer; life-spans have increased in all countries during the past few decades. Senior citizens have better health compared with several decades ago. Longevity is predicted to reach 89.1 years for women and over 84.7 years for men by 2060 in Europe (European Commission, 2014; Helbostad & Vereijken, 2016).

With the largest population in the world, China unsurprisingly has the largest aging population in the Asia-Pacific area; this demographic is forecast to increase from 110 million in 2010 to 330 million by 2050 (Suzman & Beard, 2015). As noted in the introduction, the One Child Policy, instituted in 1978 under Deng Xiaoping specified that each family could only have one child, with exceptions made for minority populations and some rural villages (Hesketh, Lu, & Xing, 2005). The impact of the policy has been to skew the population balance towards older citizens. According to Powell and Cook (2009), China's aging proportion is growing even faster than that of Japan. People in China have reasonable expectations of a longer life than the previous generation. The percentage of people over 60 in China was 13.3% in 2010 (Attané & Gu, 2014, p. 1), and the percentage keeps rising quickly. China is therefore facing a huge challenge. Powell and Cook describe the burden of the ageing population for China as 'a tiger behind and coming up soon.' (Powell & Cook, 2000) Of course, now the 'tiger' walks together with China.

Another saying reported in scholarship and policy documents is: 'growing old before becoming rich'. This saying refers to the economic impact of the aging population. The economic reforms of the 1980s brought expectation of greater prosperity. Many people did get rich (Goodman, 2008), and many more have aspired to do so. But at the same time as the population was 'artificially' manipulated by the One Child Policy, improvements in health and education were working to prolong peoples' lives. Less younger workers have meant higher wages; in addition, the rise in retirees has put unprecedented pressure on the pension system.

Several other reasons combine to exacerbate the aging population in China. Following the foundation of the Chinese Republic in 1949, people were encouraged to have large families (Uhlenberg, 2009). The group born in 1950s and 1960s are now retirees or are entering retirement. According to the *Chinese Nationwide Population Census 2010* (2011), the population reached 1.4 billion. Within China, the migration trend from less to more developed urban areas accentuates the proportion of older people living in less developed areas. Although One Child Policy has been replaced by a Two Child Policy since 2016, many people are choosing small families. With improved living conditions and access to information on the internet, people are paying more attention to health (Fredriksen-Goldsen, Kim, Shiu, Goldsen, & Emlet, 2014), from better eating habits to exercise and activities.

All these factors contribute to the dramatic 'aging' of the Chinese population, which is likely to become an economic and social problem in the next decades. For the Chinese government the question becomes one of managing the resources allocated to society in the interest of stability and economic prosperity. In the past twenty years, the so-called 'iron rice bowl' of welfare dependency has given way to a model more akin to a capitalist system in which people take personalized responsibility for their health. Ulrich Beck (2002, p.xxi) uses the term 'individualization' of society to refer to developed societies. A book entitled *iChina* (Stig & Ni, 2010) sets out how the new 'self-determining individual' operates in China. In regard to the elderly they raise two images of the elderly: the first is a burden to society; the second is a victim of modernization. Findings from Thorgensen and Ni's research (2010) in this collection on the rural elderly indicate dependency on the family. However, the focus of their research is more accurately 'fourth age'.

What has changed markedly in the past two decades in China has been peoples' access to technology—as well as a spike in the number of third agers as the Chinese baby boomers leave work. It is therefore commonplace to use terms associated with technological progress to signify how people are liberated from manual labour and time-consuming repetitive tasks. Instant messaging allows us to stay in touch, social media builds relationships, and apps deliver multiple productivity and social benefits to users. As we discuss below these 'affordances' have changed people's lives in rural China, allowing them to be more 'individualized'; that is not to say they have embraced a kind of Western-style individualism but rather have more opportunities for individual self-realisation, while at the same time being the target of online businesses selling products that are related to their stage of life. China's aging society now has a digital context and the 'third age' is where we see this trend playing out.

## THE THIRD AGE ENCOUNTERS THE DIGITAL AGE

A recent study by Park (2017) on the use of digital technologies among rural and marginalised communities in several countries, including Australia, has made use of the concept of 'digital capital'. According to Park, digital capital 'is defined as an individual's digital technology ecosystem that shapes how a user engages with digital technologies' (Park, 2017, p. 72). Digital capital is derived from Bourdieu's work on capital (economic, social and cultural). Elsewhere Bourdieu (2005) has referred to 'technological capital' in the context of firms' behaviour. Digital capital, however, in the context of a digital society, applies to an individual. The use of digital affordances among connected individuals requires certain preconditions: access to the internet, sufficient bandwidth, and sufficient literacy. The last of these varies considerably among populations and age groups. Applied to individuals, therefore, we can draw a comparison between digital capital and individualization. One's digital capital is built up over time in a similar sense to one's cultural capital (what one learns and absorbs from friends, family, society, schooling etc). With respect to the individualization thesis, the individual, here a third-age person in China, is asked to take more responsibility for their lives in a time when government services are less generous. She may choose to become connected via social media, to join in online forums, and may even become a collector of the latest apps.

### The Third Age

Let us now turn to the concept of 'third age'. This concept has seen a dramatic shift in the way scholars theorise and experience aging. In the book *A Fresh Map of Life (1991)*, Peter Laslett identifies a new stage in the life course emerging after retirement – the so called "the third age". Whereas retirement previously more or less coincided with ill health and decline, changing industrial practices and ageing demographics facilitated the emergence of a period post retirement in which individuals possessed the necessary health, vigour and attitude to realise 'personal achievement and fulfilment' (Laslett, 1991, p. 153) not possible during their working life (or second age). According to Weiss and Bass (2002) the third age is characterised by increased longevity, better health, and an increased levels of financial well-being. These characteristics in conjunction with an increase in leisure time allows for "the pursuit of new or long-latent interests, together with desired levels of sociability" (Weiss & Bass, 2002). Carr and Komp describe this age as an early stage of later life (2011).

Although Laslett is careful to avoid strict demarcations between these ages, they can be broadly summarised as following the social stages of the ageing process. For example, the first age refers more or less to childhood or a period of dependence, with the second age being a period of independence and responsibility experienced during working life. The third age typically occurs in the period leading up to or post retirement where the individual has less responsibility but maintains independence through good health and financial stability. The fourth age however "is an era of final dependence, decrepitude and death." (Laslett, 1991, p. 135).

While some theorists note that delineating the various stages of life has a long history going back at least to medieval times (Thane 2003), Laslett's concept of the third age is useful to this study because it also takes into consideration social factors such as demographics and economics in a drive for personal fulfilment. These factors work together on both a collective and whole of nation level and at the level of the individual. As Laslett explains:

*life after the second age has to last long enough for the majority of the population of that nation, and not simply for the lucky, the rich and the privileged, to expect to be able to go on to the Third Age. Which means that the third age can only appear at the time when average expectation of life begins to be high enough to allow this to happen, and when there is already a sufficiency of the whole population actually experiencing the addition to the life-course.*

The ageing society and particularly the expectation that at least half of the population will live to their 70s along with about 10% of the population being over the age of 65 contribute to the emergence of the third age on a collective level (Laslett, 1987).

Laslett (1987) focused mainly on the Western context, but did not acknowledge the unique position of people entering the third age in China, but now the situation is totally different. China entered the aging society around the turn of the twenty-first century; its population now has a reasonable expectation of living longer than previous generations. The average life expectancy in 2012 was around 75 years of age (National Bureau of Statistics of the People's Republic of China, 2012). Likewise, it is predicted that by 2050 there will be 430 million older people who are over 65 in China (Kendig, 2004).

The current third age population in China are the first to enter into this stage of the life course with extensive ICT literacy. They embrace digital technologies such as the smartphone to prolong this era of personal fulfilment and ward off the decline associated with the fourth age. The technological affordances provide more possibilities for early retirees to live a new lifestyle.

## The Mobile Only Age

In China the mobile phone is a pervasive tool in people's daily life; in 2018 mobile phone users constituted 97.5% of the total internet users in China (China Internet Network Information Center, 2018). 'Mobile use only', that is, people use the mobile phone only to go online, has become a phenomenon. People around their retired age have various motivations to use digital affordances. The rapid development of mobile internet and mobile technologies give people the possibility to live a more creative and active later life.

The development of mobile devices in China has been rapid, especially mobile phone and mobile apps. China has many very successful technology companies and mobile phones have become ever cheaper. According to Akamai, the average internet speed of China is 7.6 Mbps, which surpasses the global average connection speed 7.2 Mbps (Akamai., 2017); this represents a great achievement in a country so large. Connected devices, especially mobile phones, are an indispensable part of daily life. By 2018 there were 772 million users in China, representing 55.8% of the population, of these 753 million users are mobile phone users, or 97.5% of the total internet users. The next most popular connected devices is the computer, occupying only 53%.(China Internet Network Information Center, 2018). The phenomenon of "mobile use only" in China is therefore worth investigating.

Although the dominant internet users are aged between 10-39 years old, in 2018 the rate of internet use by people over 50 increased, from 9.4% to 10.4% (China Internet Network Information Center, 2018) and more people over 60 years old have accepted the internet and mobile apps in their daily life. In second tier cities the trend is the same. According to 2016 Internet development report of Henan province, there were more than 79 million internet users in Henan, a penetration rate of 82.8%; among these mobile phone users numbered 77.5 million (Henan Daily, 2017). "Mobile use only" has become a phenomenon among Chinese people and is more obvious for early retirees. As we discuss below the first reason is they don't use computers to deal with work tasks anymore; second, the smart phone is more accessible; third, they have accumulated significant digital capital during their working time.

Mobile apps can solve problems and improve the quality of daily life, for instance gaming apps for cognitive condition, activity apps for exercise. In addition, people use mobile apps to engage with others and connect with online communities. From a governance and policy perspective, digital technologies, especially mobile apps, are a cost-effective way to help people engage in society and support them to live a more creative and active retired life. Indeed, many international studies have explored how to use digital technologies to help aging population with various aging-related problems (Czaja & Lee, 2006; Drew & Waters, 1986; Sadana, Blas, Budhwani, Koller, & Paraje, 2016) and in the building of new lifestyles. Li and Perkins (2007) argue that the majority of senior citizens in the US view technology in a positive view and believe it will produce a better quality of life for themselves and society. In addition to the positive aspects of digital technology many studies have looked at digital divide issues (Niehaves & Becker, 2008; Brodie et al., 2000) and the disengagement of aging populations (Johnson & Mutchler, 2014; Olphert & Damodaran, 2013).

## METHOD

Many scholars in the field of aging have focused attention on first tier cities, where the 'aging population' might be seen as pioneers in using mobile devices. Less research focuses on the aging population in the second and third tier cities which have proportionally higher aging populations than the several first-tier cities in China. However, in considering the widespread use of internet and mobile phones, it is necessary to address questions of mobile use by aging populations in second or third tier cities. In addition, many researchers focus on the negative stereotype of aging, such as the medical aspect (Fredriksen-Goldsen et al., 2014), or research on the increased social expenditure as a burden for the government (Gray, 2009). Research in China follows this path. This negative image of aging population is often mentioned in reports and gives an impression that all aging populations are fragile and disempowered.

Zhengzhou, the capital of central Henan province, is a representative second tier city in China. It has the heaviest population density in China; and its economic, social and cultural development has similarities with other second-tier cities. In addition, under the context of urbanisation, many people from

Henan province have migrated to others places to work, leaving families geographically distant. For this reason, mobile use by aging population in Zhengzhou provides a useful case study.

This study is based on field work which was conducted between 6 January and 7 February 2018. The lead author, a native of Zhengzhou, made some initial contacts; she then enlisted more respondents through snow-balling. Potential participants were approached and invited to do an in-depth interview. All participants were made aware of the research and signed an ethical clearance form. The author selected participants in accordance with the third age, that is, around their retirement age. The estimated time to conduct an interview was around one hour. The interview questions included demographic information and open answer questions. The open answer questions were primarily related to the relationship between early retirees and government, early retirees and family, as well as the early retirees and smart phone and mobile apps. In addition, the interviews explored how the use of mobile apps is changing early retirees' expectations of retirement.

More than 30 participants were interviewed in Zhengzhou, and a village in Xinxiang, Henan, during January and February 2018. The discussion below focuses on 6 of these interviews. The participants are aged between 50-61. These participants either have public institutions (danwei) to retire from or they retire from a private company; they receive a retirement pension from government or commercial insurance. The gap in retirement pensions between the different two systems is not the topic of this study.

## Demographic Information

The six participants will be named A, B, C, D, E and F in this study. The basic demographic information and mobile use is shown in Table 1.

## FINDINGS AND DISCUSSION

### The Role of Mobile Use for Early Retirees

Gathering information to keep pace with the development of the world, and maintaining or developing relationships with other people are the main functions of mobile phone and apps. However, mobile use among early retirees is more than maintaining relationships. In this section, more specific and detailed uses are explored based on the interviews. These various uses are interrelated.

### Information Gathering

Gathering information is a way for retirees to engage more fully in society. Participant D is a lecturer in a university; he uses the Toutiao app (a Chinese news app), which is based on big data technology, and he is very pleased to receive information pushed by the app. Participant E also said "I use them (mobile apps) to get information and broaden my eyes…especially, the XiGua video app (an app for short videos) and Toutiao app, I use the two apps about two hours per day". Getting information on mobile apps provides a means to engage in social development and news.

*Table 1. Demographic information of participants*

| Participant | | A | B | C | D | E | F |
|---|---|---|---|---|---|---|---|
| Age | | 56 | 57 | 56 | 50 | 61 | 54 |
| Gender | | male | female | male | male | male | female |
| Education | | College degree | Bachelor degree | Bachelor degree | Bachelor degree | College degree | College degree |
| Income (RMB) | pension | 2600 | 7000 | unknown | 5000 | 4000 | 3000 |
| | salary | unknown | NA | NA | NA | unknown | NA |
| Retirement year | | 2018 | 2021 | 2020 | 2028 | 2006 | 2014 |
| Standard retirement year | | 2017 | 2011 | 2017 | 2028 | 2012 | 2014 |
| Job before retirement | | clerk of a state-owned enterprise | Lecturer in a university | Public official in provincial government | Lecturer in a university | Employee of a state-run enterprise | Typist for a state-owned enterprise |
| Job after retirement | | Support crew for the community | NA | NA | NA | Salesman for medical facilities | Ping-pong coach |
| Reasons of retirement later | | Physically good enough/ not old/ support family | Job is a way to keep positive life attitude | Too young to retire | NA | NA | NA |
| Reasons of retirement in advance | | NA | NA | NA | NA | Earn money to support family | Buy-out in advance |
| Number of adult children | | 2 | 1 | 1 | 1 | 2 | 1 |
| Mobile phone use time | | 2013 | 2010 | 2011 | 2011 | 2012 | 2014 |
| Hobby | | Writing poems on mobile phone (2 hours for one poem) | Gardening on the balcony/ baking/ meditation | Audio books/ calligraphy/ painting | Travelling/ photography/ chatting with young people | Qigong/ swimming/ play poker with friends | Ping-pong/ Playing games on phone |
| Mobile use time/day during work days | | ≥1 hour | ≈ 2 hours | 2-3 hours | ≈5 hours | ≥ 3 hours | ≈3 hours |
| Mobile use time/day during weekends | | ≈3 hours | ≈4 hours | ≈ 5 hours | ≥ 5 hours | ≥ 5 hours | ≈3 hours |

## Communicating and Maintaining Relationships

One of the most important roles of the mobile use for the early retirees is to communicate with others. This finding is consistent with a study based on Hong Kong, that the use of communication networks, including mobile phones, may enhance communication and social networking among older adults in Hong Kong (Chan, Wong, Tam, & Tse, 2014). Among my interviewees, Participant A, whose hobby is writing poems, is glad to share his poems on WeChat and Sina blog; he said "I did this for two reasons, one is I commemorate my army life time with my comrade-in-arms in the WeChat group…the other one is I want to share information with others, as well as communicate with others". Participant E says that "I use smart phone and mobile apps not only to get information to broaden my eyes, but also I need to communicate with my friends".

## Personal Trainer

Mobile devices have become a commonplace in health care settings and lead to rapid development of mobile apps for these platforms (Ventola, 2014). Exercise is a good way to maintain physical health. A study based on 726 participants (age ranged from 18-74) shows exercise apps users are more likely to exercise than those who don't use exercise apps (Litman et al., 2015). Considering the 'mobile use only' phenomenon among Chinese retirees, exercise apps play the role of personal trainer for the retirees to improve their physical health to some extent. Participants D, E and F use exercise related app and mobile devices to maintain their physical health. Participant D told me he uses WeChat Steps every day to do exercise. Similarly, Participant B told me she uses a meditation app, including Steps and the heart rate monitor function. Participant F uses Xiaomi Fitbit. She told me "I run every day to reach to the target, sometimes I run on the spot in my home to reach to the target". Similarly, studies have shown that mobile technologies and mobile apps maybe be a way to improve the health condition for aging population (Jorunn L. Helbostad et al., 2017) or eliminate some aging-related problems.

Gergen and Gergen use the term of "physical well-being" to refer to " optimal functioning of brain and body" (2001). Participant E claimed that "The benefit of the mobile apps is that they allow us enter a knowledge world, delay the time of dementia…some people are reluctant to accept new knowledge (how to use mobile apps) … this will speed up their aging process, especially dementia".

## Learning and Hobby Assistant

Learning is another way to keep positive and active retired life. Many people around retirement age are trying to learn something new based on their interests. Mobile phone and apps can help them to learn what they are interested in, and it is very economical. Participant B's hobby is gardening and baking, she said, "I learnt some knowledge, such as cooking skills, gardening knowledge and common sense of life". Participant D learnt how to use a photo editing application based on his interest in photography. Participant C expressed the view, "I use audio apps to listen to the ancient philosophers and classic books such as *Romance of the Three Kingdoms*". His interest is history. Participant F worked as a coach for several years as her hobby is playing Ping-Pong; she is now using Fitbit to maintain the habit of exercise.

## Hobby Developer

Mobile use is also a hobby developer for many retirees. Some apps provide resources to develop different hobbies. Participant E responded, "They (the aging population) can download some apps according to their interests and hobbies, I am sure they can get some knowledge, fun and happiness when they use apps." Participant D likes travelling, he uses some translation apps to help him to communicate with foreigners when he travels overseas. Participant D told me, "In my case, I can use translaton apps to improve my English communication ability when I were travelling, this will increase my chances of travelling abroad".

## Inseparable Secretary

Technological affordances allow individuals to use technology for their various specific purposes. The use of mobile apps is allowing the aging population to have more creative possibilities and to live a more interesting and diversified retired life, at least compared with the previous generation. Participant E regarded mobile phone as a personal secretary, part of his body, and part of his brain. He claimed:

*The smart phone has become part of my brain, it helps us to store valuable information, such as navigation, booking taxi, finding locations, shopping, looking for restaurants, booking hotels, selecting public utilities. Now that the apps can calculate the miles and time, you can make a decision by yourself.*

Participant A held similar ideas and said, "the smart phone is a necessity in current life. If I lose my smart phone, I would feel like I have lost some important thing: the smart phone is one part of my body". Participant E argued:

*although some people say they can live a good life without a smart phone and mobile apps, but once they know how to use it they cannot put smart phone down. If you replace my smart phone with a keyboard phone, I cannot adjust myself at all. I think I will lose at least 40% of my life quality, or even 50%".*

In her study on digital capital, Park argues that being connected is often insatiable once people know how to use such technology (Park, 2017).

## Mobile Companion

Without exaggeration, the mobile phone now plays the role of companion for many aging populations. Vincent (2006) notes the increasing emotional attachment for mobile devices during individuals' daily lives. Participant F explained how lonely she was when she lived with her daughter's family in another city:

*Now I live at 7th floor in another city, I know nobody here, sometime my daughter's mother in law comes to visit me, I have a person to communicate, otherwise, nobody. When I feel lonely, I can chat with my mother and my older sister only through WeChat.*

Participant B also said, "For me, these apps can help me, make me happier than before, healthier". Participant E responded, "The mobile apps can eliminate the retired people's loneliness". This finding shows mobile phone plays a companion role for the participants and helps to alleviate their loneliness.

## Individualization Accelerator

An unexpected finding from the research is that mobile use can accelerate the process of retirees' individualization. According to the fieldwork, some participants expressed the view that it is hopeless to rely on government, and their only adult child, to provide aged care for them in the future. This was the model of the past when government provided more services and sons and daughters were more filial. Mobile phone and apps use provide a way for government to abdicate responsibility in regard to providing retired people more choices and diversified lives. In the countryside the lack of cultural projects has become a major concern. People are turning to their phones to find their culture. There are two sides to this scenario. According to Participant E, mobile phone and apps increase his potential to live independently. As mentioned earlier, the technological affordances provided by mobile use, together with the digital capital, enable individuals to make decisions by their own. On the other hand, as mentioned above, government is providing less cultural resources for people in second and third tier cities. Phones fill the void.

## Entertainment

Mobile phone and apps supply new ways of entertainment for retirees, which is obviously good for aging population's mental status. Different individuals have different ways to find entertainment. According to participants B, she took a short video which reflect their petty bourgeois need for entertainment and stored it in her phone. Participant E said that he uses musical apps for entertainment, although he cannot sing any song. Participant D thinks using app will increase the chances of attending outdoor entertainment, he shared a story about his trip in Japan:

*I want to share the story of google map and translator app. Last year, my wife and I travelled to Japan and spent 14 days there. I downloaded two apps, which one is google map, and the other one is translation app. Both of them were suggested by my students. Can you image that me, who cannot speak English or Japanese at all, could be a guide for our trip for 14 days: with the help of the two powerful apps, our travel was unimpeded. We travelled independently.*

Participant F use video apps such as iQIYI to watch TV for entertainment. Participant F also said she played games for entertainment.

*I play Fight for Landlord online every day. The game will give you 3000 free coins every day. Once you lose all of them you cannot play the game anymore on the same day. Sometimes if I don't play well, I cannot play after two rounds. Sometimes I play well I can play longer. The limitation of 3000 free coins won't let me spend too long time on the game.*

Compared to participant F, participant D had a bad experience when he plays poker games online with unknown internet. His hardware and signal were not good enough to connect with others to play the game several years ago. He said, "I could not click my pokers (on screen) when I was playing the game. This would lead to a failure of my team, the team members cursed me". Participant D felt frustrated because of this and uninstalled the apps.

## Mood Lifter /Young Heart

Considerable research has examined positive and negative effects of mobile phones and apps to mental and mood states (Harrison et al., 2011). Gergen and Gergen (2001) uses the "sense of well-being, happiness, and optimism" to represent the positive mood of aging population. In the context of their mobile use, participant B, D and E don't think they belong to the aging population; they think they are younger than their real age, as they are still active and positive. Using mobile apps help them to keep pace with younger generation and help them engage more in society. Participant E explained the beneficial effects of mobile apps for his life. He exclaimed that "In the aspect of positive aging, I think the mobile apps are definitely good. Nowadays, without smart phone, without computer, you are isolated by the modern world. I have this feeling". Participant E also said:

*I think I belong to middle-aged people, but I do not belong to aged population. My psychological age is 50s, I never stop chasing for dream and target. I cannot get used to the aged life style. In addition, using mobile apps makes me feel younger as I can enter the new world which is full of various information... I can live harmoniously with the modern world. Another factor of using mobile apps makes me feel younger is that the funny short videos and jokes on mobile apps let me get rid of annoying things. I can adjust my mood.*

In addition, with the one-child-policy in China and the role of grandparents, some of the retirees live child-centred lives or grandchild-centred lives; they use mobile phone and apps for keeping pace with younger generation. Participant A emphasized:

*I want to upgrade my ideas, learn good things from young generation. As the time is different, I cannot just insist on my traditional values and concepts. With the changing time, I think people should be brave enough to accept new things. I try to keep pace with young generation, it will be easier to communicate. I keep myself up to date.*

Likewise, participant E has the similar opinion:

*I use mobile apps not because it is the fashion, but in order to not be isolated from the modern world. But to tell the truth, using mobile apps makes me feel younger as I can enter the new world which is full of various information.*

The rapid development of urbanization in China, as well as the one child policy, has caused more and more families to become empty nests. The findings suggest that it is more economical to maintain social ties online than offline for the aging population (Wu, 2016). These social ties promote positive moods and even feelings of youthfulness.

## Stress Producer

Of course, there are many downsides. Stressful feelings related to mobile use as well as safety problems were evident in the findings. Participant E reported "I can't operate them as well as my son... I worry about others stealing my money on WeChat". Similarly, participant F said, "I don't know how to download it (an app), I am worried about the leaking of personal information, I am afraid to operate it improperly, I am afraid to be cheated". Participant A also claimed, "I cannot use Alipay, I am worried about safety problem. You know once you click the screen, your money disappears. So, I don't connect my bank account with my smart phone".

Participant A showed a negative mental state and said "compared with them (people who can use mobile apps better), I think I am out of date. If you cannot use smart phone and mobile apps, you will be the new style of illiteracy. The gap between generations will be larger...You self development will be blocked and out of date gradually". This reflect the findings of one of the participants in Park's study who expressed similar worries—"becoming illiterate in a digitalized world" (Park, 2017).

## Eye Damage

Another negative effect of mobile use is the damage to eyesight. "The negative effect is speeding up the process of eye fatigue, and this is harmful to our health" participant D said during the interview. Several other participants expressed a similar opinion.

## CONCLUSION: MOBILE USE AND THE CHINESE THIRD AGE

In summary, this chapter has introduced the concept of third age to China and explored the emergence of Chinese "third agers". As discussed in the first section, the concept of Third Age was initially proposed by Laslett (1987). This concept of third age has provided a new perspective on the ageing population. Focusing on mainly on Western countries, Laslett (1987) and Weiss and Bass (2002) claimed that the emergence of third age was reflected in increased longevity, better health, financial wellbeing, and more leisure time. These criteria are gradually more evident in China. Laslett's study, published in 1987, did not acknowledge the people entering the third age in China. At that time the population was more normal. Now the situation is quite different. China entered the ageing society around the turn of the twenty-first century and its population now has a reasonable expectation of living longer than previous generations. As mentioned above, the average life expectancy is 75 years in 2012 according to the National Bureau of Statistics of People's Republic of China. The data on Chinese aging population indicates "longer longevity" and "better health". "Financial wellbeing", to some extent, is more evident under the dramatic economic development during the last three decades, although it still lags behind if compared with developed countries.

Within the digital world, mobile use among early retirees in China has provided more choices and potential for third agers to live a different kind of retired life. In the fieldwork, most participants expressed the opinion that they cannot rely on government and their only child to provide care for their aged life. Mobile phone and apps maybe can take some of the responsibility and role of government. Aided by the digital media, older people now can solve some of these problems by themselves and share the burden of the government. In this sense people have become more individualized; they have more autonomy, although in many cases they have no choice. The darker side of technology has not been discussed in this paper, aside from some remarks about the dangers to eyesight and financial security. Some respondents also claimed that technology was confusing and led to increased stress.

As this study shows, when we talk of the concept of third age to China, there are several notable developments to consider. First, the Chinese third age has emerged concurrently with the popularity of mobile use. Without mobile use, one of the criteria, proposed by Laslett, and Weiss and Bass, i.e. "more leisure time", is really hard to imagine. Mobile use provides an economical way for retirees to experience more leisure time. Moreover, with the popularity of mobile internet and mobile devices in China, the use of mobile phone and apps plays an important role in peoples' lives. The elderly population who are around their retirement age have begun to engage more in society and many are living a more creative and active retired life—the third age, before they enter the fourth age stage. As a result of the expanding role played by smart phone and mobile apps use an increasing number of Chinese retired people are participating are creating a new understanding of retirement. With the concept of third age, the aging population has rid itself of its stereotype, that is, peoples' sole role is to pass their remaining time looking after grandchildren.

As China is getting older and aging faster, the population of third agers is also increasing dramatically and should be explored further from social, culture, and economic aspects. Technology is just one important aspect of our changing world, and it remains to be seen what developments may occur if governments and business together accept the challenge of dealing with this social issue.

# REFERENCES

Akamai. (2017). *Akamai's State of the Internet Report*. Retrieved from https://www.akamai.com/uk/en/about/our-thinking/state-of-the-internet-report/global-state-of-the-internet-connectivity-reports.jsp

Attané, I., & Gu, B. (2014). China's demography in a changing society: Old problems and new challenges. In I. Attane & B. Gu (Eds.), *Analysing China's population: Social change in a new demographic era.* (p.1). Dordrecht: Springer Netherlands. Retrieved from https://journals.library.ualberta.ca/csp/index.php/csp/article/viewFile/28849/21140

Beck, U., & Beck-Gernsheim, E. (2002). *Individualization: Institutionalized individualism and its social and political consequences*. London: Sage.

Bourdieu, P. (2005). *The social structures of the economy*. London: Polity Press.

Brodie, M., Flournoy, R. E., Altman, D. E., Blendon, R. J., Benson, J. M., & Rosenbaum, M. D. (2000). Health information, the Internet, and the digital divide. *Health Affairs*, *19*(6), 255–265. doi:10.1377/hlthaff.19.6.255 PMID:11192412

Carr, D. C., & Komp, K. (2011). *Gerontology in the era of the third age: Implications and next steps*. New York, NY: Springer Publishing Company.

Chan, K., Wong, A., Tam, E., & Tse, M. (2014). The use of smart phones and their mobile applications among older adults in Hong Kong: an exploratory study. *GSTF Journal of Nursing and Health Care*. doi:10.5176/2345-718X_1.2.45

China Internet Network Information Center. (2018). Statistical Report on Internet Development in China of 2018 (42). Retrieved from http://cac.gov.cn/wxb_pdf/CNNIC42.pdf

Czaja, S. J., & Lee, C. C. (2006). The impact of aging on access to technology. *Universal Access in the Information Society*, *5*(4), 341–349. doi:10.100710209-006-0060-x

Deng, Z., Mo, X., & Liu, S. (2014). Comparison of the middle-aged and older users' adoption of mobile health services in China. *International Journal of Medical Informatics*, *83*(3), 210–224. doi:10.1016/j.ijmedinf.2013.12.002 PMID:24388129

Donner, J. (2008). Research approaches to mobile use in the developing world: a review of the literature. *The Information Society*, *24*(3), 140–159. doi:10.1080/01972240802019970

Drew, B., & Waters, J. (1986). Video games: Utilization of a novel strategy to improve perceptual motor skills and cognitive functioning in the non-institutionalized elderly. *Cognitive Rehabilitation, 4*(2), 26–31.

European Commission. (2014). *The 2015 Ageing Report: Underlying Assumptions and Projection Methodologies.* Retrieved from http://ec.europa.eu/economy_finance/publications/european_economy/2014/pdf/ee8_en.pdf

Fredriksen-Goldsen, K. I., Kim, H.-J., Shiu, C., Goldsen, J., & Emlet, C. A. (2014). Successful aging among LGBT older adults: Physical and mental health-related quality of life by age group. *The Gerontologist, 55*(1), 154–168. doi:10.1093/geront/gnu081 PMID:25213483

Gergen, M. M., & Gergen, K. J. (2001). Positive aging: New images for a new age. *Ageing International, 27*(1), 3–23. doi:10.100712126-001-1013-6

Gilleard, C., & Higgs, P. (2002). The third age: Class, cohort or generation? *Ageing and Society, 22*(3), 369–382. doi:10.1017/S0144686X0200870X

Gilleard, C., & Higgs, P. (2008). The third age and the baby boomers: Two approaches to the social structuring of later life. *International Journal of Ageing and Later Life, 2*(2), 13–30. doi:10.3384/ijal.1652-8670.072213

Gilleard, C., Higgs, P., Hyde, M., Wiggins, R., & Blane, D. (2005). Class, cohort, and consumption: The British experience of the third age. *The Journals of Gerontology. Series B, Psychological Sciences and Social Sciences, 60*(6), S305–S310. doi:10.1093/geronb/60.6.S305 PMID:16260712

Goodman, D. (2008). *The new rich in China: Future rulers, present lives.* London: Routledge. doi:10.4324/9780203931172

Gray, A. (2009). Population aging and health care expenditure. *China Labor Economics, 1*(10). Retrieved from http://en.cnki.com.cn/Article_en/CJFDTOTAL-ZLDJ200901010.htm

Harrison, V., Proudfoot, J., Wee, P. P., Parker, G., Pavlovic, D. H., & Manicavasagar, V. (2011). Mobile mental health: Review of the emerging field and proof of concept study. *Journal of Mental Health (Abingdon, England), 20*(6), 509–524. doi:10.3109/09638237.2011.608746 PMID:21988230

Helbostad, J. L., & Vereijken, B. (2016). *Activity app for an aging population.* Retrieved from http://www.preventit.eu/index.php/news_events/activity-app-for-an-ageing-population/

Helbostad, J. L., Vereijken, B., Becker, C., Todd, C., Taraldsen, K., Pijnappels, M., ... Mellone, S. (2017). Mobile Health Applications to Promote Active and Healthy Ageing. *Sensors, 17*(3), 622. doi:10.339017030622 PMID:28335475

Henan Daily. (2017). The release of Internet development report of Henan 2016. *Henan Daily.* Retrieved from http://www.gov.cn/xinwen/2017-05/17/content_5194606.htm

Hesketh, T., Lu, L., & Xing, Z. W. (2005). The effect of China's one-child family policy after 25 years. Mass Medical Soc. Retrived from https://www.nejm.org/doi/full/10.1056/nejmhpr051833

Johnson, K. J., & Mutchler, J. E. (2014). The emergence of a positive gerontology: from disengagement to social involvement. *The Gerontologist, 54*(1), 93–100. doi:10.1093/geront/gnt099 PMID:24009172

Kendig, H. (2004). The social sciences and successful aging: Issues for Asia–Oceania. *Geriatrics & Gerontology International, 4*(s1), S6–S11. doi:10.1111/j.1447-0594.2004.00136.x

Laslett, P. (1987). The emergence of the third age. *Ageing and Society, 7*(2), 133–160. doi:10.1017/S0144686X00012538

Laslett, P. (1991). A fresh map of life: The emergence of the third age. Massachusetts: Harvard University Press.

Laslett, P. (1994). The third age, the fourth age and the future. *Ageing and Society, 14*(3), 436–447. doi:10.1017/S0144686X00001677

Li, Y., & Perkins, A. (2007). The impact of technological developments on the daily life of the elderly. *Technology in Society, 29*(3), 361–368. doi:10.1016/j.techsoc.2007.04.004

Litman, L., Rosen, Z., Spierer, D., Weinberger-Litman, S., Goldschein, A., & Robinson, J. (2015). Mobile exercise apps and increased leisure time exercise activity: A moderated mediation analysis of the role of self-efficacy and barriers. *Journal of Medical Internet Research, 17*(8), e195. doi:10.2196/jmir.4142 PMID:26276227

National Bureau of Statistics of the People's Republic of China. (2011). *2010 The Sixth Nationwide Population Census Report*. Retrieved from http://www.stats.gov.cn/tjsj/pcsj/rkpc/6rp/indexch.htm

National Bureau of Statistics of the People's Republic of China. (2012). *The average life expectancy reaches to 74.83 in China*. Retrieved from http://www.stats.gov.cn/tjsj/tjgb/rkpcgb/qgrkpcgb/201209/t20120921_30330.html

Niehaves, B., & Becker, J. (2008). The Age-divide in e-government–data, interpretations, theory fragments. In Oya, M., Uda, R., Yasunobu, C (Ed.), IFIP International Federation for Information Processing, Volume 286; Towards sustainable society on ubituitous networks. Boston: Springer. doi:10.1007/978-0-387-85691-9_24

Olphert, W., & Damodaran, L. (2013). Older people and digital disengagement: A fourth digital divide? *Gerontology, 59*(6), 564–570. doi:10.1159/000353630 PMID:23969758

Park, S. (2017). *Digital capital*. London: Palgrave Macmillan UK. doi:10.1057/978-1-137-59332-0

Powell, J., & Cook, I. (2000). "A Tiger Behind, and Coming up Fast": Governmentality and the Politics of Population Control in China. *Journal of Aging and Identity, 5.*

Powell, J. L., & Cook, I. G. (2009). Global ageing in comparative perspective: A critical discussion. *The International Journal of Sociology and Social Policy, 29*(7/8), 388–400. doi:10.1108/01443330910975696

Sadana, R., Blas, E., Budhwani, S., Koller, T., & Paraje, G. (2016). Healthy ageing: Raising awareness of inequalities, determinants, and what could be done to improve health equity. *The Gerontologist, 56*(Suppl 2), S178–S193. doi:10.1093/geront/gnw034 PMID:26994259

Stig, T., & Ni, A. (2010). He is he and I am I: Individual and collective among China's Elderly. In M. Halskov Hansen & R. Svarverud (Eds.), *iChina: The rise of the individual in modern Chinese society*. Copenhagen: NIAS Press.

Suzman, R., & Beard, J. (2015). *Global Health and Aging. The troublesome concept of "technological affordance".* Retrieved from https://www.who.int/ageing/publications/global_health.pdf

Uhlenberg, P. (Ed.). (2009). *International handbook of population aging* (Vol. 1). Springer Science & Business Media. doi:10.1007/978-1-4020-8356-3

Ventola, C. L. (2014). Mobile devices and apps for health care professionals: Uses and benefits. *P&T, 39*(5), 356. PubMed

Vincent, J. (2006). Emotional attachment and mobile phones. *Knowledge, Technology & Policy, 19*(1), 39–44. doi:10.100712130-006-1013-7

Weiss, R. S., & Bass, S. A. (2002). *Challenges of the third age: Meaning and purpose in later life.* Oxford University Press.

World Health Organization. (2018). *Ageing and health.* World Health Organization. Retrieved from http://www.who.int/news-room/fact-sheets/detail/ageing-and-health

Wu, H. (2016). Elderly people and the Internet: a demographic reconsideration. In M. Keane (Ed.), Handbook of the cultural and creative industries in China (pp. 431–444). Cheltenham, UK: Edward Elgar Publishing. doi:10.4337/9781782549864.00041.

# Chapter 3
# Locating Gender Differences in Mobile Use and Habits

**Beliz Donmez**
*Yeditepe University, Turkey*

**Cagla Seneler**
*Yeditepe University, Turkey*

## ABSTRACT

*As an indispensable part of their lives, mobile has been playing an important role among undergraduate students in influencing their lives and academic studies. Although gender has been proven to be an important factor in distinguishing males and females in using the internet and other digital media, what are the gender differences in mobile use and habits? The authors of this chapter offer an updated answer to that question by surveying a selected sample of undergraduate students in Turkey.*

## INTRODUCTION

Today it is hard to imagine a life without a mobile phone. They have become an indispensable part of our lives. Besides, we are moving into an era when mobile devices are not just for talking and texting but also for accessing the Internet and all it has the offer (Pew Research Center, 2010). Besides, the mobile phone is no longer just a device that facilitates communication between two individuals; it is also a hybrid technology that integrates audio, video, and text with a display screen (Halder, Halder & Guha, 2014). The use of a mobile phone is not limited to speaking alone; it is being used in making a video, recording information, mobile banking and payment etc. (Halder, Halder & Guha, 2014). In contrast to traditional notions of the computer, the mobile nature of the cell phone allows these services to be accessed almost anywhere and at almost any time (Lepp, Barkley & Karpinski, 2013). The ubiquity and affordability of mobile present us with an unparalleled opportunity to improve social and economic development and positively impact lives (GSMA, 2015). Humans in this regard have become obsessed with mobile phones and Internet. It is perhaps one of the most important reasons for this obsession is to be able to handle your needs and access with only one click.

DOI: 10.4018/978-1-5225-7885-7.ch003

Gender divide in mobile and Internet usage have been widely studied in the literature. While mobile connectivity is spreading quickly, it is not spreading equally (GSMA, 2018). Most of the researches showed that males tend to use more mobile and Internet technologies compared to females.

As reported in GSMA (2017):

- Women on average are 14% less likely to own a mobile phone than a man, which translates into 200 million fewer women than men owning mobile phones.
- Even when women own mobile phones, there is a significant gender gap in mobile phone usage, which prevents them from reaping the full benefits of mobile phone ownership.
- Women in South Asia are 38% less likely to own a phone than a man, highlighting that the gender gap in mobile phone ownership is wider in certain parts of the world.
- The top 5 barriers to women owning and using mobile phones from a customer perspective are cost, network quality and coverage, security and harassment, operator/agent trust, and technical literacy and confidence. Social norms and disparities between men and women in terms of education and income influence women's access to and use of mobile technology, and often contribute to women experiencing barriers to mobile phone ownership and use more acutely than men.
- In addition to the barriers experienced by female customers above, two other key systemic barriers arose – lack of gender disaggregated data and focus on women's access to and use of technology.

Women report using phones less frequently and intensively than men, especially for more sophisticated services such as the mobile Internet (Santosham, 2015). But why? Hypothesizing that men will be more comfortable with and less anxious about Internet technology. They have proposed a variety of explanations for these predictions, often focusing on technophobia—the idea that females are more afraid of technology and therefore slower to adapt to technological advances (Shaw & Gant, 2002). And also cultural and socio-economic barriers are the most common reasons for that. Even today in some countries, women still do not have the rights that men have. Therefore, it is clear that mobile and Internet usage habits differ based on gender.

There are also significant differences between mobile and Internet usages for different groups of ages. For example, the purpose of using the mobile phone may not be the same for 65 years old and 17 years old. To be more specific, the mobile phone is generally used for calling for older people; using social media (Instagram, Facebook, Snapchat etc.) can be hard and complicated for them. Overall, as most people's capacity to adapt and memorize new things slows down when they are getting older, patience and repetition might be required in teaching, and this is often best provided by friends or relatives (Mallenius, Rossi & Tuunainen, 2007). But that does not mean this hypothesis is valid for all old people. On the other hand, a person who is at the age of 17 can easily adapt to these technologies, they use mobile phones for different purposes like texting, calling, sharing pictures on Instagram or Facebook and tweeting. Every day new apps come out, and it is not hard for teens to learn and use these applications since they have been in this new technological World from the beginning. Also, some of the researches show that adults under age of 50 are as likely as teens in mobile and Internet usage (Pew Research Center, 2010). Therefore, mobile usage has different effects depending on different age groups. Growing up with a growing technology makes easy to keep up with new trends in the technology for the younger ones. World and environment affect habits and attitudes of generations and make differences between younger and elder. To be more specific, in today's World we can say mobile phone is a must for young people and even for kids. Because they grow up with this technology, while the older ones need to accept and learn these changes. However, age has not been studied in this study due to the specific age group in the university sample.

Although gender-based mobile use has been widely studied in the literature which concluded women are more behind then men while using mobile phones, contradictory findings of these studies showed that more research is needed on these topics (Hanson, 2010; Torche, 2015; GSMA, 2018). Furthermore, there are limited studies examining the possible relations between gender and mobile use. This research discusses mainly; does gender really effect on mobile use, how frequently undergraduate students use mobile phones and what are other characteristics of undergraduate students related with mobile phones usage. In addition, this study would be very valuable to explain why we are so addicted to mobile phones and the reasons behind the over-use of mobile phones by young people.

## BACKGROUND

## Mobile Phone Evolution and Mobile Usage

Today, it is hard to find a single person who does not have a mobile phone. With the invention of the telephone in 1876, it was possible for the first time in history to have real-time conversational interaction at a distance (Katz & Aakhus, 2002). The phone, which completed its development and change day by day, was made portable in 1973 for the first time in the world, and the name of the mobile phone we used today is given. The Motorola mobile phone weighs 1 kg, and only 20 minutes of phone calls can be made while the battery is fully charged. In 1992, the world met with Nokia. The Nokia brand mobile phone has a smaller size and shorter antenna. Just 1 year after that IBM released the first smartphone, Simon. Simon has features like call and fax, as well as features like electronic mail, game, and calendar. In the early 2000s, the phone revolution moved forward with Ericsson, bringing Bluetooth to life. People on this side would now be able to share wallpapers and ringtones with each other more conveniently via Bluetooth. And 11 years ago, from the year it came out in 2017 a total of 1.2 billion sales were made, first iPhone released to the market. It was much different than all the phone models that appeared until that day. The iPhone made a big difference to its competitors with the touchscreen as well as having Internet access and running faster than other phones. In the following years, similar phones started to come from brands like Samsung, LG, HTC, etc.

Modern mobile systems such as smartphones and tablets are already an important part of our lives (Rahmati, Shepard, Zhong & Kortum, 2015). Mobile phones initially provided the luxury of immediate contact with others in one's social network at any time, any place (Forgays, Hyman & Schreiber, 2013). Mobile phones are not used only for communication, they are used for coding, playing games, listening to music, finding dates, banking, etc. Smartphone use has been changing daily routines, habits, social behaviors, emancipative values, family relations and social interactions (Samaha& Hawi, 2016). The increase in reachability has also enabled people from all social groups and ages to own mobile phones. Global sales of smartphones to end users totalled nearly 408 million units in the fourth quarter of 2017, a 5.6 percent decline over the fourth quarter of 2016 (Gartner Inc., 2018). According to Customs and Trade Ministry (2017), as cited by Cumhuriyet Newspaper, smartphone usage in Turkey has reached 84 percent of cases, the 14.7 percent cut is based on a standard mobile phone, while the non-phone cuts up to 1.5 percent. As usage area has widened, usage rate has also grown. Our mobiles became a part of our lives. In other words; the mobile phone is seen as indispensable in the daily life of individuals because mobile phone offers many facilities.

According to Joo and Sang (2013), Koreans, on average, change their cell phones every 12 months. Korean context smartphone use is affected more by cognitive and goal-oriented motivations than by habitual and less goal-oriented use motivation. Therefore, early Korean adopters appear to have purchased smartphones based mainly on their perceptions of usefulness, especially active features that allow them to optimize and customize their smartphones by selecting from a range of applications that met their needs for use.

The mobile phone also affects people's daily communication because it is a key device in the development of social relationships and communication. According to the Onedio.com's, which is the content-based social network, the survey about negative effects of mobile phones on human relations, 92 percent of 14.2 thousand people said the mobile phone has negative effects on human relations (onedio. com, 2015). Why mobile phones have a negative effect on human relations? The reasons, behind this negativity, are constantly checking mobile phones while spending time with family and friends, causing a mobile phone addiction (nomophobia) and causing asociality. Nomophobia, which abbreviation of "no mobile phone phobia", is the fear of being without the mobile phone.

In the research of Kuyucu's (2017) Use of smartphone and problematic of smartphone addiction in young people: "smartphone (colic)" university youth, it was observed that this addiction did not depend on characteristics such as gender and age, however as the age increased, the interest in mobile decreases and therefore the nomophobia ratio lessens. In this study, it was found that the probability of new generations facing the threat of nomophobia is high.

According to the research of Lee, Chang, Lin & Cheng (2014), the reasons behind the compulsive usage of smartphones are related to an external locus of control, materialism, social interaction anxiety, and the need for touch. The first two factors are found to be more influential. Because of their passive tendencies and reduced powers of self-control, individuals with an external locus of control are more likely to use their smartphones compulsively. Their findings also prove that compulsive behavior under smartphone context share similarities with other forms of compulsive behaviors such as drug and alcohol addiction. Besides the negative aspects, mobile phones have the technology that can make a human easier life by providing distance communication, people's daily needs like banking transactions, maps, e-book, mobile-learning, etc.

## Mobile Phone Usage Relying on Gender

There are a lot of researches about the gender gap in Information and Communications Technology (ICT) (Hilbert, 2011; Moghaddam, 2010; Azevedo & Mesquita, 2018). Many researchers have addressed this topic, there is no agreement on which the highest risk group for mobile phone addiction is (Hong, Chiu & Huang, 2012). The number of studies on accessing and exploiting mobile usage in the world, especially showing the differences between men and women, has increased in recent years. However, women are more likely to be dependent on mobile phones (Billieux, Linden, & Rochat, 2008), and more likely to use mobile phones (Walsh, White, Cox, & Young, 2011). Females spend more time using the phone to connect with friends and family while males use the phone to obtain information (Wei & Lo, 2006).

Joiner, Stewart and Beaney's (2015) study, which is about Gender digital divide exist and what are the explanations, shows that gender differences in how students used their mobile phones that were very similar to the gender differences they found for the Internet in 2002 and 2012. Females tended to use their mobile phones for communication. They made and received more calls than males. They sent and received more text messages and took more pictures. Males, on the other hand, used their mobile phone for entertainment. They were more likely to play games than females and watch videos on their mobile phones than females. They were also more likely to check bank accounts. But there are also researchers that argue the exact opposite of this situation, and more precisely, that men tend to use mobile more than women (GSMA, 2017).

These results can be independent of one another for many reasons, such as geographic location, development of a country, woman's right, etc. Within the same country borders, usage rates also depend on several factors such as level of education and living in urban or rural areas (AkCa & Kaya, 2016). According to GSMA (2015) report on Bridging the gender gap: Mobile access and usage in low and middle-income countries, mobile ownership and usage are driven by a complex set of socio-economic and cultural barriers negatively affecting women. Some of the findings from this research are:

- Over 1.7 billion females in low- and middle-income countries do not own mobile phones.
- Women on average are 14% less likely to own a mobile phone than men, which translates into 200 million fewer women than men owning mobile phones.
- Cost remains the greatest barrier.
- Systemic barriers, including lack of gender disaggregated data at all levels (e.g., mobile subscribers, national statistics) and unconscious biases within organisations, have kept the focus off women and sustained the gender gap in ownership and usage.

## Mobile Use and Its Impacts on Academic Performance

It is clear that, Generation Y and Generation Z born with the technology. Generation Y, which was born 1977-1994, while individuals in this generation are highly intellectual and technologically inclined. Generation Z, which was born after 1995, is grown up around complicated media and computers, the Internet is more conceptualized and internalized and more specialized than the Y generation (Kuyucu, 2017). These two generations were in this environment from the beginning of their lives. It was easy for them to adopt this changing and growing world because they were growing with it. Mobile phones, computers, Internet become part of these two generation's lives. In a sense, this age group could be considered cell phone ''natives''(Forgays, Hyman & Schreiber2013). So, they started to use their mobiles, computers, and Internet at every moment of their lives.

Much of the literature on mobile use has focused on university-aged individuals (Harley, Winn, Pemberton, & Wilcox, 2007; Walsh & White, 2007). Especially university students who are easier to adapt to change, communication and information curiosity and interest in technology, such as a smartphone with higher functionality the consumption of vehicles is increasing steadily (Demir & Çakır, 2014). Students can use their mobile to do research, to take notes in the classes. But the actual reason of use may also be independent of lessons. College students and adolescents use electronic media simultaneously with other media (e.g., checking Facebook while Instant Messaging) or during activities requiring more focused attention, like class. Students' attention can easily be dissipated by a notification on their phone. While the course is being processed, students can text with friends, check social media, or play games (Jacobsen & Forste, 2011). Mobile usage has positive impacts on academic performance as negative effects. More specifically, students may use mobile to conduct academic research, to study, and to discuss with their professors our friends about lessons.

However various studies show that mobile phones have a negative effect on academic performance (Rupert & Hawi, 2016; Karpinski, Kirschner, Ozer, Mellott, & Ochwo, 2013). Using mobile phones during the lessons can cause a distraction. And being distracted by lessons can lead decreasing of academic performance. According to the report of Tindell and Bohlander (2012) as cited by Rose, Carrier, and Cheever (2013), 91% of college students in their study had sent or received a text message in their university class and 62% felt texting should be allowed in class if it does not disturb other students.

Rosen, Carrier& Cheever (2013) reported that corroborating the work on the impact of social media on academic performance, participants who accessed Facebook one or more times during the study period had lower grade point averages. While task-switching preference was not a significant predictor of GPA, having study strategies predicted a higher GPA while being confused while studying predicted a lower GPA.

According to the finding of Human Use and Social Research Center's "Technology Usage and Dependence Photo of Turkey's Youth " research, the level of smartphone dependency and the increase in the duration of smartphone use have reduced the academic achievement level (İHH, 2015).

Study of Kibona and Mgaya (2015) shows that mobile phones have an impact on the academic performance of higher education. Students make their own choice and preference on which mobile application to use, as it is discussed above almost 48% of the respondents agreed that they tend to use mobile phones for about 5 – 7 hours per day on social communication sites (65%) like Facebook, Twitter, Instagram, WhatsApp and the like without considering that those time spent on social network could have been used on academic related works and hence yield good results at the end of semester examinations.

However, mobile phones are also used for academic reasons, for example, academic research, mobile learning. Mobile Learning (M-Learning) is an e-learning model in which learning-teaching activities are provided through mobile devices and technologies. It allows the learning activities to be avoided from a certain place and time. Educational organizations are using and supplying mobile learning to provide flexibility in education and to make student's life easier. Mobile learning helps to increase the opportunity for equality in education, facilitate communication between teacher and student.

Al-Emran, Elsherif, and Shaalan (2016) conducted a study that examined investigating attitudes towards the use of mobile learning in higher education. They found that 99% of the students have mobile devices (smartphone/tablets) while only 1% have not. This is reasonable due to the reasonable price and availability of such mobile technology devices in the market. Findings revealed that 41.5% of the students were using their mobile devices (smartphone/tablet) for surfing the Web and accessing their emails while 16.7% of them were using their mobile devices in their education. Moreover, 81.5% of the students indicated that they were using their mobile devices in their study while only 18.5% do not do so.

## METHODOLOGY

A questionnaire was administered to a class of 182 undergraduate students from different majors of Turkish university, Istanbul, Turkey. The questionnaire form of the research has been quoted from Karaaslan and Budak (2012). The reliability coefficient has been already calculated (a=0,756), so it is not recalculated again. The obtained data were tested using the SPSS statistical package. Dependent variables were analyzed by frequency distributions, and independent variables were analyzed by t-test analysis. Independent variables of the study were determined as gender, while dependent variables were determined as students' mobile phone usage habits and place & importance of mobile phone in daily communication.

In addition, unlike the quoted survey, the effect of the mobile phones on the academic life of students was investigated. At the same time, the questionnaire used was modified according to the daily technology and the research. For example, the question "Do you use mobile phones as a distraction during the class?" And "Does it affect your academic success?" were added. Besides, the question "Do you use Bluetooth?" was changed into "Do you code on your mobile phone?" to observe the technology that students use via mobile phones.

## RESULTS

### Demographic Characteristics

The average age range is between 18-23. In this research, 50.5% are females and 49.5% are males.

### Mobile Phone Brand

19.2% of the students choose their mobile phones based on its brand. 72.0% of the students are using Apple, 18.1% of students are using Samsung and 2.2% of the students are using LG brand mobile phones. While the other 7.7% are using brands such as Huawei, Asus, HTC. 44.4% of the students are choosing the brand they use if the brand has comprehensive features. 9.9% of the students choose the brand based on its price. 8.8% of them choose the brand based on its technical services' prices. 4.4% of the students' reason for choosing the brand is for prestige and influencing others. 4.4% of the students choose their mobile phones according to its appearance. Only 1.6% of the students think the warranty is important. 7.7% of the students', who chose the other option, mostly explained it with ease of use or having their phones as a gift.

### Age to Start Using Mobile Phone and Mobile Phone Experience

41.2% of the students started using the mobile phone when they were 12 years old or younger. Other 44.5% part started to use the mobile phone when they were 13 or 14 years old. 9.34% of the students started using the mobile phone when they were 15 or 16 years old. Only 4.9% started to use when they were 17 years old or older, this low rate is remarkable.

62.1% of the students have been using mobile for phones for more than 9 years. 31.9% of the students have been using mobile phones for 7-8 years. Students who have been using mobile phones for 5-6 years are 4.4%. 0.5% of the students have been using mobile phones for 3-4 years. The number of students using the mobile phone for 3-4 (0.5%) and 1-2 (1.1%) years is very low.

### Quantity of Mobile Phones That Students Have

Most of the students (97.8%) have only one mobile phone. The number of students who have two (1.1%) or three or more (1.1%) mobile phone is very low.

## Applications Used

The most commonly used (55.5%) application is Instagram. WhatsApp is used by 41.2% of the students. On the other hand, Twitter (1.1%), Snapchat (0.5%) and Facebook (0.5%) have the lowest usage rate. Other applications, which are only used by 1.1% of the students, are Google and Apple Music.

## Advertisement Effect

Above the average (59.9%) students are not influenced by the advertisements while they are buying the mobile phone. 30.8% of the students are sometimes influenced by the advertisements. While 9.3% of the students say they are influenced and advertisement can change their mind.

## Mobile Phone Replacement Frequency

Most of the students (61.0%) change their mobile phones once in 3 or 4 years. 25.8% of the students change their phones in 0-2 years. 12.1% of the students renew their mobile phones once in 5-6 years. The students who use their mobile phones for 7 years or more (1.1%) is remarkably few.

## Hours Spending on Mobile Phones

45.1% of the students use their mobile phones between 4 and 5 hours every day. 24.2% of the students use their mobile phones between 2 and 3 hours every day. 15.4% of the students are using their mobile phones for 6-7 hours. And students who use their mobile phones for 8 hours or more (13.7%) are striking. The reason why students use the mobile phone for a long time is also directly related to the applications (Instagram, Snapchat, Twitter, etc.) that they are used apart from communication. Also, only a few students use their mobile phones for less than one hour (1.6%).

## Call or Message Preference

Most of the students say that they use both texting and calling (60.5%) for communication. 22.5% students prefer texting and 17.0% of the students prefer calling when they want to communicate.

## Monthly Budget

34.1% of the students have budgets between 40 and 60 Turkish Liras (TL). Some of the students have fewer budgets than others, they prefer to or have to pay 20-40 TL (25.3%) every month. 15.4% of the students pay 100TL or more every month to their operator. 15.4% of their budgets are between 60TL and 80TL. 9.9% of the students pay 80TL-100TL every month.

## Impact on Human Communication

Most of the students (79.7%) think mobile phones are affecting daily human communication. 15.4% of the students say sometimes daily human interaction is affected by mobile phones. While the other 4.9% of the students do not believe in mobile phone effects. This finding is in line with literature. According to Campbell (2015), the mobile phone has impacted on young people's peer groups enabling a truly networked society. It has also impacted on the evolving relationships within the family; especially by the increased negotiating power the mobile phone gives to young people in regard to curfews and safety issues. Schools and educational settings report that student's mobile phone use disrupts teaching and reduces student's attention in class, resulting in negative educational outcomes.

## Students' Thought About Mobile Phone

More than average (64.8%) students says mobile phones are organizing their lives. While 35.2% of the students say mobile phones do not have an important place in their lives. Almost every student (96.7%) is aware of the harm that mobile phones can cause. 3.3% of the students do not care or be aware of its harm.

This attractive environment, which is very easy and inexpensive to access, naturally creates an addiction. The number of students (45.1%) who think they are addicted to the mobile phone is far from underestimating. 54.9% of them say that they are not addicted to their mobile phones. Most of the students' (82.4%) mobile phones are open every time, they do not close it. While 17.6% of the students close their phones once in a while. More than average (65.9%) of the students say they cannot live without their mobile phones. The other participants (34.1%) do not think mobile phones are necessary. A large majority (70.9%) of students think that the mobile phone is not an indicator of the socioeconomic level of the person. 29.1% of the students think the exact opposite; mobile phones are an indicator of the socioeconomic level of the person. A large number (70.3%) of participants do not follow the mobile phones' campaigns. Only a small group (29.7%) of students follow the mobile phones' campaigns and promotions. Mobile phones can be a distraction in the class thus, they can affect academic success. 45.6% of the participants think that way. While 54.4% of the students say mobile phones are not affecting their academic success. Almost every student (99.5%) are using Internet feature on their mobile phones. Only 0.5% of the participants which means only one person, does not have this feature on their mobile phone.

Furthermore, previous study about addiction of mobile usage (Minaz & C. Bozkurt, 2017) shows that there is not any significant difference between male and female in mobile phone addiction. In terms of gender variable, students' smart phone dependencies differ significantly. The sub-dimensions of the smartphone addiction levels of the students' averages; daily life disorders scores immediately in males and females it was nearly the same. The expectation, sense of deprivation, overuse, and abstinence the average of female students is higher than the average of male students is seen. The average of men in virtual oriented relationship size scores higher than average.

## Mobile Phone Features Used by Students

96.2% of the students listen to music from their mobile phones, 3.8% of them are not using this feature. Big group of the students (86.8%) are using banking transactions on their mobile phones.

Another feature that smartphones bring is the ability to write code. Today, students can write code from their phones, but most of the students (74.7%) are not using this feature, 3.3% of the students say, they would use it if they have it on their mobile phones. Only 5.5% of the students are using this feature. Today most of the people have an account in any of the social media. 98.9% of the participants are using social media from their mobile phones.

According to the Deloitte's research about usage of mobile phones, an important issue that arises in the research is how many times a day people look at their phones has been the answer. People in Turkey are looking at their mobile phones 78 times a day, which is above the Europe's average (48). When the daily activities about mobile phones are asked, the following results have been reached regarding the usage rates; 69% is video sharing, watching videos is 57%, reading news is 55% and playing games is 42% (Deloitte, 2017). In this research 71.4% of students are using the game feature of mobile phones, on the other hand, 87.4% of the students are watching movies on their mobile phones.

## Relationship Between Gender and the Choice of Brand

72% of 192 students prefer to use Apple brand mobile phones, which are called iPhone. Females (M=0.086, SD=0.831 N=92) prefer Apple while buying their mobile phones. On the other hand, males (M=0.137, SD=1.307, N=90) are more likely to use Samsung's mobile phones. According to Vaidya (2016), among male users Samsung is the mostly used mobile phone, followed by Apple iPhone.

## Relationship Between Gender and Reason for Choosing a Phone Brand

The reason why women choose their mobile phone's brand is comprehensive features (M=0.196, SD=1.880, N=92) that brand has when compared to other brands. While male students' (M=0.244, SD=2.316, N=90) reasons are influencing others and thinking that the mobiles are an indicator of a prestige.

## Relationship Between Gender and Advertisement Effect

Male students (M=0.054, SD=0.515, N=90) state that they certainly do not get affected by advertisements when purchasing mobile phones. On the other hand, females (M=0.068, SD=0.658, N=90) mention that advertisement of mobile phones sometimes affects them.

## Relationship Between Gender and Taking Photos or Recording Videos

There is a meaningful relationship between taking a photo and recording video on mobile phones. It is noteworthy that female students (M=0.000, SD=0.000, N=92) use taking photo and recording video features on their mobile phones more than male students (M=0.021, SD=0.207, N=90).

## Relationship Between Gender and Total Phone Usage Year

According to this research, it is seen that average mobile phone usage of females is approximately 7 to 8 years (M=0.080, SD=0.773, N=92) since they have started using mobile phones. On the contrary, mobile phone usage time is 9 years or more for male students (M=0.065, SD=0.623, N=90).

## Relationship Between Gender and Banking Operations

There is a meaningful relationship between gender and banking operations from mobile phones These days boring and difficult banking operations can be done over the mobile phones at anytime, anywhere. However, male students (M=0.028, SD=0.269, N=90) prefer using mobile application for banking operations more than females (M=0.044, SD=0.425, N=92) do. There is no such a big difference in terms of ratios, with that being said.

## Relationship Between Gender and Most Used Application

There is a meaningful relationship between gender and most used application from mobile phones. Social media is in humans' lives for a long time and is now an indispensable tool especially for university students. It was observed that Instagram application, which entered technology world in the year of 2010, was used more by females (M=0.064, SD=0.615, N=92). Male students (M=0130, SD=1.238, N=90) prefer the WhatsApp application, which is a tool for instant messaging at first and even a tool for video calling today.

## Relationship Between Gender and Usage of Notes Application

There is a relationship between gender and usage of notes application from mobile phones. This application helps people to take notes easily from the mobile phones. Findings of research show that female students (M=0.044, SD=0.422, N=92) prefer to use notes application more than male students (M=0.072, SD=0.683, N=90).

## CONCLUSION AND DISCUSSION

One of the consequences of the rapid development of technology is the intensive use of mobile phones. Thanks to the facilities that provided, mobile phones have a big place in people's lives. Previously, only computer-based operations are now done easily via mobile phones. Although, usage rates and reasons of use are not the same, from 7 to 70 people of all ages have mobile phones. As the days pass, the number of people using mobile phones increases and the user's age profile goes down into lower age groups. They are no longer only a tool for communication, they also provide banking transactions, taking photos, listening to music, shopping, GPS, etc. Smartphones with these functions have significantly affected human life. The use of mobile phones by middle and lower classes, not just those with high socio-economic levels, is another reason for the growth of this market day by day. Especially for young people, technology and innovations and developments in this area are extremely important. The widespread use of smartphones brings to mind the question of addiction. Almost everywhere, many individuals who are living their eyes without distinction from their smartphones are confronted

The gender gap in all areas of ICT is also noticeable in mobile phone use. This may be due to the socio-economic level, the level of development of the country, and the interest in technology. There are various researches on this subject, but it needs to be done more. The gender gap in phone use can be similar to the gender gap in everyday activities. For example, in daily life bank transactions are usually made by male individuals in the family, this difference is also seen in mobile bank transactions. The gender gap was based on this study and results are greeting the eye.

Students who make the sample of the research all use mobile phones and the data obtained in this study confirms these views. Contrary to other studies (Simay, 2009; Baron & Ling, 2007; Gokaliler, Aybar & Gulay, 2011) male students pointed out that they choose their mobile phones to impress other people and for prestige. Female students, on the other hand, pointed out that what they pay attention to when they pick up the phone is the comprehensive features of the mobile phone. Today Social Media has a great place in human's life and females are sharing photos or videos more frequently than men. In this research, it is observed that women use taking photos and recording videos features more than males. Also, female students start to use mobile phones for 7 and 8 years while male students start using mobile phones for 9 years or more. And the most used application by students is Instagram which is used more by female students.

The other findings of the research show that there are some differences in the mobile phone usage characteristics of the students. When the participant's daily time slot for mobile phone usage was asked, the majority of participants reported that their mobile phones are open 24 hours a day. University students have been using mobile phones since an early age. Most of the students are using mobile phones for 9 years or more (62.1%). Some of the students who consider mobile phone as an indispensable technology are not waiting for the change of mobile phones and more than half (61.0%) of the students change their mobile phone every 3-4 years. 72.0% of the students prefer to use Apple's mobile phones which are called iPhone. And the most rated reason why students choose their mobile phones' brand is that brand has comprehensive features (44.0%). Contrary to other studies (Ozascilar, 2012), students think mobile phones are necessary and they must have it.

Most of the students are aware of the mobile phones' harm. Nonetheless, students who are interested in mobile phones more than 5 hours a day and keep their phones open all the time, without taking the necessary precautions. All these findings and the fact that the phone is an essential part of their lives reveals the thought of addiction that the mobile phone creates on people. In addition, a large number of students use the mobile phone irrelevantly and distractingly in the classes, affecting their academic success. And almost half the students are aware of this situation but do not do anything to change these circumstances. Another of the technologies that the developing and changing world brings to our lives is the coding. Coding is now possible not only from the computer but also from the mobile phones. However, the vast majority of students do not use this feature and only 3.3% of the students say they would use this feature if they have it on their mobile phones.

Although the research has reached its aims, there were some unavoidable limitations. First of all, the monthly income of the students or their families was not asked in the survey, but it should be taken into consideration that the income of the majority of the students can be high due to the fact that the students are studying at the private university. Besides, there is a lack of previous studies in the literature related to this topic. Also, due to limited time, the majors of the students were not asked because they could not show an even distribution and the data might not reflect the truth. In addition, because of the limited time, research was conducted only with students from a single university. Furthermore, the age range of this study is delimited due to the fact that it had done on university students.

As a conclusion, this research helps to understand the university students mobile phone usage habits depending on gender. Despite all these conclusions and thought, as the number of students using mobile phones increases day by day, this study can be enlarged. And it can be measured that students in more universities, both the faculties and postgraduates are using mobile phones in a problematic manner while using mobile phone usage attitude, especially depending on gender.

## REFERENCES

Akca, E., & Kaya, B. (2016). The different approaches to digital divide in the concept of gender equality and it's dimensions. *Intermedia International, 3*(5). Available from: http://dergipark.gov.tr/download/article-file/399689

Al-Emran, M., Elsherif, H., & Shaalan, K. (2016). Investigating attitudes towards the use of mobile learning in higher education. *Computers in Human Behaviors [Online]*, *56*, 93–102. doi:10.1016/j. chb.2015.11.033

Azevedo, A. &Mesquita, A. (Eds.). (2018). *International Conference on Gender Research*. Porto, Portugal: Academic Conferences and Publishing International Limited. Available from: https:// books.google.com.tr/books?id=CcFWDwAAQBAJ&pg=PR6&lpg=PR6&dq=International+Conf erence+on+Gender+Research.+April+12-13+2018&source=bl&ots=7tbPObsaPt&sig=s8ALhM DV2vs-qB0SET2VCVHJGpE&hl=tr&sa=X&ved=0ahUKEwjc9OLChajbAhWGWSwKHe8pA-8Q6AEIbDAJ#v=onepage&q=International%20Conference%20on%20Gender%20Research.%20 April%2012-13%202018&f=false

Baron, N. S., & Ling, R. (2007). *Emerging Patterns of American Mobile Phone Use: Electronically-mediated Communication in Transition*. Available from: http://doczine.com/bigdata/2/1366989405_23054102ae/ emerging-patterns-ofamerican-mobile-phone-use-3.pdf

Billieux, J., Linden, M. V. D., & Rochat, L. (2008). The role of impulsivity in actual and problematic use of the mobile phone. *Applied Cognitive Psychology*, *22*(9), 1195–1210. doi:10.1002/acp.1429

Campbell. (2015). *The impact of the mobile phone on young people's social life*. Available from: https:// www.researchgate.net/publication/27465354_The_impact_of_the_mobile_phone_on_young_people's_ social_life

Deloitte. (2017). *The Place of Mobile Technologies in Our Digitalized Life, Deloitte Global Mobile User Survey* [Dijitalleşen Hayatımızda Mobil Teknolojilerin Yeri, Deloitte Global Mobil Kullanıcı Anketi]. Available from: https://www2.deloitte.com/content/dam/Deloitte/tr/Documents/technology-mediatelecommunications/deloitte_ gmcs_2017.pdf

Demir, N., & Cakır, F. (2014). *Research on determining university students' smartphone purchasing preferences* [Üniversite öğrencilerinin akıllı telefon satın alma tercihlerini belirlemeye yönelik bir araştırma]. Available from: https://www.researchgate.net/publication/304624501_Universite_Ogren-cilerinin_Akilli_Telefon_Satin_Alma_Tercihlerini_Belirlemeye_Yonelik_Bir_Arastirma

Forgays, D., Hyman, I., & Schreiber, J. (2013). Texting everywhere for everything: Gender and age differences in cell phone etiquette and use. *Computers in Human Behavior*, *31*, 314–321. doi:10.1016/j. chb.2013.10.053

Gokaliler, E., Aybar, A., & Gulay, G. (2011). The perception of Iphone branded smart phone as a status consumption symbol. *Selcuk Iletisim, 7*(1), 36-48. Available from: http://dergipark.gov.tr/josc/ issue/19023/200589

GSMA. (2015). *Bridging the gender gap 2015: Mobile access and usage in lowand middle-income countries 2015*. Available from: http://www.altaiconsulting.com/wp-content/uploads/2016/03/ GSM0001_02252015_GSMAReport_FINAL-WEB-spreads.pdf

GSMA. (2017). *GSMA Input into report on the digital gender divide*. Available from: http://www.ohchr. org/Documents/Issues/Women/WRGS/GenderDigital/GSMA.pdf

GSMA. (2018). *Mobile Gender Gap Report 2018*. Available from: https://www.gsma.com/mobilefordevelopment/wp-content/uploads/2018/02/GSMA_The_Mobile_Gender_Gap_Report_2018_Final_210218.pdf

Halder, I., Halder, S., & Guha, A. (2015). *Undergraduate students use of mobile phones: Exploring use of advanced technological aids for educational purpose*. Available from: http://www.academicjournals.org/article/article1427128975_Halder%20et%20al.pdf

Hanson, S. (2010). Gender and mobility: new approaches for informing sustainability. *Gender, Place & Culture, 17*(1), 5–23. Available from: https://www.tandfonline.com/doi/abs/10.1080/09663690903498225

Harley, D., Winn, S., Pemberton, S., & Wilcox, P. (2007). Using texting to support students' transition to university. *Innovations in Education and Teaching International, 44*(3), 229–241. doi:10.1080/14703290701486506

Hilbert, M. (2011). Digital gender divide or technologically empowered women in developing countries? A typical case of lies, damned lies, and statistics. *Women's Studies International Forum, 34*(6), 479-489. doi:10.1016/j.wsif.2011.07.001

Hong, F., Chiu, S., & Huang, D. (2012). A model of the relationship between psychological characteristics, mobile phone addiction and use of mobile phones by Taiwanese university female students. *Computers in Human Behavior, 28*(6), 2152–2159. doi:10.1016/j.chb.2012.06.020

İHH. (2015). Technology usage and dependence photo of Turkey's youth [Teknoloji kullanımı ve bağımlılığı açısından Türkiye gençliğinin fotoğrafı]. *İnsani Sosyal Araştırmalar Merkezi*. Available from: http://insamer.com/rsm/files/Teknoloji%20kullanimi%20ve%20bagimliligi.pdf

Jacobsen, W., & Forste, R. (2011). The wired generation: Academic and social outcomes of electronic media use among university students. *Cyberpsychology, Behavior, and Social Networking, 14*(5), 5. doi:10.1089/cyber.2010.0135 PMID:20961220

Joiner, R., Stewart, C., & Beaney, C. (2015). *Gender digital divide exist and what are the explanations. The Wiley Handbook of Psychology, Technology, and Society*. Oxford, UK: John Wiley & Sons, Ltd. Available from https://books.google.com.tr/books?hl=tr&lr=&id=Zb0oBwAAQBAJ&oi=fnd&pg=PA74&dq=gender+gap+and+smartphone&ots=QtV6nCtENr&sig=61o5H9gYWCF6mWsTMs1oglCQf34&redir_esc=y#v=onepage&q=gender%20gap%20and%20smartphone&f=false

Joo, J., & Sang, Y. (2013). Exploring Koreans' smartphone usage: An integrated model of the technology acceptance model and uses and gratifications theory. *Computers in Human Behavior, 29*(6), 2512–2518. doi:10.1016/j.chb.2013.06.002

Karpinski, A. C., Kirschner, P. A., Ozer, I., Mellott, J. A., & Ochwo, P. (2013). An exploration of social networking site use, multitasking, and academic performance among United States and European university students. *Computers in Human Behavior, 29*, 1182-1192. doi:10.1016/j.chb.2012.10.011

Katz, J., & Aakhus, M. (Eds.). (2002). *Perpetual Contact: Mobile Communication, Private Talk, Public Performance*. London: Cambridge University Press. Available from: https://books.google.com.tr/books?hl=tr&lr=&id=Wt5AsHEgUh0C&oi=fnd&pg=PR9&dq=mobile+usage+social+interaction&ots=YU_y_cNqkL&sig=RunBsoZK7CiToCIkEj36EvzgZ2M&redir_esc=y#v=onepage&q&f=false

Kibona, L., & Mgaya, G. (2015). Smartphones' effects on academic performance of higher learning students. *Journal of Multidisciplinary Engineering Science and Technology, 2*(4), 777–784. Available from https://pdfs.semanticscholar.org/1203/16b911f8e69ec4b79efdc5b6bda9fbf23ec6.pdf

Kuyucu, M. (2017). Use of smart phone and problematic of smart phone addiction in young people: "smart phone (colic)" university youth [Gençlerde akıllı telefon kullanımı ve akıllı telefon bağımlılığı sorunsalı: "akıllı telefon(kolik)" üniversite gençliği]. *Global Media Journal TR Edition, 7*(14). Available from: http://globalmediajournaltr.yeditepe.edu.tr/sites/default/files/mihalis_kuyucu_-_genclerde_akilli_telefon_kullanimi_ve_akilli_telefon_bagimliligi_sorunsali_akilli_telefonkolik_universite_gencligi.pdf

Lee, Y., Chang, C., Lin, Y., & Cheng, Z. (2014). The dark side of smartphone usage: Psychological traits, compulsive behavior and technostress. *Computers in Human Behavior, 31*, 373–383. doi:10.1016/j.chb.2013.10.047

Lepp, A., Barkley, J., & Karpinski, A. (2013). *Computers in Human Behavior.* Kent, OH: Elsevier. Available from https://www.sciencedirect.com/science/article/pii/S0747563213003993?via%3Dihub

Madden, M., Lenhart, A., Duggan, M., Cortesi, S., & Gasser, U. (2013). *Teens And Technology.* Available from: http://www.pewinternet.org/Reports/2013/Teens-and-Tech.aspx

Mallenius, S., Rossi, M., & Tuunainen, V. (2007). *Factors affecting the adoption and use of mobile devices and services by elderly people – results from a pilot study.* Available from: https://www.researchgate.net/publication/228632076_Factors_affecting_the_adoption_and_use_of_mobile_devices_and_services_by_elderly_people-results_from_a_pilot_study

Minaz, C. B. (2017). *Investigation of university students smartphone addiction levels and usage purposes in terms of different variables.* Available from: http://dergipark.gov.tr/download/article-file/352363

Moghaddam, G. (2010). Information technology and gender gap: Toward a global view. *The Electronic Library, 28*(5), 722–733. doi:10.1108/02640471011081997

Newspaper, C. (2017). Smart phone usage rate in Turkey [İşte Türkiye'de akıllı telefon kullanım oranı]. *Cumhuriyet.* Available from: http://www.cumhuriyet.com.tr/haber/ekonomi/816129/iste_Turkiye_de_akilli_telefon_kullanim_orani.html

Onedio.com. (2015). *Akilli Telefon Bagimliliginin Insan Iliskilerine Etkisini Gozler Onune Seren Reklam Kampanyasi.* Available from: https://onedio.com/haber/akilli-telefon-bagimliliginin-insan-iliskilerini-nasil-etkiledigini-gozler-onune-seren-fotograflar-550175

Ozascilar, M. (2012). Mobile phone usage and personal security of young individuals: University students' usage of mobile for personal security [Genç bireylerin cep telefonu kullanımı ve bireysel güvenlik: üniversite öğrencilerinin cep telefonunu bireysel güvenlik amaçlı kullanımları]. *Journal of Sociological Research.[Online], 15*, 1. Available from http://dergipark.ulakbim.gov.tr/sosars/article/view/5000093023

Rahmati, A., Shepard, C., Zhong, L., & Kortum, P. (2015). Practical context awareness: Measuring and utilizing the context dependency of mobile usage. *IEEE Transactions on Mobile Computing, 14*(9), 1932–1946. doi:10.1109/TMC.2014.2365199

Rosen, L., Carrier, L., & Cheever, N. (2013). Facebook and texting made me do it: Media-induced task-switching while studying. *Computers in Human Behavior, 29*(3), 948–958. doi:10.1016/j.chb.2012.12.001

Rupert, M., & Hawi, N. (2016). Relationships among smartphone addiction, stress, academic performance, and satisfaction with life. *Computer in Human Behavior. [Online], 57*, 321–325. doi:10.1016/j.chb.2015.12.045

Samaha, M., & Hawi, N. (2016). Relationships among smartphone addiction, stress, academic performance, and satisfaction with life. *Computers in Human Behavior, 57*, 321–325. doi:10.1016/j.chb.2015.12.045

Santosham, S. (2015). Closing the gender gap in mobile phone access and use. *Better Than Cash.* Available from: https://www.betterthancash.org/news/blogs-stories/closing-the-gender-gap-in-mobile-phone-access-and-use

Shaw, L., & Gant, L. (2002). Users divided?Exploring the gender gap in internet use. *Cyber Psychology & Behavior, 5*(6), 517–527. doi:10.1089/109493102321018150 PMID:12556114

Simay, A. E. (2009). Mobile Phone Usage and Device Selection of University Students. *Symposium for Young Researchers, 20-22*, 185-193.

Torche, F. (2015). *Gender differences in intergenerational mobility in Mexico.* Available from: http://www.ceey.org.mx/sites/default/files/adjuntos/dt-011-2015_si.pdf

Vaidya. (2016). *Mobile phone usage among youth.* Available from: https://www.researchgate.net/publication/299540610_Mobile_Phone_Usage_among_Youth

Walsh, S. P., & White, K. M. (2007). Me, my mobile, and I: The role of self- and prototypical identity influences in the prediction of mobile phone behavior. *Journal of Applied Social Psychology, 37*(10), 2405–2434. doi:10.1111/j.1559-1816.2007.00264.x

Walsh, S. P., White, K. M., Cox, S., Young, R., & Mc, D. (2011). Keeping in constant touch: The predictors of young Australians' mobile phone involvement. *Computers in Human Behavior, 27*(1), 333–342. doi:10.1016/j.chb.2010.08.011

Wei, R., & Lo, V.-H. (2006). Staying connected while on the move: Cell phone use and social connectedness. *New Media & Society, 8*(1), 53–72. doi:10.1177/1461444806059870

Wunmi, B., & Rob, M. (2018). *Gartner says worldwide sales of smartphones recorded first ever decline during the fourth quarter of 2017.* Gartner, Inc. Available from: https://www.gartner.com/newsroom/id/3859963

# Chapter 4
# Mobile Dating:
## A Fancy Way to Fall in Love?

**Lu Sun**
*Communication University of China, China*

## ABSTRACT

*In addressing the question as to whether mobile dating is a fancy way to fall in love, the author of this chapter ignited further interest in examining the issue by reviewing up-to-date studies on mobile dating, examining mobile dating applications and individual users' motivations at different stages of mobile dating. Furthermore, the author also included LGBTQ studies in her investigation. Finally, the author identified various problems related to mobile dating that invite further examination and solutions.*

## INTRODUCTION

Mobile dating, also known as cell phone dating, enables people to chat, flirt, meet, hook-up and probably start potential romantic relationships by using text messaging, voice calling, video chatting, and the Internet. Mobile dating applications (MDAs) are designed to make interactions easier for mobile dating based on mobile devices, especially for smart phone users. The concept of "mobile dating" stems from "online dating" which is different from conventional face-to-face offline dating and refers to the practice of using dating websites or mobile devices for the purpose of finding potential partners. Online dating has been developed for several decades since websites are prevalent in daily life. Toma (2015) described the mechanism of online dating: dating websites operate by requesting users to compose self-descriptive profiles and then connecting them with databases of potential partners. Several factors facilitate the booming of online dating. In fact, social and commercial institutions that facilitate courtship and marriage are diverse and long-standing (Ahuvia & Adelman, 1992). Matchmaking and introductory intermediaries, especially for facilitating marriage, play a role in the marriage-courtship market long before the beginning of online dating. Furthermore, computer-mediated communication (CMC) has been used for romantic matching for over 60 years (Finkel et al., 2012). Thanks to the development of technology, video-dating emerged in the 1980s, which was popular at that time for mate seeking and encouraged users to provide profile descriptions and photographs and participate in a video-based chat (Ahuvia & Adelman, 1992; Woll & Cozby, 1987).

DOI: 10.4018/978-1-5225-7885-7.ch004

Meanwhile, "cyber love", an emerging fancy word, was regarded as as "a romantic relationship consisting mainly of CMC" (Ben-Ze'ev, 2004, p. 4). Li and Chen (2005) made it more clearly by classifying cyber love into two main categories. (1) Cyber love I-a: romantic relationships developing mainly through the Internet, but the people involved had actually met each other face-to-face before; Cyber love I-b: relationships in which two people meet each other online and later form offline romantic relationships; (2) Cyber love II: two people form an online romantic relationship and do not meet face-to-face.

"Online dating" has been described "as an adventure" (Lawson and Leck, 2006, p. 197). They viewed online dating as a gambling activity where users are supposed to take risks on the premise of social exchange theories. As social interaction is an exchange process, individuals should balance costs and rewards to obtain profit to the most. In this case, risks are one of the costs in online dating, they indicated (Lawson and Leck, 2006).

Finkel et al. (2012) did a comprehensive study on online dating from the perspective of psychological science. They asserted that online dating is drastically different from traditional offline dating and it is hard to say whether online dating promotes better romantic outcomes than conventional offline dating. Accessing, communication and matching are three main services online dating offered (Ahuvia & Adelman, 1992; Finkel et al., 2012).

(1) Access to a potential partner who is exposure to the same platform. (2) Communication with various forms of CMC to interact with specific potential partners through the dating site before meeting face-to-face. (3) Matching by the use of a mathematical algorithm to select compatible partners for users. (Finkel et al., 2012, p. 3)

Although online dating has some similar characteristics with mobile dating, the use of mobile devices makes mobile dating more convenient and technical. Individuals' daily life is entangled with digital media, especially mobile devices (Goggin, 2006), and this involved into to sexual acts and intimate relationships (Light, 2014). With the proliferation of smart phones, individuals prefer mobiles over PCs in recent years when dating online. Because it is more convenient to use a portable smart phone all day and wherever you like. According to ASDF survey, young generations prefer mobile dating apps to matrimonial sites (Diary, 2018). By incorporating user's real-time location and intelligently connecting to social networks, mobile dating systems could give more appropriate options due to the potential partner's proximity, interests, priorities.

## BACKGROUND

Mobile dating alters the conventional way to find a mate by offering numerous options rather than getting to know each other naturally by chance. The digital data pool gives users a large amount of opportunities to find potential love. At the same time, targeted selection processes make it easier to find appropriate partners when set searching standards with filtering characteristics such as location, age and hobbies etc. For instance, music is helping connect Tinder users with individuals harboring a similar taste in singers and genres. In addition, mobile dating can also meet specific users demands such as LGBTQ groups (lesbians, gays, bisexual, transgender and queer people) as well. Furthermore, individuals are able to use MDAs to chat everywhere and every time in a cheapest way and without barrier of distance. In addition, mobile dating business makes essential contribution to economic profits. Dating websites and MDAs services that support the search for romantic and sexual partners are increasingly developed and became a substantial part of the burgeoning "app economy" (Goldsmith, 2014).

Various MDAs can be divided into several categories.

1.  Geo-social proximity and social networks-based apps. Tinder, one of the first location-based real-time dating (LBRTD) applications, is representative. Instead of creating a new account, subscribers are welcome to link Tinder with Facebook accounts to alleviate phony-profile problems and enhance authenticity. This sort of MDA usually recommends potential love information by location, shared interests and common friends. Tinder, one of the most popular MDAs, has attracted users in 196 countries, counting more than 10 billion matches worldwide (Dredge, 2015) and attaining one of the highest popularities matching 26 million pairs of daters per day (Zhang& Yasseri, 2016). Users can swipe right (like) or swipe left (dislike) to express their attitude towards potential daters. At this moment, they make their judgment based on profiles from Facebook, a short self-introduction, as well as a linked Instagram or Spotify account as optional additions. When both swipe right, they are allowed to chat which means that there is a chance of a success to match.

2.  Live Video Chat apps. Live video chats may prelude a new era for MDAs. Software like Skype and FaceTime make mobile dating more convenient. In North America, platforms like Dating.fm, Flikdate, Video Date, View N Me and Instamour adopt to this tech in order to help daters save time and money with a video first date. MoMo, a real-time video supplier, is prevailing in China with around 75 million active users (MoMo Inc., 2016). Other popular MDAs in China are TanTan, Tinder, Datetix, Zoosk, Lovestruck, Coffee Meets Bagel, and Skout etc.

3.  Niche MDAs designed for targeted audiences. For example, JDate is a flagship ethnic dating service for Jewish people. Similarly, Christian Mingle has become a leading online community for Christian singles. In addition, resources are available for LGBTQ groups. For instance, Grindr has become to be the largest social networking app for the LGBTQ community. This app also offers need user's location to find partners nearby. Besides, LGBTQutie, Chappy, Scruff, Growlr and Scissr are also aimed to foster meaningful relationships based on similar interests.

For instance, Dattch, as the lesbians-only MDA was redesigned with a focus on longer-term social interaction and culture, rather than short-term attraction based on geographic proximity. The company altered its strategy by looking insights into data analysis generated by customers using behavior (Murray & Sapnar, 2016)

4.  Extra-technic applied and high-qualified MDAs. In those specific designed apps, more information is required for achieving better matching quality. Extensive personality questionnaires and partner tests are taken in apps such as EliteSingles and eDarling to try to connect singles with similar personal beliefs and values. Also, customers usually asked to pay for specially designed service.

## A WORLDWIDE PICTURE OF MOBILE DATING

According to Pew Research Center's report, nearly 15% of young adults use online dating or mobile applications; meanwhile, 41% of Americans know someone who uses online dating when forming romantic relationships (Smith & Anderson, 2016).

According to Statista's Global Consumer Survey (Statista, 2018a) conduced in 2017, 63.6% of digital dating users are male. 25-34-year-old users are nearly in half of the population, taking a share of

42.2%. 35-44-year-old customers are approximately in one quarter of the total (24.9%). 18-24 years old subscribers come next with a percentage of 18. 45-54-year-old and 55-64 years-old ones are the least of two, with a share of 9.6% and 4.6% respectively (Statista, 2018 a). Meanwhile, a share of 44.7% of users comes from high-income groups with a distinct gap between low-income (26.7%) and medium-income ones (28.5%) (Statista, 2018a). As for users' penetration in the world, the top 5 countries comparatively look similar. The Netherlands takes the first place with rate of 12.5%. Next four following countries are South Korea (12.4%), the United States (12.3%), the United Kingdom (11.6%) and Canada (11.5%) in subsequence (Statista, 2018a).

The progressive technology and prevalent accessibility of smart phone contribute to the burgeoning industry of mobile dating. Taking an aerial view globally, five top countries in digital dating revenue (in million US dollars) are the United States ($590 m, ranking first and far ahead of others), China ($129m), the United Kingdom ($97 m), Germany ($74 m), Australia ($46 m) (Statista, 2018b).

Statista's investigation (2018c) illustrates the most popular online dating apps in the United States by audience size. Tinder ranks first by having 8.2 million users. One reason why Tinder holds such a significant place is that it established early at 2012 and innovatively linked with social networks (Facebook). Plenty of fish just follows with a 6.7 million audience. Both Match.com and OK Cupid attract 5.1 million subscribers respectively. Grindr and Bumble have a market share of 1.9 and 1.4 million customers. Bumble is specialized for women to initiate the conversation in heterosexual matches while neglecting to same-sex matches.

Match.com and eHarmony are subscription-oriented brands, which typically charge between $30.00 and $60.00 per month. There are also so-called free dating brands like OKCupid. Advertising contributes to the majority of its revenue. Meanwhile, customers are also recommended to choose premium membership options for better service. Overall, the proportion of online dating rises to 31.0% of the total dating market revenue and is projected to grow continuously (IBISWorld, 2017).

Mobile dating industry consists of numerous giant multinational entities, in which Match.com Group is one of the major representatives. Match.com Group belongs to a leading media and Internet company called IAC, which includes platforms such as Match.com, OkCupid, Tinder, Plenty of Fish and others. In 2017, the revenue reached to 1.28 billion U.S. dollars. In August 2018, the Match.com Group announced that Tinder has over 3.7 million paid subscribers, with a salient 81 percent increase during the same quarter in 2017(Emily, 2018).

In conclusion, the digital dating market is still promising. Revenue in 2018 reaches to $1,383 million US dollars and will result in a larger market volume of $1,610 million dollars by 2022; at the same time, user penetration is expected to increase from 5.7% in 2018 and hit 6.3% by 2022 globally. (Statista, 2018b)

## THEORIES ABOUT MOBILE DATING

There are several theoretical frameworks related to mobile dating taken by researchers to articulate individuals' mobile dating behaviors.

Erving Goffman (1959) put forward the concept of impression management by depicting of human interaction as a theatre stage. Impression management refers to a process in which individuals attempt to influence others' perceptions by regulating and controlling information in social interaction (Goffman, 1959). Impression management happens with self-presentation synonymously which is not only applied to face-to-face communication, but also expands to digital interactions such as computer-mediated communication (CMC) and MDAs. Although impression management strategy might take place differently online and offline, strategic self-presentation acts similarly in directing identities towards individual's ideal self which the potential partner would like to see rather than one's real self (Ellison et al., 2006; J. Rosenberg & Egbert, 2011).

Uses and gratifications (U&G) theory could also be an explanation on the use of MDAs which concerning as follows:

The social and psychological origins of (2) needs, which generate (3) expectations from (4) the mass media or other sources, which lead to (5) differential patterns of media exposure (or engagement in other activities), resulting in (6) need gratifications and (7) other consequences, perhaps mostly unintended ones. (Katz, Blumler, & Gurevitch, 1973, p. 20)

Aiming to illustrate sexual motivations, Sexual Strategies Theory (SST, Buss & Schmitt, 1993, 2016) divides human mating strategies into two categories: short-term mating and long-term mating. Long-term mating involves extended courtship, pair-bonding emotions and dedication of resources over time, while short-term mating refers to more fleeting sexual encounters (Buss & Schmitt, 1993, 2016; Botnen et al., 2018).

Uncertainty reduction theory (URT) indicates that a leading goal of interactants during communication is to reduce uncertainty by enhancing the predictability of others' behaviors. While decreasing uncertainty is definitely a major motivation for many communicators, the unique features of MDAs create an emerging context that can alter how people perceive, evaluate, and act upon uncertainty when forming interpersonal communications (Corriero & Tong, 2016).

Previous studies concerning with information-seeking and self-disclosure in web-based online dating were explained by URT (Gibbs et al., 2010). Whereas, the use of MDAs adds a new layer of complexity to interpersonal uncertainty, so that URT cannot explain well in some circumstances. The afterwards research adopts uncertainty management theory (UMT; Brashers, 2001) as primary framework. UMT no longer perceives uncertainty intrinsically negative, instead, whether beneficial or harmful of uncertainty is determined by users' cognitive appraisals. Several factors influence appraisals such as the use's goals, the specific contextual circumstances involved, and their assumed controllability (Brashers & Hogan, 2013; Brashers et al., 2000). Later, Rains and Tukachinsky (2015) assumed that information-seeking behavior was also a pivotal predictor of desire for uncertainty, and they found that uncertainty was not significantly related to uncertainty appraisal.

## Three Phases in Mobile Dating Relationships

Finkel et al. (2012) concluded a nine-step process for online dating which is also thought provoking for the study of mobile dating.

*(1) Seek information about one or more dating sites. (2) Register for one or more dating sites. (3) Create a profile on one or more dating sites—and, where relevant, complete a matching questionnaire. (4) Browse others' profiles (optional). (5) Initiate contact through the dating site (optional). (6) Receive contact through the dating site. (7) Engage in mutual mediated communication. (8) Meet face-to-face. (9) Develop an offline relationship. (pp. 13-18)*

Afterwards, some scholars use three significant periods to depict for establishing relationships in mobile dating: the profile stage, matching stage and discovery stage. (Markowitz, Hancock, & Tong, 2018)

1.  The profile period. The profile stage comes first, which refers to the stage when mobile dating apps subscribers create their personal accounts (some can link other social networks). Individual usually develops a profile to write an autobiography to the best or share own story to draw affection. In addition, other characteristics are optional (e.g., job, interests, graduating university etc.). Profile gives others the first impression, which especially matters at the beginning. Toma (2015) categorized three types of dating profile data: physical attributes, personal interests, and photos.

Interestingly, males are more likely to be visual oriented than females. Thus, there are tricks or even deception during profile period. At the meantime, women tend to post their "perfect" good-looking profiles after highly modifying them by mobile software. Phony profiles bring problems. "Is that the lady I chat with? " This inner ticking question depicts the frustration of a man's first off-line date (Markowitz et al., 2018). It is not a single case and even happens quite frequently. Although users are supposed to pay attention to their profiles (as well as other pictures), over-editing should be prohibited. Furthermore, this authenticity issue may cause trust loss and eventually the relationship ending.

2.  The matching period. Both potential lovers make a judgment based on profile information right after the profile stage. Daters will soon decide to accept or reject a potential partner by swiping right or left (as most MDAs are designed like Tinder).
3.  The discovery period. The third phase begins after two users express mutual connections in the profile stage and show interest during the matching period, they can then start interpersonal communication. This is called the discovery phase, because it represents a time when communication plays a crucial role in the decision to meet the other partner face-to-face (Markowitz et al., 2018).

## Motivations to Use MDAs: Sexuality Comes First

Several professors identified user's motivations and gratifications via using MDAs. As a worldwide popular application, Tinder draws a focusing attention by researchers.

Sumter et al. (2016) collected 46 items covering the physical, sexual, and psychosocial motivations for using Tinder. "The results of their exploratory factor analysis revealed six categories of dominant motivations for using Tinder: love, casual sex, ease of communication, self-worth validation, thrill of excitement, and trendiness" (Solis & Wong, 2018, p.16).

Ranzini & Lutz (2017) did an online survey through Amazon Mechanical Turk participated by 497 respondents in United States (in 2016). Selected items regarding narcissism (by Narcissism Personality Inventory (NPI)-16 Scale), self-esteem (by Rosenberg Self-Esteem Scale) and loneliness (by De Jong Gierveld Scale) were measured and respectively. Based on data analyzing via using structural equation modeling, Ranzini & Lutz would like to assess how Tinder users present themselves, the impact of their personality characteristics while self-presenting, their demographics and motives of use. Self-esteem is the most pivotal psychological predictor in this experience, enhancing authentic self-presentation while weakening deceptive self-presentation. There are six motives of use also matter to authentic and deceptive self-presentation: hooking up/sex, friendship, relationship, traveling, self-validation, and entertainment.

Entertainment is the most pronounced motivation for the respondents and self-validation the weakest. Demographic characteristics and psychological antecedents influence the motives for using Tinder, with gender differences being especially pronounced. Women use Tinder more for friendship and self-validation, while men use it more for hooking up/ sex, traveling, and relationship seeking. (Ranzini& Lutz, 2017, p.88)

Two causal factors probably affect the motivation for having sex: the gender of individuals and their mating strategy (Buss & Schmitt, 1993). Men and women are supposed to deal with different adaptive problems through human evolution where gender difference appears (Buss, 1998). For instance, related to mating and parental investment. Men are likely to take "lower minimum parental investment" strategy (Trivers, 1972). Men are expected to short-term mating strategies with devotion of a larger proportion of their total mating effort such as energy and resources (Buss & Schmitt, 1993). One reason is that men can take advantage of fitness benefits by having numerous sex partners comparing to woman. Because once a woman is pregnant, it takes great amount time and energy for her to give birth to a baby. Thus, in this theory, women are more serious about having sex. Consequently, comparing to women, men will (1) more likely to desire short-term partners; (2) desire a larger number of short-term partners; (3) require less time before consenting to sex or desire to have sex with an attractive partner (Schmitt, Shackelford, & Buss, 2001, Botnen et al., 2018). Evidence suggests those gender differences are exist and expected to be universal (e.g., Lippa, 2009; Schmitt, 2005). Moreover, within each gender, individual differences and personal preference for short-term mating relationships will influence relevant sexual behavior (Botnen et al., 2018).

Solis & Hong (2018) measured motivations and risks as predictors of outcomes in the use of mobile dating applications in China. The results showed that sexuality was the only predictor of the reasons that people use MDAs to meet people offline for dates and casual sex. MDA users in China were selected in this study random from the Sojump database (www.sojump.com) in December 2016. After a filtering process, there are 433 valid respondents in this experiment. All their respondents came from the main-land of China (57.5% males & 42.5% females). Beijing and Shanghai ranked top two places for current residence (10% and 8%, respectively). The mean age of these MDA users was about 30 years (ranging from 11 to 58 years old). About 89% of the users used three or less MDAs at a time. Respondents used MDAs in different frequency: 30% individuals used three times a week; another 30% used daily. 82.2% of the respondents used MoMo and 42.3% used TanTan. 67% percent of the respondents met MDA matches in real life rather than just chatting online. 55% had gone on an off-line date and about 25% had sex with strangers after matching process (Solis & Wong, 2018).

Botnen et al. (2018) found individual differences in sociosexuality predict picture-based mobile dating app (PBMDA) use. The research recruited 641 Norwegian university students (aged ranging from 19 to 29 years old, 55.8% women) to answer a questionnaire and it was analyzed within the framework of sexual strategies theory (SST; Buss & Schmitt, 1993, 2016). As a result, Botnen et al. (2018) indicated that women and men considered differently for reasons of using PBMDAs. Men are more eager to establish contact with dates, are more positive to more potential partners, and seek to have short-term sexual encounters with matches to a larger degree than women are (effect sizes ranged from d=0.45 to d=1.08) (Botnen et al., 2018). This was also coherent with research outside of digital dating (Buss & Schmitt, 2016). Men appeared to pursue more actively sexual actions that were not contingent upon relationship commitment. Despite women and men spending quite equal time daily on dating apps in this study, men seemed to spend more time on mate relevant behaviors, as they were likely to be engaged more in the actual and concrete hooking-up activities (Botnen et al., 2018).

Whereas, women regarded self-affirmation (to feel good) more valuable when they would like sexual actions and when they want a committed relationship (Botnen et al., 2018). Consequently, PBMDA users are not primarily looking for a long-term relationship, while the desire of seeking for a romantic partner may be a secondary following after primary short-term aims.

Choi& Chan (2013) conducted a sociological research on "why and how men in Hong Kong use QQ to chase women in Mainland China". QQ, a popular Chinese application, was developed as an instant messaging and a popular dating platform since users could to make friends. The feeling of solitude in the "real world"(Hong Kong) cannot be eliminated by sexual pleasure alone (approaching sex workers). They can also get gratification deriving from their online partners. The intimate relationship between Hong Kong men and mainland women also involved cross-border communication. Understanding other cultures is a way to cure ethnocentrism, which assumes that one's own culture is superior to all other cultures (Giddens & Griffiths, 2006). In general, Hong Kong's economical development is superior to Mainland China and "love is not purely biological but also sociological"(Choi& Chan, 2013, p. 9). Thus, the men in Hong Kong would like to chase mainland women rather than foreign women or Hong Kong women because they could feel a sense of superior to woman from Mainland China. They think that by plugging themselves into the online world, the men are trying to disengage from the "real" world and enjoy themselves in another world (Choi & Chan, 2013, p. 158). According to Choi and Chan, the "chasing game" involves the exchange of fake personal information and deceptive self-presentation. In addition, it would bring unpleasant consequence: "personal belongings being stolen, being cheated by bar-sales or clothes-sales, [and] being infected by sexually transmitted diseases and the risks of non-reciprocation" (2013, p. 146; Solis& Wong, 2018).

## HOOK-UP CULTURE CHALLENGES TRADITIONAL LOVE

With the prevalent of hook-up culture, mobile dating overturns the traditional cognition of love. Citizens get to know their potential partner by systematic selection software rather than having meeting for falling in love by chance. Face-to-face communication pales in comparison to chatting via cell phone. Thus we redefine intimate relationship. Individuals are willing to use Emoji, symbol of mobile dating sub-cultural language, instead of lyrics of poems to express their emotions. Also, as mobile becomes ubiquitous, users are prone to be immersive and even addicted so that cannot tell the real world and virtual society. Worsley (2015) assumes that mobile dating has made romance too easy by lamenting that the "slow, exquisite torture of love" from the age of Jane Austen was dead; the interpersonal dynamics of mobile dating is "a networked self and love" (Worsley, 2015).

Youngsters are more open to sexual issues even in conservative Asian countries like China and India. An increasing number of young people in China are more openly to talk about sexual experiences in public by using MDAs especially for post-1990 generation (Tatlow, 2013). In Tatlow's study, a sexual freedom pioneer MA Jiajia said: "Sex, should be openly talked about, 'fun' and 'funny', not 'dirty' or 'depressing' " which changes drastically comparing with Chinese traditional values"(2013). Also, Tatlow (2013) found that MDAs, often location-based and with a hookup element, are popular among millions of Chinese youngsters. Brubaker et al. (2016) perceived that apart from "finding the right kind of person", hook-up apps with real time geo-location function could also help users to "categorize and structure themselves in spaces where others can find them.

Nearly at the same time, ASSOCHAM Social Media Foundation (ASDF, 2018) did a random survey of 20-30-year-old people in 10 cities of India (including around 1,500 individuals). Majority (55%) of the total respondents said they had used a dating app for casual dating, meaningful relationship outside the prescribed norms and traditions. While 20% of people said they were using MDAs as they were more serious about finding a life long partnership. Nearly 10% use MDAs to interact socially or for networking other than dating. The rest had no idea about MDAs.

"Pick-up Artists" (PUA) movement goes to extreme of hook-up culture. It is a counter-feminist movement of males whose goal is seduction and sexual success with and access females, is notorious but not so uncommon in MDAs. PUA focuses on direct sexual seducing and very explicit metaphors of hunting and gaming (Almog & Kaplan, 2015).

Unfortunately, the popularity of hookup culture arouses concerns on interpersonal communication and even leads to public moral panics. For example, users blame the increase of sexually transmitted infections on Tinder. A report from United Nations said an upsurge in HIV infections among teenagers in the Asia Pacific region might be due to casual sex after using mobile dating apps (Clark, 2015). MDAs like Grindr, Tinder, and Growlr increase the possibility of spontaneous casual sex within users live nearby.

## RESEARCH ON LGBTQ GROUPS

LGBTQ Groups are not neglected by social scientists, while the majority of those programs are focused on Grindr, a popular all-male, location-based MDA.

Corriero & Tong (2016) engaged in exploring users' motivations and concerns of Grindr under the framework of uncertainty management theory (UMT). Their study is designed to understand experience and desire for uncertainty and how uncertainty would influence their behavior of seeking information. In their first experiment, by using online snowball-sampling survey on publicly linked on Facebook, Twitter, and Reddit, 62 self-identified Grindr users expressed six main concerns as follows: (1) Misrepresentation of personal/social information (30.3%); (2) misrepresentation of health/serostatus information (10.1%); (3) virtual privacy (12.7%); (4) physical privacy (11.4%); (5) recognition (11.4%); (6) social stigma/judgment (e.g., slut-shaming; 7.6%). An additional 1% of the data were categorized as "other" (e.g., racism) and 16% was N/A.

Previous study basing on web platforms found that perceptions of risk facilitate sexual encounters. Users regard uncertainty for potential partners as undesirable and would like to reduce it. In turn, such desire will also influence their information-seeking behavior. Consequently, there are various information-seeking and self-disclosure strategies to reduce risk, "it also seems that a certain degree of uncertainty was tolerated, and even sought, as integral part of the fulfillment of their sexual goals" (Couch and Liamputtong, 2007). Thus, Corriero &Tong indicated that the risk accompanying with casual sex might also increase daters' excitement within Grindr (2016).

There's always the danger that the next guy you hook up with may be a rapist, pedophile, an axe murderer. All that I am given is a picture (which 99% of the time is not of their face or even of their body) and location. (Corriero &Tong, 2016, p.131)

Openly recognizing sexual identity on MDAs can also come to an issue for those who are not "out" to the public (admitting their homosexual identity). The sense of "slut-shaming" still existed but was less intense comparing to previous study. Whereas, as Grinder was viewed a hook-up platform, sometimes user was perceived "sleazy" by others.

In their second study, 326 participants were recruited aged 18 and older (M = 24.33, SD = 5.63) by using the same method as study 1. The sample consisted of users as follows: Caucasian (82.2%), Hispanic/Latino (5.9%), Asian (5.4%), African American (1.7%), and other (4.8%). self-identities of respondents were as gay (87%), bisexual (12.1%), or other (0.8%). Concerns of (1) (2) (3) referred in study one were significantly related to desire for uncertainty. While there were concerns about recognition, physical harm, and social judgment were not related to desire for uncertainty. In addition, casual sex goals greatly increased desire for uncertainty. "The second study confirmed that a specific set of user goals and concerns predicted daters' desire for uncertainty, which in turn predicted information-seeking behavior" (Corriero &Tong, 2016).

## PROBLEMS ON MOBILE DATING

### Deceptive Self-Presentation

Deceptive self-presentation is significantly influenced by self-esteem, education, and sexual orientation. Thus, the reasons for deceptive self-presentation can vary from more explicit strategic considerations (to attract sexual partners) to more implicit emotional motives (to get self-validation).

Men often exaggerate their height to be more attractive to women and women often underestimate their weight to be more attractive to men (Toma, 2015; Hancock, & Ellison, 2008; Markowitz & Hancock, 2018). Furthermore, women are more likely to show pictures appearing younger and men prefer to display wealth and social status (Hancock & Toma, 2009; Markowitz & Hancock, 2018).

Interestingly, Ranzini& Lutz (2017) find that individuals who use Tinder might be more honest in their self-presentation rather than a deceptive one because of their long-term perspective and the likelihood that deceptive self-presentation could be harmful to them. This would be coherent with previous scientists' findings on dating sites (Ellison et al., 2012; Hancock et al., 2007).

Markowitz & Hancock (2018) try to figure out how deception works before daters meet. They use a content analysis method by analyzing message conversations in MDAs. There are two goals for daters to use impression management during the discovery period: (1) self-presentation enhancement (less for physical attributes and more concerning romantic availability and desirability) and (2) availability (when to meet, avoid or continue to develop closer relationship). Their two studies found that 66% of lies were driven by the two goals above in impression management. In addition, nearly 7% of messages were deceptive, and there is a correlation between one's lying rate and the partner's.

### Digital Dating Abuse

Mobile dating also could give rise to "digital dating abuse" (DDA). DDA reflected a set of actions, including monitoring someone's activities and whereabouts, hostility, and pressuring for sexual behavior using the Internet or cell phones (Futures without Violence, 2009; Reed, Tolman & Ward, 2016). DDA also referred actions in the context of digital media or using of digital media to harass, pressure, threaten, coerce, or monitor a dating partner (Reed et al., 2018).

Reed et al., (2018) did a research on DDA in adolescent dating relationships, which explored the role of stereotypical gender and dating beliefs (SGDBs) in shaping DDA perpetration. This study collected

703 valid surveys in heterosexual high school students (from a suburban area of Southeast Michigan) with dating experience investigated the role of gender beliefs in DDA perpetration. Structural equation modeling (SEM) was used in this study. It turned out that girls were prone to perpetration of some types of DDA, and boys showed greater endorsement of stereotypical gender and dating beliefs. Recipient reported the behavior on "pressured my dating partner to sex" (6.3% of girls and 22.2% of boys); as for the issue of "pressured my dating partner to have sex or do other sexual activities" 5.8% of girls and 18.8% of boys admitted (Reed et al., 2018). The study also indicated that high school adolescents took different DDA tactics to control in their relationships. "Boys are more likely to use aggressive and sexually coercive tactics. Girls are more likely to use more passive monitoring behaviors, perhaps as a means of exercising possessiveness and ensuring fidelity" (Reed et al., 2018). This research also suggested that societal beliefs about gender and dating might lead to the problems on digital media in dating relationships.

## Privacy Problems

Individuals get gratification for love and sex but still take risk for privacy problems. Mobile dating entails customers sharing more private information via filling blanks and chatting (including age, education, height and weight, family, job, as well as continuous location data). Admittedly, privacy risks are inherently located in those apps especially when neglecting confidential protection.

Some scientists did a series of experiments to test 18 MDAs (such as ChatOn, Grindr, MeetMe, Plenty of Fish, Tinder, MoMo) on whether they send sensitive HTTP traffic, include the current location of users, send the actual distance to other users, use static links, etc. (Patsakis et al., 2015) Studies show that there are significant security holes that could be vulnerable to attacks even by inexperienced attackers to obtain very sensitive hidden personal information such as telephone numbers, e-mails, degree of interaction between users, etc. without the users' consent. Another team of experts tested 9 popular Android MDAs and they all have potential privacy risk for users. (Farnden et al., 2015)

As for privacy policies for location-sharing information of Grindr and Tinder, both of the companies provide lengthy regulations in general, but neither could offer clear and detailed instructions in helping users to manage the shared geo-information by themselves (Albury et al., 2017). Tinder's Privacy Policy says:

We automatically collect information from your browser or device when you visit our Service. This information could include your IP address, device ID and type, your browser type and language, the operating system used by your device, access times, your mobile device's geographic location while our application is actively running, and the referring website address. (Tinder, 2017)

Meanwhile, Grindr illustrates as follows in terms of how and why geo-information is stored (Grindr also has the functions for users to disable real-time location):

When you use the Grindr App, we will collect your location to determine your distance from other users through the GPS, Wi-Fi, and/or cellular technology in your device . . . Your last known location is stored on our servers for the purpose of calculating distance between you and other users. (Grindr, 2018)

Barreneche (2012) illustrated that sophisticated forms of "geo-demographic profiling" were accumulation in data pool to segment users and bring inferences about them. This sort of data carried immense potential commercial value, especially related to location-based advertisement and data analysis where partial revenue of Grindr and Tinder came from.

## CONCLUSION

This chapter has explored the up-to-date researches of mobile dating by discussing related theories, demographics of worldwide users, varies sorts of applications, individual's motivations, different stages, LGBTQ studies and problems etc.

Coming back to the previous question: "is mobile dating a fancy way to fall in love? " It is complicated to give a clear response and the answer might be yes and no. Mobile dating, accompanying with profile phase, matching stage and discovery period, brings convenience and filtered matching data pool for individuals. The sprouting MDAs (targeted different audiences) attract numerous users and gain promising giant market. For instance, as previously mentioned, Tinder has attracted 8.2 million users and 3.7 million paid subscribers. The revenue of digital dating in 2018 reaches to $1,383 million US dollars and is predicted to grow continuously in the future decades. In some sense, mobile dating shifts our way of acquainting potential lovers.

Whereas, statistics show that users are prone to take short-term mating strategies when based on geographic proximity MDAs. Fewer people are serious about finding a life long partnership by using MDAs. As a consequence, hook-up culture is emerging and individuals who do not even care about encountering with casual sex, just desire to release their sexual motivation via mobile dating. The sense of uncertainty could make them excited but might also lead to transmission of sexual diseases at the same time. Moreover, UN has reported HIV infections in Asian areas might due to casual sex after using MDAs. LGBTQ groups are also able to participate in specifically designed MDAs, such as Grindr. Whereas, users take divergent strategies on self-disclosure and some will be sensitive to admit their homosexual identity.

Problems on mobile dating cannot be ignored. Deceptive self-presentation is not uncommon in mobile dating. In profile stage, individuals would like to make a good impression by editing profile pictures and beautifying themselves towards ideal standards. Unfortunately, over exaggeration is beyond understandable behaviours but a deceptive action. In addition, teenagers are prone to be digital abuse victims by being pressured to have sex with partner. According to a survey in this chapter, boys are more likely to use aggressive and sexually coercive method while girls are more likely to use more passive behaviour. Furthermore, privacy risk is another issue. Scientists have found that several MDAs have significant security holes so that users' personal information could be unintentionally obtained.

Admittedly, this chapter still have some limitations. Although the author would like to present up-to-date situations of mobile dating globally, current studies are mostly concentrated on North America, Europe and Asia (China and India mainly). Materials about mobile dating are hardly found in Africa, East Asia and some other areas where are relatively less developed. One possible reason is that the proliferation of mobile dating demands a prevalence of high-tech portable digital devices. Unfortunately, those areas mentioned might not reach the standard. There might be another digital gap in global sphere. Meanwhile, traditional values and religious beliefs also matter a lot. For Muslims, mobile dating might violate their customs and conventions and cannot be acceptable. Also, according to limited academic surveys (and few studies pay attention to these group of people), seniors are not so active in mobile dating in comparison to young generations. They might not so familiar with the use of MDA or unpleased to accept such a new way of dating. In conclusion, mobile dating brings a new way of mate finding experience, while related issues such as ethnics, behaviours, technics and solutions still deserve to explore.

# ACKNOWLEDGMENT

This research was supported by China Scholarship Council [grant number 201707050010].

The author would also express great appreciation to Prof. Junhao HONG from State University of New York at Buffalo (U.S.A.) and Prof. Li ZHANG from Communication University of China (China).

# REFERENCES

Ahuvia, A. C., & Adelman, M. B. (1992). Formal intermediaries in the marriage market: A typology and review. *Journal of Marriage and the Family, 54*(2), 452–463. doi:10.2307/353076

Albury, K., Burgess, J., Light, B., Race, K., & Wilken, R. (2017). Data cultures of mobile dating and hook-up apps: Emerging issues for critical social science research. *Big Data & Society, 4*(2), 2053951717720950. doi:10.1177/2053951717720950

Almog, R., & Kaplan, D. (2015). The nerd and his discontent: The seduction community and the logic of the game as a geeky solution to the challenges of young masculinity. *Men and Masculinities*. doi:10.1177/1097184X15613831

Barreneche, C. (2012). Governing the geocoded world: Environmentality and the politics of location platforms. *Convergence (London), 18*(3), 331–351. doi:10.1177/1354856512442764

Ben-Ze'ev, A. (2004). *Love online: Emotions on the Internet.* Cambridge, UK: Cambridge University Press. doi:10.1017/CBO9780511489785

Botnen, E. O., Bendixen, M., Grøntvedt, T. V., & Kennair, L. E. O. (2018). Individual differences in sociosexuality predict picture-based mobile dating app use. *Personality and Individual Differences, 131*, 67–73. doi:10.1016/j.paid.2018.04.021

Brashers, D. E. (2001). Communication and uncertainty management. *Journal of Communication, 51*(3), 477–497. doi:10.1111/j.1460-2466.2001.tb02892.x

Brashers, D. E., & Hogan, T. P. (2013). The appraisal and management of uncertainty: Implications for information-retrieval systems. *Information Processing & Management, 49*(6), 1241–1249. doi:10.1016/j.ipm.2013.06.002

Brashers, D. E., Neidig, J. L., Haas, S. M., Dobbs, L. K., Cardillo, L. W., & Russell, J. A. (2000). Communication in the management of uncertainty: The case of persons living with HIV or AIDS. *Communication Monographs, 67*(1), 63–84. doi:10.1080/03637750009376495

Brubaker, J. R., Ananny, M., & Crawford, K. (2016). Departing glances: A sociotechnical account of 'leaving' Grindr. *New Media & Society, 18*(3), 373–390. doi:10.1177/1461444814542311

Buss, D. M. (1998). Sexual strategies theory: Historical origins and current status. *Journal of Sex Research, 35*(1), 19–31. doi:10.1080/00224499809551914

Buss, D. M., & Schmitt, D. P. (1993). Sexual strategies theory: An evolutionary perspective on human mating. *Psychological Review, 100*(2), 204–232. doi:10.1037/0033-295X.100.2.204 PMID:8483982

Buss, D. M., & Schmitt, D. P. (2016). Sexual strategies theory. In T. Shackelford & V. Weekes-Shackelford (Eds.), *Encyclopedia of evolutionary psychological science*. Cham, Switzerland: Springer. doi:10.1007/978-3-319-16999-6_1861-1

Chan, L. S. (2018). Ambivalence in networked intimacy: Observations from gay men using mobile dating apps. *New Media & Society*, *20*(7), 2566–2581. doi:10.1177/1461444817727156

Choi, M. K., Chan, K. B., & Chan, K. (2013). *Online dating as a strategic game: why and how men in Hong Kong use QQ to chase women in Mainland China*. Retrieved from http://ebookcentral.proquest.com

Clark, J. (2015). Mobile dating apps could be driving HIV epidemic among adolescents in Asia Pacific, report says. *BMJ: British Medical Journal*, *351*. https://doi-org.gate.lib.buffalo.edu/10.1136/bmj.h6493

Corriero, E. F., & Tong, S. T. (2016). Managing uncertainty in mobile dating applications: Goals, concerns of use, and information seeking in Grindr. *Mobile Media & Communication*, *4*(1), 121–141. doi:10.1177/2050157915614872

Couch, D., & Liamputtong, P. (2007). Online dating and mating: Perceptions of risk and health among online users. *Health Risk & Society*, *9*(3), 275–294. doi:10.1080/13698570701488936

Diary, O. (2018). *Youth prefers mobile dating apps over matrimonial sites/ads*. Retrieved from http://link.galegroup.com.gate.lib.buffalo.edu/apps/doc/A527756418/STND?u=sunybuff_main&sid=STND&xid=da2f1060

Dredge, S. (2015, May 7). Research says 30% of Tinder users are married. *The Guardian*.

Ellison, N., Heino, R., & Gibbs, J. (2006). Managing impressions online: Self presentation processes in the online dating environment. *Journal of Computer-Mediated Communication*, *11*(2), 415–441. doi:10.1111/j.1083-6101.2006.00020.x

Ellison, N. B., Hancock, J. T., & Toma, C. L. (2012). Profile as promise: A framework for conceptualizing veracity in online dating self-presentations. *New Media & Society*, *14*(1), 45–62. doi:10.1177/1461444811410395

Emily, M. (2018, Aug. 8). Tinder Sends Match Earnings Blazing Past Estimates. *Bloomberg News*.

Farnden, J., Martini, B., & Choo, K. K. R. (2015). *Privacy risks in mobile dating apps*. arXiv preprint arXiv:1505.02906

Finkel, E. J., Eastwick, P. W., Karney, B. R., Reis, H. T., & Sprecher, S. (2012). Online dating: A critical analysis from the perspective of psychological science. *Psychological Science in the Public Interest*, *13*(1), 3–66. doi:10.1177/1529100612436522 PMID:26173279

Futures without Violence. (2009). *"That's not cool" initiative background and development research*. Retrieved from http://www.thatsnotcool.com/tools/index.asp?L1.1

Gibbs, J. L., Ellison, N. B., & Lai, C. H. (2010). First comes love, then comes Google: An investigation of uncertainty reduction strategies and self-disclosure in online dating. *Communication Research*, *38*(1), 70–100. doi:10.1177/0093650210377091

Giddens, A., & Griffiths, S. (2006). *Sociology* (5th ed.). Cambridge, UK: Polity Press.

Goffman, E. (1959). *The presentation of self in everyday life*. New York, NY: Doubleday.

Goggin, G. (2006). *Cell Phone Culture: Mobile Technology in Everyday Life*. London: Routledge.

Goldsmith, B. (2014). The smartphone app economy and app ecosystems. In G. Goggin & L. Hjorth (Eds.), *The Routledge Companion to Mobile Media* (pp. 171–180). New York, NY: Routledge.

Grindr. (2018). Retrieved from https://www.grindr.com/privacy-policy/(accessed Nov. 2018)

Hancock, J. T., Toma, C., & Ellison, N. (2007). The truth about lying in online dating pro-files. In *Proceedings of the SIGCHI Conference on Human Factors in Computing Systems* (pp. 449–452). New York, NY: ACM. 10.1145/1240624.1240697

Hancock, J. T., & Toma, C. L. (2009). Putting your best face forward: The accuracy of online dating photographs. *Journal of Communication, 59*(2), 367-386. doi:.1460-2466.2009.01420.x doi:10.1111/j

IBIS World. (2017). Retrieved from https://www.ibisworld.com/industry-trends/market-research-reports/other-services-except-public-administration/personal-laundry/dating-services.html

Jayson, S. (2014, Jan. 26). Latest trend in digital dating: Live video chat dates. *USA Today*. Retrieved from https://www.usatoday.com/story/tech/personal/2014/01/26/dating-mobile-phone-video/4674651/ (accessed Oct. 3rd 2018)

Keegan, V. (2007, Dec. 6). Dating moves from the PC to the mobile. *The Guardian*. Retrieved from https://www.theguardian.com/technology/2007/dec/06/digitalvideo.mobilephones(accessed Oct. 3rd 2018)

Lawson, H. M., & Leck, K. (2006). Dynamics of Internet dating. *Social Science Computer Review, 24*(2), 189–208. doi:10.1177/0894439305283402

Li, L., & Chen, T. (2005). Cyberlove and its ethical issues: A clarification. *Studies in Ethics, 1*(1), 72–74.

Li, M., Cao, N., Yu, S., & Lou, W. (2011). Findu: Privacy-preserving personal profile matching in mobile social networks. In *INFOCOM, 2011 Proceedings IEEE*. IEEE. doi:10.1109/INFCOM.2011.5935065

Light, B. (2014). *Disconnecting with Social Networking Sites*. Basingstoke, UK: Palgrave Macmillan. doi:10.1057/9781137022479

Lippa, R. A. (2009). Sex differences in sex drive, sociosexuality, and height across 53 nations: Testing evolutionary and social structural theories. *Archives of Sexual Behavior, 38*(5), 631–651. doi:10.100710508-007-9242-8 PMID:17975724

Markowitz, D. M., & Hancock, J. T. (2018). Deception in mobile dating conversations. *Journal of Communication, 68*(3), 547–569. doi:10.1093/joc/jqy019

Markowitz, D. M., Hancock, J. T., & Tong, S. (2018). Interpersonal dynamics in online dating: Profiles, matching, and discovery. In Z. Papacharissi (Ed.), A networked self and love (pp. 50–61). New York, NY: Routledge.

MoMo Inc. (2016). *MoMo announces unaudited financial results for the second quarter 2016* [Press release]. Retrieved from http://media.corporate-ir.net/media_files/IROL/25/253834/2016/Momo2016Q2-final.pdf

Murray, S., & Sapnar, A. M. (2016). Lez takes time: Designing lesbian contact in geosocial networking apps. *Critical Studies in Media Communication*, *33*(1), 53–69. doi:10.1080/15295036.2015.1133921

Patsakis, C., Zigomitros, A., & Solanas, A. (2015). Analysis of privacy and security exposure in mobile dating applications. In *International Conference on Mobile, Secure and Programmable Networking* (pp. 151-162). Springer. 10.1007/978-3-319-25744-0_13

Rains, S. A., & Tukachinsky, R. (2015). An examination of the relationships among uncertainty, appraisal, and information-seeking behavior proposed in uncertainty management theory. *Health Communication*, *30*(4), 339–349. doi:10.1080/10410236.2013.858285 PMID:24905910

Ranzini, G., & Lutz, C. (2017). Love at first swipe? Explaining Tinder self-presentation and motives. *Mobile Media & Communication*, *5*(1), 80–101. doi:10.1177/2050157916664559

Reed, L. A., Tolman, R. M., & Ward, L. M. (2016). Snooping and sexting: Digital media as a context and tool for dating violence among college students. *Violence Against Women*, *22*(13), 1556–1576. doi:10.1177/1077801216630143 PMID:26912297

Reed, L. A., Ward, L. M., Tolman, R. M., Lippman, J. R., & Seabrook, R. C. (2018). The association between stereotypical gender and dating beliefs and digital dating abuse perpetration in adolescent dating relationships. *Journal of Interpersonal Violence*. doi:0886260518801933

Rosenberg, J., & Egbert, N. (2011). Online impression management: Personality traits and con- cerns for secondary goals as predictors of self-presentation tactics on Facebook. *Journal of Computer-Mediated Communication*, *17*(1), 1–18. doi:10.1111/j.1083-6101.2011.01560.x

Ruggiero, T. E. (2000). Uses and gratifications theory in the 21st century. *Mass Communication & Society*, *3*(1), 3–37. doi:10.1207/S15327825MCS0301_02

Sager, R. L., Alderson, K. G., & Boyes, M. C. (2016). Hooking-up through the use of mobile applications. *Computer Communication & Collaboration*, *4*(2), 15–41.

Schmitt, D. P. (2005). Sociosexuality from Argentina to Zimbabwe: A 48-nation study of sex, culture, and strategies of human mating. *Behavioral and Brain Sciences*, *28*(02), 247–275. doi:10.1017/S0140525X05000051 PMID:16201459

Schmitt, D. P., Shackelford, T. K., & Buss, D. M. (2001). Are men really more oriented toward short-term mating than women? A critical review of theory and research. *Psychology Evolution & Gender*, *3*(3), 211–239. doi:10.1080/14616660110119331

Smith, A. (2016). *15% of American adults have used online dating sites or mobile dating apps*. Pew Research Center.

Smith, A., & Anderson, M. (2016). 5 facts about online dating. *Pew Research Center*. Retrieved from http://www.pewresearch.org/fact-tank/2016/02/29/5-facts-about-online-dating/(accessed July 3rd 2018)

Solis, R. J. C., & Wong, K. Y. J. (2018). To meet or not to meet? Measuring motivations and risks as predictors of outcomes in the use of mobile dating applications in China To meet or not to meet? Measuring motivations and risks as predictors of outcomes in the use of mobile dating applications in China. *Chinese Journal of Communication*, 1–20. doi:10.1080/17544750.2018.1498006

Statista. (2018a). Retrieved from https://www.statista.com/outlook/372/100/online-dating/worldwide#market-age

Statista. (2018b). Retrieved from https://www.statista.com/statistics/449390/quarterly-revenue-match-group/(accessed May 1st 2018)

Statista. (2018c). Retrieved from https://www.statista.com/statistics/826778/most-popular-dating-apps-by-audience-size-usa/(accessed May 2018)

Sumter, S. R., Vandenbosch, L., & Ligtenberg, L. (2017). Love me Tinder: Untangling emerging adults' motivations for using the dating application Tinder. *Telematics and Informatics, 34*(1), 67–78. doi:10.1016/j.tele.2016.04.009

Tatlow, D. K. (2013, July 24). *Apps offer Chinese a path to the forbidden.* Retrieved from http://www.nytimes.com/2013/07/25/world/asia/25iht-letter25.html(accessed July 3rd 2018)

Tinder. (2017). Retrieved from https://www.gotinder.com/privacy

Toma, C. L. (2015). Online dating. In C. Berger & M. Roloff (Eds.), *The international encyclopedia of interpersonal communication* (pp. 1–5). Hoboken, NJ: John Wiley & Sons, Inc. Retrieved from https://onlinelibrary.wiley.com/doi/pdf/10.1002/9781118540190.wbeic118

Toma, C. L., & Hancock, J. T. (2012). What lies beneath: The linguistic traces of deception in online dating profiles. *Journal of Communication, 62*(1), 78-97. doi:.1460-2466.2011.01619.x doi:10.1111/j

Toma, C. L., Hancock, J. T., & Ellison, N. B. (2008). Separating fact from fiction: An examination of deceptive self-presentation in online dating profiles. *Personality and Social Psychology Bulletin, 34*(8), 1023–1036. doi:10.1177/0146167208318067 PMID:18593866

Trivers, R. (1972). Parental investment and sexual selection. In B. Campell (Ed.), *Sexual selection and the descent of man: 1871–1971* (pp. 136–179). Chicago: Aldine-Atherton.

Woll, S. B., & Cozby, C. P. (1987). Video-dating and other alternatives to traditional methods of relationship initiation. In W. H. Jones & D. Perlman (Eds.), Advances in personal relationships (Vol. 1, pp. 69–108). Greenwich, CT: JAI.

Worsley, L. (2015). Dating apps have killed romance, says historian. *Times.* Retrieved from https://www.bbc.com/news/technology-34455738

Zhang, J., & Yasseri, T. (2016). *What Happens After You Both Swipe Right: A Statistical Description of Mobile Dating Communications.* arXiv preprint arXiv:1607.03320

# Chapter 5
# My Little Joy in Life:
## Posting Food on Instagram

**Wan Chi Leung**
*University of Canterbury, New Zealand*

**Anan Wan**
*Georgia College & State University, USA*

## ABSTRACT

*To post food on social media has become a frequent source of fun and joy in life for many mobile users. In investigating such a common scene on Instagram among its young users, the authors of this chapter investigated the relationship between social activity, personal traits like narcissism and shyness, and uses and gratifications from posting food photos on Instagram. Uses of Instagram for posting selfies were also examined for comparison. Results showed that while posting food photos were associated with social activity, posting selfies were associated with shyness. Narcissists were more likely to involve in posting both food photos and selfies. Implications of the results in explaining the generation of visual contents on social media are discussed.*

## INTRODUCTION

Food, as an important part of daily life, is likely to be universally welcomed by everyone. The joy from food not only comes with its taste, but with its visual appeal as well. With the development of technology in recent years, the joy from the appearance of food can be visually recorded and shared anytime, anywhere. One of the important apps for photo sharing is Instagram, which was launched in October 2010, and has become popular since then for its visual components and the hands-on creative features. Instagram is particularly popular among young adults. As of October 2018, 31% of the global Instagram users were aged 18-24, compared with 27% of the global Facebook users in the same age group (Statista, 2018). The gender distribution of young adult Instagram users aged 18-24 was more balanced than Facebook: 15% of the global Instagram users were female and 16% were male, versus 11% of the global Facebook users were female and 16% were male. It shows that Instagram were popular among both female and male young adults.

DOI: 10.4018/978-1-5225-7885-7.ch005

Among various categories of photos on Instagram, food was identified as one of the top eight popular photo categories (Hu, Manikonda, & Kambhampati, 2014). Over 30% of the users in Hu et al.'s (2014) study posted more than two photos about food in their accounts. As of November 2018, there have been more than 300 million posts using the hashtag #food on Instagram, and more than 180 million posts using the hashtag #foodporn, indicating the popularity of sharing food photos on Instagram.

Instagram's focus on visual arts and the App-embedded filters makes it a suitable and convenient platform for photo sharing. Comparing to other social media platforms such as Twitter and Facebook, Instagram has lowered the requirements in artistic and photography skills, so almost anyone can enjoy producing attractive photos via Instagram. Verbal descriptions on how visual and gustatory attractive the food is might be difficult, but showing a photo of the food is a more convenient way to convey the same messages.

Although posting food photos has been popular for years (Hu et al., 2014), little is known about how people could gratify from posting food and why they would especially like to post food on social media. In view of the popularity of Instagram among young adults, the purpose of this exploratory study is to investigate young adults' gratifications from posting food photos (photos with food as the main theme) on Instagram, and how these gratifications are related to the use of Instagram for posting food photos. Also, how young adults' social activity and personal traits including narcissism and shyness are associated with posting food photos on Instagram is examined.

This study provides significant findings in indicating how the ubiquitous use of visual images of food for communication can satisfy young adult users. In other words, food consumption, is no longer only about nutritional needs, but the visual aspects of food can serve as a communication tool with Instagram as a platform. A number of studies have been conducted about Instagram selfie uploaders, (e.g., Al-Kandari, Melkote & Sharif, 2016; Williams & Marquez, 2015; Dhir et al., 2016; Kim et al., 2016), but not much have been done studying the motives of a large amount of Instagram users who have posted food photos. This study fills the research gap by investigating how young Instagram users satisfy their needs through posting food, an important aspect of everyday life.

Past research suggested that common Instagram users who focus more on posting food photos still like to post other categories of photos as well, which is quite different from the "selfies-lovers" who prefer posting self-portraits on Instagram exclusively (Hu, Manikonda, & Kambhampati, 2014). While both food photos and selfies can be easily posted anytime, anywhere, the two types of visual contents differ greatly by their levels of self-disclosure. In view of this, we also investigate young adults' use of Instagram for posting selfies (a self-portrait photo taken by oneself, which can include an individual alone or an individual with any other persons) on Instagram to see how the differences in visual appeals lead to different Instagram use. By comparing the use of Instagram in posting food photos and selfies among narcissistic and shy individuals, this study provides important implications on the relationship between the visual contents and individual characteristics.

## LITERATURE REVIEW

### Gratifications of Posting Food on Instagram

The uses and gratifications approach has been employed by researchers to study audiences' active consumption of mass media contents for more than half a century. It assumes the audiences actively select the media to meet their social and psychological needs and expectations (Katz et al., 1973). The term "gratifications set" refers to multiple possibilities for audiences to form and re-form the basis of their media-related interests, needs or preferences (McQuail, 2005). As Lindlof and Schatzer (1998) suggests, the difference between computer-mediated communication and other media use is that the former is more transient and multimodal, with fewer codes of conduct governing use, and allows for a higher degree of "end-user manipulation of content." Nowadays, the uses and gratifications approach has been applied to study the users of social media to generate contents. Leung's (2009) study indicates the motives of online user-generated contents include recognition needs, cognitive needs, social needs and entertainment needs, and their civic engagement.

Today, a smart phone is both a personal medium and a multipurpose device converging with the internet (Humphreys, Von Pape, & Karnowski, 2013). Wei's (2008) study found that the use of the mobile phone for news-seeking and web-surfing was driven by instrumental use motives, while playing video games via the mobile phone was driven by the motive of passing time. Gerlich et al. (2015) conducted an exploratory analysis of the uses and gratifications sought of mobile apps. Findings suggested that reasons for using mobile apps included engagement/disengagement, passing time, gaining knowledge and education, and social uses. In this study, we focus on Instagram, a popular mobile social media app which relies largely on photos as a visual component for information sharing. Lee, Lee, Moon, and Sung (2015) suggest that Instagram users have five social and psychological motives to share and review photos on Instagram: social interaction, archiving, self-expression escapism, and peeking. The activity of food blogging involves the creation and production of photos, and the processes of selecting and editing images, which make food blogging time consuming but very enjoyable (Cox & Blake, 2011).

In view of the literature using the uses and gratifications approach, the following hypothesis is proposed:

**H1:** The more gratifying the Instagram users find from posting food photos, the higher their personal involvement in posting food photos.

## Social Activity

Rubin and Rubin (1982) associated television use and the concept of "contextual age", measured by various factors including social activity, as an important indicator for assessing life-position and communication behavior. While motives for posting photos on Instagram can be very different from television usage, social activity can be an important factor associated with Instagram use. Social activity indicates to what extent a person is socially active. A socially active person displays little affinity with the media and feels more comfortable with interpersonal interaction than a less mobile person. Instagram allows users to share photos and network with other users via visually-attractive photos, and the online visual expression can be a self-disclosure of social activity in their real lives (Rubin & Rubin, 1982; Sheldon & Bryant, 2016). A research question is proposed to examine how social activity is associated with the food photo posting behaviors:

**RQ1.1:** How is social activity associated with the personal involvement in posting food photos?

Previous visual studies have shown that human faces are powerful ways to communicate non-verbally (Takeuchi & Nagao, 1993). In online social media contexts, visuals with human faces also have more power in engaging and interacting with other users, for example, photos with faces are more likely to receive likes and comments (Bakhshi, Shamma & Gilbert, 2014). A study has shown that self-portraits were the most popular photo category on Instagram (Hu, Manikonda, & Kambhampati, 2014). Al-Kandari et al.'s (2016) survey found that the need for visual self-expression was the strongest predictor of the self-disclosure use of Instagram. As posting food photos and selfies involve different kinds of self-disclosure in social activity, a research question is proposed to compare the different behaviors:

**RQ1.2:** How is social activity associated with the personal involvement in posting selfies?

## Narcissism

In recent years, a number of researchers have studied the role of narcissism in predicting social media use. According to Campbell and Foster (2007), "Individuals with narcissistic personality possess highly inflated, unrealistically positive views of the self. Often-times, this includes strong self-focus, feelings of entitlement, and lack of regard for others. Narcissists focus on what benefits them personally, with less regard for how their actions may benefit (or harm) others" (p.115). Social media are different from the traditional media, in a sense that they allow users to generate the contents on their own accounts – meaning that narcissists now have their platform to build up their images of the narcissistic self, which can be "positive, inflated, agentic, special, selfish, and oriented toward success" (Campbell and Foster, 2007, p.118).

Poon and Leung (2011) have found that narcissistic individuals reported more frequent online content production. Davenport et al.'s (2014) study indicated that narcissism was a stronger predictor of Facebook friends than Twitter followers. Ong et al.'s (2011) study found that narcissistic people are more likely to engage in features presenting self-generated content such as profile picture rating, status update frequency. Sheldon and Bryant (2016) surveyed college students and found a positive relationship between narcissism and using Instagram for the purposes of surveillance and being cool. The authors argued that Instagram appeared to be cool, accommodating narcissists' wish to be perceived in a positive light. Therefore, the following hypothesis is proposed:

**H2.1:** The higher the level of narcissism of the Instagram users, the higher their personal involvement in posting food photos.

Paramboukis, Skues and Wise's (2016) online survey suggested that uploading photos of one's physical appearance was associated with grandiose narcissism (traits such as exhibitionism and aggression). The following hypothesis is proposed:

**H2.2:** The higher the level of narcissism of the Instagram users, the higher their personal involvement in posting selfies.

## Shyness

Shyness is defined as "one's reaction to being with strangers or casual acquaintance's: tension, concern, feelings or awkwardness and discomfort, and both gaze aversion and inhibition of normally expected social behavior." (Cheek & Buss, 1981, p.330; Buss, 1980). To shy individuals, the development of asynchronous computer-mediated communication such as social media has been important - Asynchronous computer-mediated communication not only reduces the need to interpret peripheral communicative behaviors such as body languages and tones of voice, but also gives shy individuals more control of the interaction and reduces the effects of situation-specific cues such as unexpected interruptions (Chan, 2011).

Past research has shown that social media remove the divide between shy and non-shy individuals. Baker and Oswald's (2010) survey found that among relatively shy individuals, greater Facebook use predicted satisfaction, importance, and closeness with Facebook friends, and increased social support received from friends. Stritzk, Nyugen and Durkin's (2004) experiment indicated that shy individuals differed from non-shy individuals in terms of rejection sensitivity, initiating relationships, and self-disclosure in an offline context, but they were not significantly different on these three domains in the online context. Sheldon's (2013) study suggested that while shyness was negatively correlated with self-disclosure to a face-to-face friend, it was not correlated with time spent on Facebook and self-disclosure to a Facebook friend. The following hypotheses are proposed:

**H3.1:** The higher the level of shyness of the Instagram users, the higher their personal involvement in posting food photos in posting food photos.

**H3.2:** The higher the level of shyness of the Instagram users, the higher their personal involvement in posting selfies.

Finally, in order to explore the uses of Instagram in posting food photos, the following research question is proposed:

**RQ2:** How do demographics, narcissism, shyness, social activity, general Instagram use, and gratifications sought predict a) personal involvement in posting food photos on Instagram; b) frequency of posting food photos on Instagram?

# METHOD

## Data Collection

To investigate Instagram use by young adults, a survey was administered to undergraduate students who had posted food photos on Instagram in the past. Following approvals by the Institutional Review Board in the U.S. and the Human Ethics Committee in New Zealand, a focus group of undergraduate students and a pilot test of the questionnaire, data were gathered through an online survey on Qualtrics and the distribution of paper questionnaires in 2016 in two public universities in the United States and in New Zealand respectively. The online survey link was snowballed from the undergraduate students, and the paper questionnaires were distributed to undergraduate students in class. A total of 373 respondents took part in the survey. After eliminating responses who indicated no experience in posting food photos on Instagram and incomplete responses, 223 responses (59.8%) were eligible for further analysis.

Among the analyzed responses, 136 (61.0%) were from the United States, and 87 (39.0%) were from New Zealand. 151 (67.7%) respondents were female, 53 (23.8%) were male, and 19 (8.5%) had their gender not revealed or undefined. The majority of the respondents were White/Caucasian (n = 181, 81.2%). The mean age of the respondents was 21.6, with 189 (84.8%) respondents aged 25 or below.

## MEASUREMENTS OF MAJOR VARIABLES

### Gratifications for Posting Food Photos on Instagram

To assess the gratifications sought from posting food photos on Instagram, respondents were asked to indicate their opinion for 34 items, on a 7-point Likert scale ranging from 1 (strongly disagree) to 7 (strongly agree). The items were adapted from previous studies on Instagram uses and social media use in general (Sheldon & Bryant, 2016; Lee et al., 2016; Whiting & Williams, 2013; Malik, Dhir & Nieminen, 2016). A pilot test with 15 Instagram users was conducted to eliminate the ambiguous and irrelevant items. The final survey instrument consists of 34 statements about gratifications sought from posting food photos.

An exploratory factor analysis of 34 items, with principal components and Varimax rotation, was conducted. Six factors were yielded using following criteria: eigenvalue greater than 1.0, and all factor loadings greater than 0.4. Six factors emerged, which explained 79.76% of the total variance. Eleven items were deleted due to low factor loadings.

Factor 1 was labeled "self-promotion," containing six items (Cronbach's alpha = .94). Factor 2, "escapism", contained four items (Cronbach's alpha = .94). Factor 3 was named "information sharing" (Cronbach's alpha = .93). Factor 4, "archiving," has a Cronbach's alpha at .80. Factor 5 was named as "creativity" (Cronbach's alpha = .90). Factor 6 was "self-disclosure" (Cronbach's alpha = .86). The items of all factors can be found in Table 1. The scores of the items under each factor were summated and averaged to become a single variable.

### Personal Involvement in Posting Food Photos and Selfies on Instagram

Respondents' level of personal involvement in posting food photos on Instagram was assessed by their perceived importance of posting food photos on Instagram. Respondents indicated their opinion toward three items taken and adapted from the five items in Mittal's (1995) Product Category Involvement Scale (PCIS), which captures people's involvement in products. In this study the measurement used is a 7-point Likert scale, ranging from 1 (strongly disagree) to 7 (strongly agree), with three items: 1) Posting food photo on Instagram is important to me; 2) Posting food photo on Instagram means a lot to me; 3) Posting food photo on Instagram is valuable to me. The three items gave a high reliability (Cronbach's alpha = .957), and the scores of the three items were summated and divided by three to give the final score of the variable (M = 3.19, SD = 1.65).

Three similar items were used to measure the personal involvement in posting selfies on Instagram. The three items also gave a high reliability (Cronbach's alpha = .941), and the scores of the three items were summated and divided by three to give the final score of the variable (M = 2.77, SD = 1.65). The difference between the personal involvement in posting selfies and food photos were then computed.

### Frequency in Posting Food Photos on Instagram

Respondents were asked to indicate their frequency of posting food photos by a 7-point Likert scale, ranging from 1 (Almost never), 2 (Rarely), 3 (Sometimes), 4 (Quite often), to 5 (Very often) (M = 2.60,

*Table 1. Factor analysis of gratifications sought from posting food photos on Instagram*

| I post food photos on Instagram… | Component | | | | | | Mean | S.D. |
|---|---|---|---|---|---|---|---|---|
| | **1** | **2** | **3** | **4** | **5** | **6** | | |
| **Factor 1: Self-promotion** | | | | | | | | |
| 1. To gain attention | .863 | | | | | | 3.66 | 1.79 |
| 2. To be more popular | .857 | | | | | | 4.10 | 1.76 |
| 3. To get more likes | .845 | | | | | | 4.14 | 1.85 |
| 4. To get more comments | .844 | | | | | | 3.69 | 1.79 |
| 5. To be noticed by others | .724 | | | | | | 3.76 | 1.77 |
| 6. Because sharing food photos on Instagram is trendy | .612 | | | | | | 4.07 | 1.74 |
| **Factor 2: Escapism** | | | | | | | | |
| 7. To forget about troubles | | .903 | | | | | 2.61 | 1.67 |
| 8. To avoid loneliness | | .878 | | | | | 2.52 | 1.67 |
| 9. To escape from reality | | .847 | | | | | 2.65 | 1.68 |
| 10. To relax | | .780 | | | | | 3.19 | 1.90 |
| **Factor 3: Information sharing** | | | | | | | | |
| 11. To share something useful about food | | | .861 | | | | 3.46 | 1.72 |
| 12. To share something informative about food | | | .860 | | | | 3.59 | 1.77 |
| 13. To share something important about food | | | .823 | | | | 3.49 | 1.77 |
| **Factor 4: Archiving** | | | | | | | | |
| 14. To record my traces (e.g., trip) via photomap | | | | .796 | | | 3.92 | 1.94 |
| 15. To remember special events | | | | .743 | | | 5.25 | 1.51 |
| 16. To take fancy food photos and save them online | | | | .675 | | | 4.09 | 1.82 |
| 17. To depict my life through photos | | | | .587 | | | 4.58 | 1.74 |
| **Factor 5: Creativity** | | | | | | | | |
| 18. To create visual art | | | | | .798 | | 4.41 | 1.86 |
| 19. To show off my photography skills | | | | | .793 | | 4.12 | 1.78 |
| 20. To produce attractive visual content | | | | | .770 | | 4.70 | 1.76 |
| **Factor 6: Self-disclosure** | | | | | | | | |
| 21. To express my actual self (who I really am) | | | | | | .778 | 4.15 | 1.78 |
| 22. To disclose happenings around me | | | | | | .695 | 4.41 | 1.70 |
| 23. To share my personal information with others | | | | | | .695 | 3.64 | 1.78 |
| Eigenvalues | 10.47 | 2.52 | 2.10 | 1.23 | 1.03 | 1.00 | | |
| Variances explained | 45.53 | 10.94 | 9.13 | 5.34 | 4.46 | 4.36 | | |
| Cronbach's Alpha | .94 | .94 | .93 | .80 | .90 | .86 | | |
| Extraction Method: Principal Component Analysis. Rotation Method: Varimax with Kaiser Normalization. | | | | | | | | |
| a. Rotation converged in 6 iterations. | | | | | | | | |

N = 223.

SD = .776). Respondents also answered the question "On average, how long do you spend on taking and editing a food photo before posting it for others to see on Instagram?" by indicating a number in minutes (M = 6.36, SD = 7.62).

## General Instagram Use

Respondents' general Instagram use were measured by the number of followers they had on Instagram (M = 5.19, SD = 461), and the number of Instagram accounts they were following (M = 4.51, SD = 328).

## Social Activity

The items of social activity, modified from Rubin and Rubin (1982), were measured by three items on a 7-point scale: 1) I often travel, vacation, or take trips with others; 2) I often visit friends, relatives, or neighbors in their homes; 3) I often participate in games, sports, or activities with others. The three items give an acceptable reliability (Cronbach's alpha = .763), and they were summated and divided into the final score (M = 5.11 SD = 1.25).

## Narcissism

Narcissism was measured by four items on a 7-point Likert scale, adapted from the Hypersensitive Narcissism Scale (HSNS) (Hendin & Cheek, 1997): 1) I can become entirely absorbed in thinking about my personal affairs, my health, my cares or my relations to others; 2) I dislike sharing the credit of an achievement with others; 3) I feel that I have enough on my hands without worrying about other people's troubles; 4) I feel that I am temperamentally different from most people. The Cronbach's alpha of the four items was .741, showing they were fairly reliable. The scores of the four items were summated and divided by four to become the final score of narcissism (M = 3.86, SD = 1.21).

## Shyness

Four items, adapted from the shyness scale by Cheek and Buss (1981), were set up to measure shyness: 1) I am socially somewhat awkward; 2) I don't find it hard to talk to strangers; 3) I feel tense when I'm with people I don't know; 4) I am often uncomfortable at parties and other social functions. Respondents were asked for their opinion toward the four items, on the 7-point scale ranging from 1 (strongly disagree) to 7 (strongly agree). The four items gave a very high reliability (Cronbach's alpha = .911). The scores were summated and divided by four as the final score of the variable (M = 3.49, SD = 1.61).

## Demographics

Respondents were asked to indicate their age, gender, and ethnicity. Age was indicated by a number in years. Options for gender were Male; Female; and Others (please specify). Options for ethnicity were Hispanic or Latino; American Indian or Alaska Native; Asian; Black or African or African American; Native Hawaiian or Other Pacific Islander; White or Caucasian or European; Maori; Arab or Middle Eastern; Others (please specify).

## RESULTS

H1 proposed that the more gratifying the Instagram users find from posting food photos, the higher personal involvement in posting food photos on Instagram. To test the hypothesis, a hierarchical regression was run. The first block of predictors were users' demographic variables, including age, race and nation of residence. Race and nation of residence were entered as dummy variables, with 0 = White or Caucasian or European, 1 = Others, and 0 = the U.S., 1 = New Zealand respectively. The six gratifications were entered as the second block of the predictors ($F = 21.71$, $p < .001$, adjusted R square = .509). The multicollinearlity was acceptable (VIF < 2.30). Escapism ($\beta = .154$, $p < .05$), information sharing ($\beta = .310$, $p < .001$), archiving ($\beta = .164$, $p < .05$) were found to be significant predictors of personal involvement in posting food photos on Instagram. H1 was supported.

RQ1.1 asked how social activity is associated with the personal involvement in posting food photos. To answer a research question, a hierarchical regression with demographics controlled as the first block and the social activity as the second-block predictor was run ($F = 3.22$, $p < .01$, adjusted $R^2 = .052$). Social activity was found to be a significant predictor ($\beta = .154$, $p < .05$) of the personal involvement in posting food photos on Instagram. The result of the regression was shown in the first column of Table 2.

RQ1.2 asked how social activity is associated with the personal involvement in posting selfies. A similar hierarchical regression was run, with demographics controlled. The regression equation was insignificant.

H2.1 proposed that the higher the level of narcissism of the Instagram users, the higher their personal involvement in posting food photos on Instagram. A hierarchical regression with demographics controlled ($F = 4.62$, $p < .001$, adjusted $R^2 = .082$) indicated that narcissism was a significant predictor of the personal involvement in posting food photos ($\beta = .224$, $p < .001$). H2.1 was supported. The result of the regression was shown in the second column of Table 2.

*Table 2. Hierarchical regressions of personal involvement in posting food photos on Instagram*

| DV | Personal Involvement in Posting Food Photos | |
|---|---|---|
| **Predictors** | β | β |
| **Block 1: Demographics** | | |
| Age | .085 | .095 |
| Race | .025 | .068 |
| Gender | -.062 | -.074 |
| Nation | -.151* | -.175* |
| *Adjusted* $R^2$ | .036 | .036 |
| **Block 2: Narcissism and shyness** | | |
| Social activity | .154* | |
| Narcissism | | .224*** |
| *Incremental* $R^2$ | .016 | .046 |
| Total adjusted $R^2$ | .052 | .082 |

Note: * p < .05, ** p < .01, *** p < .001; N = 223.

H2.2 proposed that the higher the level of narcissism of the Instagram users, the higher their personal involvement in posting selfies on Instagram. A hierarchical regression with demographics controlled (F = 3.81, $p < .001$, adjusted $R^2 = .077$) indicated that narcissism significantly predicted personal involvement in posting selfies ($\beta = .266, p < .001$). H2.2 was supported. The result of the regression was shown in the first column of Table 3.

H3.1 proposed that the higher the level of shyness of the Instagram users, the higher their personal involvement in posting food photos on Instagram. A hierarchical regression with demographics controlled (F = 3.03, $p < .05$, Adjusted $R^2 = .048$) indicated that shyness was insignificant in predicting of the personal involvement in posting food photos. H3.1 was not supported.

H3.2 proposed that the higher the level of shyness of the Instagram users, the higher their personal involvement in posting selfies on Instagram. A hierarchical regression with demographics controlled (F = 2.06, $p < .05$, Adjusted $R^2 = .031$) indicated that shyness was a significant predictor of personal involvement in posting selfies ($\beta = .160, p < .05$). H3.2 was supported. The result of the regression was shown in the second column of Table 3.

RQ2a asked for demographics, narcissism, shyness, social activity, general Instagram use, and gratifications sought predicted personal involvement in posting food photos on Instagram. To answer this research question, a hierarchical regression was performed. The first block of predictors were users' demographic variables. The second block of predictors were narcissism and shyness, followed by the number of followers on Instagram and the number of following Instagram accounts as the third block of predictors. The six factors of gratifications sought from posting food photos on Instagram were entered as the fourth block of predictors. Personal involvement in posting food photos on Instagram was entered as the dependent variable. The regression equation was significant (F = 13.670, $p < .001$), with an adjusted R square at .495. Results showed that significant predictors included nation of residence ($\beta$

*Table 3. Hierarchical regressions of personal involvement in posting selfies on Instagram*

| DV | Personal Involvement in Posting Selfies | |
|---|---|---|
| **Predictors** | **β** | **β** |
| **Block 1: Demographics** | | |
| Age | -.025 | -.036 |
| Race | .035 | .035 |
| Gender | -.194* | -.180* |
| Nation | .011 | -.017 |
| *Adjusted* $R^2$ | .011 | .011 |
| **Block 2: Narcissism and shyness** | | |
| Narcissism | .266*** | |
| Shyness | | .160* |
| *Incremental* $R^2$ | .066 | .20 |
| *Total adjusted* $R^2$ | .077 | .031 |

Note: * p < .05, ** p < .01, *** p < .001; N = 223.

= -.164, $p < .01$), escapism ($\beta = .16$, $p < .05$), information sharing ($\beta = .277$, $p < .001$), and archiving ($\beta = .174$, $p < .01$). Detailed results are shown in Table 4.

RQ2b was proposed to examine how demographics, narcissism, shyness, social activity, general Instagram use, and gratifications sought predicted the frequency of posting food photos on Instagram. Predictors were entered into the hierarchical regression equation with the same order as before, with the

*Table 4. Hierarchical regressions of personal involvement in posting, frequency of posting, and time spent editing food photos on Instagram*

| DV | Personal involvement | Frequency |
|---|---|---|
| Predictors | β | β |
| **Block 1: Demographics** | | |
| Age | .022 | .000 |
| Race | .004 | -.097 |
| Gender | -.035 | -.161* |
| Nation | -.164** | -.167* |
| Adjusted $R^2$ | .034 | .027 |
| **Block 2: Narcissism and shyness** | | |
| Narcissism | .005 | .099 |
| Shyness | .025 | -.168 |
| Incremental $R^2$ | .086 | .057 |
| **Block 3: Social activity** | | |
| Social activity | .005 | .033 |
| Incremental $R^2$ | .012 | .003 |
| **Block 4: General Instagram use** | | |
| Followers | .042 | -.123 |
| Following a/c | -.024 | -.052 |
| Incremental $R^2$ | .014 | -.07 |
| **Block 5: Gratifications sought** | | |
| Social interaction | .119 | -.043 |
| Escapism | .160* | .226** |
| Information sharing | .277*** | .296*** |
| Archiving | .174** | .121 |
| Creativity | .102 | .089 |
| Self-disclosure | .079 | .034 |
| Incremental $R^2$ | .383 | .0253 |
| Total adjusted $R^2$ | .495 | .333 |

Note: * $p < .05$, ** $p < .01$, *** $p < .001$; N = 223.

frequency of posting food photos as the dependent variable (F = 7.445, $p < .001$, adjusted $R^2 = .333$). Results showed that significant predictors included gender ($\beta = -.161$, $p < .05$), nation of residence ($\beta = -.167$, $p < .05$), escapism ($\beta = .226$, $p < .01$), and information sharing ($\beta = .296$, $p < .001$). Detailed results are also shown in Table 4.

## DISCUSSION

Findings in this study offer meaningful implications for the theoretical approaches in uses and gratifications. In this exploratory study, six gratifications sought from posting food photos on Instagram by young adults were found. They were self-promotion, escapism, information sharing, archiving, creativity and self-disclosure. Among the six gratifications, escapism, information sharing, and archiving were found to be significant predictors of personal involvement in posting food photos on Instagram. Motives including self-promotion and self-disclosure were not significant predictors of personal involvement and frequency of posting food photos on Instagram. Al-Kandari et al.'s (2016) survey showed the need for visual self-expression was the strongest predictor of the self-disclosure use of Instagram. Sharing food photos without human faces reduces the extent of self-disclosure, which may not satisfy the need for self-expression on social media as other types of photos, such as selfies.

However, posting food photos does satisfy Instagram users' psychological needs by escapism. Unlike selfies that may lead to criticisms on the uploader's appearance, taking photos of food allows the individual to reduce the stress caused by criticisms, by showing the gustatory attractive food without any engagement of a personal appeal. Distress may also be eased physiologically with the consumption of food after taking the food photo. Sharing food photos on Instagram brings ordinary joy that can be shared by everyone, regardless of physical appearances and physical conditions.

Food photos are also informative, as shown by the gratifications of information sharing and archiving. Food photos contain important information about eating, which is always an important aspect of human life. The information on a food photo can include a person's lifestyle, a recipe, a restaurant, and so on, and is worth sharing. Archiving did not predict the frequency of posting food photos, but significantly predicted the personal involvement of posting food. While food is something that appears in everyone's life frequently, it seems everyday food does not need to be archived. Archiving with food photos may only be needed under special circumstances such as remembering an event, celebrating important dates, visiting a famous restaurant, traveling to another place, etc.

Many of the past studies treated content generation on social media as a single, holistic behavior, but our analysis suggests the underlying psychological mechanism behind posting different types of contents can be different. Our findings explore young adults' uses of social media in different ways, by comparing different psychological factors' associations with personal involvement of posting food photos and posting selfies. The comparison enriches our understanding of the content generation behavior on social media.

Social activity significantly predicted the personal involvement in posting food photos, but not selfies. In other words, more socially active people involve more in posting food photos. It can be because posting food photos is a "non-invasive" type of self-disclosure. Food involved in social activity, such as food during gatherings and travels, are usually more visually appealing and can initiate the motives of sharing them online. The mobile nature of Instagram allows food photos to be shared in these social activities easily, but keeping the privacy of the users at the same time because their own appearances are not shown in these photos. On the contrary, posting selfie is a kind of self-disclosure closely related to self-expression (Al-Kandari et al., 2016). The act of taking and posting selfies may not involve any types of social interactions, and thus it does not necessarily associate with social activity.

This study also disclosed the relationships between narcissism and the involvements of Instagram users in posting food photos and selfies among young adults. The more narcissistic the Instagram user, the more involved they were in posting both food photos and selfies on Instagram. Such a finding is in line with previous research (e.g., Poon & Leung, 2011; Kim et al., 2016). Posting either food or selfies on Instagram can be a way to satisfy the narcissists' needs to promote themselves, attract attentions from others, and share personal information with others. After all, regardless of food or selfies, posting photos on Instagram is a form of self-disclosure. In addition to sharing something with others, Instagram is a perfect platform to equip normal social media users with hands-on mobile photography skills to produce and share visually-attractive creative works with others. To narcissistic individuals, food can be visual extension of themselves, that serves a similar communication function with their selfies.

Shy individuals were found to be more involved in posting selfies. For shy individuals, face-to-face interactions with strangers or casual acquaintances can be a source of awkwardness and discomfort (Cheek & Buss, 1981), so expressing themselves by posting selfies on the social media may reduce the stress caused by self-expressions in real life. On the contrary, posting food photos is less associated with self-expressions and is not associated with tensions built up in face-to-face interactions among shy individuals. This piece of finding shows clearly that the nature of the visual content posted on social media is an important factor influencing users' communication patterns. Food, without any visual self-disclosure, is a more universal way of communication across people with different levels of shyness.

Our sample covers college students from both U.S. and New Zealand, and findings indicated that Instagram users in the U.S. were more involved in posting food photos. One reason may be the differences in eating habits between the two countries. For example, the per capita food expenditure away from home in 2014 in the U.S. was USD 2,293, while the per capita expenditure eating out in New Zealand was much lower at NZD 1,688 (approximately USD 1,182); The share of consumer expenditures on food that were consumed at home in 2015 was much higher in New Zealand (14.9%) than in the U.S. (6.4%) (United States Department of Agriculture, 2016). Food that are non-homemade (offered by restaurants, cafes, etc.) are usually visually more appealing, initiating greater desires to share on Instagram, which may explain the national difference.

In addition, results showed that women involved more than men in posting selfies on Instagram, but there was no gender difference in posting food photos. This is consistent with previous research findings, for example, Dhir et al. (2016) found that women were more likely to take personal and group selfies and post compared to men. Food photos, without showing any physical appearances, on the other hand, engaged men and women equally. Women significantly posted food photos more frequently than men, but both genders involved in posting food photos similarly.

While race was not a significant predictor in the personal involvement in posting selfies, Williams and Marquez's (2015) semi-structured interviews suggested White social media users had an aversion to selfies, whereas Black and Latino users generally approved of selfies. Contrary to selfies, food photos do not disclose the uploaders' race. As we argued before, the joy of posting food photos on Instagram can be shared by everyone, regardless of their gender and race. While selfies involve the visual appeal of the photographer and are more popular among people with good appearances, food photos can be uploaded by all demographics, because literally, everyone consumes food every day.

## LIMITATIONS AND FUTURE STUDY

Although this exploratory study provides several promising insights for understanding young adult Instagram users' behaviors of posting food and selfies, it does have several limitations. First, all participants were recruited from college campuses through snowball sampling. Future study should use a more diverse and representative sample outside the university. Then, differences were discovered from comparing the sample from different countries of residence, the United States and New Zealand, but both countries are dominated by western cultures. Future study can explore young adults' use of Instagram in countries with other cultures, and investigate whether and how culture influences their mobile photography preferences and behaviors. Finally, this study only examined food-and-selfie-posting behaviors on Instagram, so future study would investigate similar topic on other social media platforms such as Facebook, Twitter, Snapchat etc. to understand more about users' self-disclosure of visual contents under the specific characteristics

## REFERENCES

Al-Kandari, A., Melkote, S. R., & Sharif, A. (2016). Needs and Motives of Instagram Users that Predict Self-disclosure Use: A Case Study of Young Adults in Kuwait. *Journal of Creative Communications*.

Baker, L. R., & Oswald, D. L. (2010). Shyness and online social networking services. *Journal of Social and Personal Relationships*, *27*(7), 873–889. doi:10.1177/0265407510375261

Bakhshi, S., Shamma, D. A., & Gilbert, E. (2014, April). Faces engage us: Photos with faces attract more likes and comments on Instagram. In *Proceedings of the 32nd annual ACM conference on Human factors in computing systems* (pp. 965-974). ACM. 10.1145/2556288.2557403

Buss, A. H. (1980). Shyness and sociability. *Journal of Personality and Social Psychology*, *41*, 330–339.

Campbell, W. K., & Foster, J. D. (2007). The narcissistic self: Background, an extended agency model, and ongoing controversies. *Self*, 115–138.

Chan, M. (2011). Shyness, sociability, and the role of media synchronicity in the use of computer-mediated communication for interpersonal communication. *Asian Journal of Social Psychology*, *14*(1), 84–90.

Cheek, J. M., & Buss, A. H. (1981). Shyness and sociability. *Journal of Personality and Social Psychology*, *41*(2), 330–339. doi:10.1037/0022-3514.41.2.330

Cox, A. M., & Blake, M. K. (2011, March). Information and food blogging as serious leisure. In P. Willett (Ed.), ASLIB proceedings (Vol. 63, No. 2/3, pp. 204-220). Emerald Group Publishing Limited. doi:10.1108/00012531111135664

Davenport, S. W., Bergman, S. M., Bergman, J. Z., & Fearrington, M. E. (2014). Twitter versus Facebook: Exploring the role of narcissism in the motives and usage of different social media platforms. *Computers in Human Behavior*, *32*, 212–220. doi:10.1016/j.chb.2013.12.011

Dhir, A., Pallesen, S., Torsheim, T., & Andreassen, C. S. (2016). Do age and gender differences exist in selfie-related behaviours? *Computers in Human Behavior*, *63*, 549–555. doi:10.1016/j.chb.2016.05.053

Gerlich, R. N., Drumheller, K., & Babb, J. (2015). App Consumption: An Exploratory Analysis of the Uses & Gratifications of Mobile Apps. *Academy of Marketing Studies Journal, 19*(1), 69.

Hendin, H. M., & Cheek, J. M. (1997). Assessing Hypersensitive Narcissism: A Re-examination of Murray's Narcissism Scale. *Journal of Research in Personality, 31*(4), 588–599. doi:10.1006/jrpe.1997.2204

Hu, Y., Manikonda, L., & Kambhampati, S. (2014, June). What We Instagram: A First Analysis of Instagram Photo Content and User Types. ICWSM.

Humphreys, L., Von Pape, T., & Karnowski, V. (2013). Evolving mobile media: Uses and conceptualizations of the mobile internet. *Journal of Computer-Mediated Communication, 18*(4), 491–507. doi:10.1111/jcc4.12019

Katz, E., Blumler, J. G., & Gurevitch, M. (1973). Uses and gratifications research. *Public Opinion Quarterly, 37*(4), 509–523. doi:10.1086/268109

Kim, E., Lee, J. A., Sung, Y., & Choi, S. M. (2016). Predicting selfie-posting behavior on social networking sites: An extension of theory of planned behavior. *Computers in Human Behavior, 62*, 116–123. doi:10.1016/j.chb.2016.03.078

Lee, E., Lee, J. A., Moon, J. H., & Sung, Y. (2015). Pictures Speak Louder than Words: Motivations for Using Instagram. *Cyberpsychology, Behavior, and Social Networking, 18*(9), 552–556. doi:10.1089/cyber.2015.0157 PMID:26348817

Leung, L. (2009). User-generated content on the internet: An examination of gratifications, civic engagement and psychological empowerment. *New Media & Society, 11*(8), 1327–1347. doi:10.1177/1461444809341264

Lindlof, T. R., & Shatzer, M. J. (1998). Media ethnography in virtual space: Strategies, limits, and possibilities. *Journal of Broadcasting & Electronic Media, 42*(2), 170–189. doi:10.1080/08838159809364442

Malik, A., Dhir, A., & Nieminen, M. (2016). Uses and Gratifications of digital photo sharing on Facebook. *Telematics and Informatics, 33*(1), 129–138. doi:10.1016/j.tele.2015.06.009

McQuail, D. (2005). *Mass communication theory* (5th ed.). London: Sage.

Mittal, B. (1995). A comparative analysis of four scales of consumer involvement. *Psychology and Marketing, 12*(7), 663–682. doi:10.1002/mar.4220120708

Ong, E. Y., Ang, R. P., Ho, J. C., Lim, J. C., Goh, D. H., Lee, C. S., & Chua, A. Y. (2011). Narcissism, extraversion and adolescents' self-presentation on Facebook. *Personality and Individual Differences, 50*(2), 180–185. doi:10.1016/j.paid.2010.09.022

Paramboukis, O., Skues, J., & Wise, L. (2016). An Exploratory Study of the Relationships between Narcissism, Self-Esteem and Instagram Use. *Social Networking, 5*(02), 82–92. doi:10.4236n.2016.52009

Poon, D. C. H., & Leung, L. (2011). *Effects of narcissism, leisure boredom, and gratifications sought on user-generated content among net-generation users.* Academic Press.

Rubin, A. M., & Rubin, R. B. (1982). Contextual age and television use. *Human Communication Research, 8*(3), 228–244. doi:10.1111/j.1468-2958.1982.tb00666.x

Rubin, R. B., Rubin, A. M., Graham, E., Perse, E. M., & Seibold, D. (2010). *Communication research measures II: A sourcebook*. Routledge. doi:10.4324/9780203871539

Sheldon, P. (2013). Voices that cannot be heard: Can shyness explain how we communicate on Facebook versus face-to-face? *Computers in Human Behavior, 29*(4), 1402–1407. doi:10.1016/j.chb.2013.01.016

Sheldon, P., & Bryant, K. (2016). Instagram: Motives for its use and relationship to narcissism and contextual age. *Computers in Human Behavior, 58*, 89–97. doi:10.1016/j.chb.2015.12.059

Statista. (2018). *Statistics and Market Data on Social Media & User-Generated Content*. Accessed at https://www.statista.com/markets/424/topic/540/social-media-user-generated-content/

Stritzke, W. G., Nguyen, A., & Durkin, K. (2004). Shyness and computer-mediated communication: A self-presentational theory perspective. *Media Psychology, 6*(1), 1–22. doi:10.12071532785xmep0601_1

Sung, Y., Lee, J. A., Kim, E., & Choi, S. M. (2016). Why we post selfies: Understanding motivations for posting pictures of oneself. *Personality and Individual Differences, 97*, 260–265. doi:10.1016/j.paid.2016.03.032

Takeuchi, A., & Nagao, K. (1993, May). Communicative facial displays as a new conversational modality. In *Proceedings of the INTERACT'93 and CHI'93 Conference on Human Factors in Computing Systems* (pp. 187-193). ACM. 10.1145/169059.169156

United States Department of Agriculture. (2016). *Food expenditures*. Retrieved from https://www.ers.usda.gov/data-products/food-expenditures/food-expenditures/#Expenditures%20on%20food%20and%20alcoholic%20beverages%20that%20were%20consumed%20at%20home%20by%20selected%20countries

Whiting, A., & Williams, D. (2013). Why people use social media: A uses and gratifications approach. *Qualitative Market Research, 16*(4), 362–369. doi:10.1108/QMR-06-2013-0041

Williams, A. A., & Marquez, B. A. (2015). Selfies| The Lonely Selfie King: Selfies and the Conspicuous Prosumption of Gender and Race. *International Journal of Communication, 9*, 13.

# Chapter 6
# Smartphone Usage, Gratifications, and Addiction:
## A Mixed-Methods Research

**Hafidha S. AlBarashdi**
*The Research Council, Oman*

**Abdelmajid Bouazza**
*Sultan Qaboos University, Oman*

## ABSTRACT

*Combining a survey and focus groups as a mixed-methods research, the authors of this chapter examined the functions, types, and motivations of smartphone usage and gratifications. Furthermore, the authors also investigated the rates, symptoms, and reasons of smartphone addiction. Still another achievement of the authors was to look into the relationship between smartphone usage, gratifications, and addiction with academic achievements among college students. On top of identifying three levels of addiction, the authors also located distinctive traits of these levels. This chapter provided an interesting example of how mixed-methods research can be employed to investigate mobile use, gratifications, and addiction. It is expected from the editor that this chapter would lead to more comparative studies between or among countries or cultures using by using a mixed-methods research to triangulate and/or complement findings of using different research methods. The inclusion of this chapter in this volume is also meant to invite further studies to investigate the gap between what mobile users want and what they actually get from using mobile as well as its related experience on both the normative and the empirical sides.*

## INTRODUCTION

The Smartphone is one of the most prominent types of information and communication technology (ICT) device, which has shown the most spectacular development (Chóliz, 2012). It entered human life and affected directly and indirectly many aspects of human relationships and interactions and changed most aspects of everyday life (Szpakow, Stryzhak, & Prokopowicz, 2011). It has been changing the ways in which people communicate with others, find information, have fun and manage their everyday lives.

DOI: 10.4018/978-1-5225-7885-7.ch006

Furthermore, the recent developments of new operating systems, abundant applications and competition between vendors have facilitated a remarkable growth in the number of users (Park, Kim, Young & Shim, 2013).

As a result, Smartphone use is rapidly spreading worldwide. In fact, the number of smartphone users is predicted to grow from 2.1 billion in 2016 to around 2.5 billion in 2019, with smartphone penetration rates increasing as well. Just over 36 percent of the world's population is projected to use a smartphone by 2018, up from about 10 percent in 2011(statista.com, 2018). Oman is no exception to this trend. Oman's mobile phone subscriber base crossed 7 million at the end of April 2017, with a penetration rate of more than 150 per cent (timesofoman.com, 2018).

Today, the wide spread of Smartphones can be attributed to many factors, including their being fast, light, strong and ever more convergent. Likewise, Smartphones come with different features that attract users, such as a digital camera, movie camera, diary, phone book, GPS, radio, MP3 player, web browser, data storage device, encyclopedia, alarm clock, Dictaphone, personal organizer, flashlight and many more. All of these factors have encouraged the majority of people in developed as well as developing countries to use mobile phones (Takao, Takahashi & Kitamura, 2009). Moreover, since the introduction of the iPhone, mobile phones with touchscreens have begun to dominate the Smartphone market (Henze, Rukzio, & Boll, 2012)

While Smartphone use has been increasing in all economic and age sectors, university students have been considered as one of the most important target markets and the largest consumer group for Smartphone services (Head & Ziolkowski, 2012). Likewise, Bianchi and Phillips (2005) revealed that problematic mobile phone usage was most widespread among younger users; hence, it is arguable that mobile phone addiction is most likely to occur among this group. Hong, Chiu and Huang (2012) argued that mobile phones are popular among university students because they increase their social communication and expand their opportunities for making social relationships.

## OBJECTIVES OF THE STUDY

The chapter aims to provide a comprehensive understanding of the Smartphone usage, gratifications, and addiction among undergraduates using UGT as a framework. In addition, it examines the relationship between Smartphone usage, gratifications, addiction and undergraduates' academic achievement.

### Significance of the Study

The expected contribution of this study is as follows:

1. To broaden our knowledge about Smartphone usage, gratifications, and addiction among undergraduates.
2. To provide exploratory insights into the nature of the Smartphone addiction problem: The prevalence rate of different Smartphone addiction levels, the symptoms of Smartphone addiction and the effects of Smartphone addiction on academic achievement among SQU undergraduates.
3. To determine the motives behind Smartphone use and addiction, as well as the types of Smartphone usage behavior associated with Smartphone addiction among undergraduates. Consequently, the

study will increase the awareness of students, parents and teachers about risk factors associated with Smartphone addiction.

4.  The findings of this study maybe useful to various parties, including university students, parents, educators, researchers and policy-makers.

## LITERATURE REVIEW

This chapter conducts a review of the body of literature relating to Smartphone usage, gratifications and addiction, as follows:

### Smartphone Usage Behavior Among University Students

Despite the recent huge rise in popularity of Smartphone use and their impact on users' lifestyles, few studies have been conducted on Smartphone usage patterns (Ahn, Wijaya, & Esmero, 2014). Therefore, this chapter will explore the types of Smartphone usage behavior and their association with Smartphone gratifications and addiction. Therefore, it is important to describe these usage types and their relationship with Smartphone gratifications and addiction by reviewing past studies. In this regard, Hooper and Zhou (2007) correlated six types of mobile phone usage behavior (addictive, compulsive, dependent, habitual, voluntary and mandatory) with seven underlying motivations for mobile phone usage (social interaction, dependency, image/identity, safety, job-related, freedom and gossip). Likewise, Shambare, ugimbana, and Zhowa (2012) identified the types of behavior associated with mobile phone usage among university students. They tried to determine whether students exhibited one type of behavior more than another or a set of behavior types more than others. They also attempted to categorize mobile phone usage according to the typologies that are commonly identified in the literature (addictive, compulsive, dependent, habitual, voluntary and mandatory).

Lee, Chang, Lin, and Cheng (2014), on the other hand, examined links between psychological traits and compulsive behaviors of Smartphone users, and looked further into the stress caused by those compulsive behaviors. The results suggested that compulsive usage of Smartphone and techno-stress were positively related to certain psychological traits, including locus of control, social interaction anxiety, materialism and the need for touch.

According to Fukuda, Asai, and Nagami (2015, October) users are gradually but steadily adopting WiFi at home, in offices, and public spaces over these three years. The majority of light users have been shifting their traffic to WiFi. Heavy hitters acquire more bandwidth via WiFi, especially at home. The percentage of users explicitly turning off their WiFi interface during the day decreases from 50% to 40%. In South Korea, Jeong, Kim, Yum, and Hwang (2016) examined the user characteristics and media content types that can lead to addictive behavior. With regard to user characteristics, results showed that those who have lower self-control and those who have greater stress were more likely to be addicted to smartphones. For media content types, those who use smartphones for SNS, games, and entertainment were more likely to be addicted to smartphones, whereas those who use smartphones for study-related purposes were not.

Fullwood, Quinn, Kaye, and Redding (2017) found that using Smartphones to alleviate boredom had become habituated for some users. Findings also imply that users may not be attached to the device itself, but rather the affordances on offer. However, for Yang (2018) using a combination of Extended

Technology Acceptance Model (TAM2) and consumer behavior theories can provide a better understand the planning, execution, and assessment of multi-platform advertising campaigns that affect users' usage behavior.

To conclude, various previous studies described the types of usage behavior that are associated with Smartphone use (e.g Ahn et al., 2014; Fukuda et al., 2015; Fullwood et al., 2017; Jeong et al., 2016; Lee et al., 2014; Shambare et al., 2012; Yang, 2018). However, many previous focused on the most used Smartphone functions and applications that can lead to negative usage behavior (Jeong et al., 2016; Lee et al., 2014). In line with these findings, the current study articulated the following questions:

**Q1:** What are the main types of Smartphone usage behavior exhibited by SQU undergraduates based on their underlying motivation?

**Q2:** What are the most used Smartphone functions and applications among SQU undergraduates?

## Smartphone Gratifications Among University Students

A growing number of studies have applied UGT in investigating the gratifications of Smartphone usage among university students. Interestingly, Walsh (2007) investigated the factors underlying mobile phone use and the indicators of addiction, applying UGT as a framework. This study identified three mobile phone gratification motives: Self, social and security. Social and self-gratifications predicted the level of use and addictive tendency, with self-gratifications exhibiting the greatest impact on the three addiction indicators. Likewise, Grellhesl and Punyanunt-Carter (2012) analyzed the most highly sought gratifications for using SMS text messaging among university students. They identified seven gratifications: Immediate access and mobility, relaxation and escape, entertainment, information seeking, coordination, socialization and affection, and status.

In Malaysia, Balakrishnan and Loo (2012) explored mobile phone and Short Message Service (SMS) usage among urbanized university students. By applying UGT, they explored mobile phone purchasing factors, reasons to use mobile phone and SMS, usage pattern and behavioral issues related to mobile phone and SMS. The results identified the following motives for using a mobile phone: Socializing, privacy, status symbol and safety. On the other hand, motives to use SMS included: To make/cancel appointments, gossip and maintain relationships, provides privacy, cheap, easy to use and quick.

In Korea, Kim and Shin (2013) examined Smartphone users' motivation and gratification based on UGT. They also examined the relationship between Smartphone usage motives and the level of gratification. The findings identified five motives: Accessibility, entertainment, social status, portability, and problem-solving capacity. The study also identified five factors of gratification: Entertainment, social homogeneity, quick information search, usefulness, and convenience. It is notable that entertainment and social homogeneity were the first gratification factors. Besides, among five using motives, influential orders were entertainment, social, usefulness, quickness. However, Sheldon and Bryant (2016) investigates motives for Instagram use. These were "Surveillance/Knowledge about others," "Documentation," "Coolness," and "Creativity." Recently, Reid and Thomas (2017) indicated that the primary gratifications received from smartphone usage differed for each gender; for males, gratification from smartphone usage was more evenly spread across different gratification types, whereas females placed an overwhelming emphasis on social gratification. While both groups use smartphones primarily for their social connectedness, females value this more so. These findings highlight the diversity of uses and gratifications that the smartphone can satisfy.

From above listed studies, it clear that a growing number of studies have applied UGT to investigate gratification motives of mobile phone usage among university students (e.g. Balakrishnan & Loo, 2012; Grellhesl & Punyanunt-Carter, 2012; Kim & Shin, 2013; Reid & Thomas, 2017; Sheldon & Bryant, 2016; Uys et al., 2012; Walsh (2007).Therefore, the current study suggested this question:

**Q3:** What Smartphone motives do SQU undergraduates gratify from Smartphone usage and addiction?

## Smartphone Addiction Among University Students

Since this chapter aims to explore the nature of Smartphone addiction among SQU undergraduates, it is first important to discuss how previous studies identified this type of addiction, classified its various levels, and how they determined the symptom.

Park (2005) measured mobile phone based on seven indicators of dependency: Tolerance, withdrawal, unintended use, cutting down, time spent, displacement of other activities and continued use. Results showed that mobile phone users grow tolerant of mobile phones despite the fact that they may cause problems such as high phone bills and public annoyance.

Krajewska-Kułak et al. (2012) showed that students usually used mobile phones for sending text messages, taking photos and accessing the Internet. However, more Polish students than Belarusian respondents knew that mobile phone users could become addicted. Almost 1/5 of Polish students and 1/10 of the Belarusians had symptoms of mobile phone addiction.

In Korea, Kwon et al. (2013) developed the first scale of Smartphone addiction. In this study, each subject group was assessed and the respondents were divided into high-risk, low- to medium-risk, and general groups. The findings showed that Smartphone addiction rates of the high- risk and low- to medium-risk groups were 2.2 and 9.3% respectively in adolescents and 1.0 and 6.7% respectively in adults. Based on the factor analysis results, there were six subscales for Smartphone Addiction Scale (SAS): Daily-life disturbance, positive anticipation, withdrawal, cyberspace-oriented relationship, over-use and tolerance. While, Park (2014) investigated differences between Smartphone users with high and low addiction tendencies among Korean college students. The results revealed that more females than males exhibited high addiction tendency. Highly addicted Smartphone users have a greater level of motivation for chatting, caring for others, and accessibility to others than the low addictive users. Addicted Smartphone users tend to prefer particular Smartphone activities, such as voice calls, social networking programs and chatting.

Moreover, Valderrama (2014) developed the problematic Smartphone use scale (SPUS). He found strong positive relationships between the Internet Addiction Test (IAT) and the (SPUS), providing evidence of convergent validity. Moreover, evidence of good-to-excellent internal consistency was found for two of the scale's factors: Problematic use and mood modification. These results supported the use of the PSUS as a measure for problematic Smartphone use.

Demirci, Orhan, Demirdas, Akpinar, and Sert, (2014) found that 13.3% of the students considered themselves as addicted to their Smartphone, while 60.5% of the students considered themselves to be not addicted, and 26.2% of the students were unsure. In addition, total scores on Smartphone addiction were significantly higher for females than for males. Likewise, the average scale scores were the highest in users who used Smartphones for over 16 hours.

In Holland, Bolle (2014) suggested that Smartphone addiction could develop through habit. The result revealed that younger users are more vulnerable to develop this type of addiction, especially when they have higher levels of social stress, are weak at self-regulation, and extensively use their Smartphones for social and process purposes. Moreover, females were somewhat more sensitive to develop addiction because of their higher rates of social stress and social usage. However, in Turkey, Gökçearslan, Mumcu, Haşlaman, & Çevik (2016) investigated the roles of smartphone usage, self-regulation, general self-efficacy and cyberloafing in smartphone addiction. The results showed that both the duration of smartphone usage and cyberloafing positively affected smartphone addiction. The effect of self-regulation on smartphone addiction was negative and significant. Likewise, Pearson and Hussain (2017) investigated the relationship between smartphone use, narcissistic tendencies and personality as predictors of smartphone addiction. The results revealed that 13.3% of the sample was classified as addicted to smartphones. Higher narcissism scores and neuroticism levels were linked to addiction. Three themes of social relations, smartphone dependence and self-serving personalities emerged from the qualitative data. Interpretation of qualitative data supports addiction specificity of the smartphone. It is suggested smartphones encourage narcissism, even in non-narcissistic users. Similarly, Cha, and Seo (2018) reported that 30.9% of students were classified as a risk group for smartphone addiction. However, Alhassan, Alqadhib, Taha, Alahmari, Salam, and Almutairi (2018) found that 19% of the sample can considered as non-addicted to smartphone usage, 64% can considered as slightly addicted to smartphone usage and only 17% were probably addicted.

To conclude, several studies highlighted the negative effects of Smartphone addiction among university students. These studies identified the nature of this type of addiction by indicating its symptoms, classifying its levels and developing tools to measure it (e.g. Bolle, 2014; Demirci et al., 2014; Gökçearslan et al., 2016; Krajewska-Kułak et al., 2012; Kwon et al., 2013; Park, 2014; Pearson & Hussain, 2017; Valderrama, 2014). Given that, the following questions were suggested:

**Q4:** What are the prevalence rates of the various Smartphone addiction levels among SQU undergraduates?
**Q5:** What are the main symptoms of Smartphone addiction among SQU undergraduates?
**Q6:** What are the reasons behind Smartphone addiction among SQU undergraduates?

## The Relationship Between Smartphone Usage, Gratifications, and Addiction

There is some controversy among researchers regarding the relation between Smartphone usage, gratification and addiction. Whilst some researchers have found no relation between them (Song, Larose, Eastin, & Lin, 2004), others have claimed that enjoyable gratification seeking leads to habit and, finally, to addiction through operant conditions (Bolle, 2014). However, this study aims to apply UGT to provide better understanding of Smartphone usage, gratifications, and addiction and relationship between them. In addition, the UTG can improve our understanding of gratifications obtained from social networking via a Smartphone, since it considers university students as motivated users. Particularly, to examine why and how university students use the Smartphone for social networking in order to gratify a need, the underlying motivational purposes, and what leads them to addiction.

Recently there has been a revival of UGT research in the form of a growing number of UGT studies to investigate the gratification motives of mobile usage among university students (e.g. Balakrishnan & Loo, 2012; Grellhesl & Punyanunt-Carter, 2012). In this regard, Bolle (2014) indicated that the difference between internet and Smartphone addiction is in the usage gratifications and usage context of the two. Smartphones have different gratifications or features that can produce strong positive reinforcements, such as a pleasurable experience. Therefore, the current study suggested this hypothesis:

**H1:** There are significant relationship between Smartphone usage, gratifications, and addiction among SQU undergraduates.

## The Relationship Between Smartphone Usage, Gratifications, Addiction and Academic Achievement

Since this study is partially concerned with examining relationship of Smartphone usage, gratifications, and addiction with academic achievement, it is important to discuss both the positive and negative effects of Smartphones on academic achievement. In this regard, some previous studies have highlighted the positive role of Smartphones in advancing students' learning. For example, Cheon, Lee, Crooks and Song (2012) reported that advancements in mobile technology are rapidly widening the scope of learning in areas outside formal education by allowing flexible and instant access to rich digital resources.

However, Casey's (2012) found that the lower the grade of the students, the higher the likelihood that they would become addicted to Smartphones. In addition, Hong et al. (2012) found that female university students showed a high level of mobile phone addiction, which affected their academic achievement, including time management and other related problems. In order to decrease their addiction, the study recommended that those students should engage in fun activities with others. Likewise, Kuznekoff and Titsworth (2013) found that students who were not using their mobile phones during the lecture wrote down 62% more information in their notes. They also took more detailed notes, were able to recall more detailed information from the lecture, and scored a full letter grade and a half higher on a multiple-choice test than those students who were actively using their mobile phones during the lectures.

Recently, some researchers have focused on examining how social media sites addiction through Smartphone chatting affects university students' academic achievement. In this regard, Daffalla and Dimetry (2014) pointed out that in 7 out of 10 students, academic achievement was affected negatively by their use of social networking programs, especially Facebook and WhatsApp. Similarly, Bijari, Javadinia, Erfanian, Abedini, and Abassi (2013) reported a negative relationship between SNs usage and students' GPA. Likewise, Kibona and Mgaya (2015) found out the impact of smartphones on academic performance of higher learning students. However, Tavakolizadeh, Atarodi, Ahmadpour, and Pourgheisar (2014) found no significant relationship between excessive mobile phone use and students' academic achievement.

Some previous studies highlighted the gratifications of SNs usage and how they affect academic achievement among university students. For example, Muriithi and Muriithi (2013) found that connecting to classmates, sending messages, opinions and updates, socializing, chatting and updating their profiles were the most common SNs gratifications among university students. Conversely, those students might be expected to use SNs to raise their academic achievement. However, Bröns, Greifeneder and Støvring (2013) produced a different result, reporting that students seek gratifications of information sharing, as well as to express their opinions, in particular for academically related purposes.

However, Samaha and Hawi (2016) explored whether satisfaction with life mediated by stress and academic achievement facilitates smartphone addiction. The results showed that smartphone addiction risk was positively related to perceived stress, but the latter was negatively related to satisfaction with life. Additionally, a smartphone addiction risk was negatively related to academic achievement, but the latter was positively related to satisfaction with life. Recently, according to Sage and Burgio (2018), some neurodevelopmental and neurobehavioral changes due to exposure to wireless technologies, such as symptoms of retarded memory, learning, cognition, attention, and behavioral problems.

Hence, quite a number of studies correlated mobile phone use with a decrease in students' academic achievement (e.g., Bijari et al., 2013; Bröns et al., 2013; Casey, 2012; Cheon et al., 2012; Daffalla & Dimetry, 2014; Hong et al., 2012; Kibona & Mgaya, 2015; Kuznekoff & Titsworth, 2013; Muriithi & Muriithi, 2013; Sage & Burgio, 2018; Samaha & Hawi, 2016). However, Tavakolizadeh et al. (2014) found no significant relation between excessive mobile phone use and academic achievement. It worth noting here that no previous studies specifically investigated relationship between Smartphone usage, gratifications, and addiction and the academic achievement of university students, and the literature merely explored this topic in general. In line with these studies, current study suggested the following hypotheses:

**H2:** There are significant relationships between Smartphone usage, gratifications, and addiction and SQU undergraduates' academic achievement.

## RESEARCH METHODOLOGY

A mixed-approach investigation were applied in order to achieve the objectives of the study. A stratified sample of 5% (n=849) of SQU undergraduates were selected to collect the data through the study questionnaires. In addition, four discussion sessions with two focus groups were conducted using a sample of (16) students. The description of the study tools were as follows:

## Smartphone Addiction Questionnaire (SPAQ)

To assess the extent to which SQU undergraduates are addicted to Smartphone use, previous questionnaires were adapted and modified to measure Smartphone addiction among SQU undergraduates. This questionnaire was adapted from those used by Walsh (2009) and Casey (2012). The new questionnaire cronbach's alpha Coefficient was (0.76). It consisted of 39 items and utilized a five- point Likert-type scale. The questionnaire have four subscales, as follows:

1. **Smartphone Information:** Seven questions were included to collect information regarding the behavioral characteristics of the respondents. These included questions about the number of phones they owned, kind of Smartphone, usage hours, money spent on Smartphone bill, etc.
2. **Smartphone General Usage Rate:** The questionnaire included five items regarding Smartphone use, including the average number of calls sent, calls received, texts sent and texts received every day, in order to measure the level of use as identified by Walsh (2009). The purpose of these questions was to identify the amount of time the students allocated to the use of Smartphones and the amount of money they spent on the performance of a number of functions through Smartphones use (e.g. "How many calls would you make on your Smartphone per day?").
3. **Smartphone Functions and Applications Level of Usage:** Seventeen items were included to cover the level of usage to Smartphone different functions and applications. The purpose of these items was to identify the most used Smartphone's functions and applications among users (e.g., "How often do you use your Smartphone to do the following: Send WhatsApp\SMS messages, use the internet, set the alarm or reminders, take pictures, chat via social networking programs, listen to audio clips, use the calendar, voice calls, ....etc.)".

4.  **Smartphone Addiction Symptoms Appearance Rate:** This subscale included seventeen items to cover symptoms of Smartphone addiction, distributed in a five-factor Smartphone addiction profile as identified by Casey (2012). These factors were: Disregard of harmful consequences (items: 1, 2, 3, 4), preoccupation (items: 7, 9, 10), inability to control craving (items: 5, 6, 11, 12), productivity loss (items: 13, 14, 15) and feeling anxious and lost (items: 8, 16, 17). The purpose of these items was to identify the frequency of the appearance of Smartphone addiction symptoms among users.

## Smartphone Usage Behavior Questionnaire (SPUBQ)

To obtain quantitative data to explore the Genes (Gratification) and the types of Smartphone usage behavior, a questionnaire was developed to measure these Genes and types, using the UGT as a framework. The questionnaire benefited from Hooper and Zhou, (2007), Glaser, (2010), Grellhesl and Punyanunt-Carter (2012) instruments. The questionnaire, which consisted of 42 items and utilized a five- point Likert-type scale, was designed to provide information illustrating the Smartphone gratification as well as the types of usage behavior that reflect these motives. This questionnaire measured both the gratifications and the types of Smartphone usage behavior. The cronbach's alpha Coefficient for the questionnaire was (0.91). Below is a description of the questionnaire's various subscales:

1.  **Smartphone Gratifications:** This subscale measures six gratifications that influence Smartphone users' usage behavior and each one gratifies seven motives (items) as follows:
    a.  **Social Interaction:** Seven items were included to cover the use of Smartphone for purposes of social interaction, such as to stay in touch with their friends and families) e.g., "I use my Smartphone to activate ongoing communication with others"). These were items (1, 2, 3, 4, 5, 6, and 7).
    b.  **Information Sharing and Entertainment:** Seven items were included to cover the use of Smartphone for the purposes of information sharing and entertainment. These items explained the role of Smartphone as an essential tool in students' life for seeking information, sharing experiences and collaboration with others. Moreover, they also reflected the acquisition of Smartphones for study-related reasons (e.g. "I need my Smartphone to exchange information and experiences with colleagues". They also reflected how the students used their phones for entertainment purposes, for instance, to play games, listen to music, watch movies...etc. (e.g., "I use the entertainment programs in my Smartphone, for example: Video player, audio player and games to get rid of my boredom and to have fun."). These were items (8, 9, 10, 11, 12, 13, and 14).
    c.  **Self-Identity and Conforming:** Seven items were included to cover the use of Smartphone for the purpose of self-identity and conforming. These items illustrated how some students see Smartphones as giving status or conforming group identity. In this regard, the Smartphone and its usage help students personalize their phones, express themselves and conform to belonging to a particular group of friends (e.g. "I change my Smartphone constantly because it makes me feel special among my friends"). These items were (15, 16, 17, 18, 19, 20, and 21).
    d.  **Self-Developing and Safety:** Seven items were included to cover the use of Smartphone for the purpose of self-developing and safety. These items reflected the reasons mentioned for purchasing a Smartphone, which included to enhance self-completeness and to feel safe, deal with conflict situations, increase self-confidence, parents' belongingness and ideas diffusion

(e.g., "I bought a Smartphone to help me in emergency situations."). These items were (22, 23, 24, 25, 26, 27, and 28).

e.  **Freedom and Privacy:** Seven items were included to cover the use of Smartphone for the purpose of freedom and privacy. These items were (29, 30, 31, 32, 33, 34, and 35). They reflected the role of Smartphone in providing privacy and giving the student freedom to contact people at any time, which must be associated with responsibility and respect of others' privacy) e.g., "My Smartphone gives me the freedom to contact any person at any time").

f.  **Self-Express and Gossip:** Seven items were included to cover the use of Smartphone for the purpose of self-express and gossip. These items were (36, 37, 38, 39, 40, 41, and 42). They clarified the role of Smartphone in enabling students to express their feelings and opinions and to keep in touch, but also to engage in more extended gossiping with friends and families (e. g., "Smartphone apps offer me the chance to meet new friends and exchange views with them.").

2.  **Types of Smartphone Usage:** This subscale measures six types of Smartphone usage behavior, as follows:

a.  **Addictive Behavior:** Seven items were included to cover the use of Smartphone as an addictive type of behavior. These items were (20, 29, 32, 34, 35, 37, and 39). These items reflected the users' excessive usage of Smartphone, with continued use in spite of negative outcomes (e.g., "I use my Smartphone to escape from the worries of everyday life.").

b.  **Compulsive Behavior:** Seven items were included to cover the use of Smartphone as a compulsive type of behavior. These items were (16, 17, 18, 19, 21, 31, and 40). These items reflected the users' irrational need to use Smartphone, often despite negative consequences and it is usually periodic (e.g., "I usually ignore the harmful consequences of spending too much time talking on my Smartphone simply because I cannot live without it.").

c.  **Dependent Behavior:** Seven items were included to cover the use of Smartphone as a dependent type of behavior. These items were (8, 9, 12, 13, 14, 36, and 42). These items reflected the users' reliance on Smartphones to contact others and perform other functions (e.g., "I rely on my Smartphone to share knowledge with others in my specialty.").

d.  **Habitual Behavior:** Seven items were included to cover the use of Smartphone as a habitual type of behavior. These items were (1, 2, 5, 6, 15, 33, and 38). These items reflected users' automatic and routine behavior that they repeat, because it is easy, comfortable or rewarding (e.g., "I am used to carrying my Smartphone in my hand, even when I do not need it.").

e.  **Voluntary Behavior:** Seven items were included to cover the use of Smartphone as a voluntary type of behavior. These items were (3, 4, 7, 10, 26, 30, and 41). These items reflected the users' reasoned behavior, driven by specific motivations like social and personal benefits (e.g., "Because of my Smartphone I was able to develop my social relationships and gain popularity among my friends.").

f.  **Mandatory Behavior:** Seven items were included to cover the use of Smartphone as a mandatory type of behavior. These items were (11, 22, 23, 24, 25, 27, and 28). These items reflected behavior which is required to be done, followed, or complied with by the user, usually because it is driven or prompted by environmental consequences (e.g., "I bought a Smartphone so my parents can check up on me at any time.")

## Focus Group Discussion Guide

The researchers designed a focus group discussion guide comprising a series of focusing statements and open-ended questions, to initiate a discussion among a small sample of SQU undergraduates. Topics included their points of view with regard to the reasons behind Smartphone addiction and explaining the relationships between Smartphone usage, gratifications, and addiction among SQU undergraduates.

## FINDINGS

### The Main Types of Smartphone Usage Behavior Exhibited by SQU Undergraduates

The means and standard deviations for each type of Smartphone usage behavior were calculated, table 1.

The results in table 1 show that voluntary behavior was the most common type of Smartphone usage behavior among SQU undergraduates (3.68), followed by dependent behavior (3.65) and mandatory behavior (3.29), while compulsive behavior was the least common type of Smartphone usage behavior among SQU undergraduates (2.82).

### The Most Used Smartphone Functions and Applications Among SQU Undergraduates

The means and standard deviations for each of Smartphone functions and applications were calculated. Then based on the means, these functions and applications were organized under categories from the most common to the least common among SQU undergraduates as follows: Casual usage (1-2.29), Moderate usage (2.30-3.59), and Heavy usage (3.60-5.00), table 2.

The results in table 2 show the levels of usage to Smartphone functions and applications categorized into three levels based on their means. It is clear that at the heavy usage level, sending messages was the heaviest Smartphone function among SQU undergraduates (4.53), followed by using the Internet (4.44), while making voice calls was the least. At the moderate level of usage, watching video clips was the

*Table 1. Means (M) and standard deviations (SD) of the types of Smartphone usage behavior among SQU undergraduates (N=849)*

| The Main Types of Smartphone Usage Behavior | M | SD |
|---|---|---|
| 1. Voluntary Usage Behavior | 3.68 | 0.61 |
| 2. Dependent Usage Behavior | 3.65 | 0.64 |
| 3. Mandatory Usage Behavior | 3.29 | 0.49 |
| 4. Habitual Usage Behavior | 3.00 | 0.59 |
| 5. Addictive Usage Behavior | 2.88 | 0.77 |
| 6. Compulsive Usage Behavior | 2.82 | 0.71 |

*Table 2. Means (M) and standard deviations (SD) of Smartphone functions and applications usage among SQU undergraduates*

| Smartphone Functions and Applications | M | SD | Level of Usage |
|---|---|---|---|
| 1. Send WhatsApp messages | 4.53 | 0.79 | Heavy usage |
| 2. Use the Internet | 4.44 | 0.91 | |
| 3. Set the alarm or reminders | 4.32 | 1.04 | |
| 4. Take pictures | 4.25 | 0.98 | |
| 5. Chat via social networking programs (Facebook, Twitter, etc.) | 4.08 | 1.14 | |
| 6. Listen to audio clips | 3.82 | 1.16 | |
| 7. Use the Calendar | 3.68 | 1.16 | |
| 8. Voice calls | 3.66 | 1.17 | |
| 9. Watch video clips | 3.51 | 1.25 | Moderate usage |
| 10. Use the notepad | 3.29 | 1.38 | |
| 11. Receive or send e-mails | 3.18 | 1.33 | |
| 12. Use the calculator | 3.13 | 1.37 | |
| 13. Use the games | 2.97 | 1.38 | |
| 14. Download ringtones, games and programs | 2.71 | 1.12 | |
| 15. E-shopping | 2.22 | 1.39 | Casual usage |
| 16. Use the GPS maps and GRS | 2.27 | 1.33 | |
| 17. Vote for television programs and competitions | 1.49 | 1.08 | |

most used function (3.51), while download ringtones, games and programs was the least used activity (2.71). Finally, electronic shopping, use the GPS maps and GRS and voting for television programs and competitions were at the casual level of Smartphone usage.

## Smartphone Motives Underling SQU Undergraduates Smartphone Usage and Addiction

The means and standard deviations for each Smartphone Gene were calculated, table 3.

The results in table 3 reveal that the seeking information and entertainment was the most common Smartphone gratification among SQU undergraduates (3.92), followed by social interaction (3.48), and then by self-developing and safety (3.19). This was followed by self-express and gossip, and self-identity and conforming (2.95) together.

## The Prevalence Rates of the Various Smartphone Addiction Levels Among SQU Undergraduates

Smartphone addiction average scores for SQU undergraduates were calculated by assessing the Smartphone addiction indicators collected from (22 items) in the Smartphone addiction questionnaire. Based on their overall average score on Smartphone addiction SQU undergraduates were classified under three groups- casual level, moderate level and heavy, table 4.

*Table 3. Means (M) and standard deviations (SD) of the Smartphone Gratifications among SQU undergraduates (N=849)*

| The Genes of Smartphone | M | SD |
|---|---|---|
| 1. Seeking information and entertainment | 3.92 | .80 |
| 2. Social interaction | 3.48 | .52 |
| 3. Self-developing and safety | 3.19 | .49 |
| 4. Self-express and gossip | 2.95 | .49 |
| 5. Self-identity and conforming | 2.95 | .82 |
| 6. Freedom and privacy | 2.83 | .62 |

*Table 4. Prevalence rates of Smartphone addiction levels among SQU undergraduates (N=849)*

| Smartphone Addiction Levels | Addiction Score | | Prevalence Rate | |
|---|---|---|---|---|
| | SD | M | N | % |
| Casual (1-2.29) | 0.56 | 1.73 | 43 | 5.1 |
| Moderate (2.30-3.59) | 0.78 | 2.95 | 525 | 61.8 |
| Heavy (3.60-5.00) | 0.66 | 3.97 | 281 | 33.1 |

Note: M=Means, SD=Standard deviation, N= Number of size

The results in table 4 reveal that moderate Smartphone addiction was the most prevalent level among SQU undergraduates (61.8%), followed by the heavy level (33.1%) and finally the casual level (5.1%). Figure (13) summarizes the Smartphone addiction percentages among SQU undergraduates according to these three addiction levels.

## The Main Symptoms of Smartphone Addiction Among SQU Undergraduates

The means and standard deviations were calculated for each factor of Smartphone addiction symptoms, table 5.

The results in table 5 show the main symptoms of Smartphone addiction among SQU undergraduates, organized based on their means. It clear that the most common symptom was "disregard of harmful

*Table 5. Means (M) and standard deviations (SD) of Smartphone addiction factors (N=849)*

| The Factors of Smartphone Addiction Symptoms | M | SD |
|---|---|---|
| 1. Disregard of harmful consequences | 3.76 | .84 |
| 2. Productivity loss | 3.52 | .95 |
| 3. Inability to control craving | 3.39 | .88 |
| 4. Feeling anxious and lost | 3.26 | 1.08 |
| 5. Preoccupation | 3.02 | 1.08 |

consequences" (3.76), followed by "productivity loss" (3.52), then "inability to control craving" (3.39), then "feeling anxious and lost" (3.26) and finally "preoccupation" (3.02). To sum up, "Disregard of harmful consequences" was the most common symptom, while "preoccupation" was the least common.

## The Reasons Behind Smartphone Addiction Among SQU Undergraduates

All focus group participants offered seven reasons behind Smartphone addiction among SQU under-graduates. Five of these reasons related to students' Smartphone overuse, while two reasons related to Smartphone manufacturers. The reasons that related to students included: (1) Using Smartphones extensively to entertain themselves in order to escape from academic pressure; (2) Using Smartphones for self-expression, especially through SNs; (3) Over depending on certain Smartphone functions and Apps to accomplish their academic work; (4) The negative desire for excellence through experiencing new devices and apps before others; and (5) Addict chatting via SNs to maintain and develop social relationships. In sequence, comments illustrating each reason are provided as follows:

- In my opinion, with respect to Smartphone addiction among SQU undergraduates, I think one of the reasons is the need for entertainment. Because of the academic pressure, it appears that some students seek entertainment through listening to music or video games, whereas others prefer communication functions such as talking and texting. (P14)[1]

- Some SQU undergraduates become addicted to Smartphone usage because they enable them to express themselves. Shy students can easily express their opinions, feelings and emotions by chatting through social media networking sites. They prefer to interact in the virtual world via Smartphone more than the real world. (P7)

- Most students depend on Smartphones' functions and applications to organize their daily work, but some students are addicted to using these devices because they depend on them too much, for example, they cannot finish their assignments without using their Smartphones. (P10)

- Another reason behind Smartphone addicted among the students was the desire for excellence among peers through experiencing new devices and applications before others, which makes them seek continually to discover more apps. (P8)

- Chatting via social media networking apps using Smartphones is the easiest way to maintain friendships and increase social interactions. Despite that, some students chat online mainly to keep in contact with family members and friends. However, many students become addicted to chatting in a way that distracts them from their normal life and affects their academic achievement. (P13)

Likewise, the focus group participants presented two reasons relating to Smartphone manufacturers: 1) Continuous upgrading of Smartphone devices, and 2) Attracting young consumers by developing new apps. Two of the participants clarified these reasons:

- I think that the Smartphones manufacturers play an important role in making people addicted to Smartphone use. For example, they continue to upgrade these devices and develop new functions and applications to increase their benefits, which forces the consumer to change their devices con-stantly to follow these changes. As a result, the customers' usage rate continues to increase. (P 5)

- Smartphones manufacturers gain benefits by attracting young customers through developing new applications in a way that keeps customers busy with activities to download and experience using the latest applications. (P 6)

## The Relationship Between Smartphone Addiction, Gratifications, and Usage Types Among SQU Undergraduates

First, Pearson Correlation Coefficients between Smartphone addiction average score and the Smartphone gratifications were calculated, table 6.

The results in table 6 reveal significant correlation between Smartphone addiction and three of the Smartphone gratifications: Self-express and gossip, self-developing and safety, and self-identity and conforming. However, there were no significant correlation between Smartphone addiction average score and three other Smartphone gratifications: Social interaction, freedom and privacy and seeking information and entertainment.

Second, Pearson Correlation Coefficients between Smartphone addiction average score and the types of Smartphone usage behavior were calculated, Table 7.

The results in table 7 show that there were significant relationships between Smartphone addiction average and three types of Smartphone usage behavior: Addictive, compulsive and habitual. Nevertheless, there were no significant differences between Smartphone addiction average score and three other types of Smartphone usage behavior: Dependent, voluntary and mandatory.

*Table 6. Results of correlations between Smartphone addiction and Smartphone gratifications*

| Dependent Variables | Correlations | P |
|---|---|---|
| Smartphone Addiction × Social Interaction | -.04 | .245 |
| Smartphone Addiction × Freedom and Privacy | .05 | .165 |
| Smartphone Addiction × Self-Express and Gossip | .15** | .01 |
| Smartphone Addiction × Self-Developing and Safety | .16** | .01 |
| Smartphone Addiction × Self-Identity and Conforming | .12** | .01 |
| Smartphone Addiction × Seeking Information and Entertainment | .01 | .75 |

*Table 7. Results of correlations between Smartphone addiction and types of Smartphone usage behavior*

| Dependent Variables | Correlation | P |
|---|---|---|
| Smartphone Addiction × Addictive Behavior | 0.16** | 0.01 |
| Smartphone Addiction × Compulsive Behavior | 0.15** | 0.01 |
| Smartphone Addiction × Dependent Behavior | 0.03 | 0.46 |
| Smartphone Addiction × Habitual Behavior | 0.12** | 0.00 |
| Smartphone Addiction × Voluntary Behavior | 0.04 | 0.20 |
| Smartphone Addiction × Mandatory Behavior | -0.01 | 0.80 |

## The Differences in Smartphone Usage, Gratifications, and Addiction Among SQU Undergraduates Related to Academic Achievement

First, the means and standard deviations for each type of Smartphone usage behavior among SQU undergraduates were calculated. Then, One-Way ANOVA testing was conducted to test the effect of the academic achievement levels- high, middle and low (Table 8).

According to the results in table 8 no significant differences were observed among SQU undergraduates in any types of Smartphone usage behavior according to their academic achievement levels.

Second, the means and standard deviations as well as One-Way ANOVA test were calculated for each Gene of Smartphone usage behavior among SQU undergraduates according to their academic achievement. The students' academic achievement was classified into three levels (high, middle and low) in order to test their effect on the Smartphone gratifications among them, table 9.

The results in table 9 demonstrate no significant differences in Smartphone gratifications among SQU undergraduates according to the students' academic achievement.

Third, the means and standard deviations for each Smartphone addiction indicator were calculated according to the students' academic achievement, which was divided into three levels (high, middle and low). Then One-Way ANOVA testing was applied to test the effect of academic achievement on the Smartphone addiction indicators, table 10.

*Table 8. Results of One-Way ANOVA for types of Smartphone usage behavior among SQU undergraduates related to their academic achievement*

| Dependent Variable | Academic Achievement | N | M | SD | f-value | P |
|---|---|---|---|---|---|---|
| Addictive Behavior | High | 181 | 2.80 | .78 | 1.26 | 0.28 |
| | Middle | 583 | 2.89 | .77 | | |
| | Low | 84 | 2.95 | .73 | | |
| Compulsive Behavior | High | 181 | 2.76 | .75 | 0.66 | 0.51 |
| | Middle | 583 | 2.83 | .70 | | |
| | Low | 84 | 2.82 | .67 | | |
| Dependent Behavior | High | 181 | 3.71 | .64 | 1.36 | 0.25 |
| | Middle | 583 | 3.62 | .63 | | |
| | Low | 84 | 3.64 | .63 | | |
| Habitual Behavior | High | 181 | 2.99 | .62 | 0.19 | 0.82 |
| | Middle | 583 | 3.00 | .57 | | |
| | Low | 84 | 3.03 | .61 | | |
| Voluntary Behavior | High | 181 | 3.69 | .62 | 0.78 | 0.45 |
| | Middle | 583 | 3.68 | .60 | | |
| | Low | 84 | 3.60 | .65 | | |
| Mandatory Behavior | High | 181 | 3.28 | .50 | 1.24 | 0.28 |
| | Middle | 583 | 3.28 | .50 | | |
| | Low | 84 | 3.37 | .42 | | |

Note: M=means, SD=Std. Deviation

*Table 9. Results of One-Way ANOVA for the gratifications of Smartphone among SQU undergraduates related to their academic achievement*

| Smartphone gratifications | Academic Achievement | N | M | SD | F | P |
|---|---|---|---|---|---|---|
| Social Interaction | High | 181 | 3.47 | .52 | 1.01 | .36 |
| | Middle | 583 | 3.49 | .50 | | |
| | Low | 84 | 3.40 | .57 | | |
| Freedom and Privacy | High | 181 | 3.99 | .76 | 1.17 | .31 |
| | Middle | 583 | 3.88 | .81 | | |
| | Low | 84 | 3.93 | .75 | | |
| Self-express and Gossip | High | 181 | 2.92 | .85 | .11 | .89 |
| | Middle | 583 | 2.95 | .80 | | |
| | Low | 84 | 2.96 | .80 | | |
| Information Seeking and Entertainment | High | 181 | 3.16 | .50 | 1.75 | .18 |
| | Middle | 583 | 3.19 | .49 | | |
| | Low | 84 | 3.28 | .45 | | |
| Self-Developing and Safety | High | 181 | 2.78 | .62 | 1.12 | .33 |
| | Middle | 583 | 2.83 | .62 | | |
| | Low | 84 | 2.89 | .56 | | |
| Self-Identity and Conforming | High | 181 | 2.91 | .54 | .69 | .51 |
| | Middle | 583 | 2.96 | .46 | | |
| | Low | 84 | 2.94 | .49 | | |

Note: M=means, SD=Std. Deviation.

The results of one-way ANOVA in table 10 indicate that there were significant differences in the following Smartphone addiction indicators: Calls received and messages sent among SQU undergraduates related to their academic achievement. Hence, Post Hoc testing, using the Scheffe test, was conducted for multiple comparisons (Table 11).

The results of Post Hoc tests in table 11 reveal that there were significant differences in calls received among SQU undergraduates, in favor of higher-grade students more than among middle and low-grade students. Moreover, there were significant differences in messages sent among SQU undergraduates, in favor of low and middle-grade students more than among higher-grade students.

## DISCUSSION

### The Type of Smartphone Usage Behavior Among SQU Undergraduates

The results of the current study showed that voluntary behavior was the most common Smartphone usage behavior among SQU undergraduates, followed by dependent behavior and mandatory behavior, while compulsive behavior was the least common type of Smartphone usage behavior among SQU

*Table 10. Results of One-Way ANOVA for Smartphone addiction among SQU undergraduates related to their academic achievement*

| Smartphone Addiction Indicators | Academic Achievement | N | M | SD | f-value | P |
|---|---|---|---|---|---|---|
| Money Spent | High | 181 | 1.31 | .62 | 1.32 | .27 |
| | Middle | 583 | 1.26 | .53 | | |
| | Low | 84 | 1.20 | .46 | | |
| Calls Sent | High | 181 | 1.60 | .93 | 1.73 | .18 |
| | Middle | 583 | 1.49 | .78 | | |
| | Low | 84 | 1.41 | .85 | | |
| Calls Received | High | 181 | 2.04 | 1.26 | 7.08*** | .01 |
| | Middle | 583 | 1.71 | .93 | | |
| | Low | 84 | 1.76 | 1.01 | | |
| Messages Received | High | 181 | 4.36 | 1.21 | 1.24 | .29 |
| | Middle | 583 | 4.51 | 1.05 | | |
| | Low | 84 | 4.42 | 1.11 | | |
| Messages Sent | High | 181 | 4.33 | 1.25 | 7.68*** | .01 |
| | Middle | 583 | 4.66 | .95 | | |
| | Low | 84 | 4.70 | .85 | | |
| Addiction Symptoms | High | 181 | 2.62 | .81 | 1.32 | .27 |
| | Middle | 583 | 2.56 | .69 | | |
| | Low | 84 | 2.47 | .69 | | |
| The Average Score | High | 181 | 2.71 | .49 | .36 | .70 |
| | Middle | 583 | 2.69 | .44 | | |
| | Low | 84 | 2.66 | .49 | | |

Note: M=means, SD=Std. Deviation

*Table 11. Results of Post Hoc tests for Smartphone addiction among SQU undergraduates related to their academic achievement*

| Smartphone Addiction Indicators | (I) GPA | (J) GPA | Mean Difference (I-J) | P | Post Hoc |
|---|---|---|---|---|---|
| Calls Received | High | Middle | 0.33** | 0.01 | High>middle |
| | | Low | 0.28 | 0.12 | |
| | Middle | Low | -0.05 | 0.92 | |
| Messages Sent | High | Middle | -0.33** | 0.01 | middle > High |
| | | Low | -0.37** | 0.02 | Low > High |
| | Middle | Low | -0.05 | 0.93 | |

undergraduates. In specific terms, Smartphone usage among SQU undergraduates could be regarded more as voluntary, dependent or mandatory behavior rather than habitual, addictive or compulsive. This phenomenon could possibly relate to the fact that SQU undergraduates were voluntary engaged in Smartphone use for seeking information and entertainment -as the present study results revealed previously. Another contributory reason might be that many SQU undergraduates do not study in their hometown, so they seek to establish new contacts and social relationships. Thus, it is a conscious decision to use their Smartphone to acquire positive results. It suits their lifestyle and core values.

As mentioned in the literature review, there is little previous research on the topic of types of Smartphone usage behavior. Hence, the current study provides rich new information regarding the types of Smartphone usage behavior among SQU undergraduates. Nevertheless, the current finding differs from the Shambare et al. (2012) result that suggested that mobile phone usage is dependent, habitual and addictive. Likewise, the current finding differs from some previous studies that associated Smartphone use with negative usage behavior types (e.g. Jeong et al., 2016; Lee et al., 2014) as these studies revealed that Smartphone compulsive behavior has been regarded as the core of addiction.

## Smartphone Usage Among SQU Undergraduates

The current results showed that sending messages, especially via WhatsApp, was the most used Smartphone activity among SQU undergraduates, while voting for television programs and competitions was the least used Smartphone activity. This relates to the fact that messages can be composed and sent in a short time, from virtually anywhere. University students prefer WhatsApp messages because the service is quick, cheap and convenient. They also use text messaging to coordinate with both friends and family. Interestingly, the focus group discussion results supported these explanations and highlighted the five following specific reasons regarding why SQU undergraduates heavily use WhatsApp messenger:

The simplicity of WhatsApp messenger use, the low cost, the numerous positive uses, the program's good features, such as group chatting, sending photos, video, location, and contacts. Finally, the possibility of broadcasting WhatsApp messages to various contacts. Moreover, this finding tends to support Lecturer et al.'s (2014) and Daffalla and Dimetry (2014) observation that university students were addicted to WhatsApp messenger usage.

## Smartphones Gratifications Among SQU Undergraduates

The current findings indicated that seeking information and entertainment was the most common Smartphone gratification among SQU undergraduates, which relates to the fact that SQU undergraduates had a strong need to acquire information for learning and entertaining purposes. Besides, SQU undergraduates from different fields of study need information to write their term papers, assignments, and update their knowledge. Therefore, they found that Smartphones were very useful tools for seeking information and sharing it with friends and classmates. Although there is little previous research on this topic, the present study finding tends to agree with results obtained by other studies regarding mobile phone gratifications (e.g. Balakrishnan & Loo, 2012; Grellhesl & Punyanunt-Carter, 2012; Kim & Shin, 2013; Reid & Thomas, 2017; Sheldon & Bryant, 2016). These studies indicated that seeking information and entertainment, and social interaction were the most common mobile phone gratifications among university students. Yet,

UGT explained the motivations to use technologies, especially it explain why consumers use media to satisfy their social and psychological needs. Therefore, according to UGT students use Smartphones to gratify their needs for information and entertainment.

## Smartphone Addiction Levels Among SQU Undergraduates

The current results regarding Smartphone addiction levels among SQU undergraduates revealed that the moderate level of Smartphone addiction was the most prevalent, followed by the heavy level and finally the casual level.

This result showed that the majority of students were not frequently involved in heavy addiction. However, the (33%) of heavy addiction among these SQU students represents a high percentage. This result can be explained by the fact that Smartphones increase students' social communication and expand their opportunities to establish social relationships. Furthermore, Smartphones are equipped with various features that facilitate communication and entertainment for their users. Another reason is the availability of Smartphone's devices. However, the focus group results regarding Smartphone addiction reasons among SQU undergraduates specified two types of reasons: reasons related to student and reasons related to Smartphone manufacturers. First, the reasons related to students were as follows: escape from academic pressure; negative self-expression; over dependence on Smartphone to accomplish academic work; the negative desire for excellence by experiencing new devices and apps before others; addicted chatting via SNs. These reasons reflected the psychological needs of Smartphone addicts. Additionally, other reasons related to Smartphone manufacturers who make people addicted to Smartphone use: continuous upgrading of Smartphone devices and attracting young customers by developing new applications. Consequently, this study suggested that the transition from casual to moderate or heavy Smartphone addiction occurs when the user views the usage as an important mechanism to release academic pressure, relieve stress and anxiety.

Comparing to previous studies, the percentage of students with heavy Smartphone addiction is slightly similar to that obtained by such as Tavakolizadeh et al. (2014), who reported (36.7%) of mobile phone addicted students. Our result was also similar to Cha and Seo (2018) who reported that 30.9% of south Korean students were classified as a risk group for smartphone addiction. However, our result was differed from that obtained by Alhassan et al., (2018) which show that 17% of student were probably addicted to smartphone. Likewise, Pearson and Hussain (2017) results revealed that 13.3% of the sample was classified as addicted to smartphones.

## The Relationship Between Smartphone Usage, Gratifications, and Addiction Among SQU Undergraduates

There is some controversy in research on the relation between Smartphone addiction, types of usage, and gratifications. However, the current study results found significant relationships between Smartphone addiction and three types of Smartphone usage behavior: Addictive, compulsive and habitual, whereas no significant differences were observed between Smartphone addiction and dependent, voluntary and mandatory usage behavior. In other words, it seems that there were significant relationships between Smartphone addiction and negative types of Smartphone usage behavior.

Moreover, the focus group discussions confirmed the above explanation and provided other evidence regarding the existence of relationships between Smartphone addiction and addictive, compulsive, and habitual usage behavior. First, Smartphone addiction usually leads to negative usage behavior. Second, use of Smartphone as habit turns into addiction with time. Third, addiction is usually linked to passive types of Smartphone usage behavior. Finally, Smartphone addicts lose control of their Smartphone use.

The focus group discussions confirmed previous explanations and added other explanations as follows:

1.    Addicts need to use their Smartphones for self-expression;
2.    Addicts try to satisfy their psychological needs through their Smartphones;
3.    Addicts seek compatibility with friends through use of social networking programs;
4.    Absence of the family's role in satisfying the addict's psychological needs, and
5.    Addicts feel shame and social phobia.

Furthermore, the present results identified significant relationships between all types of Smartphone usage behavior and Smartphone gratifications among SQU undergraduates, except in the cases of dependent behavior and self-identity and conforming. In addition, this study identified no relationship between habitual behavior and information seeking and entertainment.

According to UGT, a potential explanation for this result is that SQU undergraduates displayed the various types of Smartphone usage behavior to gratify the same motives, except for students who displayed dependent usage behavior, because they were not following self-identity and conforming motives. The explanation could be that self-identity and conforming motives appear more with independent usage behavior.

## The Differences in Smartphone Usage, Gratifications, and Addiction Among SQU Undergraduates Related to Academic Achievement

According to student's academic achievement, there were significant differences in Smartphone addiction, while there were no significant differences in terms of Smartphone usage and gratifications between SQU undergraduates. Specifically, the current study results revealed that low-academic achievement students were addicted to sending messages, while high-academic achievement students received more calls than other groups. This difference may be attributable to three main reasons. First, it seems that low-academic achievement students waste their precious time by keeping themselves busy with writing and sending useless messages in a way that leads to poor academic achievement. Secondly, this also may be because low-academic students lack time management skills and the ability to balance between personal and practical lives. As a result, they tend to use their Smartphone extendedly, which affects their academic performance. Thirdly, Smartphone addicts grow tolerant of Smartphone use and become very anxious and irritated when the phone is not available, despite the fact that their addiction may cause many academic problems, which leads to a decrease in their academic achievement level. According to Kim (2013), young Smartphone users with poor academic achievement usually receive less respect from surrounding people. Besides, poor academic achievement might be associated with low self-esteem and other behavioral problems, such as sleep disorders, aggression or depression. Those kinds of feelings and isolation would cause these users to go online in a search for feelings of belonging and self-satisfaction. Comparing to previous studies, this result consist with number of studies that correlated mobile phone use with a decrease in students' academic achievement (e.g., Bijari et al., 2013; Bröns et al., 2013; Casey, 2012; Cheon et al., 2012; Daffalla & Dimetry, 2014; Hong et al., 2012; Kibona & Mgaya, 2015; Kuznekoff & Titsworth, 2013; Muriithi & Muriithi, 2013; Sage & Burgio, 2018; Samaha & Hawi, 2016). However, this result disagree with Tavakolizadeh et al. (2014) which found no significant relation between excessive mobile phone use and academic achievement

## CONCLUSION

This study considered novel and noteworthy phenomena: Smartphone usage, gratifications and addiction. Little research has cast light on these issues, despite the increasing negative influence of this addiction on university students. One of the theoretical contributions of this study is that it explained the relationship between Smartphone usage, gratifications, and addiction among undergraduates in the light of UGT. It also provides comprehensive insights into the nature of the Smartphone addiction problem, the prevalence rate of different Smartphone addiction levels, the symptoms of Smartphone addiction and the effects of Smartphone addiction on academic achievement among SQU undergraduates. Moreover, it determines the motives behind Smartphone use and addiction, as well as the types of Smartphone usage behavior associated with Smartphone addiction among undergraduates. Consequently, the study will increase the awareness of students, parents and teachers about risk factors associated with Smartphone addiction. The findings of this study maybe useful to various parties, including university students, parents, educators, researchers and policy-makers.

Therefore, the first suggestion for further research would be to investigate the characteristics of SQU undergraduates with heavy levels of Smartphone addiction by employing more in-depth study. Another suggestion would be to investigate the phenomenon of Smartphone addiction within other research communities, such as school students and staff in companies and institutions.

Moreover, the present study results indicated that further research is needed on the role of academic stress in Smartphone addiction, as the relation between them still unknown. It could be that Smartphone use has an effect on academic stress among university students and vice versa. Likewise, there is a need for deeper investigation of the relationship between social gratifications and Smartphone addiction. Although this study found no significant relationship, previous ones reported that addicts use their Smartphones excessively for social gratification. This variation in results calls for a deeper look.

The research investigated the role of certain demographic factors in Smartphone usage, gratification, and addiction but it would also be interesting to enquire into Smartphone impacts on users' personality. Therefore, a further research is recommended to investigate the psychological characteristics of Smartphone addicts, such as social anxiety, loneliness, and depression.

Yet further research is also recommended to examine Smartphone effects on cognitive and cultural factors among undergraduates, and how these affect their perceptions. Indeed, other factors could also be examined that relate to questions of students' heath in general. In addition, another further research is recommended to explore the relationship between Smartphone use and car accidents.

Finally, the present study's results suggest that further research is needed to investigate other types of technology addiction among Omani students, such as SNs addiction especially WhatsApp addictions.

# REFERENCES

Ahn, H., Wijaya, M. E., & Esmero, B. C. (2014). A Systemic Smartphone Usage Pattern Analysis : Focusing on Smartphone Addiction Issue. *International Journal of Multimedia and Ubiquitous Engineering*, 9(6), 9–14. doi:10.14257/ijmue.2014.9.6.02

Alhassan, A. A., Alqadhib, E. M., Taha, N. W., Alahmari, R. A., Salam, M., & Almutairi, A. F. (2018). The relationship between addiction to smartphone usage and depression among adults: A cross sectional study. *BMC Psychiatry*, 18(1), 148. doi:10.118612888-018-1745-4 PMID:29801442

Balakrishnan, V., & Loo, H.-L. (2012). Mobile Phone and Short Message Service Appropriation, Usage and Behavioral Issues among University Students. *Journal of Social Sciences*, 8(3), 364–371. doi:10.3844/jssp.2012.364.371

Bianchi, A., & Phillips, J. G. (2005). Psychological predictors of problem mobile phone use. *Cyberpsychology & Behavior*, 8(1), 39–51. doi:10.1089/cpb.2005.8.39 PMID:15738692

Bijari, B., Javadinia, S. A., Erfanian, M., Abedini, M., & Abassi, A. (2013). The Impact of Virtual Social Networks on Students' Academic Achievement in Birjand University of Medical Sciences in East Iran. *Procedia: Social and Behavioral Sciences*, 83, 103–106. doi:10.1016/j.sbspro.2013.06.020

Bolle, C. (2014). *Who is a Smartphone addict? The impact of personal factors and type of usage on Smartphone addiction in a Dutch population* (Unpublished Master's Thesis). University of Twente, Enschede, The Netherlands.

Bröns, P., Greifeneder, E., & Støvring, S. (2013). How Facebook Promotes Students' Academic Life. *Zeitschrift für Bibliothekskultur, 3*, 116-126.

Casey, B. M. (2012). *Linking Psychological Attributes to Smartphone Addiction, Face-to-Face Communication, Present Absence and Social Capital* (Unpublished Master's thesis). The Chinese University of Hong Kong, Hong Kong, China.

Cha, S. S., & Seo, B. K. (2018). Smartphone use and smartphone addiction in middle school students in Korea: Prevalence, social networking service, and game use. *Health Psychology Open, 5*(1).

Cheon, J., Lee, S., Crooks, S. M., & Song, J. (2012). An investigation of mobile learning readiness in higher education based on the theory of planned behavior. *Computers & Education*, 59(3), 1054–1064. doi:10.1016/j.compedu.2012.04.015

Chóliz, M. (2012). Mobile-phone addiction in adolescence: The Test of Mobile Phone Dependence (TMD). *Progress in Health Sciences, 2*(1), 33–44.

Daffalla, A., & Dimetry, D. A. (2014). The Impact of Facebook and Others Social Networks Usage on Academic Performance and Social Life among Medical Students at Khartoum. *International Journal of Scientific & Technology Research*, 3(5), 3–8.

Demirci, K., Orhan, H., Demirdas, A., Akpinar, A., & Sert, H. (2014). Validity and reliability of the Turkish Version of the Smartphone Addiction Scale in a younger population. *Bulletin of Clinical Psychopharmacology*, 24(3), 226–235. doi:10.5455/bcp.20140710040824

Fukuda, K., Asai, H., & Nagami, K. (2015, October). Tracking the evolution and diversity in network usage of smartphones. In *Proceedings of the 2015 Internet Measurement Conference* (pp. 253-266). ACM. doi:10.1145/2815675.2815697

Fullwood, C., Quinn, S., Kaye, L. K., & Redding, C. (2017). My virtual friend: A qualitative analysis of the attitudes and experiences of Smartphone users: Implications for Smartphone attachment. *Computers in Human Behavior*, *75*, 347–355. doi:10.1016/j.chb.2017.05.029

Gökçearslan, Ş., Mumcu, F. K., Haşlaman, T., & Çevik, Y. D. (2016). Modelling smartphone addiction: The role of smartphone usage, self-regulation, general self-efficacy and cyberloafing in university students. *Computers in Human Behavior*, *63*, 639–649. doi:10.1016/j.chb.2016.05.091

Grellhesl, M., & Punyanunt-Carter, N. M. (2012). Using the uses and gratifications theory to understand gratifications sought through text messaging practices of male and female undergraduates. *Computers in Human Behavior*, *28*(6), 2175–2181. doi:10.1016/j.chb.2012.06.024

Head, M., & Ziolkowski, N. (2012). Understanding student attitudes of mobile phone features: Rethinking adoption through conjoint, cluster and SEM analyses. *Computers in Human Behavior*, *28*(6), 2331–2339. doi:10.1016/j.chb.2012.07.003

Henze, N., Rukzio, E., & Boll, S. (2012, May). Observational and experimental investigation of typing behaviour using virtual keyboards for mobile devices. In *2012 ACM annual conference on Human Factors in Computing Systems (CHI '12)* (pp. 2659-2668). New York: ACM. 10.1145/2207676.2208658

Hong, F.-Y., Chiu, S.-I., & Huang, D.-H. (2012). A model of the relationship between psychological characteristics, mobile phone addiction and use of mobile phones by Taiwanese university female students. *Computers in Human Behavior*, *28*(6), 2152–2159. doi:10.1016/j.chb.2012.06.020

Hooper, V., & Zhou, Y. (2007, June). *Addictive, dependent, compulsive? A research of mobile phone usage*. Paper presented at the 20th Bled e-Conference e-Mergence: Merging and Emerging Technologies, Processes and Institutions, Bled, Slovenia.

Jeong, S. H., Kim, H., Yum, J. Y., & Hwang, Y. (2016). What type of content are smartphone users addicted to?: SNS vs. games. *Computers in Human Behavior*, *54*, 10–17. doi:10.1016/j.chb.2015.07.035

Kibona, L., & Mgaya, G. (2015). Smartphones' effects on academic performance of higher learning students. *Journal of Multidisciplinary Engineering Science and Technology*, *2*(4), 777–784.

Kim, T.-Y., & Shin, D.-H. (2013). The Usage and the Gratifications About Smartphone Models and Applications. *International Telecommunications Policy Review, 20*(4). Retrieved on March 2, 2013 from: SSRN: http://ssrn.com/abstract=2373428

Krajewska-Kułak, E., Kułak, W., Stryzhak, A., Szpakow, A., Prokopowicz, W., & Marcinkowski, J. T. (2012). Problematic mobile phone using among the Polish and Belarusian University students: A comparative research. *Progress in Health Sciences*, *2*(1), 45–50.

Kuznekoff, J., & Titsworth, S. (2013). The Impact of Mobile Phone Usage on Student Learning. *Communication Education*, *62*(3), 233–252. doi:10.1080/03634523.2013.767917

Kwon, M., Lee, J.-Y., Won, W.-Y., Park, J.-W., Min, J.-A., Hahn, C., ... Kim, D.-J. (2013). Development and Validation of a Smartphone Addiction Scale (SAS). *PLoS One, 8*(2), 1–7. doi:10.1371/journal.pone.0056936 PMID:23468893

Lecturer, J. Y., Dominic, G., & Lecturer, E. (2014). The Impact of WhatsApp Messenger Usage on Students Performance in Tertiary Institutions in Ghana. *Journal of Education and Practice, 5*(6), 157–164.

Lee, Y., Chang, C., Lin, Y., & Cheng, Z. (2014). The dark side of Smartphone usage: Psychological traits, compulsive behavior and techno stress. *Computers in Human Behavior, 31*, 373–383. doi:10.1016/j.chb.2013.10.047

Lin, Y-H., Chang L-R., Lee, Y-H., Tseng, H-W., Kuo, T. B. J., & Chen, S-H. (2014). Development and Validation of the Smartphone Addiction Inventory (SPAI). *PLOS One, 9*(6), 1-5. doi:10.1371/journal.pone.0098312

Muriithi, M. K., & Muriithi, I. W. (2013). Student's motives for utilizing social networking sites in private universities in Dar Es salaam, Tanzania. *Academic Research International, 4*(4), 74–83.

Park, N. (2014). Nature of Youth Smartphone Addiction in Korea Diverse Dimensions of Smartphone Use and Individual Traits. *Journal of communication research, 51*(1), 100–132.

Park, N., Kim, Y., Young, H., & Shim, H. (2013). Factors influencing Smartphone use and dependency in South Korea. *Computers in Human Behavior, 29*(4), 1763–1770. doi:10.1016/j.chb.2013.02.008

Park, W. K. (2005). Mobile Phone Addiction: Mobile Communications. *Computer Supported Cooperative Work, 31*(3), 253–272.

Pearson, C., & Hussain, Z. (2017). Smartphone use, addiction, narcissism, and personality: A mixed methods investigation. In Gaming and Technology Addiction: Breakthroughs in Research and Practice (pp. 212-229). IGI Global. doi:10.4018/978-1-5225-0778-9.ch011

Reid, A. J., & Thomas, C. N. (2017). A Case Study in Smartphone Usage and Gratification in the Age of Narcissism. *International Journal of Technology and Human Interaction, 13*(2), 40–56. doi:10.4018/IJTHI.2017040103

Sage, C., & Burgio, E. (2018). Electromagnetic fields, pulsed radiofrequency radiation, and epigenetics: How wireless technologies may affect childhood development. *Child Development, 89*(1), 129–136. doi:10.1111/cdev.12824 PMID:28504324

Samaha, M., & Hawi, N. S. (2016). Relationships among smartphone addiction, stress, academic performance, and satisfaction with life. *Computers in Human Behavior, 57*, 321–325. doi:10.1016/j.chb.2015.12.045

Shambare, R., Rugimbana, R., & Zhowa, T. (2012). Are mobile phones the 21st century addiction? *African Journal of Business Management, 6*(2), 573–577.

Sheldon, P., & Bryant, K. (2016). Instagram: Motives for its use and relationship to narcissism and contextual age. *Computers in Human Behavior, 58*, 89–97. doi:10.1016/j.chb.2015.12.059

Song, I., Larose, R., Eastin, M. S., & Lin, C. A. (2004). Internet gratifications and internet addiction: On the uses and abuses of new media. *Cyberpsychology & Behavior*, 7(4), 384–394. doi:10.1089/cpb.2004.7.384 PMID:15331025

Statista.com. (2018). *Number of smartphone users worldwide from 2014 to 2020 (in billions)*. Retrieved from: https://www.statista.com/statistics/330695/number-of-smartphone-users-worldwide/

Szpakow, A., Stryzhak, A., & Prokopowicz, W. (2011). Evaluation of threat of mobile phone – addition among Belarusian University students. *Progress in Health Sciences*, 1(2), 96–101.

Takao, M., Takahashi, S., & Kitamura, M. (2009). Addictive personality and problematic mobile phone use. *Cyberpsychology & Behavior*, 12(5), 501–507. doi:10.1089/cpb.2009.0022 PMID:19817562

Tavakolizadeh, J., Atarodi, A., Ahmadpour, S., & Pourgheisar, A. (2014). The Prevalence of Excessive Mobile Phone Use and its Relation With Mental Health Status and Demographic Factors Among the Students of Gonabad University of Medical Sciences in 2011 – 2012. *Razavi International Journal of Medicine*, 2(1), 1–7. doi:10.5812/rijm.15527

timesofoman.com. (2018). *Oman's mobile phone subscriber base crosses 7 million*. Retrieved from: http://timesofoman.com/article/110822/Oman/Oman%27s-mobile-phone-subscriber-base-crosses-7-million

Valderrama, J. A. (2014). *Running head: Problematic Smartphone use scale development and validation of the problematic Smartphone use scale* (Unpublished PhD dissertation). Alliant International University, San Francisco, CA.

Walsh, S. P., White, K. M., & Young, R. M. (2007, July). Young and connected: Psychological influences of mobile phone use amongst Australian youth. In *Proceedings Mobile Media* (pp. 125-134), University of Sydney.

Yang, K. C. (2018). Understanding How Mexican and US Consumers Decide to Use Mobile Social Media: A Cross-National Qualitative Study. In Multi-Platform Advertising Strategies in the Global Marketplace (pp. 168-198). IGI Global.

## ENDNOTE

[1]    (P n)= Participant number in the focus group discussion.

# Chapter 7
# Mobile Phone Usage and Its Socio–Economic Impacts in Pakistan

**Sadia Jamil**
*University of Queensland, Australia*

## ABSTRACT

*Through examining use of mobile in Pakistan's Sindh province, the current chapter presents a unique and interesting case of the socio-economic impacts of mobile use on users' lifestyles. Although there exists an obvious divide between urban and rural areas in terms of impacts of mobile use, the case of Pakistan could serve as an alert to scholars that why mobile use remains limited in narrowing the gap between urban and rural areas against a backdrop of mobile being widely believed to be able to play a big role in narrowing the social and economic gap between urban and rural areas. The author of this chapter found that mobile use was also gender-biased in rural areas, resulting in a gap between males and females as far as social and economic impacts of mobile use on their lifestyles.*

## INTRODUCTION

The proliferation of mobile phone technologies has impacted the people's lives in terms of socio-economic progress across the globe (Madden & Savage, 2000; Sinha, 2005). It is widely recognized that mobile phones are not just a source of communication and networking, but their usage is seen as having a catalyst affect in prompting social and economic changes and development (Banks & Burge, 2004). International organizations such as World Trade Organization (WTO) and International Telecommunication Union (ITU) acknowledges the significant role of telecommunication services in the economic development of the countries and its social impacts on lives of common individuals especially in developing countries (Sridhar & Sridhar, 2007; Abraham, 2007).

There is a considerable penetration of Information and Communication Technology (ICT) infrastructure and an increase in mobile phone subscribers within the South Asian region during the last one decade.

DOI: 10.4018/978-1-5225-7885-7.ch007

"This has brought significant changes in the socio-economic environment of the SAARC (South Asian Association for Regional Cooperation) countries" (Chowdhury, 2015). Especially, Pakistan's growth in terms of telecommunication and mobile infrastructure has gone beyond all expectations because the country's mobile phone subscription has reached to 148 million by the end of January 2018 (Pakistan Telecommunication Authority, 2018)[1]. The Pakistani people, living either in rural or urban areas, are now using mobile phones for multiple purposes. The case of Pakistan is certainly one good example of how a technology can become an integral part of the public's life.

There are many country- or region-specific international studies that have analysed the impacts of mobile phones usage on socio-economic development in Africa (Goodman, 2005; Donnar, 2006; Frempong, Essegbey & Tetteh, 2007; Aker, 2008; Ilahiane & Sherry, 2012; Krone, Donnenberg & Nduru, 2016); in India (Kathuria, Uppal, & Mamta, 2009; Ansari & Pandey, 2011); in Sri Lanka (de Silva & Ratnadiwakara, 2008); in Bangladesh (Bayes, von Braun & Akhter, 1999; Cohen, 2001; Bairagi, Roy & Polin, 2011; Chowdhury, 2015); in India, Brazil, China, Korea, Lithuania and the United Kingdom (Kushchu, 2007).

The aforementioned prior studies have revealed that the use of mobile phone improves the performance of agriculture system and helps farmers to pursue their agricultural activities effectively in India (Ansari & Pandey, 2011) and in Africa (Aker, 2008). Some other studies in Africa (Donnar, 2006; Frempong, Essegbey & Tetteh, 2007; Ilahiane & Sherry, 2012) have shown that small business entrepreneurs are able to expand their business and manage commercial activities by using mobile phones. In Bangladesh, a number of studies have unpacked that people are able to attain micro credits using their mobile phones, which has helped in poverty reduction to quite some extent (Cohen, 2001; Bayes, von Braun & Akhter, 1999). From a sociological perspective, studies in Bangladesh (Bairagi, Roy & Polin, 2011) and in Sri Lanka (de Silva & Ratnadiwakara, 2008) have suggested that the use of mobile phone helps in an increased level of social networking, connectivity and results in stronger family relationships.

In Pakistan, there are some studies that have analysed mobile phone use among youth and in relation to economic and agricultural developments of the country (Malik, Chaudhry & Abbass, 2009; Shaukat & Shah, 2014; Kayani, 2015; Shahzad, Ahmed, Shahzad, Hussain & Riaz, 2015; Maqsood, 2015). These studies mainly focus on cities and rural areas of Punjab province and Islamabad. A very few academics have carried-out comparative analysis of the public's use of mobile phone and its effect on their socio-economic development in Pakistan's Sindh province (Chhachhar, Chen, & Jin, 2017). Therefore, drawing on the theory of technological determinism, this study aims to explore the public's experience of mobile phone usage in Pakistan's Sindh province, thereby to examine the socio-economic impacts of mobile phone use on people's lives and whether it has changed their lifestyle in urban and rural areas of the province.

Findings reveal that Sindh's urban residents use mobile phone for diverse reasons regardless of their age and gender and without any major problem except of a few, namely: the government's taxes on mobile phone subscription, the cost of mobile internet packages, mobile phone snatching, occasional network failure and the government's suspension of mobile phone services for security reasons during religious events (such as twelfth Rabi-ul-Awwal and tenth Moharram)[2]. The study highlights that male residents of rural Sindh areas use mobile phone for economic and social purposes more than rural women. Findings suggest that rural women have less and at times no access to mobile phone due to patriarchal culture and poverty. A majority of rural women are not literate and they do not have basic skills, which are necessary to operate mobile phone. Thus, to explain these findings in detail, this chapter firstly reviews past studies into the socio-economic impacts of mobile phone use and briefly reviews literature relating to the theory of technological determinism. The chapter goes on discussing the study's results and finally presents its conclusion.

## LITERATURE REVIEW

### Socio-Economic Impacts of Mobile Phone Use

The advancement in Information and Communication Technology (ICT) infrastructure has resulted in the proliferation of mobile phones in developing world, which has significant socio-economic implications. Past international studies have shown that mobile phone is much convenient and less costly way of communication as compared to the traditional networks of landline phone in different regions of the world. The smart features of mobile phone enable people to communicate and access to the internet anywhere. In this way, it serves as an alternate of desk-top computer and laptop for poor and to some extent illiterate people (Geser, 2005). The author thinks that instant and low-cost communication and internet access are very basic socio-economic benefits of mobile phone. Several previous studies have addressed much wider economic advantages of mobile phone use at macro-level. For example, Sridhar & Sridhar (2006) have examined the use of mobile phone in economic growth of developing countries. Their study reveals that the use of mobile phone contributes to increase the annual GDP level of many developing countries.

Furthermore, there are a number of prior micro-level studies that have highlighted how the use of mobile phone support in improving agricultural system and practices (Aker, 2008; Islam & Gronlund, 2008; Ansari & Pandey, 2011; Shaukat & Shah, 2014; Maqsood, 2015; Krone, Donnenberg & Nduru, 2016); in generating self-employment (Baro & Endouware, 2013); in providing market information (Jensen, 2006; Shimamoto, Yamada & Gummert, 2015); in facilitating fiscal transactions, remittances and mobile financial services[3] (Talbot, 2008; The Boston Consulting Group, 2011); in attaining micro-credits (Cohen, 2001) and in helping small business entrepreneurs to develop their business (Donner, 2006; Jagun, Heeks & Whalley, 2007; Frempong, Essegbey & Tetteh, 2007; Ilahiane & Sherry, 2012). Much of these studies emphasize that the use of mobile phone provides people access to information and connectivity, which is essential for economic activities and thus results in better economic outcomes.

Moving beyond the economic implications, the use of mobile phone has significant social impacts as well. Several past sociological studies have revealed that mobile phones foster social networking, cohesion and family relationships (Goodman, 2005; Kwaku Kyem & LeMaire, 2006; De Silva, Ratnadiwakara & Zainudeen, 2009; Bairagi, Roy & Polin, 2011). Similarly, one prior study in Taiwan has suggested that using "mobile phone strengthen users' family bonds, expands their psychological neighbourhoods, and facilitates symbolic proximity to the people they call by streamlining communication" (Wei & Lo, 2006, p. 53). The social implications of mobile phone use for rural residents are more noteworthy. In rural areas, where there is no infrastructure of landline phones, the use of mobile phone enables people to communicate easily with their family members and friends in urban areas and overseas on a comparatively low or reasonable cost. Nevertheless, Baro & Endouware (2013), in their case study of Niger Delta region of Nigeria, highlight the potential problems associated with the mobile phone usage in rural areas such as "network failure, non-availability of recharge cards, unreliable or complete absence of power supply to charge batteries, high charges by Network Service Providers, stealing of mobile phones, and unskilled persons repairing phones in rural areas". The author opines that these challenges are not specific to the case of Nigeria and rural population may face these obstacles in many other developing countries.

Some previous sociological inquiries into mobile phone usage have specifically focused on the women's use of mobile phone. For example, Rakow's study (1992) has suggested that the use of mobile phone lessens women's loneliness and boredom by assisting them to deal with confinement better at home. This gender aspect of mobile phone use is very relevant to mention here because women are desalinated from mainstream socio-economic life in many conservative and Islamic societies due to cultural and religious restrictions (such as women's use of veil) and they are deprived of face-to-face direct social networking and interaction. Hence, mobile phone use has an empowering and liberating effect on the social and economic lives of women in developing countries (like Pakistan), where they experience cultural and at times religious barriers and thus they have very less opportunities to share their perceptions and to work for improving their financial situation. In this regard, Lee (2009) suggests:

Features unique to mobile phones, such as portability, text messaging and data downloading, may also allow women to more easily participate in the labour force by giving them more timely and accurate market information and greater flexibility of communication. In addition, unlike other ICT devices, mobile phones do not require literacy or sophisticated skills that many women lack. Direct access to phones may empower women by increasing opportunities for communication and obtaining information. (p.9)

The author partly agrees with Lee's argument because some studies have revealed that a lack of literacy and basic skills restrain the women's ability to use mobile devices for socio-economic purposes. For example, Maqsood (2015) in her recent study, highlights that many Pakistani rural women, who work in farms, are not able to use mobile phone because of a lack of awareness to use it and hence they are not able to contribute for the expansion of agriculture system in the country's Punjab province. When reviewing the case of Pakistan, there is no substantial literature that addresses the women's use of mobile phone in the country. However, there are studies that have analysed the pattern of mobile phone usage among university students (Shahzad, Ahmed, Shahzad, Hussain & Riaz, 2015) and socio-economic impacts of mobile phone usage (Abbass, 2009; Malik, Chaudhry & Abbass, 2009; The Boston Consulting Group, 2011; Shaukat & Shah, 2014; Kayani, 2015). All of these studies focus only on areas of Punjab province and Islamabad in Pakistan. Noticeably, very limited research is available on the economic use of mobile phone in rural areas of Pakistan's Sindh province. For example, Chhachhar, Chen & Jin (2017) have examined the level of efforts for advancing the usage of mobile phone among the farmers of rural Sindh province in order to improve their agricultural practices and economic outputs. However, their study does not offer any insights into the social benefits of mobile phone use in Sindh's urban and rural areas. Therefore, this study offers a more comprehensive analysis of the public's experiences of mobile phone usage in rural parts and urban cities of the province, thereby to assess its socio-economic impacts on their lives.[4] For this purpose, this study uses the theory of technological determinism to evaluate how and to what extent mobile technology reshapes the lifestyle of public in Pakistan's Sindh province. Thus, the next section articulates the theory of technological determinism and its application in this study.

## Theory of Technological Determinism

The theory of technological determinism (TD) primarily emphasizes the important effects of technology on human lives. The theory has "informed" many analyses of changes in socio-economic structures: "the transition from feudalism to capitalism, changing occupational and skill structure of the labour force in the 20th century, the emergence of post-industrialism in the post-World War II era, the subsequent emergence of the 'information society,' 'post-Fordism,' and globalization" (Adler, 2006, p. 1). Raymond Williams, in his conception of technological determinism, acknowledges the progressive elements of the theory and asserts that new technologies facilitate social change and thus they are seen as "potential drivers of progress" in any social and cultural setting (Miller, 2011, p.3).

Categorically, the theory has been conceptualized in two forms. Firstly, hard determinists see technology as shaping individuals' lives powerfully using only those social aspects that help in its expansion. Thus, hard determinists conceptualize technology as a dominant factor in shaping human lives and social change beyond other natural and social elements (Ellul, 1964). On the other hand, soft determinists suggest that technology helps to foster industrial growth and overall development, but it is not the 'only determining factor' in bringing change and individuals can make decisions for possible or desirable outcomes. Beyond the individuals' decisions, some critics of technological determinism assert that technology depends on socio-economic structure within which it is implemented or operated (Winner, 2004). The author also thinks that technology cannot be considered as a 'solo determinant' of change and progress, and its consequences are shaped by social, political, cultural and economic settings.

When analysing the case of mobile technology and its impacts on the public's life and their experiences, many factors may appear as crucial in shaping the implications of mobile technology in Pakistan. Thus, this study attempts to address whether mobile technology has changed the public's lifestyle in the country's Sindh province and whether its use is dependent on certain factors for inducing change in their everyday life.

## METHODOLOGY

The study uses mix-method sequential explanatory design, which involves the collection and analysis of quantitative survey data followed by the collection and analysis of qualitative data (i.e., in depth interviews). In this study, there are three reasons to use mix-methods. Firstly, it has helped to explain, interpret or contextualize survey findings in detail by incorporating results from in-depth interviews. Secondly, it has enabled the author to validate the findings of survey and in-depth interviews data by using the method of triangulation[5]. And finally it has helped to provide a comprehensive analysis of gathered data from quantitative and qualitative perspectives both.

In this study, firstly the quantitative method of survey has been used to gather the data. The survey includes 100 respondents from seven urban cities of Sindh province (including Karachi, Hyderabad, Sakkhar, Dadu, Nawabshah, Larkana and Badin) and 100 rural respondents who live in adjacent rural areas of these cities. To ensure the diversity of feedback, survey sample includes fifty male and fifty female respondents of age ranging between 25 to 65 years. The selection of respondents has been made using the 'purposive sampling'[6] so as to ensure the representation of people belonging to seven major cities of Sindh province and the adjacent rural areas of these cities.[7] The bi-lingual survey questionnaire (in English and Urdu languages) examines the public's experience of mobile phone usage across nine themes, namely: (i) communication, (ii) social networking, (iii) recreation and entertainment (audio-visual use, mobile games and apps etc), (iv) mobile learning, (v) information seeking, (vi) community and health services, (vii) travel, (viii) business and employment opportunities, (xi) security (access to police and law enforcement agencies). The survey data has been analysed thematically and by using 'relative frequency statistics'[8] in order to present the respondents' response in the form of percentage.

The study also incorporates eight interviews of representatives belonging to <u>franchise outlets</u> of four local mobile phone companies including Mobilink, U-fone, Telenor and Zong. The eight interviewees include representatives from sales and customer care sections at franchise outlets of aforementioned mobile companies in Karachi. The purpose of including the feedback of mobile companies' representative is to unpack their point-of-views regarding the proliferation of mobile phone infrastructure, the public's use of mobile phones and challenges related to its usage in Pakistan. To ensure the confidentiality, the names of interviewees have been replaced by alphabets (A-H).

## FINDINGS AND DISCUSSION

### The Public's Use of Mobile Phone in Rural Sindh Province and Its Socio-Economic Impacts on Their Lifestyle

Survey result, in this study, reveals that the public's experience of mobile phone use is less diverse among Sindh's rural population since 79 per cent of male rural respondents use it for communication and recreational purposes only. Findings suggest that 48 per cent male rural respondents (of age between 30 and 50 years) use mobile phone for business and employment purposes and for their utility bills' payment (through mobile Easypaisa Service)[9]. Moreover, 42 per cent male and only 2 per cent female rural respondents use mobile phone to seek information respectively. Interestingly, rural women's use of mobile phone is not very high and their mobile phone usage is mainly restricted to communication purpose. This study finds no public's use of mobile phone for learning, security, travel, community and health purposes in rural areas of Sindh province. Figure 1, below, explains the mobile phone usage among the residents of rural Sindh province (survey data).

Interviewees' feedback suggests that there is a considerable penetration of mobile phone infrastructure in rural Sindh, which has led to an increased level of awareness among rural masses for the usage of mobile phone for communication and other purposes. The participating representatives of mobile phone companies, in this study, view a moderate level of mobile-laden change in the lifestyle of Sindh's rural residents. For instance, a male customer care representative of Zong states:

*In Sindh's rural areas, people are now using mobile phones to a large extent like other provinces of Pakistan. Earlier, there were many rural and remote areas, where there was no infrastructure of landline*

*Figure 1. The mobile phone usage among the residents of rural Sindh province*

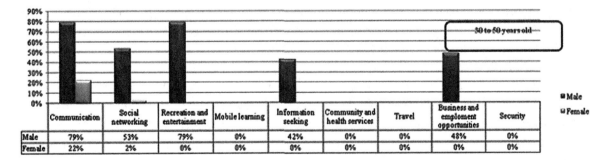

| | Communication | Social networking | Recreation and entertainment | Mobile learning | Information seeking | Community and health services | Travel | Business and employment opportunities | Security |
|---|---|---|---|---|---|---|---|---|---|
| Male | 79% | 53% | 79% | 0% | 42% | 0% | 0% | 48% | 0% |
| Female | 22% | 2% | 0% | 0% | 0% | 0% | 0% | 0% | 0% |

*phone, and people had to face a lot of problems for communication with their relatives living in urban areas or in foreign countries............Actually, mobile phone is a new and handy source of communication, entertainment and recreation especially for rural residents. There are hardly proper cinemas, play areas and parks in rural areas. In such environments, people can entertain themselves by using audio and video content, mobile games and social media. So, people's connectivity and recreational level have increased at a very affordable price because of the proliferation of mobile infrastructure. More importantly, the use of mobile phone has helped rural men to manage their financial matters in much better ways than before. Farmers are now able to get market information, weather updates and correspond with their buyers and sellers more easily. People, either in urban or rural areas, can search jobs online and they can do many other tasks such as online payments of their utility bills, e-banking and online business using their mobile phones. (Interviewee A)*

Chhachhar, Chen & Jin (2017), in their recent study, also confirm that most of the farmers in rural Sindh province have their own mobile phone and "more than half around 64 per cent of the respondents directly call the buyers and negotiate to sell their goods" (p.1). Agriculture is backbone of Pakistan's economy and it is the main source of people's income in most of the rural areas of the country. The use of mobile technology can definitely bring significant and positive changes in socio-economic lives of rural masses in Sindh province. However, interviewees' response, in this study, validates the survey data and highlights that mobile phone penetration has moderately changed the lifestyle of rural residents because they face certain day-to-day problems for frequent use of mobile phone. For instance, a male sales representative of Telenor unpacks the thought-provoking situation of Interior Sindh that is a bit similar to the case of Niger Delta of Nigeria (Baro & Endouware, 2013). He states:

*Power failure and electricity shortage are major problems in Sindh and rather in entire Pakistan. How people can charge the mobile phones' batteries without electricity? No one can use mobile phone effectively with six or eight or some time more than ten hours of electricity load shedding. In many small villages, there are no shops for repairing of mobile phone and its credit recharge and people have to travel either to nearby towns and cities for buying mobile credits, its charger and other accessories. Also, mobile phone snatching is very common in Interior Sindh, apart from other safety risks...............*
*It is very nice to read the growing statistics of mobile phone users in Pakistan, but either government or mobile phone companies are not paying attention to solve these nitty gritty issues that people confront for using their mobile phones especially in villages. (Interviewee F)*

Some other interviewees reveal that rural men use mobile phone more as compared to rural women in the province. They collectively suggest that factors of poverty, low literacy rate and patriarchal culture cannot be ignored for a woman's use of mobile phone in rural areas because still many women cannot have a mobile phone and only male members of family can use it. According to a male sales representative of Mobilink:

*Poverty is the foremost factor that impinges on the adoption of mobile phone. Not every person in a rural family can afford to have a mobile hand-set due to very low income and limited economic resources. Life is very miserable in interior Sindh.......................... In many rural areas of Pakistan, people do not have access to clean water, food, basic infrastructure, and health and education facilities. So, people's first priority is of course their food, shelter and clothing.............................The country's rural women are largely restricted due to patriarchal customs. In most of the rural areas (including Sindh province), women do not have mobile phone and only male members of the family use it. Although mobile phone is a good alternative for face-to-face communication and women can interact and socially engaged themselves by using it. However, a majority of women do not have mobile phone and they lack literacy and essential skills to operate it. This is an unfortunate situation because many rural women can develop contacts for managing their small businesses (such as handicrafts and homemade foods) and those who work in farms can improve their knowledge relating to agriculture. (Interviewee D)*

The factor of literacy is very important for the infusion of mobile technology in any context. In the case of Pakistan, the overall literacy rate is low in the country and rural masses lack basic skills that are necessary for mobile phone adoption. That is why; the rural population is actually deprived of full-blown socio-economic benefits of mobile phone technology, which can reshape their lives in many aspects.

Moreover, the author thinks that the liberating and empowering effects of mobile phone use on the Pakistani women's lives are largely ignored by researchers. Particularly, the country's rural women face cultural barriers and restrictions and they have usually a fewer opportunities to raise their concerns on pressing issues and to pursue their career. The accessibility of mobile phone, without gender-discrimination, can empower women by giving them better access to social services (such as mobile health and community services) and economic opportunities for jobs and business. The use of mobile phone can also encourage rural women to report domestic violence and threat of karo-kari[10] that is very common in Sindh province. In this regard, a female employee of Mobilink states:

*Rural women need to be skilled in order to use mobile phone for multiple purposes. Unlike men, who are comparatively more skilled, a majority of rural women are not educated and often cannot operate hand-sets due to a lack of skills, language and educational barriers. In Sindh province, Sindhi and Urdu are the common languages. How can they operate different mobile functions in English being illiterate? And at times, women are even unable to read Urdu language................................Domestic violence and Karo-kari are common in the province. Women can seek help from police immediately by using their mobile phone. Nevertheless, this never happens because either women do not have mobile phone or they cannot operate it skilfully.......................... In rural areas, a lot of women die every year during pregnancy due to a lack of awareness around health issues. Ironically, provincial government does not pay attention to promote the use of mobile health and mobile security service among rural women. (Interviewee E)*

The author believes that mobile technology can transform the way healthcare is provided to rural masses in Pakistan. Information and awareness regarding different disease can be disseminated to the public through text messages. Mobile help-lines can be established for those seeking advice from doctors. Not only healthcare, mobile help-lines can promote fast security, community and other services for the rural residents. These are some ways through which mobile technology can significantly reshape the public's lifestyle in rural Sindh province, provided if the problems of illiteracy, electricity failure/or shortage and cultural restrictions are effectively dealt by the government.

## The Public's Use of Mobile Phone in Urban Sindh Province and Its Socio-Economic Impacts on Their Lifestyle

In urban areas of Sindh province, male and female respondents (of age between 25 and 65 years) reveal a very high use of mobile phone for diverse purposes, namely: communication; recreation and entertainment; social networking; travel (for online bookings of flight, Uber and Careem and other taxi services); security; health and community services; mobile learning and information seeking; employment and business. These findings substantiate the interview data that reveals the growth of 3G/4G/LTE features-enabled mobile phones in major urban cities of Sindh province with a speedy rise in the public's use of mobile internet for communication; networking; information and employment seeking; access to travel, health, community and security services. Figure 2, below, illustrates the mobile phone usage among the residents of urban Sindh province (survey data).

Findings reveal that all surveyed urban residents in Sindh province (i.e., 100 per cent) use mobile phone for communication purpose. This is not surprising due to massive proliferation of mobile infrastructure in major urban cities of the province. Survey result indicates that 98 per cent and 89 per cent male and female urban residents use mobile phone for social networking and recreational reasons respectively. Interviewees' feedback, in this study, confirms these findings and suggests that urban dwellers in Sindh province substantially rely on mobile phone for communication, social networking and entertainment purposes.

This study highlights that urban residents with traditional mind-set consider social networking through mobile phone as an ill practice, which can influence community relationships. According to a male customer care representative of Zong, "many people think that the use of mobile phone alienates people socially and affects the conventional patterns of interaction, for instance face-to-face communication and public gatherings" (Interviewee I). Contrary to the aforementioned apprehensions, some studies have shown that the use of mobile phone makes people more socially engaged and help them in coping with isolation. For example, Hampton, Goulet, Ja Her & Rainie (2009), in their study, suggest that the use of mobile phone helps people to participate in online discussions and expands their social networks through a variety of social media platforms (p.3). The author thinks that the penetration of mobile phone has altered the pattern of social interaction as now it is more online instead of face-to-face communication. For instance, in this study, survey data shows that almost all urban respondents use their mobile phones for communication and social networking purposes, indicating a possibility of people's less physical or face-to-face engagement in social interaction, activities and gatherings (See Figure 2).

*Figure 2. The mobile phone usage among the residents of urban Sindh province*

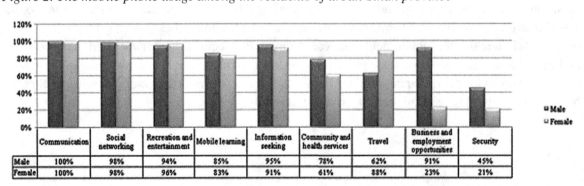

| | Communication | Social networking | Recreation and entertainment | Mobile learning | Information seeking | Community and health services | Travel | Business and employment opportunities | Security |
|---|---|---|---|---|---|---|---|---|---|
| Male | 100% | 98% | 94% | 85% | 95% | 78% | 62% | 91% | 45% |
| Female | 100% | 98% | 96% | 83% | 91% | 61% | 88% | 23% | 21% |

This implies that people could be physically isolated in many cases, but they are well-tied technologically regardless of time and space.

In addition, survey findings suggest that 95 per cent and 91 per cent male and female urban residents use mobile phones for information seeking, business and employment purposes respectively. These findings indicate that mobile technology impacts on socio-economic lives of urban residents positively. Interviewees' feedback, in this study, confirms these findings. For example, a female sales representative of U-fone states:

*Pakistan's mobile phone industry has witnessed a rapid growth during the past ten years. Most of the mobile companies have well-developed infrastructure in almost all major and small urban cities of Pakistan. Communication, information seeking, social networking and entertainment are a few key uses of mobile phones. People are also using it for online banking, generating employment opportunities and running their business ..............................Mobile learning among university students is becoming popular in Sindh's major universities. Students can now access their university's website and they can seek information through their mobile phones much easily and without using desk-top computers and laptops. Women, especially in Karachi, are using mobile phone for mobile taxi services (such as Uber and Careem) and community health services for their personal consultation and their children's vaccinations..................................There is also a growing trend of women's use of mobile phones for online shopping and tracking of their shopping orders through text messages. Interestingly, people frequently use mobile phone for food delivery at their homes especially in Karachi and Hyderabad, which are the biggest cities of Sindh...................Markedly, local political leaders have substantially used mobile phone for their election campaigns and for communicating with voters in Pakistan's 2018 election...............
......................................Hence, mobile technology is not only transforming people's life, but also it is reshaping the political trends and business model of Pakistan's fashion and food industry. (Interviewee B)*

This study unpacks that the penetration of mobile phone infrastructure has substantially changed the public's lifestyle in urban cities of Sindh province as compared to the rural residents of the province. These findings are contrary to some past local studies that have analysed the socio-economic impacts of mobile phone growth in Pakistan's Punjab province. For instance, Malik, Chaudhry & Abass (2009) suggest that people in Punjab's rural areas use mobile phone more than the urban dwellers of the province. Malik et al. (2009), in their study, reveal that rural residents of Punjab province use mobile phone to establish social contacts; to keep themselves informed about local markets, national developments and political situation. Malik, Chaudhry & Abass (2009), in their study, do not consider those factors that may affect the use of mobile phone among rural residents such as: affordability of mobile phone, electricity failure and shortage, mobile phone snatching, a lack of knowledge and basic skills to operate mobile phone. Also, their study does not explain the women's use of mobile phone that could offer gender-based insights into the impacts of mobile phone usage on rural and urban residents of Punjab province.

Furthermore, in addition to highlighting the benefits of mobile phone usage, some interviewees reveal the problems associated with its usage in Sindh urban areas. For example, according to a male customer care representative of Telenor:

*Mobile services are suspended in Karachi and other major urban cities of Sindh province on Eid-Milad-un-Nabi (12th Rabi-ul-Awwal), Ashura and Moharram's processions[11] for security reasons. Landline phone is the only way of communication during these religious dates. This is just one problem. Mobile snatching is very common in Karachi and other parts of the province. Many people then prefer to keep low cost mobile hand-sets instead of smart phones that are equipped with more significant features of internet access and other mobile applications. (Interviewee C)*

The author argues that suspension of mobile phone services cannot ensure security from any terrorist attack and rather it causes inconvenience to the public in several ways. Suspension of mobile phone services not only affects communication between people, but economic and other activities are halted as well. Therefore, mobile services should not be suspended on any occasion. Despite the aforementioned issues, one of the interviewees mentions some more important aspects that affect the public's use of mobile phone in Pakistan. According to a male sales representative of U-fone:

*Government's taxes on mobile phone and internet packages are making the affordability of mobile phone difficult especially for students and middle and lower-middle class people. Another issue is network failure that is mostly happened because of either any major technical fault or rough weather. By enlarge, the use of mobile phones has reshaped urban life not only in Sindh, but in other provinces of Pakistan too. (Interviewee H)*

Thus, this study has manifested the positive implications of mobile phone use among the urban residents of Sindh province more than the rural dwellers. This does not mean that rural people are not able to gain benefits through the proliferation of mobile phone infrastructure in the province, but certain factors (such as mobile snatching, poverty, illiteracy, language and cultural barriers) affect their effective use of mobile phone.

## CONCLUSION

This study reveals that the penetration of mobile phone infrastructure has largely altered the public's life style in urban cities of Sindh province. Interview and survey data suggest that urban residents use mobile phone for varied purposes regardless of their age, religion and gender and without any major obstacle except of a few such as: the government's taxes on mobile phone subscription, mobile snatching, network failure, the government's suspension of mobile phone services during religious events and the cost of mobile internet packages. In urban areas, the use of mobile phone has enabled people to keep themselves well-informed about political condition and election campaigns' update and they are socially well-connected and able to pursue their economic activities (such as business and search for employment) in much easier way than before. And thus, mobile technology can be seen as a potential source of social and economic change in urban areas of Pakistan.

Moreover, this study highlights that the public's experience of mobile phone usage is shaped by multiple factors of age, gender, literacy rate and cultural practices in rural areas of Sindh province. Rural men (of age between 30 and 50 years) use mobile phone more for economic activities as compared to rural men (of age between 55 and 65 years). A majority of rural women, who have participated in this study, are unable to use mobile phone because of a lack of skills, poverty, low literacy rate and patriarchal culture. Rural women, being illiterate, are unable to understand or seek information through their mobile phones. And they are not capable to access security, healthcare and community services that can help them to improve their lives in many ways. For instance, women's mortality rate during pregnancy, post-delivery infants' mortality rate, patients' disability and deaths caused due to other disease (such as cancer, stroke, polio, hepatitis B and C) are very high in rural areas of Sindh province. Unfortunately, the national healthcare system in Pakistan is still not operating well to deliver easy and affordable access to healthcare facilities to the public. Considering the operational and financial limitations, mobile health services can help in strengthening the public's healthcare awareness and facilities not only in rural Sindh, but also in other provinces of the country.

The author believes that the penetration of mobile phone network in rural areas is not only enough, but mobile technology is dependent on cultural factors as well to trigger broader socio-economic impacts of mobile phone usage (such as mobile community, health and other commercial services). And hence there is a pressing need for promoting literacy, social awareness and rural women's access to mobile technology in Pakistan.

Last but not least, the author recognizes the limitation of this study as it focuses only on one province of Pakistan. Also, it does not explore the negative social implications of mobile phone use in the country. Future research into the social effects of mobile phone can help in addressing many questions. For example, does mobile phone addiction among youth affects their attitudes towards studies and develops negative habits (like mobile teasing)? How public's dependence on mobile phone alter the traditional patterns of communication, family relations and social engagement? And has the public's use of landline phones reduced due to the proliferation of mobile phone infrastructure in Pakistan? These are just a few recommended research questions and a wide range of aspects can be explored particularly in relation to the scope of mobile learning, mobile travel, mobile healthcare and mobile finance services in the country.

## REFERENCES

Abraham, R. (2007). Mobile phones and economic development: Evidence from the ashing industry in India. *Information Technologies and International Development, 4*(1), 5–17. doi:10.1162/itid.2007.4.1.5

Aker, J. C. (2008). *Does digital divide or provide? The impact of mobile phones on grain markets in Niger (Working paper No. 154)*. Berkeley, CA: Department of Agriculture and Resource Economics, University of California. Retrieved at http://are.berkeley.edu/aker/cell.pdf

Ansari, M. A., & Pandey, N. (2011). *Assessing the potential and use of mobile phones by the farmers in Uttarakhand (India): A special project report*. Pantnagar, India: G.B. Pant University of Agriculture and Technology.

Bairagi, A., Roy, T., & Polin, A. (2011). *Socio-Economic Impacts of Mobile Phone in Rural Bangladesh: A case Study in Batiaghata Thana, Khulna District*. Retrieved at www.ijcit.org/ijcit_papers/vol2no1/IJCIT-110738.pdf

Banks, K., & Burge, R. (2004). *Mobile Phones: An Appropriate Tool for Conservation and Development*. Cambridge, UK: Fauna & Flora International.

Bayes, A., von Braun, J., & Akhter, R. (1999). *Village pay phones and poverty reduction: Insights from a Grameen Bank initiative in Bangladesh* (ZEF Discussion Papers on Development Policy No. 8). TeleCommons Development Group. Retrieved at http://www.telecommons.com/villagephone/ Bayes99.pdf

Boston Consulting Group. (2011). *The Socio-Economic Impact of Mobile Financial Services: Analysis of Pakistan, India, Bangladesh, Serbia and Malaysia*. Retrieved at https://www.telenor.com/wp-content/uploads/2012/03/The-Socio-Economic-Impact-of-Mobile-Financial-Services-BCG-Telenor-Group-2011.pdf

British Broadcasting Corporation. (2011). *What is Ashura?* Retrieved at www.bbc.com/news/world-middle-east-16047713

Chhachhar, R., Chen, C., & Jin, J. (2017). Performance and Efforts Regarding Usage of Mobile Phones among Farmers for Agricultural Knowledge. *Asian Social Science, 13*(8), 1–11. doi:10.5539/ass.v13n8p1

Chowdhury, M. (2015). Socio-economic penetration of mobile penetration in SAARC countries with special emphasis on Bangladesh. *Asian Business Review, 5*(2), 66–71. doi:10.18034/abr.v5i2.56

Cohen, N. (2001). *What Works: Grameen Telecom's Village Phones?* A Digital dividend Study by The World Resources Institute. Retrieved at http://www.digitaldividend.org/pdf/grameen.pdf

De Silva, H., & Ratnadiwakara, D. (2008). *Using ICT to reduce transaction costs in agriculture through better communication: A case-study from Sri Lanka*. Retrieved at http://lirneasia.net/ wp-content/uploads/2008/11/transactioncosts.pdf

De Silva, H., Ratnadiwakara, D., & Zainudeen, A. (2009). Social influence in mobile phone adoption: Evidence from the bottom of pyramid in emerging. *Asia*. doi:10.2139srn.1564091

Dean, S., & Illowsky, B. (2010). *Sampling and Data: Frequency, Relative Frequency and Cumulative Frequency*. Retrieved at https://www.saylor.org/site/wp-content/uploads/2011/06/MA121-1.1.3-3rd.pdf

Donnar, J. (2006). The use of mobile phones by microentrepreneurs in Kigali, Rwanda: Changes to social and business networks. *MIT Information Technologies and International Development, 3*(2), 3–19. doi:10.1162/itid.2007.3.2.3

Ellul, J. (1964). *The Technological Society*. New York: Vintage Books.

Flick, U. (2000). Triangulation in Qualitative Research. In U. E. Flick, E. V. Kardoff, & I. Steinke (Eds.), *A Companion to Qualitative Research* (pp. 178–183). London: Sage Publications.

Frempong, G., Essegbey, G. O., & Tetteh, E. O. (2007). *Survey on the use of mobile telephones for micro and small business development: The case of Ghana*. Retrieved at https://www.idrc.ca/en/project/survey-use-mobile-telephone-micro-and-small-business-development-ghana

Frost & Sullivan. (2006). *Social impact of mobile telephony in Latin America*. Retrieved at http://www.gsmlaa.org/ªles/content/0/94/Social%20Impact%20of%20Mobile%20Telephony%20in%20Latin%20America.pdf

Goodman, J. (2005). *Linking mobile phone ownership and use to social capital in rural South Africa and Tanzania* (Vodafone Policy Paper Series, 2, pp. 56-65). Retrieved at https://www.vodafone.com/content/dam/vodafone/about/public_policy/policy_papers/public_policy_series_2.pdf

Gray, P. S., Williamson, J. B., Karp, D. A., & Dalphin, J. R. (2007). *The research imagination: An Introduction to qualitative and quantitative methods.* New York, NY: Cambridge University Press. doi:10.1017/CBO9780511819391

Green, N. (2002). On the move: Technology, mobility, and the mediation of social time and space. *The Information Society*, *18*(4), 281–292. doi:10.1080/01972240290075129

Hampton, K., Goulet, L. S., Ja Her, E., & Rainie, L. (2009). *Social Isolation and New Technology.* Retrieved at http://www.pewinternet.org/2009/11/04/social-isolation-and-new-technology/

Ilahiane, H., & Sherry, J. W. (2012). The problematics of the "Bottom of the Pyramid" approach to international development: The case of micro-entrepreneurs' use of mobile phones in Morocco. *Information Technologies and International Development*, *8*(1), 13.

Islam, M. S., & Gronlund, A. (2007). Agriculture market information e-service in Bangladesh: A stakeholder-oriented case study. In M. A. Wimmer, H. J. Scholl, & A. Gronlund (Eds.), *Electronic Government* (pp. 167–178). Berlin: Springer-Verlag Berlin Heidelberg. doi:10.1007/978-3-540-74444-3_15

Jagun, A., Heeks, R., & Whalley, J. (2007). *Mobile telephony and developing country micro-enterprise: A Nigerian case study.* Retrieved at http://itidjournal.org/itid/article/view/310

Jensen, R. (2007). The digital provide: Information (technology), market performance, and welfare in the South Indian fisheries sector. *The Quarterly Journal of Economics*, *122*(3), 879–924. doi:10.1162/qjec.122.3.879

Kathuria, R., Uppal, M., & Mamta. (2009). *An econometric analysis of the impact of mobile* (Vodafone Policy Paper Series, 9, pp. 5-20). Retrieved at http://www.icrier.org/pdf/public_policy19jan09.pdf

Krone, M., Dannenberg, P., & Nduru, G. (2016). The use of modern information and communication technologies in smallholder agriculture: Examples from Kenya and Tanzania. *Information Development*, *32*(5), 1503–1512. doi:10.1177/0266666915611195

Kushchu, I. (2007). *Positive Contributions of Mobile Phones to Society.* Publication of the Mobile Government Consortium International UK. Retrieved at http://www.kiwanja.net/database/document/report_positive_impact.pdf

Kwaku Kyem, P. A., & LeMaire, P. K. (2006). *Transforming recent gains in the digital divide into digital opportunities: Africa and the boom in mobile phone subscription.* Retrieved at http://www.ejisdc.org/ojs2/index.php/ejisdc/ article/viewFile/343/189

Madden, G., & Savage, S. (2000). Telecommunications and economic growth. *International Journal of Social Economics*, *27*(7-10), 893–906. doi:10.1108/03068290010336397

Malik, S., Chaudhry, I., & Abbass, Q. (2009). Socio-Economic Impact of Cellular Phone Growth in Pakistan: An Empirical Analysis. Pakistan. *Journal of social Sciences*, *29*(1), 23–37.

Maqsood, M. (2015). *Use of mobile technology among rural women in Pakistan for agriculture extension information.* Retrieved at https://d.lib.msu.edu/.../USE_OF_MOBILE_TECHNOLOGY_AMONG_RURAL_WO

Miller, V. (2011). *Understanding Digital Culture*. London: Sage.

Murphie, A., & Potts, J. (2003). *Culture and Technology*. London: Palgrave. doi:10.1007/978-1-137-08938-0

Patel, A., & Gadit, A. (2008). Karo-kari: A form of honour killing in Pakistan. *Transcultural Psychiatry*, *45*(4), 683–694. doi:10.1177/1363461508100790 PMID:19091732

Rakow, L. F. (1992). *Gender on the line: Women, the telephone, and community life*. Urbana, IL: University of Illinois Press.

Shaukat, R., & Shah, I. (2014). Farmers Inclinations to Adoption of Mobile Phone: Agriculture Information and Trade System in Pakistan. *Journal of Economics and Social Studies*, *4*(2), 191–220. doi:10.14706/JECOSS11428

Shazad, M., Shazad, N., Ahmed, T., Hussain, A., & Riaz, F. (2015). Mobile phones addiction among university students: Evidence from twin cities of Pakistan. *Journal of Social Sciences*, *1*(11), 416–420.

Shimamoto, D., Yamada, H., & Gummert, M. (2015). Mobile phones and market information: Evidence from rural Cambodia. *Food Policy*, *57*, 135–141. doi:10.1016/j.foodpol.2015.10.005

Sridhar, K. S., & Sridhar, V. (2007). Telecommunications Infrastructure and Economic Growth: Evidence from Developing Countries. Applied Econometrics and International Development. *Euro-American Association of Economic Development*, *7*(2), 37–61.

Talbot, D. (2008). *Upwardly mobile*. Retrieved at https://www.technologyreview.com/s/411020/upwardly-mobile/

Wei, R., & Lo, V. H. (2006). Staying connected while on the move. *New Media & Society*, *8*(1), 53–72. doi:10.1177/1461444806059870

Winner, L. (2004). Technology as Forms of Life. In Readings in the Philosophy of Technology (pp. 103-113). Rowman & Littlefield Publishers, Inc.

## ENDNOTES

[1]   See Pakistan Telecommunication Authority's statistics at http://www.pta.gov.pk/en/telecom-indicators

[2]   Twelfth Rabi-ul-Awaal is the day of birth of Holy Prophet Muhammad (peace be upon him). "Tenth of Muharram is the first month of the Islamic lunar calendar. It is marked with mourning rituals and passion plays re-enacting the martyrdom" (British Broadcasting Corporation, 2011).

[3]   "Mobile financial services include two broad categories: *branchless banking via mobile phones* and *mobile banking* as a channel for financial services. With branchless banking, users can take advantage of services allowing them to make basic payments—utilities and other bills—and domestic and international remittances. These transactions become fast, easy, and cost-effective through MFS. Users can also participate in savings, credit, and insurance programs. Such services

drive the financial inclusion of the unbanked through m-wallet solutions, micro-loans, and micro health and crop-failure insurance" (The Boston Consulting Group, 2011, p.6).

4    The author has limitations for direct access to urban and rural residents in other provinces of Pakistan. Therefore, this study focuses only on Sindh province of the country.

5    'Triangulation' refers to the analysis of a research problem from at least two different perspectives or aspects (Flick, 2000).

6    The purposive sampling refers to the "selection of certain groups or individuals for their relevance to the issue being studied" (Gray, Williamson, Karp & Dalphin, 2007, p. 105).

7    "A relative frequency is the fraction of times an answer occurs. To find the relative frequencies, divide each frequency by the total number of students in the sample - in this case, 20. Relative frequencies can be written as fractions, percents or decimals" (Dean & Illowsky, 2010, p. 1).

8    In Pakistan, Easypaisa Bill Payment service helps people to pay their electricity, gas, telephone, water and internet bills through your Easypaisa mobile account.

9    Karo-Kari is a sort of honour killing, which is usually practiced in rural and tribal areas of Sindh, Pakistan. The murderous acts are mainly committed against women who are thought to have brought dishonour to their family by engaging in illegitimate pre-marital or extra-marital relations. In order to restore this honour, a male family member must kill the female in question or doubt (Patel & Gadit, 2008).

10   Eid Milad-un-Nabi is the day of birth of Holy Prophet Muhammad (peace be upon him). It is a public holiday in Pakistan. Sunni Muslims celebrate Milad-un-Nabi on 12th Rabi-ul-Awwal, which is the third month of the Islamic calendar.

11   "The day of Ashura is marked by Muslims as a whole, but for Shia Muslims it is a major religious commemoration of the martyrdom at Karbala of Hussein, a grandson of the Prophet Muhammad. It falls on the 10th of Muharram, the first month of the Islamic lunar calendar. It is marked with mourning rituals and passion plays re-enacting the martyrdom. Shia Muslim men and women dressed in black also parade through the streets slapping their chests and chanting" (British Broadcasting Corporation, 2011).

# Chapter 8

# Commercial Use of Mobile Social Media and Social Relationship:
## The Case of China

**Li Zhenhui**
*Communication University of China, China*

**Dai Sulei**
*Communication University of China, China*

## ABSTRACT

*China is well known for its wide and increasing commercial use of mobile social media for various purposes in different areas, ranging from online shopping to social networking. Such a popular commercial use was insightfully examined in relation to social relationship in the age of mobile internet, which enables people of either weak or strong connections to socialize anywhere anytime, leading to scenarios where mobile social media can be leveraged for profits. In what way can user experiences be guaranteed while platforms' value-added targets be achieved at the same time? In addressing that question, the authors of this chapter examined the commercial use of mobile social media in the context of complicated social networks. It is expected from the editor that further studies are to be carried out to comprehensively and comparatively examine the same topic in different countries or cultures.*

## INTORDUCTION

In the Internet age, especially in the age of mobile Internet, online social networks are making the connection, interaction, and relationship among people even more complicated. The social relationship of the younger generation is also being made more complicated, in which individual roles are being constantly transformed (Luo, 2017). In the past, mechanistic logic and reductionism were applied to analyze the problems of social networks from social sciences perspectives. In fact, the strength of weak ties (Granovetter, 1973) could play a more important theoretical role in explaining social networks in the

DOI: 10.4018/978-1-5225-7885-7.ch008

Internet age, which argued that the weak ties in social networks could satisfy some social needs, with those who have favorable resources play a key connecting role as bridges.

Before studying changes in social networks, technical factors should be first taken into consideration. Information technology is clearly the fundamental factor for social reshaping, and in an era where strength and efficiency surpass any source of power, the technical logic has begun to replace the functions of social regulation and cultural traditions in certain areas, which changes people's cognitive and action frameworks. To make people's connection more convenient and meet people's social needs, mobile social media are growing more mature with technology, and developers spare no efforts to strive to occupy every single market. Even though mobile users are so different in their perceptions and expectations of mobile social media, mobile social media have gradually been transformed a communication tool to a living necessity. This development has been driven by business and user traffic, regardless of platforms, channels or media, which in turn would attract more attention and capital. When mobile social media and commercial capital are becoming more maturely integrated, interacted, interconnected, or even interdependent, how mobile social media have been commercially leveraged in the context of the mobile Internet and in the presence of the strength of weak ties of social networking. That is an imperative topic to be fully investigated from mixed perspectives in relation to communication and economics studies.

## RESEARCH PURPOSE

Based on the view of strength of weak ties of social networking, the research on mobile social software, driven by Internet technology, is deconstructing the power structure and communication pattern. And when audiences have more power, they will have their own commercial value and they are likely to pay for channels and contents.

Mobile Internet, firstly, is to deconstruct the power structure of traditional society. Fei (2006) proposes 'Differential mode of association' of agricultural society and he thought blood relationship is the basis of agricultural social relations where egoism occupies personal emotions and there is on obvious distinction between public and private, also, violent ruling, without democracy; Industrial society appears *'Group pattern'*, where nation controls rare resources and builds a new organizational framework with production materials, employment position and living space, so that it can eliminate the differential mode of association based on the blood relationship. Internet society has brought elimination of those power structures. The decentralization and fragmentation of state power, and the opening and connection of the Internet have changed the scarcity of resources, and the mobile Internet has made information sharing easier, and even the marginal cost of surplus social resources is close to zero. The essence of *''sharing'* makes the Internet burst out with greater energy. Sun (1993) argues that the basic unit of social control and resource allocation gradually loses the power to monopolize social resources and to control social relations. The society becomes a relatively independent source of resources and opportunities, and individual dependence on the state is significantly weakened (Sun, 1993).

Mobile Internet then deconstructs the dependence of traditional media on content and channels and the content is not the key point anymore and channels are not the only choice. In the internet system, Yu (2016) proposes a new developing direction which is 'Relationship Empowerment'. He argues that 'relationship empowerment gives the public right and ability to discuss and participate in public affairs, by stimulating individual value and relationship networking. Hence, the environment and pattern of social governance are undergoing unprecedented changes (Yu, 2016). This way of empowerment could also be seen as a paradigm for the reconstruction of the value of the media in the Internet age. The Six Degree Separation theory nowadays can even become to Three Degree. It is very important for every transformation of media form whether the vital hint which hided among everyone could be accurately controlled and stimulated or spread forms could be strengthened with activation and scene technology.

Another important reason for supporting the theory of strong-weak ties theory is the technology and means provided by the Internet, which increases the scenes of users. When new scenes appear, people can play different roles in a variety of scenes. For example, virtual reality technology can create a versatile field to make people enter a scene that connects many people's social connections and feelings to achieve the immersive experience. Once these technologies are widely used, users can generate emotional experiences, and personal emotions are fully driven. This is the best mobilization and utilization of relational resources and emotional resources. The emergence of multiple situations is a key point which can become a strong relationship between social media and audience.

Information technology surpasses the supreme power source in efficiency and intensity, showing its own logic, and even in some areas has been or has replaced the functions of social regulation and cultural traditions, reshaping people's cognitive, behavior, and perception frameworks. Therefore, the social relationship is being reshaped, and the allocation system of social resource is transformed. Mobile social applications are clearly closely related to users. And user relationship is vital to social media.

## LITERATURE REVIEW

User-centric relationship network covers various application scenes. SoLoMo (Social, Local, Mobile) has deeply changed internet user's habit to obtain information. In internet age, mobile internet changes the whole relation structure from the bottom of society, and various mobile social media nearly have occupied all social networking nodes of people. Developers are also constantly occupying market gaps, and the business model of social applications is being promoted by capital. Now, the business model of mobile social media is mainly built from two ways. The one is that application itself can realize cash storage or income and another one is to attract flow to external applications for cash achieving. No matter which way, it is reusing the weak ties which are already gathered. From the strong-weak ties theory, interpreting the social relationship changes in the era of mobile internet and the meaning behind it, and paying attention to the commercial phenomena appearing in social applications, to support some viewpoints are the purpose of this chapter.

### Research on Strong-Weak Ties of Social Relationships

The relationship between the strengths and weaknesses of the audience is related to the changes in social relations. 'Free flow resource' and 'Free Space' both promote structural differentiation between societies. For the performance of the audience on social media, social structure differentiation is the underlying reason. Sun (1993) describes the manifestation of structural differentiation in his article. Firstly, he argues that society becomes a relatively independent source of resources and opportunities and the dependence of individuals on the state are evidently weakened. Then, relatively independent social power would be developed and formed. Lastly, the intermediary organization would appear, which is a civil organization between the state and family and these organizations do not target social services and profitability. Because of the 'Free flow resource' and 'Free space', the re-division and structuring between the state and society has brought far-reaching significance to all aspects of Chinese society. For individuals, changes in identity and status will naturally lead to more independence.

Yu (2016) mentions one concept called 'Relationship Empowerment' in his paper that the most prominent trait of the Internet society is that information technology has become the basic power shaping society. In internet society normal people are given by some power which transcends any age and their (internet celebrities and opinion leaders) value and influence hardly came from administration, capital or force (Yu, 2016). Internet, especially social media, not only gives individuals the speaking and executive power, but also meets individuals needs of social resources and materials in basic survival and longer-term development. Personal internal needs and value system are being re-arranged.

Based on the development of technology, the emergence of mobile terminals actually realizes the switching of scenes at any time. Peng (2015) argues that the mobile Internet includes three areas that is content, social and service, and mobile media has made a leap in the three directions of content media, relationship media and service media. She also mentions that in the analysis and application of mobile scenes, the current focus is on the location and significance of users here and now, but in the long run, the analysis and application of mobile scenes need to involve three stages (Peng, 2015). In addition to here and now, mobile scenes also need to extend to two different space-times of 'Before here and after here'. Our research on social software is to explore all aspects of the various situations, such as causes, conditions, and impacts. Once the audience have become dependent on the scene, it means that the fixed relationship is highlighted, or briefly, the strong relationship is established.

Actually, defining the transformation of strong-weak ties in social media can be described in terms of very vivid words, which is from 'Masked Internet' to 'Face-seeing internet'. Masked internet brings great freedom based on anonymity. Face-seeing internet is constantly updating social platform functions of social media. Social attributes lead to a strong tie with a limited scope, but this transformation needs many activate factors, such as multiple interactions which brought by social users' common concerns and providing real-time communication and advancement of sticky communication mobile platforms. This layer of conversion implementation would make the Internet more and more close to interpersonal communication, and of course, it would be the basis for mobile social media to expand business applications.

## Research on the Business Application of Social Mobile Media

The development of mobile technology brought about the rapid growth of mobile data. As of December 2017, the number of mobile Internet users in China reached 753 million (CNNIC, 2018). The report (2018) points out the proportion of Internet users using mobile terminals s increased from 95.1% in 2016 to 97.5%. Moreover, there is a change about mobile terminals. Smart devices represented by mobile phone have become the basis of 'Internet of Everything'. Smart lights for cars and home appliance have begun to enter personalized and intelligent application scenarios. This is also consistent with 'scenarios that bring strong connections'. Business application research for the mobile social industry has gradually become a hot topic. Data output is mainly concentrated in industry reports. Some scholars interpret the attributes of mobile social media from the perspective of communication. Some scholars analyze the innovation of social media from the field of marketing.

On the whole, the business model on mobile social media has few refining results. All of them are based on case studies. Cheng (2014) uses SWOT analysis to expound the commercialization of Tencent WeChat. He proposes that the WeChat business model components include proposition, network, maintenance and realization of value. Liu (2015) conducts a business form analysis on social networking sites, Weibo, and instant messaging. But it lacks systematic summaries and improved views. Zhao and Luo (2015) use the 'street-side network' as an example to obtain the optimization steps of social network application software through the research methods of questionnaires. It includes 'person', 'machine', 'material', 'law' and 'ring'. In five aspects, the product is iterated. But providing personalized services, securing user information and improving the user experience can provide a precedent for other social media. Li (2014) analyzes that WeChat's profit points include user payment, advertising revenue, value-added services, profit sharing, e-commerce and game revenue. It points out that WeChat's profit direction is mainly focused on value-added services, marketing platforms, games and e-commerce platforms. Zhao (2014) points out that the core business model of WeChat is 'platform business model'. But the content of the discussion still focuses on four types of modules: value-added, games, marketing and e-commerce.

In terms of review, the scholars' research focuses on the case study of the representation. The power structure and relationship network behind the mobile social platform are not mentioned. The expression of the viewpoint is not refined enough. The discussion on the profit model only stays at present, not highlighting the changes of mobile social which resulted from the changes of mobile internet. So what are the changes of advertising, value-added, game and e-commerce in future? Will there be new profitable revenue points? This chapter attempts to propose a new profit model.

## Research Methods and Research Problem Design

This chapter mainly adopts the investigation method with a systematic understanding of case, industry data and existing research results. It also conducts text analysis on the survey data. It shows the current development status and prospects of the deep logic and profit model of current mobile social media by reading the relevant documents. At the same time, in order to make the arguments more sufficient, the case analysis method is introduced to analyze the commercialization method of mobile social software.

## DATA ANALYSIS AND DISCUSSION

### Mobile Social Media From the Perspective of Social Relations

#### Technological Change Makes Mobile Internet Become the Dominant Technology

Mobile Internet is the technological change firstly. It breaks through the barriers of time and space from the technical level. It also brings the possibility of sending and receiving information anytime and anywhere. The individual has become a small tower. With the continuous development of technology, the speed of information dissemination and the expansion of information storage, the audience has gradually gained a new understanding of their own roles. Under the drive of subjective initiative, they have made adjustments in the era of mobile internet. The ''empowerment'' function of mobile internet technology even surpasses the Internet. Because the conditions for changing the scene at any time are realized, people become more active actors. The emergence of scenes means the deconstruction of traditional rights structure, as well as society redistribution of resources.

Based on technologies such as Internet of Things, big data, cloud computing and high-frequency information transmission, on one hand, mobile internet inherits the characteristics of PC terminal interaction and the advantage of crossing the information gap. On the other hand, it breaks down the barriers of time and space. It adds meaningful 'instant' communication to users. In addition, in the diversity of information, it is more creative. It can meet the transmission of various types of information such as text, voice, pictures, video, etc. With the help of high-frequency communication technologies such as 4G, 5G and Wi-Fi, users can even make up for the shortcomings of mass communication - 'lacking of interpersonal communication interaction' under the carrier of the Internet APP.

Especially on various smart phone applications, it covers almost every corner of the user's life, such as social, shopping, entertainment and knowledge sharing by satisfying the needs of users at all levels. All things that users can think of and hope to use are developed. It's better to say that the mobile Internet is connected by multiple small fields instead of a large of field. It is precisely because of the technology that the relationship between these small fields is complicated. In the case of social software, the crowds gathered on the same social software. It means that the recognition of the software interaction, whether it is interpersonal communication between people, or group communication within a small group, or the organization and dissemination of social software are gathered on this platform. Individuals become the center of platform maintenance. Multiple individuals are connected together to provide a continuous source of power for social software. It provides cumulative users and further profitability.

## Mobile Social Media's Role Changes

From a personal point of view, it is a change in personal roles. Internet pioneer Negroponte (1997) once predicted that digital survival naturally has the essence of 'empowerment', which would lead to positive social changes. And in the digital future, people will find new dignity. We can see that this prophecy has become a reality today. The Internet empowers ordinary people more than any era. It has dealt a blow to any source of power. Both the opinion leaders in Internet and people who influence the surrounding people pass on their information to others. When some creative and influential mobile Internet users use those apps with non-social attributes, they will make them socialized. It can be seen that under the level of technological innovation, the way of social interaction has changed significantly. Specifically, individuals have such performance in the mobile Internet.

Individuals become the direct productivity of social networks. In other words, the wisdom of individuals provides the raw materials for social networks. The operational logic of Web 2.0 makes everyone a relatively independent 'propagation base station (Peng, 2013). On the basis of data, individuals can maintain the passive acceptance of information in the past, as well as actively control the data production. Through the shaping of data, Internet people have gradually formed a way to survive and perform in this era. In the traditional sense, the 'background' has been moved to the 'front desk'. People are no longer constrained by whether they need to take into account the image, but play a role in the massive wave of information. Based on the 'performance' and 'relationship' of data, individuals are gradually exposed to form a 'data memory' about this era. As far as social networks are concerned, based on the use and satisfaction of individual contributions, attention resources, knowledge reserves and user data support for social platforms can even become the basic driving force for the existence of the platform. Zhihu, as a platform for user knowledge collection, the way of rationally treating problems has become a gathering place for a large number of outstanding intellectuals. It is precisely because this platform provides a relatively rational communication environment that it can stand out among many social platforms. The power of users brings together a steady stream of content for the platform.

Individuals assume the responsibility of spreading nodes. And the network of individual connections becomes a new place for public opinion. When the technical conditions satisfy the individual's ability to act as a small mobile tower, we can regard the individual as a node in the communication network. Of course, this node will be based on the individual's grasp of the information, the social status of the individual and the reason of the opinion leader. But the common point is that the entire process of 'coding, decoding, decoding' can be completed. Mobile social software is very easy to become a new 'public opinion generation and fermentation platform' when it acts as a life communication assistant. In the public opinion event, the individual is continuously coding as an information node, releasing the code at a relatively fast speed and then decoding. Sending and promoting the process of public opinion events has become a new place for public opinion fermentation.

The habits of personalized usage are obvious. Mobile social software is also tending to introduce personalized customization services. Different people have different software usage habits. People's information 'experience domain' is different in size. So the performance is different in the process of socialization. Users use mobile social media to obtain information and experience knowledge. They use it as a medium to interact with people in the society to generate information flow. On the one hand, it is to understand the surrounding environment. It can be called the 'environmental radar' function of Lasswell's three functions. On the other hand, it is better to integrate into society and complete the socialization process. With the advancement of Web3.0 technology, social software has also showed different needs in the process of socialization between people, paying more attention to personalized customization and creating private scenes.

## 'Mobile Community': Scenarioization of Mobile Social Media Mimicry Environment

The Internet platform combines the breadth of mass communication and the depth of interpersonal communication. The mobile social application restores the communication advantages of interpersonal communication. It includes the comprehensive meaning of transmitting information, the strong two-way, the high feedback, and the flexible method. It also broadens the types, time and space of interpersonal communication on a technical level. On the one hand, it is no longer limited by time and place. People who are not in the same situation can also interact. On the other hand, it developed multiple ways of paralleling sound and picture in the ways of communication. Then a private imaginable discourse space can be built.

'Relationship' plays a fundamental role in mobile social applications and it embodies in two aspects. First of all, strong connections are based on the users' trust. Mobile social software has numerous complex functions and has accumulated a large amount of users' information. Besides, it can record users' normal life and even analyze such behavioral characteristics like reading and consumption habits. In this case, if it is not built on a high degree of trust between users, it will not be easy to obtain data. In other words, the essence of the existence of mobile social networks is based on the users' trust in sharing information, including sharing their hobbies, interests, status, activities and locations online. Network topology varies, and mobile social network is a tool to connect nodes of users. Interpersonal trust manifests the interactive generation of value conception, attitude, mood, and even personal charm between the users. The word ''network trust'' has initiated many research by scholars when it was put forward in the 20th century. Lu (2003) has concluded network trust into three categories: the trust relationship between the users and the websites in the electronic commerce activities, the 'trust system' of technology, the trust relationship in the process of online interpersonal communication and the network trust among the users in the mobile social networking platform.

In addition, the relationship between the platform and users is also worth studying. Unlike WeChat as a social software, there are also many platforms maintaining a light connection with users and rarely being used unless when they are needed. This kind of trust is based on the theory of 'Uses and Gratifications', the theory shows that whether in mental or in action, once the platform has the characteristics which the users need and could bring a certain degree of satisfaction for users, it may obtain basic trust. What's more, if users feel the platform is coordinated with their own using habits, the reliability will be enhanced. This is also why the platforms will make a difference in the competition. To focus attention on the commercial realization of mobile social application, only if users are confident with security and privacy of the platform may the consuming behavior or the 'payment' action can be generated, which means users confident with the platform and willing to pay attention and money for support.

From the perspective of social scenes, all communication activities are carried out under specific scenes. Merowitz(2010) believes that it takes a long time for the traditional development of society to form a universal connotation of any scene. Electronic media also create a scene in the development and become a situational factor which will influence the human behaviors. Lippmann(2010) believed that the communication behavior of modern people is not in the real environment, but in the 'pseudo-environment' rendered by mass communication. In English, words like situation, context, settings and field are used to express the semantics of a situation, and in Chinese, there are also synonyms words like situation, background, and environment, which are not easy to distinguish.

From the perspective of space-time dimension, Goffman (2009) believes that a scene is 'a place that is limited to a certain extent by the perceptible boundary'. In a specific time and place, only people who face each other in the same three-dimensional space can perceive the same information. Merowitz (2010) and Goffman (2009) also stressed from the situational perspective, 'places create information system of live communication, and other channels create many other types of situation.' One of the most conspicuous signs is the birth of the television which created a new situation for people directly. People sitting in front of the TV to watch the content on the screen and their mood going up and down with the plot as if they are one part of the virtual world.

Some scholars believe that the elements of scene in the mobile era include space and environment, users' real-time state, their habits and social atmosphere based on the perspective of compound latitude of new era. These scenes become the entrance of data for mobile media.

Despite of the dimensions, the emergence of the scene is ultimately oriented to the deconstruction of the traditional power structure. In a macroscopic view, information technology disintegrates the operation mechanism of power structure in traditional society. First of all, the interaction between different scenes become more common when a new social scene has emerged, and the word 'decentralization' can best represent the disperse of traditional social rights. In the past, the privileged shapes their authority through clearly defined scenes which could sustain their uniqueness and mystique. Network platform becomes more open and full of information hybridity. Due to Internet technology, network platform has greatly filled the chasm between ordinary people and power center caused by information asymmetry, so that people from different social classes are placed in a unified scene. French philosopher Foucault (2012) proposed that information asymmetric is equivalent to a low-cost and efficient tool for social governance, just like the pyramid prison in ancient Rome: prisoners are kept in different cells and the jailer can monitor them at the top meanwhile the other prisoners could not see him. It is known as 'Panopticon' while Internet technology has generated a new social structure described as ''common view prison'' which is completely relative to 'Panoticon'. The later, as an 'onlooker' structure, concentrates on many-to-one model. In this way, information is relatively symmetrical. The information in this time is relatively symmetrical, managers will no longer have absolute control of the information resources and almost every manager are in the surveillance of ordinary citizens thus the role has exchanged between the two. Quantitatively, managers are less outnumbered, and moreover, the protections which new information technology has brought to the democracy have become more rampant, such as anonymous, hint, group-behavior-infection.

Microscopically, the mechanism of power granting has broken through the previous mode of "institutional appointment". Toffler (2006) believes that force, wealth, and knowledge constitute the triangle cornerstone of different power frames based on the criterion of the evolution of human society. The early human society need force to penalize but with the rise of the capital market, a part of power has changed from force to wealth, money can make a clear distinction between reward and penalty and more flexible; As the industrial civilization declines, knowledge became the dominant force and people from all social stratification has an opportunity to grasp, it could violate the violence if applied appropriately so that it has been called 'high-grade right'.

However, there is no one such as web celebrity, influential WeChat official accounts, Taobao celebrities could obtain the position from international agencies for it comes from their own relationship resources. The social status in today's market-oriented society will be relatively easy to obtain when every individual and every institution has channels to obtain attentions and financial support.

A deeper reason is that those groups such as marginal groups and isolated individuals who are easily neglected in traditional society, are also 'empowered' in the era of mobile Internet. Different from the superposition of order in the period of institutional empowerment and the pursuit of maximization of economic interests in the period of industrial capitalism, however, mobile Internet is different, it has changed the paradigm of individual empowerment and fundamentally changed the rules of the game of power. Those who have traditionally been kept out of the empowerment sphere have deservedly stepped into the center of the stage. Moreover, there is no upper or lower level of empowerment. For the web celebrities and their fans, they are all the consensual subjects participated and no one is forced. 'Mutual benefit, mutual respect and mutual identity' are necessary conditions for the existence of cooperation mechanism. After scaling up to the entire society, interpersonal cooperation will be enhanced exponentially instead of growing layer by layer in the pattern of hierarchy.

For social resources, it also breaks through the dominance of 'scarce resources' in the past and advocates 'sharing economy' now. The characteristics of internet such as open, interactive and complex have changed the endorsement of trust among various subjects. Nowadays the core of trust construction is the capacity of connecting, integrating and applying relational resources. As Tencent Charity for example, in August 2017, it launched a charity activity called 'one yuan purchasing a painting', it chose to advertise in the WeChat moments, and turning the paintings created by mental disorder children into electronic ones selling on 1 yuan. Each work has a simple introduction and summary of status of the children. People could scan the QR code to pay and sending the blessings, writing the messages meanwhile. This activity spread widely on WeChat platform. Different from the past form of donating money on the spot and large amount of the remittance, the form of 'Internet and commonweal' has made charity no longer just a process of mobilization, dissemination of information, donating money, but can widely obtain the effect of emotional resonance among the public. The conception of 'charity is nearby' and 'everyone could become a commonweal' will become more down to earth. In this micro public welfare platforms, common strangers are connected with each other due to consensus and trust. Together, each person's puny effect will confluence into huge power.

The change that the mobile internet brings to the social environment, is connecting everyone in every corner of the world, letting them abandon their suffering in the real life and be connected together. It gives confidence to everyone to integrate into the society, because common resources are shared and various ethnic groups, all kinds of demands are included within a framework to interact, so that people could be harmonious yet different.

## Changes in Communication Content and Forms of Mobile Social Media

After mobile network successfully enter the market, all kinds of applications have showed carrying out by smart phones. Among numerous of mobile applications, instant-messaging software is on the top of downloads ranks in mobile application market for replacing text function and integrating social circle. For now, this software is mature after development and the problems of adhesiveness of users has been resolved, the only problem left is how to break through the bottleneck problem of user experience. Besides, there is no doubt that for the emergence of new social software, the requirement of attracting attention is still necessary.

Though the content of mobile social media still follows the mode of user-generated content, it is developing towards the direction of professional, depth and precision. Meanwhile, many content producers that are widely concerned have emerged. With the era of live broadcasting and short video has come, more ordinary people become a popular star. Those content provides a new subculture environment for the new generation of young Internet users who admire the secular pleasure and their spiritual satisfaction are related to the short-time value, so that they could compete with the mainstream culture. Of course, with the further evolution of the Internet, those subcultures will eventually be integrated after the process of compromise. From the view of professional communicators, precision is the main feature for the users can get information exactly right what they need, and the analysis tools are needed to produce accurate delivery and production. In this way, under the support of multiple forms of contact, that fragmented information actually has a rule to follow, which is accorded with information personalized and customized.

In terms of communication form, due to the flexibility of mobile terminals, the form setting must conform to the reading characteristics of mobile intelligent devices, which is, flexible switching, short text, light reading, multi-text and so on. Compared with traditional social media, the function of browsing and searching of mobile social media are less important. However, in recent years, public accounts and other community functions have been opened and searching functions have been added to expand the functions of social media. It has changed the 'treelike' information flow mode of traditional media into the 'network' information flow mode of the new media era and increased the subjective initiative of the audience, finally strengthened the 'interaction' and 'feedback' effect. With the further development of interactive media, the future media environment will be more immersive.

## Theoretical Basis of Commercialization of Mobile Social Media

Mobile social media owns a huge number of users, it is also a way of communication based on social relations. This means that mobile social media holds huge social capital and has the ability to carry out business transformation. Massive users accumulated by strong relationships can provide a great marketing space. By using aggregation effect, information can be aggregated, and the essence of attention economy can be shown.

## Three Degrees of Separation- Strong and Weak Ties- Structural Hole- SoLoMo

The theory of 'Six Degrees of Separation' is a well-known social theory, also known as the 'Small World Theory'. The theory shows that weak relationships are ubiquitous in society and play a very powerful role. At the beginning, the theory is closely linked to the Internet, and then lead to emergence of 'social software' that support people to establish a close and mutual trust relationship. These are also what we are familiar today as the social software. Blog took the lead in popularity at the initial period, thanks to its equipped features of posting feelings, communicating with ease and personalized displaying. After a while of gradual evolution, there came these chatting software with purpose of making friends which were built on direct social relations. And more innovative social interaction cases were then springing up. Domestically, platforms such as Renren, Qzone, Douban.com and Zhihu were all role models, who knew well how to use games, knowledge discussions as ways to better communicate with their users. With development of mobile network, 'Six Degrees of Separation' was upgraded into 'Three Degrees of Separation'. That's where mobile social network was built on the basis of a 'strong relationship', as compared to the Internet, mobile web can bring people much closer to burst out more concentrated capabilities. Hence in an era of mobile social media, people often find that it is easier to get acquainted with strangers who share common interests or hobbies, or whom they are more willing to get to know. And the function of location sharing makes users feel much closer.

Based on Granovetter (1973) 's 'The Strength of Weak Ties' theory, American sociologist Burt(1995) proposed the theory of 'Structural Holes' in his study on what hinders interpersonal communications. A structural hole is understood as a gap between two individuals who have complementary sources to information. The theory suggests that some individuals in a social network hold certain positional advantages/disadvantages from how they are embedded in neighborhoods or other social structures. And if we see the network as a whole, it seems that there are gaps in the network structure (Dong & Li, 2011, p.40-43). Coleman (1988)'s theory of social capital has the greatest impact on the theory of structural holes. Only by participating in social groups and establishing group ties can people gain social capital. Heterogeneity is significant among weak ties and the feasibility of sharing scarce resources between the two sides is also greater. People who do not have a strong relationship with each other are not able to communicate smoothly because of objective or subjective barriers. The hypothesis of weak ties can be regarded as the foundation of structural holes theory. On the contrary, strong ties are shaking this foundation. It believes that if people are closely connected in mobile social networks, the holes will shrink or even disappear.

The concept of mobile social networking that based on strong ties is SoLoMo, which means Social, Local, Mobile. Since the concept was put forward by John Doerr, a partner of a well-known venture capital firm in February 2011, technology companies started to consider it as the development trend of future Internet marketing. In fact, this is indeed a keyword of mobile internet. In order to obtain real services in the virtual Internet, people can achieve their goals through shifting. LBS applications (positioning services) have been pushed into people's vision. Virtual networks can react to the real human society through this application technology. LBS applications (Location service) have been pushed into people's vision. Virtual networks can react to the reality through this technology. Mobile social media knows that it's not enough for social users to just record where they've been. What's more important is how different LBS applications can make, that is, 'geographic location information can provide what services'. In November 2010, Renren became the biggest LBS service provider in China after the launch of the 'Everyone Check-in' product based on mobile Renren's clients. Users are willing to 'check in' because the product integrates the real social relationship in Renren, which makes it possible to increase the opportunities for interaction between friends. Businessmen are always sensitive, and advertisers always pay attention to what users care about. In order to promote the new flavor of beverage, Tingyi (Cayman Islands) Holding Corp launched campaign through LBS services, 'Everyone Check-in' and 'Sina Wei Territory', which brought direct profit growth offline.

## Attention Economy and Community Shared Economy

'...In an information-rich world, the abundance of information means the lack of information consumption. Now, the object of information consumption is the attention of its recipients. The abundance of information leads to a lack of attention. Therefore, attention needs to be allocated effectively in an excessive amount of consumable information resources,' said Herbert Simon, the Nobel Prize Winner for Economics in 1978. Because attention is a psychological concept, therefore, a psychologist named Thorngate put forward 'attention economy'. Zhang (2009) mentions that the so-called attention economy is human interaction mode of production, processing, distribution, exchange and consumption of attention resources. In the Internet era, the attention economy is more obvious, the group of internet celebrities is the example. In the era of mobile social networking, information tends to be more fragmented, so it is hard to gather up the attention from audience. But at the same time, delivering information accurately becomes easier. Users can customize the information based on their preferences. Mobile social media has huge marketing potential in allocating users' attention.

Felson and Spaeth (1978) mentioned the concept of 'Collaborative Consumption'. But sharing economy did not become popular until recent years. Under the background of mobile social networking, the essence of sharing economy is based on the interpersonal relationship, and economic value is realized by detonating a certain scale of user groups. The greatest value of mobile internet lies in the network effect. Sharing economy is more like a community economy on mobile social software. People stop hiding their name in social network communities, mutual trust starts to increase, and then value got generated from it, together to form a self-operating, self-cycling economic system. In these community systems, any need or interest may eventually evolve into a business purpose. And both extensive networking resources and business development are worth noting and exploring. In a many-to-many relationship, information, resources and creativity are stimulated by the interaction, and then they will reproduce content and value. Therefore, social interaction will be self-operating and self-enhancing once it develops to a certain extent.

## Social Virtual Currency and Fan Economy

According to mUserTracker (2017), a monitoring product of iResearch, a provider of online audience measurement and consumer insights in China, the monthly number of mobile devices that contain social APPs reached nearly 590 million in May 2017. Among the mobile netizens, the social communication APP usage rate is the highest, namely 91.8%.(Figure 1) There are many ways to use the interpersonal relationship in the mobile social era for marketing. Based on some common understandings, social platforms produce many virtual gifts to replace money for fans to interact with their idols, which is essentially word-of-mouth communication. From the Stealing Vegetables game in the Qzone.com to giving out gifts during live-streaming, those actions are all aimed at increasing the activeness of fans through using social virtual currencies on mobile social platforms. From a certain point of view, it is the

*Figure 1. China mobile social app monthly independent device number trend map during June 2016-May 2017*

Note: China's social network includes independent network communities like Baidu's social product Baidu Tieba, social products of portal websites like Sina Weibo and Qzone, and excludes instant messaging APPs.

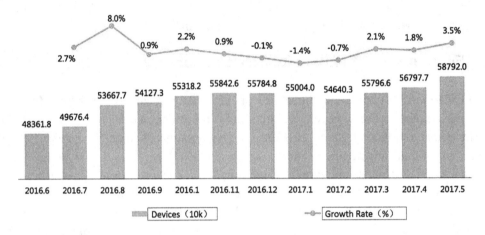

monetization of fans, apart from letting fans to consume directly, it also involves the deep interaction between individuals and brands. There is an important feature of social currency, which is the ability to arouse users' emotional resonance by presenting interesting content, and in turn stimulating their sharing behaviors. That is a typical communication process of transforming the influence from thoughts to actions. With the development of live-streaming, the scale of users has increased dramatically, and social dividend has increased. Under the external push of technology support and capital boost, the scale of live-streaming industry keeps expanding.

*Figure 2. The utilization rate of China netizens' mobile application in May 2017*

Source: mUserTracker 2016.11. Based on data of 4 million mobile phones and tablet mobile device software on a daily basis, as well as communication data from over 100 million mobile devices plus joint computing research.

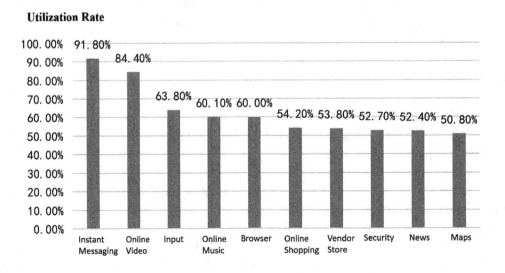

Mobile social networking offers the opportunities for people to have the decision-making power, to release them from the shackles of society. Social virtual currency becomes the carrier of time consumption in social networking, it also carries emotions and thoughts. It is a common asset held by consumers and enterprises, which helps to enhance user's recognition and loyalty to the brand, and also helps to realize the rapid marketing promotion of the brand in a short time, that is, word of mouth.

## Ways to Commercialize Mobile Social Media: A Case Study of Communication Social Media

Mobile social media is a combination of mobile internet, smart terminal and media service. CIC(2015) introduced the different categories that mobile social media applications fall into, including Weibo/blog, dating networking, entertainment social networking, instant messaging, community social networking, anonymous social networking and workplace social networking. In the above categories, instant messaging applications can play a role comparable to mobile phone in people's daily life. With the largest cumulative user traffic, mature user experience and greater stickiness of users, the foundation for cash revenue is becoming increasingly solid. This is also the focus of this chapter.

## The Current Profit Model of Mobile Social Media

The commercial practice on mobile social media in recent years are mainly focus on internal cash flow, drainage to external APP and advertising.

## Turing Internal Value-Added Flow Into Cash

Mobile social media can usually bring a group of core users together. They are able to maintain daily activity and overlap the communication between users invisibly. Under the background of three-degree segmentation, it is less difficult to connect two strangers. For platforms, it has the basis to user operation and the ability to bring value-added products through internal flow.

The most important way of realizing internal flow is to use social virtual currency to complete the value-added closed-loop within the media. As mentioned in the previous article, virtual currency is sold by catering for people's psychological social needs. And with the development of social software algorithms, these virtual currencies are dressed up with aesthetics. From QQ space decoration in the early years to sending presents in live-streaming, they have become the media of user communication. Taking anonymous social media Momo as an example, Momo announced its unaudited financial statements for the third quarter of 2017. Quarterly growth in paid subscribers brought in total revenue of value-added services of $26.3 million, up 45% year-on-year, mainly concentrated on members' subscription revenue and virtual gift revenue. The products ordered by members include recording visitors, quiet viewing, voice self-introduction, group online reminder, exclusive membership logo and personalized information pages, etc.

Value-added service means mobile social media transforming their social attributes and user resources into real wealth. It also creates a more interesting social environment for users. In addition, there is also the form of liquidation which is developing downstream products within mobile social media.

Ding Talk is a working APP, the main functions are providing smart office phone, recording office attendance, etc. The registered users climbed after the APP becoming an internal office communica-

*Table 1. Momo 2017 First three quarters earnings*

| | Total net revenue ($ one hundred million) | Gross Revenue of Value-added Services ($ Ten thousand) | Mobile Game Revenue ($ Ten thousand) | Mobile Marketing Revenue ($ Ten thousand) |
|---|---|---|---|---|
| 2017 Q3 | 3.545 | 2630 | 800 | 1740 |
| 2017 Q2 | 3.122 | 2460 | 910 | 1900 |
| 2017 Q1 | 2.652 | 2290 | 1160 | 1790 |

tion platform. In order to highlight its workplace function, it started to focus the function of recording attendance. By accumulating users online and developing hardware products offline, the company used the platform to publicize and sell, and achieve its goal to revenue.

## Using Internal Flow to Attract External Flow

Flow has great potential for adding values. Social media will naturally attract capital attention on the basis of such huge traffic and share with other offline enterprises. On the one hand, it can achieve a win-win situation on using resource; on the other hand, it can add a new way to create cash flow. This is also a widely used way of making profits in mobile social media

## Social Game Mode: Rich Benefits, Promising Prospects

Adding the game section in social media to trigger multi-user participation has been commonly used in early non-mobile social media. It is represented by various small games in QQ space. Fast, interactive, interesting, simple and relational attributes make these social games unique and make people happy. Mobile social media also choose this way to enhance the user experience, and it takes more effort in game form and page design. After the 'small program' was launched, WeChat allowed the company to develop its own small program based on WeChat's huge user base. It was equipped in WeChat platform and directly entered the program page, which not only directly promoted its own brand and products, but also realized user value conduction.

'Jumping' is a very representative WeChat applet developed by WeChat team. When it was first launched on December 28, 2017, the WeChat team introduced a mandatory launch page and recommended it to the huge WeChat users. This little game with smart voices, simple operations, and challenging gameplay allows people to 'jump up' in an instant and make competition in the circle of friends. A continuous influence of 'jumping' game decryption and being able to send to the group to hold an invitational tournament can easily help people kill time. By setting up an exclusive game springboard for the brand, WeChat has also achieved the function of advertising, in order to obtain advertising fees, and gain more.

Lao Yue Gou is a social APP that uses games to attract users. In May 2017, Lao Yue Gou announced that it had 45 million registered users in China. The main ways to promote themselves are using joint live-streaming platforms and the fan communities. Inside the APP, user's behavior data are generated around the game. Through the settings of different game types, users can get a more authentic display of their self-state. For a game social platform, it will have a higher matching efficiency if it can present user portraits more realistic and comprehensive, compare with the settings that matching friends with strangers. In the process of mining user's value, Lao Yue Gou has developed value-added services such as providing accompanies, opening self-owned internet cafes and table games shops. All those actions form a closed-loop with online games and other value-added services. Lao Yue Gou's revenue reached 40 million yuan in 2017 (36Kr, 2017).

## 'Social Media + E-Commerce' Model

This is a reliable way to mature development and to commercialize at present, and transaction sharing can be a significant profit point for social media. First of all, for the e-commerce platforms such as Taobao, Tmall, and JD.com, cooperation with social media can introduce sufficient traffic into e-commerce. In addition, they can use the user data obtained from social media to enhance fan marketing and content marketing, so as to make the shopping mall recommendation more accurate and improve the shopping experience. JD-WeChat shopping, mobile QQ shopping and other social e-commerce provide businesses and consumers with another platform to directly establish a trust relationship, which is a new entry for consumers to mobile shopping. Secondly, for social media, e-commerce has a payment logistics system that is complementary to the social media O2O closed-loop and realizes commercialization by diverting traffic to the e-commerce platform.

In 2014, Sina weibo announced the launch of Alipay's payment tool, "Weibo payment" on its platform. It opened a receiving function to the users who is has enterprise qualification certification, this means that the platform where the fans gather is going to employ fan economy. After Sina Weibo announced its cooperation with Taobao in 2013, Weibo started to imported flow to Taobao and obtained transaction share. It has taken another step on the road of commercialization. By using marketing strategy with social elements and the word-of-mouth to spread the brand and to improve brand recognition, it changed the weak ties in the past into strong ties which have multi-direction links. It shows that the 'marriage' of e-commerce and social media can bring win-win relation.

## Secondary Sale Model

The secondary sale we know is to sell time and space from TV to advertisers, the essence is to sell the audience's attention to advertisers. When audience's attention shifts to internet platform, internet advertising has gradually replaced television advertising, and capital tends to be attracted to most people's attention. As audiences divert their attention to mobile social media, the profit-seeking nature of capital completes for the secondary sale. The products are still the audience's attention, and in the era of mobile internet, audience's attention is more focused on the content than in the era of television.

Social media has an advantage which cannot be found in the traditional content era. It can be well targeted at a particular type of consumers to improve the conversion rate of advertisements. The big data provide accurate information on age, gender, income, and interest, and individualized environment creates a sense of personal identity. The connection between the strong relationship and the weak relationship also invisibly constructs the user's social identity, thereby distinguishing the complete virtual anonymity of the internet era, so that advertisers are more targeted in the promotion to provide better services.

In 2016, the scale of social advertising in China was 23.96 billion yuan. China is also following the global trend of rapid growth of social networks, with the development of advertising technology, the original information stream advertisement, video advertisement, H5 advertisement and soft text marketing advertisement based on social media have pushed social advertisement to a high-speed development period.

## The Efficiency of Social Media's Cash Conversion Under Social Scene

Although mobile social networking has more flexibility and business opportunities, there are also shortcomings in various ways of cash conversion.

## Internal Cash Conversion is Easy to Reach Saturation Point

Each social media has its own characteristics and target users, which limits the main direction of one social media in the process of turning flow into cash. Both social and media attributes require social media to consider both the strength and weakness of the relationship between users. It relies on the relationship chain for business development, but also need to convey certain information to users, whether through pictures, videos, or text.

Being the one-to-one dating APP, Momo was initially turning flow into cash through ads. Now it put video social networking as its main business and most of its revenue comes from live-streaming services. According to Momo, in the third quarter of 2017, the proportion of live-streaming business increased from 80% in the first quarter to 85.36%. However, making money from live-streaming subscribers has entered into a stagnation period, which means that Momo has encountered the predicament of saturation. The solution is either to maintain the existing cash conversion way and enhance media content, or to continue to seek the next business opportunity to develop new business.

## The Limited Ability of Turning Flow Into Cash

However, consumers are difficult to fully adjust to. For different types of advertisements, the level of audience attraction varies. Therefore, advertising has become a relatively mature one in all commercialization channels, but it cannot monopolize all channels. The advertisers will use the benefit as the standard, and the audience will also pay attention to some advertisements.

WeChat's 'Moments' function is a form of precise delivery of advertisement based on big data analysis. However, from the perspective of existing WeChat advertising technology, it is impossible to achieve the target of investment, let alone precise marketing. In the context of WeChat 'Moments', information tends to be redundant and complicated. People rarely discuss topics in income, consumption, and living standards in 'moments'. Even if so, the revealed information could be misleading, such as the act of showing off wealth, and the desire to seek weird ideas. The strategy of topic marketing in social media is nothing more than a notion. The hot topics can always attract people's attention and have the advantage of viral spreading. But there are also problems, such as poor controllability, fast propagation speed, and rapid update of public opinion trends. There are not many brands that can accurately predict public opinion. Hence, this kind of topic marketing has its limitations and faces difficulties.

## The Strong Ability to Attract External Flow and Turning It Into Cash

Shareaholic (2015), a foreign content marketing platform published a report announcing that social media has become the biggest source of recommendation traffic for websites, targeting the huge traffic that Facebook and Twitter have brought into blogs.

Taking shopping as an example, although social media has constructed social and shopping scenarios, it would be friction if the entrance is not packaged and publicized to cultivate users' habit of using scenario entrance. It attracts users by using discounts. Meanwhile, it strengthens users' memory of scene entrance by triggering scene memory, such as sending short messages and APP messages.

## The Social Game Model Works Well

For the user experience in the social context of strong-weak ties, scene is the most emotional alternative to enable users to have a stronger immersive experience. Games can compensate for simple social media attributes and provide users with a perfect scene. By providing tips and plots to stimulate users' willingness to consume in the game, using interesting social functions to improve users' activities, game developers, advertisers and online payment systems can all benefit from the industry chain.

"Face to Lite" is a typical APP that integrates games into mobile strangers' social networking. After users' registration, there will be a various types of dating game. These functions can create a romantic scene, which can offer opportunities to chat with strangers at any time and can also transfer weak ties between strangers into a strong connection.

## CONCLUSION

To analyze the strong and weak relationship in the context of a mobile social era is aimed to solve social relationship development behind the technical level. It doesn't matter if it's a weak connection among strangers created by network attributes, or it's a strong connection built upon social communities and scenes. Both are based on people's instinct to socialize anywhere and anytime. This is also the basis for social media to develop into multiple branches and make profits. In what way can user experiences be guaranteed while platforms' value-added targets be achieved at the same time? This chapter attempts to summarize commercial application rules of mobile social media for communication.

## Build a Scene to Maintain User Activity

Attracting users to retain is the primary condition for communication social media to survive as a functional software. The drawback of the tool will be remembered by users only when certain scenes are present. For example, all mobile phones have a call and text message function, which leads to the condition that the tool solves the user's just need problem. But it is still limited by the frequency of use, which requires further active users.

The reason why WeChat is regarded as a 'dependency' by people is its multi-faceted scene setting. First of all, group scenes are very easy to stimulate the gathering of people. The first group breaks through the limits of time and distance and regroups together. The collection of occasional groups can also be guaranteed. The community plays the role of the media, releasing information and exchanging information, it becomes a distribution center for information. Secondly, the circle of friends is a big social platform. Self-information sharing at any time and the function of comment, like, sharing makes it a place for information diffusion. It also has the saying that 'We can know the world in the circle of friends'. These functions are all creating an image scene for the user, allowing the user to experience the feeling of 'a lot of information and high social frequency' in WeChat. Construction of scenarios help to open up space for new products, and also means there is commercial potential.

## Games and E-Commerce Access to Increase Profit Points

Communication-based social media is targeting the communications space, but it can also increase profit margins by accessing games and e-commerce. The game can realize the scene substitution, and the scene of socialization is the part that every enterprise is trying to build. In the past, the relationship between enterprises and users was not equal. The majority of users were eager for getting quality content but it's not available, and there was almost no relationship between users. In this context, Tencent's QQ, which does not create content, has made great strides. The most successful part of this social software is established a perfect membership system that allows users to create themselves through interaction, when most companies focus on production content. The more active Zhihu (Chinese network Q&A community) and Douban (Chinese community website) in the Chinese market also follow this principle. QQ has developed the user's payment habits, and the game sector in Tencent has also brought real benefits to the company. In 2003, Tencent released a series of games prefixed with QQ, 'QQ Fantasy' and 'QQ Tang' became the childhood memories of the older generation of QQ. In 2006, Tencent officially entered the field of online games and embarked on the development of highways. Until now, it has accounted for half of China's game industry. Tencent Games, which carries the QQ scene, has given itself a well-performing advertisement and helped Tencent possess a stable revenue channel.

The traditional e-commerce platform relies on its own platform to carry users, and also uses traditional advertising methods in publicity and promotion. In May 2016, think tank Analysys and JD.com jointly released 'the China Mobile Social E-Commerce Development Special Research Report'. It showed that domestic mobile e-commerce transactions reached 2.07 trillion yuan in 2015. This is not only a quantitative accumulation, but also a qualitative change. Looking at the development of e-commerce platforms over the years, there is a place worthy of attention: the cooperative relationship with social media.

In 2014, JD.com teamed up with instant communication social media like QQ's mobile phone app and WeChat to tap into mobile e-commerce industry. Among all the social media, those in the communication category have the largest user base and are used most frequently as they fit better into mobile social communication's features. Therefore, they become the e-commerce platforms' first choice.

For e-commerce platforms, the first move to realize their purpose of gaining traffic is promotion. Social media has such advantages in concentrating dispersed traffic and improving users' engagement. It's the most reliable way to reach a large number of users in a short time through mobile social media. Most consumers will refer to other buyers' reviews to make purchases. But such relation between consumers themselves and other users belongs to the 'weak tie' category. Sharing information among friends based on social media platforms is able to enhance users' trust in products. Moreover, the opening of mobile social e-commerce platforms also improves overall operational efficiency of both sides. Take JD.com as an example, after joining hands with WeChat and QQ, these two mobile social applications have launched many public platforms in addition to instant messaging and social entertainment. It has also opened interfaces to third parties. All these have enhanced users' consumption experiences and provided richer user data for companies, making it easier for precision marketing.

## Invest in High Quality Advertising to Attract Attention

The significance of advertising is to attract attention. The advantage of mobile social media is that it can reach large-scale users. In recent years, social marketing has continued to develop, and advertising creativity combined with life is also easily accepted by the public, but the disadvantage of social media is that technology is not enough to achieve precision marketing.

Mobile social users have obvious likes and dislikes of advertising forms, which results in completely different recommendation effects of various advertisements. Those attracting the highest attention are video advertising, QR code advertising, feeds and APP recommendations. While those with the most selective attention are screen ads or pop-up advertising. All these stems from users' experiencing effects. Based on the theory of Usage and Satisfaction, users seek attention according to their own needs. Mobile social users pay more attention to product performances introduced by advertisements, while also have the desire to further understand high-quality audio-visual effects of advertisements and advertising discount information.

For mobile social platforms, high-quality advertising means having the effect of reaching users, but also stimulating users' desire to view and pay attention. The goal of all advertising is to motivate public action. The Moment advertisements in WeChat are collected through open appraisal to reach the attention information. The convergent advertisement method is close to the user's habits, but there are often cases of inaccurate promotion. A good social advertisement can instantly capture the user's eyes, trigger the desire to explore, allow users to think that they have a real need for such information or items from their own perspective. Or it pushes public join the purchase list in the future, and make people actively talk about brands with friends to form a share path, not simply conduct rude closure or rushing through.

## REFERENCES

Burt, R. S. (1995). *Structural Holes: The Social Structure of Competition*. Cambridge, MA: Harvard University Press.

Cheng, J. F. (2014). *Research on business models of mobile social media* (Master Thesis). Northwestern University.

CNNIC. (2018). *China Social Media Overview' released by CIC*. Available at: http://www.cac.gov.cn/2018zt/cnnic41/index.htm

Coleman, J. S. (1988). Supplement: Organizations and Institutions: Sociological and Economic Approaches to the Analysis of Social Structure. *American Journal of Sociology*, *94*, S95–S120. doi:10.1086/228943

Dong, X. M., & Li, F. Y. (2011). Research on rural private lending on structural hole theory. *South China Finance*, *8*, 40–43.

Fei, X. T. (2006). *Local China*. Shanghai People's Publishing House.

Felson, M., & Spaeth, J. L. (1978). Community Structure and Collaborative Consumption: A Routine Activity Approach. *The American Behavioral Scientist, 21*(4), 614–624. doi:10.1177/000276427802100411

Feng, L. (2004). Social scene: Psychological field of the communication subject. *China Communication Forum.*

Foucault, M. (2012). *Discipline and Punishment: the birth of prison.* Life, Reading and New Knowledge SanLian Bookstore Press.

Goffman, E. (2009). The presentation of self in everyday life. *Threepenny Review, 21*(116), 14–15.

Granovetter, M. S. (1973). The Strength of Weak Ties. *American Journal of Sociology, 78*(6), 1360–1380. doi:10.1086/225469

Harwit, E. (2017). WeChat:social and poitical development of China's dominant messaging app. *Chinese Journal of Communication, 10*(3), 312–327. doi:10.1080/17544750.2016.1213757

Lippmann, W. (2010). Public opinion and the politicians. *National Municipal Review, 15*(1), 5–8. doi:10.1002/ncr.4110150102

Liu, C.Z. (2015). Discussion on profit models of mobile social media. *News World*, (8),188-189.

Lu, X. H. (2003). *Network reliance: Challenges between virtual and reality.* Shanghai: Southeast University Press.

Luo, J. D. (2017). *Complex: Connections, Opportunities and Layouts in the Information Age.* CITIC Publishing Group Co., Ltd.

Meyrowitz, J. (2010). Shifting worlds of strangers: Medium theory and changes in "them" versus "us". *Sociological Inquiry, 67*(1), 59–71. doi:10.1111/j.1475-682X.1997.tb00429.x

Negroponte, N. (1997). *Being Digital.* Hainan Press. doi:10.1063/1.4822554

Peng, L. (2013). Evolution of ''Connection'': The Basic Clue of the Development of Internet. *Chinese Journal of Journalism & Communication*, (2), 6-19.

Peng, L. (2015). Scene: new elements of media in the mobile age. *Journalism Review*, (3), 21-27.

Sun, L. P. (1993). Free flow resources and free activity space --- China's social structure changes during its reform. *Probe*, (1): 64–68.

Sun, L.P.(1996). Relations, social network and social structure. *Sociological Studies*, (5), 20-30.

Yu, G. (2016). *Social currency- The road to business monetizing in the era of mobile social networking.* Posts and Telecom Press.

Yu, G.M. (2009). Media revolution: From panorama prison to shared-scene prison. *People's Tribune, 8*(1), 21.

Yu, G.M. (2016). Reconstruction of media influence under the paradigm of relationship empowerment. *News and Writing*, (7), 47-51.

Yu, G.M., & Ma, H. (2016). New power paradigm in digital era: Empowerment based on relation network in social media --- Social relation reorganization and power pattern dynamics. *Chinese Journal of Journalism & Communication,* (10), 6-27.

Yu, G.M., & Ma, H. (2016). Relationship empowerment: a new paradigm of social capital allocation --- The logical change of social governance under network reconstruction of social connection. *Editorial Friend*, (9), 5-8.

Yu, G.M., Zhang, C., Li, S., Bao, L.Y., & Zhang, S.N. (2015). The era of individual activation: Reconstruction of communication ecology under the logic of the Internet. *Modern Communication*, (5), 1-4.

Zhang, L. (2009). *The research of the western attention economy school.* China Social Sciences Press.

# Chapter 9
# Use of Mobile Apps and Creator–Audience Matchmaking:
## The Case of India

**Biplab Lohochoudhury**
*Visva-Bharati, India*

## ABSTRACT

*With a bird's-eye-view of the journey of exploring mobile use and its impacts in India, one of the biggest mobile markets in the world, the author of this chapter provides an analytical and insightful review of earlier studies on mobile communication development and utility-driven usage. Although country-specific, this chapter provides a very interesting, insightful, and invaluable model to the world, that is, the model of creator-audience matchmaker. As demonstrated by the two cases in this chapter, the creator-audience matchmaker model can be used as a heuristic tool for understanding mobile app development and success in India.*

## INTRODUCTION

India' tryst with the mobile communications system has been a little over twenty years. In the interim, the nation has been witness to changes galore from the point of view of both hardware and software technologies. India has emerged globally as one of the biggest markets for mobile phones. From a total wireless subscription base count of 346.89 million till December 2008, the figure now stands at a staggering 1166.90 million (as per the latest available TRAI Telecom Subscriptions Report) till August 2018. With introduction of the 4G technology and the entry of new service providers, the rates of both voice calling and data usage have taken a significant dip.

Number of internet users has increased fast in India from 2012 to 2015 piggybacking the mobile phone of various dispensations. Indian Cellular Association (ICA) claimed that in 2018 India has emerged as the second largest mobile phone producer after China. Production increased from 3 million units in 2014 to

DOI: 10.4018/978-1-5225-7885-7.ch009

11 million units in 2017, ICA claimed. Mobile handset price has gone down while number of functional features in phones increased making available several information and communication services. Data has become more affordable especially after entry of Jio brand in mobile telephony and data communication segments last year. The number of mobile phone users is on rise rapidly, and the market is expanding fast in Rural India. With 2/5$^{th}$ of population under 18 aspiring to have one phone which is becoming more affordable day by day due to home production in the country, mobile phone is a huge impact.

The 'Digital India' vision of the Indian government appears to have given a significant boost to the mobile technology. "The emphasis by the government on its [sic recte: its] flagship 'Digital India' program coupled with the upcoming NTP – 2018 which is expected to include cable TV networks and satellite communications including DTH for broadband penetration is expected to take broadband to the last mile." (Ernst & Young LLP & FICCI, 2018, p. 105). In 2008, 7 percent of Indian households could afford a smart phone, whereas in January 2018, it is 27 percent (approximately 66 million).

Thus, prima facie it seems that the Indian mobile sector is advancing at a fast pace and is likely to eclipse the other mediums in the user-consumer count race. While it must be conceded that the other mediums are also introducing various innovations in their respective technologies to stay alive in the game, perhaps the 'mobile' is the best-suited to adapt to the changing praxis of the new-age media-market mechanism which has itself undergone significant transformation in the digital age. This has come with emergence of mass self-communication (Castells 2011), a communication mode which has obliterated the centre-periphery relation between producer-manufacturer and consumer-citizen dual existence of audience. The reality of Prosumer (producer consumer role reversal) thrown open by New Media is challenging the existing notion of market. Mobile communication is in the forefront of this enormous change.

## THE PHENOMENON AND THE PROBLEM

Advanced mobile technology opens population to possibility of border-less information and entertainment causing gradual alteration in communication spectrum (LohaChoudhury, 2001, p. 194) of every community. When this observation is combined with the present Indian reality of increasing purchasing capacity of villagers enabled by cash supply from Mahatma Gandhi National Rural Employment Guarantee Act ensured 100 days lean season local work and other governmental schemes, as well as sharp fall in the price of mobile phone, the fast expansion of mobile communication in India is understandable. The mobile communication literacy being the easiest among all Information and Communication Technologies (ICT), even-spread of mobile communication creates opportunity for choice of consumption, exchange and income generation. 'From Home' work advantage, localization and networking are driving towards the mobile-first and fast-mobile business ecosystem whose effect on economy, social system, community dynamics and democracy is only perfunctorily studied from perspectives developed before the self-mass communication such as mobile communication emerged.

How all these enablers help evolving the phenomenon of mobile communication in a country of huge diversity like India needs multiple studies of different aspects of the phenomenon.

The present study is only a base study which informs about the mobile communication expansion drivers in the country. Without understanding this, research to inform policy and practice would remain perfunctory. To understand the same in Indian context, suitability of perspectives and methods used so far for mobile communication studies is examined.

## HOW THE STUDY IS DONE

For such a country-level study, logic of choosing some documents over others and putting some instances ignoring others must be supported by the perspective behind phenomenon. Therefore, this study needs to have a clear perspective to form research questions. For the same one has to either take help of existing theory, frameworks, constructs, or even literature existing elsewhere, or create a new construct in case of inadequacy of existing ones for the purpose.

Pearce (2013) suggested five theoretical approaches to guide mobile studies in developing countries whose inadequacy for our research purpose are discussed first. Our study does not fit 'Diffusion of Innovation' theory of how, why, and at what rate new ideas and technology spread through social systems (Rogers, 2003) for two reasons. First, it's not a study about changes in social system. Second, the scope of the study does not allow to observe behavioral and normative changes in human agency for usage of mobile phones.

Haddon (2003) suggested 'Domestication' theory which "concentrates on how individuals go through the process of discovering, purchasing, and integrating devices into their lives, and helps to account for how individuals judge others' use of the devices as well as the social consequences of the device." Our study does not aim at any of such questions. We are not going to study either linkages between kinds of motivation for choice of mobile phone or needs and gratifications of mobile phone usage. Therefore, the study excludes 'Uses and Gratification' perspective.

The authors exclude 'Digital Divide' as the guiding term for data selection and explanation as the study does not look into socio-economic gap between mobile phone users and non-users, and the implication of the gap for the country.

The study is not about how "Mobile phones can affect social ties and cohesion. Through small rituals enabled by mobiles throughout the day, individuals maintain social cohesion that was not possible before" (Ling, 2004, 2008; Rice & Hagen, 2010). Therefore, the study excludes 'Social cohesion and maintenance' approach identified by Pearce in several mobile communication studies.

The study, to be guided for its goal (to inform about mobile communication expansion drivers in India) has drawn its own perspective from the current reality. To explain the nature of the unfolding phenomenon, "Creator-Audience Matchmaker" (LohaChoudhury, 2001, p. 194) model for media product development has been utilized.

## Construct: Media Is Market

The new millennium is witness of completely changed media scenario in India. From domination of passive medium of mass communication, the media has gradually shifted into active mass-self communication mode courtesy New Media on the wing of ICT. It has emerged as the prime catalyst in Indian economic scenario expanding from the market support role (advertising and publicity). During last millennium media had primary role to inform, entertain and educate. Thus, the 'media' and 'market' went their individual ways.

Slowly and steadily, the opportunity of media as the potential platform for market manipulation in addition to market support roles opened up. Simultaneously, the media also started taking cognizance of the role and importance of the market forces for its survival and proliferation. This gradually led to the mutual realization that both the media and the market can actively complement the furtherance of each other's interests. Thus, came inevitable commercialization of the media and the 'market' began to actively employ the 'media' in its market operations. The 'media' emerged as a 'market' component working hand in gloves with production, marketing and consumer components.

From being a mere ally of the consumer market, the 'media' overarched the 'market' itself and in the process, the 'media' emerged as the 'market'. Mobile telephony, in emerging economies like India has quickly democratized the use of new media as the bridge in digital divide. Now every smart mobile phone is a market- from display of goods and services (private or government) to order to payment to delivery tracking and grievance re-dressal. It changed the concept of market as physical space, and goods as tangible to be touched and experienced. It has also made time irrelevant for the purpose of purchase. Present study uses this construct 'media as market' to establish the base for future studies.

Figure 1 represents the transforming nature of the media-market relation.

Be it as virtual market with huge display or with manipulative text, or advertisements which directly try to influence the consumers or as recruiter of agents or the portrayer of stereotype characters-signs-symbols in media productions to influence the masses to embrace a particular market-savvy cultural orientation, the media holds the key. The web media entrepreneurs' click, visit and download offer for real money has given rise to a home job segment. In this scenario, mobile medium presents the netizens with the necessary flexible platform to emerge as effective content and process creators in sync with the expectations of consumer-citizen dual existence, thus, fulfilling the role of "Creator-Audience Match-maker" (Figure 2)(LohaChoudhury, 2001, p. 194) which is pivotal to the success of any communication programme.

## Creator-Audience Matchmaker

The model of Creator –Audience Matchmaker was developed in 2001 to suggest how the media industry would behave amid increasing convergence of media technology, information technology and communication technology for product development and marketing. The model came out from broad range of suggestions received from media company owners and managers in India. A close look into mobile application software development and marketing over last few years attests the efficacy of the model. Participatory way of mobile App development and improvement, which itself integrates marketing mode with functionality modes is driving the expansion of mobile communication in India.

In building a consumer-enticing and citizen-serving milieu, the media needs to employ a persuasive strategy which essentially mandates resorting to both short-term and long term influencers. While all the media forms have the potential to build and promote such a consumer-enticing environment, the mobile has emerged as the front-runner in the new millennium by a considerable margin from all its peers for its sheer presentation platform and accommodation abilities. This is helped largely by consumer's need driven mobile application software (App in short) development mainly in banking, entertainment, shopping, travel and security. Around 50,000 software developers were engaged in mobile application software development (2016).

*Figure 1. Media-Market evolution*

**'Media and Market'** ⟹ **'Media as Market'** ⟹ **'Media is Market'**

*Figure 2. Creator-Audience Matchmaker Mechanism(LohaChoudhury, 2001, p. 194)*

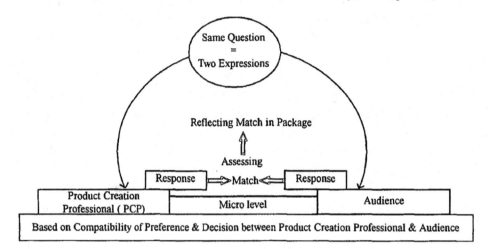

## FINDINGS

This qualitative work connects to the 'media as market' construct by descriptive analysis to provide better understanding of the unfolding mobile telephony phenomenon in India. For the same case study and reporting from national level documents/reports are resorted to from reliable industry bodies and specialist organizations.

Creator-audience matchmaking is employed to study specific cases for matchmaking between the creator and the user population.

### Mobile App: Media Matches Need

Global mobile analytics firm Appflyer survey between January 2017 to January 2018 shows that India tops mobile apps in average download of 40 apps per month, surpassing the US for the highest number of non-organic app downloads. During one year, a three-fold increase in the average number of installs per App is reported while Indian consumers spend nearly 170 mins per day on Apps. As per the survey findings, India has seen over one billion App installs, four billion App opens, 950 Apps and $400 million in-App revenue in the last one year (State of App Marketing in India Report, March 2018).

The share of buying users is relatively high in India with 7 percent of installers making in-app purchases within 30 days. However, about 32 percent of installed apps are deleted within 30 days and only about 5 percent of users remain active 30 days after installing an app. A case below illustrates how even the smallest bank account holder in the country downloads uses App with increasing ease signalling success of audience match by app developers.

The case studies how mobile application in financial transaction is gradually changing the way citizens bank in India where ease of services matters the most along with security. The study Bharat Interface for Money (BHIM) mobile App is pertinent for this reason.

By 2016, India's digital push reached considerable height. So when world's biggest demonetisation exercise was taken up by GoI in November,2016, GoI's BHIM App was introduced immediately. National Payments Corporation of India (NPCI) statistics (October,2018) shows 35.5 million BHIM App downloads for Android and 1.7 million for IOS by 30th September,2018 with more than INR 8206.37 crores transacted by 18th October,2018 from a mere INR 1.85 crores in December,2016. Immense popularity of this App stems from few consumer-centric advantages such as connecting to multiple bank accounts, using Hindi besides English and three-factor authentication for high security.

Net banking, mobile wallets or plastic cards practice were existent even before demonetisation of high value currency on November,2016. However, BHIM App of just two features- one to send or ask for money and the other to check account balance or transactions has the least complication along with very high security. It works as mobile wallet with security of net banking and cards.

BHIM is the uppermost layer software on UPI common library — a piece of code that NPCI made and gave to every bank to be embedded into their net banking application so that it connects and transacts between an account and bank. Its encryption, first factor of authentication used by the App to communicate with the payment's server is the same for all e-wallets. Net banking and credit or debit cards use two-factor authentication.

During first use of BHIM, the application automatically binds to user's unique device ID and unique phone number with SIM card- preventing use of same UPI from two phones or any phone without SIM card. Thus, BHIM works only with both the unique device and the active number. Use of any of the two may be masked during fraud, but masking both at once is almost impossible. This is the second factor.

The UPI PIN, set by the user, is needed for every BHIM transaction. This is the third security factor. Once everybody gets a Virtual Payment Address, the need to upload money into a wallet from one's bank stops. With the BHIM App installed in mobile, the user can select bank out of more than a hundred listed banks. The application matches the mobile number with the chosen bank's data base to automatically detect the account whose Know Your Customer details is already filled in while opening the bank account.

BHIM has eliminated third party transfer time of minimum 20 to 30 minutes even in the same bank through Net Banking while giving users facility to access the bank account from anywhere. It has also eliminated some problems of using plastic card such as limited Point of Sales terminals, chance of card data compromise at ATM counters and logistics of card delivery. This advantage is revolutionizing banking experience; the transaction time has come down to few seconds.

## Mobile Phone as the Frontrunner

The latest Indian Media and Entertainment Industry Report (March 2018) observes, "The mobile phone is now clearly the primary mode of communication, news and entertainment for a large number of Indians." (Ernst & Young LLP & FICCI, 2018, p. 105)

Many reasons can be attributed to this. Some of the significant developments witnessed in the Indian mobile sector presently that have been highlighted in the latest M&E report (March 2018) and the TRAI Annual Report (2016-2017) are as follows:

1. Increase in subscriber base of general mobile users
2. Increase in subscriber base of mobile internet users
3. Growth in 4G network connectivity
4. Increase in low-cost smartphone options
5. Significant increase in smartphone usage
6. Significant fall in data usage charges
7. Significant increase in data consumption-video, audio, news, social media

8.  Increase in both rural and urban teledensity
9.  Increase in Mobile Number Portability (MNP) requests
10. Increase in digital advertising share and subscription

Direct fall out of these key developments is the emerging clout of the mobile as the preferred medium of content access in terms of information, education and entertainment among the Indian masses. The rapid advancements in both hardware (low-cost smartphone options) and software (4G network connectivity) have resulted in increase in the subscriber base of both general mobile users and mobile internet users. The increase in subscriber base of mobile internet users has partly been encouraged by the greater availability of low-cost smartphone options, and partly been boosted by the significant fall in data usage charges due to the entry of new market players with deep pocket such as Reliance Jio. While the new entrants have demolished the practice of exorbitant data usage charge in India, the fidelity test for quality mobile connectivity still needs to be passed with flying colours. The increase in Mobile Number Portability (MNP) requests augurs well for the Indian mobile sector considering that mobile service providers can no longer afford to sit idly on consumer service concerns. However, concerns remain over poaching of consumers by rival companies with lucrative tariff offers. Perhaps the best take away from the recent trends is the increase in both rural and urban mobile phone density. While the mobile sector in India was urban-centric, the rural India is also rising to the call of the digital age. Digital India programme of Government of India (GoI) has catalysed huge availability of key board mobiles at low price for poorer section. Huge number of Indian entertainment Apps drive this segment to entertainment mode mostly.

Smooth sail for the mobile sector is evident from substantial increase in subscription. Internet and Mobile Association of India estimated that from around 33% in 2012, the use of internet on mobile phone reached 60% by 2015. By that time 65% of internet traffic was mobile traffic on 2G and 3G. As reported in the latest Indian Media and Entertainment Industry Report (March 2018), "India reached a telecom subscriber base of 1.19 billion in 2017...While dual-SIM phones exist, and many SIMs remain inactive, the actual number of cell phone users is estimated to be 650 million in 2017. This is expected to reach 1 billion unique mobile subscribers In India by 2020."

The hardware and software up gradation in the mobile technology have solved many a problems encountered in advertising through mobile platform. The increase in digital advertising share attests this. Internet on mobile attracted INR 7.7 billion digital market advertisement expenditure by 2015 (FICCI-KPMG 2015) when the technology was 2G and 3G only. With coming of 4G, the report predicts the ad spending to increase to INR 27.1 billion by 2019.

## India's Vision for Mobile Usage

However, the assessment of any medium remains incomplete unless a due account of its impact on the lives of the common masses is taken. That the mobile technology can be a very potential medium for fostering people's development was envisaged way back in 2004 by Dr. A. P. J. Abdul Kalam, the President of India in following observations:

"When we provide electronic connectivity to the rural areas with the broadband, satellite and wireless connectivities for last mile access, it paves the way for knowledge enablement to the rural people via tele-education. Through this, we can take the knowledge to their doorsteps with the latest innovations, digital library, experiences in value-addition, state-of-the-art practices, system oriented approach, entrepreneurial management skills, technical skills, business knowledge with the help of Universities,

R&D organizations, Industries, Management and Technical institutions…The infusion of technology is needed to make these systems affordable for every one of our panchayat and our villages, so that the knowledge can reach the common man in all the corners of the nation…The Internet based technology is a connection less network and because of its use of IP; it is amenable to even for communicating with small form factor mobile devices. These are good for one to few interactive interactions." (Kalam, 2004)

It has been more than a decade since Dr. Kalam had made his observations highlighting the immense scope of utilising the mobile technology in people-centric development initiatives and fulfilling the vision of a knowledge society. Thus, the true test of the success of the mobile communication experience in India lies in its contribution to initiatives towards making a positive change in the lives of the needy masses. In the next section, we shall see some success stories of India's experience with the mobile medium as a catalyst of people's development.

In the last decade, a lot of emphasis has been laid on utilizing mobile technology as a medium for initiating and promoting development in India. GOI initiated the 'Digital India' programme "with a vision to transform India into a digitally empowered society and knowledge economy" (Ministry of Electronics & Information Technology, GoI, 2018).One of the primary pillars of this programme is the thrust area of 'Universal Access to Mobile Connectivity' which "focuses on network penetration and filling the gaps in connectivity in the country"(Ministry of Electronics & Information Technology, GoI, 2018). The Digital India programme is centred on three key vision areas (Figure 2): (Ministry of Electronics & Information Technology, GoI, 2018)

## Evidences From Grassroot

In this section, we shall look into some studies. Perhaps the earliest assessment study on the impact of mobile technology in empowering the rural masses was recorded by Jensen (2007) who reported the findings of a micro level survey (1997) conducted to understand the impact of the introduction of mobile technology on the fishermen and wholesalers of "fifteen beach markets in northern Kerala" (p. 881). The Jensen study of the period 1997 to 200 revealed that "the adoption of mobile phones by fishermen and wholesalers was associated with a dramatic reduction in price dispersion, the complete elimination of waste, and near-perfect adherence to the Law of One Price" (Jensen, 2007, p. 879) and eventually, "both consumer and producer welfare increased" (ibid.). Since then, many studies have been conducted to assess the impact of various mobile technology experiments in India.

Two significant studies in this regard have been attempted by Srivastava and Sen (2016) and Razvi et al (2016). Srivastava and Sen (2016) have attempted to present a compilation of more than 100 mobile

*Figure 3. Three key vision areas of digital India programme (Courtesy: MEITGoI)*

technology based initiatives that have sought to further the attainment of the Millennium Development Goals in India. Razvi et al (2016) have focused specifically on reviewing 14 such case studies from 12 Indian states (CGNetSwara, Mobile Vaani, GPower, BridgeIT, GIS@School, Learn Out of The Box, Arogyashreni, eMamta, Hamari Ladli, Mobiles for Mothers, mSakhi, Mobile Kunji, ReMiND, Vatsalya Mandla). Both the studies have classified the various case studies into three broad categories based on their utility: information dissemination, monitoring and tracking, and support to frontline workers. The classification of the case studies by Razvi et al (2016) is reproduced in Table 1.

These earlier works have already presented significant data on the relevance of the mobile technology in addressing peoples' concerns in different geographical regions of India from three different utility aspects. The present work will highlight one recent case study from India based on the utility aspect of mobile technology in disaster management. This case demonstrates the relevance of the mobile medium as a suitable creator-audience matchmaker platform at all the four stages of peoples' concerns:

1. Information dissemination
2. Monitoring and tracking
3. Support to frontline workers
4. Disaster management

## Amrita Kripa: Disaster Management

"Amrita Kripa" is a mobile application developed by the Amrita Vishwa Vidyapeetham (Amrita University) for disaster management during inland flooding. The application was developed by a team from the Amrita Vishwa Vidyapeetham in response to the devastating floods in Kerala in August 2018. It works on android phones. Designed primarily "for effective and timely management of relief and rehabilitation

*Table 1. Categorisation and geographical spread of select mobile-based interventions in India*

| S/N | Mobile-Based Model | Category | Geographical Spread |
|---|---|---|---|
| 1. | CGNetSwara | Information dissemination | Chhattisgarh, Madhya Pradesh |
| 2. | Mobile Vaani | Information dissemination | Bihar, Jharkhand |
| 3. | GPower | Monitoring & tracking | West Bengal |
| 4. | BridgeIT | Information dissemination | Harayana, Tamil Nadu &Andhra Pradesh |
| 5. | GIS@School | Monitoring & tracking | Madhya Pradesh |
| 6. | Learn Out of The Box | Information dissemination | Assam |
| 7. | Arogyashreni | Monitoring & tracking | Karnataka |
| 8. | eMamta | Monitoring & tracking | Gujarat |
| 9. | Hamari Ladli | Monitoring & tracking | Madhya Pradesh |
| 10. | Mobiles for Mothers | Information dissemination | Jharkhand |
| 11. | mSakhi | Support to frontline workers | Uttar Pradesh |
| 12. | Mobile Kunji | Support to frontline workers | Bihar |
| 13. | ReMiND | Support to frontline workers | Uttar Pradesh |
| 14. | Vatsalya Mandla | Monitoring & tracking | Madhya Pradesh |

(Source: Razvi et al, 2016, pp. 43-44)

efforts during inland flooding" (Amrita Vishwa Vidyapeetham, 2018), the application offers assistance to both rescue-seekers and rescue-volunteers. Besides providing basic instructions on flood situation awareness to inform other situations such as snake-bite, the application offers the following features: (ibid.)

1. Ability to request for rescue, medical help, supplies such as food, clothing, medicines, etc., shelter, and services such as water, electricity, telephone services, etc. by the disaster victims
2. Ability to offer rescue, medical help, supplies such as food, clothing, medicines, etc., services, and shelter by the relief providers (individuals, organizations and government)
3. Ability to report people missing and people found orphaned either conscious or unconscious

All these features are offered through a set of high-performance, user-friendly mobile and web applications in multiple languages. Even though AmritaKripa is being developed to deal with inland flooding situations, it can be used in any disaster scenario. The use of smart phones with internet data services is widespread especially in rural areas of Kerala. Even though the mobile network may get disrupted during a disaster, it gets restored fairly quickly. Though the phone lines may get busy, the data network is seen to be more resilient. By building a robust, durable suite of applications that can be readily deployed in these situations, Amrita Kripa seeks to improve the effectiveness, traceability and scalability of such crowd-sourcing efforts manifold. (ibid.)

Regarding the application, Dr. Maneesha Sudheer, Director, Amrita Center for Wireless Networks and Applications (Amrita WNA) observes, "By directly linking the help-seekers and providers in a post-disaster environment using smart phones and the Internet, we are helping expedite the relief-and-rescue operations and improve their effectiveness by making them targeted at the individual level. We want this to be a worldwide app. The Kerala floods were its first deployment, but it is ready to be used in future disasters as well." (Sharma, 2018)

Sethuraman Rao, Team Leader at AmritaWNA further adds, "The use of smartphones with data services is widespread, even in rural areas. Even though mobile networks may get disrupted during a disaster, they also get restored rather quickly. Phone lines may get busy, but the data network stays resilient. In disaster situations, pinpointing locations of survivors is the key. In the AmritaKripa app, location data is picked up automatically based on user location using real-time GPS data. Users can also enter data to the nearest landmark location, and the app has the capability to automatically identify the user's location. By building a robust, durable suite of applications that can be readily deployed in these situations, Amrita seeks to improve the effectiveness, traceability and scalability of such efforts." (ibid.)During the 2018 Kerala floods, the Amrita Kripa "helped locate, rescue and provide relief to over 12,000 people stranded in floods all over Kerala". (ibid.)

## CONCLUSION

The present chapter has sought to present a bird's-eye-view of the mobile communication development in India driven by a clear national vision. In doing so, a new perspective, the 'media' as 'market' is drawn. The discussions guided by the perspective pertains to the remarkable journey of the mobile medium and its rising popularity among the Indian masses which seems well on course to establish the former as the front-runner of communication mediums in the new age India. 'Creator-Audience Matchmaker' model is used as a heuristic tool for understanding development of mobile app and their success. The first case discussed is that of a simple banking App BHIM which could offer solution to security as well as time wastage problem faced by bank account holders in India. Second case has been proposed for consideration of recognition of the relevance of the mobile medium as a suitable "creator-audience matchmaker" platform at all stages of peoples' concerns by presenting a recent case study that has proved the worth of the mobile medium in crisis situations. Of course, concerns shall always remain on the over-dependence on any technology. Yet, such is nature of all technologies that each of them is destined to enjoy unprecedented success at some age. The digital age well and truly belongs to the mobile technology and India is affording rightly to recognize the same.

## REFERENCES

Amrita Vishwa Vidyapeetham. (2018). *Amrita Kripa*. Retrieved November 21, 2018, from https://play.google.com: https://play.google.com/store/apps/details?id=edu.amrita.awna.floodevac&hl=en_IN

Appflyer. (2018). *State of App Marketing in India Report*. Author.

Ernst & Young LLP. (2018). Re-imagining India's M&E Sector. Kolkata: Ernst & Young LLP.

Jensen, R. (2007). The Digital Provide: Information (Technology), Market Performance, and Welfare in the South Indian Fisheries Sector. *The Quarterly Journal of Economics*, *122*(3), 879–924. doi:10.1162/qjec.122.3.879

Kalam, A. P. (2004, June 30). *Address at the Technology Day Award Function, Pragati Maidan, New Delhi*. Retrieved November 21, 2018, from http://abdulkalam.nic.in: http://abdulkalam.nic.in/sp300604.html

LohaChoudhury, B. (2001). *Media Performance: The Experience of Calcutta Media* (PhD thesis). Silchar, Assam: Assam University.

Ministry of Electronics & Information Technology. (2018a). *Introduction to Digital India*. Retrieved November 21, 2018, from http://digitalindia.gov.in: http://digitalindia.gov.in/content/introduction

Ministry of Electronics & Information Technology. (2018b). *Universal Access to Mobile Connectivity*. Retrieved November 21, 2018, from http://digitalindia.gov.in: http://digitalindia.gov.in/content/universal-access-mobile-connectivity

Ministry of Electronics & Information Technology. (2018c). *Vision and Vision Areas*. Retrieved November 21, 2018, from http://digitalindia.gov.in: http://digitalindia.gov.in/content/vision-and-vision-areas

Pearce, K. E. (2013). Phoning it in: Theory in mobile media and communication in developing countries. *Mobile Media & Communication*, *1*(1), 76–82. doi:10.1177/2050157912459182

Razvi, S., Srivastava, R., & Halder, B. (2016). Mobile Phone: A Public Tool (Analysing the Use of Mobile Technology in Civic Participation, Education & Health). New Delhi: Digital Empowerment Foundation & UNICEF.

Sharma, N. C. (2018, September 8). *Mobile app saves 12,000 flood victims in Kerala*. Retrieved November 21, 2018, from https://www.livemint.com: https://www.livemint.com/Politics/56R0AnrHfdsP2lagpTGrpJ/Mobile-app-saves-12000-flood-victims-in-Kerala.html

Srivastava, R., & Sen, A. (2016). Mobile Phones for Social and Behaviour Change + *Initiatives in India*. New Delhi: Digital Empowerment Foundation.

# Section 2
# Investigating the Impacts of Mobile Experience

*Behind each mobile use lies mobile experience as each use of mobile generates each unique mobile experience, which can be defined as an outcome of actors' mobile use in mobile-related activities. The seven chapters in this section deal with seven different topics with one common theme, that is, mobile experience and its impacts.*

*As the world is moving towards experience economy, consumers are paying more and more attention to memorable and fun experience beyond a product or service. In the context of experience economy, learners are also expecting the same, especially when learning goes mobile. After identifying mechanisms to measure and evaluate mobile learning experience in Chapter 10, Danielle McKain reviewed what mobile learning resources could be leveraged to enhance mobile learning experience, followed by recommendations for further studies. In their Chapter 11, Rong Hu and Xiaoge Xu reviewed earlier studies on mobile experience in learning Chinese as a second language and recommended dimensions and directions for further studies, inviting further efforts to compare mobile learning experience in general by leveraging features and functions of mobile technology.*

*In their Chapter 12 on mobile translation experience, Nancy Xiuzhi Liu and Matthew Watts have identified a tightly intertwined relationship between mobile translation and machine translation. They have also found that the technological side is more dynamic than the user side in the case of mobile translation and machine translation, which may lead to a gradual reduction of people learning foreign languages and a possible loss of professional translators and language specialists. When it comes to contextual and textual translation, however, human translators currently outperform mobile or machine translators. Although human contribution will be determined by translation scenarios or specific translation tasks, the human-mobile/machine interaction in translation deserves further studies.*

*In Chapter 13, Pooja Tabeck and Anurupa B. Singh presented the landscape of contemporary experience through mobile phones in the field of economic value creation, social value and health among bottom of pyramid in India while in Chapter 14, Shixin Zhang located similarities and differences in mobile experience in war and conflict reporting among professional journalists, citizen journalists, governments, militaries, rebels, NGOs, activists, and communities. After exploring the changes and trends related to mobile experience in war and conflict reporting, she also offered specific dimensions and directions for further studies.*

*In Chapter 15, using a mobile experience index (Xu, 2018), Wendi Li and Xiaoge Xu located and also explained the gap between news consumers' expectations and news apps experience-rich features by comparing mobile news apps of News York Times and The Guardian. In Chapter 16, Natalia Menezes, Belem Barbosa, Carolina Barrios Laborda and Dayana R Pinzón Callejas identified the benefits and impacts of mobile use to tourists and their experiences. Besides locating similarities and differences in using mobile for tourism, they confirmed that mobile empowers tourists to get more from their vacations and to have more flexible planning, resulting in satisfaction and accomplishment.*

*For illustrations of dimensions and directions of further studies of mobile use and experience, the concluding chapter of this volume, examined mobile features and mobile journalism on the one hand and presented a new approach to explanation and prediction of mobile experience for further studies and also explained how mobile experience can be maximized to secure sustainable development.*

# Chapter 10
# Mobile Learning Experience:
## Resources and Review

**Danielle McKain**
*Robert Morris University, USA*

## ABSTRACT

*As the world is moving towards experience economy, consumers are paying more and more attention to memorable and fun experience beyond a product or service. Learners are the same, especially when learning goes mobile. Mobile learning has been examined in different areas ranging from forms and formats to features and functions. Mobile experience in learning, however, has not yet fully examined. After identifying mechanisms to measure and evaluate mobile learning experience, this chapter reviewed what mobile learning resources could be leveraged to enhance mobile learning experience, followed by recommendations for further studies.*

## INTRODUCTION

Mobile learning (m-learning) started as electronic learning (e-learning). Before smartphones and tablets, computers and the Internet allowed information to be shared and learning to take place, but typically one had to be stationed at a computer. Now smartphones and tablets enable learning to take place virtually anywhere. If information is stored, some resources can even be accessed without an Internet connection. This is often referred to as ubiquitous learning (u-learning) as it can take place anytime and anywhere. Mobile learning is often viewed as informal while e-learning can be considered more formal. Furthermore, e-learning is sometimes considered to be used for more in-depth topics, whereas, m-learning can be more for brief lessons. Learning management systems were typically websites that allowed access to course content for e-learning. Many of these systems now have apps that can be used to access course content for m-learning. The two classifications are often generalized as online learning and used interchangeably without considering the difference. To complicate things further, the notation of the word e-learning is debated as eLearning, E-Learning, and e-Learning are all variations of the word depending on the resource. There has been a considerable amount of research on online learning, electronic learning, and mobile learning, but in the quickly changing market, it is difficult to measure

DOI: 10.4018/978-1-5225-7885-7.ch010

and evaluate success. This is complicated further by the reason for learning taking place. Learning can be intended for K12 education, higher education, individual purposes, or employment.

Interconnecting the different practices is a common thread, that is, mobile learning experience, which can be broadly defined in this chapter as a memorable, beneficial, and enjoyable interaction between learners and learning content or services. After a brief but critical review of earlier studies on mobile learning, this chapter provides resources that can be leveraged to enhance mobile learning experience. Furthermore, this chapter also offers recommendations on how mobile learning experience can be further enhanced by drawing up on the findings of earlier studies on mobile learning experience.

## RESOURCES FOR ENHANCING MOBILE LEARNING EXPERIENCE

Mobile learning continues to grow and more options are available making it difficult to evaluate and make recommendations. Recommendations from one study can completely change with new technology; thus it is important to look at mobile learning as a whole and take general recommendations from research that can be applied to multiple resources.

Traxler (2007) establishes categories of mobile learning as technology-driven, miniature portable, eLearning, connected classroom learning, informal personalized, situated, mobile learning, mobile training/performance support, and remote/rural/development mobile learning. Furthermore, he explains the challenge of evaluating mobile education is that it is difficult to classify "good" characteristics of mobile learning. Traxler (2007) provides possible attributes that would make a good evaluation as rigorous, efficient, ethical, proportionate, appropriate, consistent, authentic, and aligned, but states there are problems with the epistemology, ethics, and gathering and analyzing data for mobile learning evaluations.

A critical review of the challenges in eLearning in developing countries found 30 challenges that were broken down into four categories: (a) courses, (b) individuals, (c) context, and (d) technology (Andersson & Gronlund, 2009). Although all of these challenges were also faced in developed countries, the challenge of technology was more common in developing countries. As a result of their research, Andersson and Gronlund recommend creating a conceptual framework that could be used as a guide for eLearning issues faced in both developed and developing countries. Bhuasiri, Xaymoungkhoun, Zo, Rho, and Ciganek (2012) identified factors that influence eLearning success in developing countries. Curriculum design, technical knowledge, motivation, and learner behavior were found to be necessary for successfully implementing eLearning.

ELearning allows instruction to reach more trainees at a fraction of the price. According to the 2015 Brandon Hall Group Study, the five main reasons for switching Learning Management Systems are to improve user experience, improve administrative experience, enhanced reporting, integration of systems, and need for mobile capabilities. They recommended the five most important priorities of learning technology as social and collaboration tools, mobile delivery, data analysis, virtual classrooms, and content management (ELearning Market Trends and Forecast, 2016).

Recommendations for mobile learning are difficult without knowing the reason for the learning and learning goal. To better determine which recommendations to follow, it is essential to consider what the mobile learning is being used for and how it is being used. The following section provides an overview of a variety of resources for mobile learning organized by authoring software, K12, Learning Management Systems, Employee Training, Online Courses, and additional resources.

The Global ELearning market was worth over $165 billion in 2015 and is expected to exceed $275 billion by 2022 (Global ELearning Market, 2017). There are, however, a growing number of mobile learning resources that are available for free. The number of available resources can be overwhelming, and it is difficult to know where to start. Authoring software can be used by instructors to create online content and courses. Many K12 resources provide pre-made content. Learning Management Systems (LMS) provide a platform to organize existing course content or instructor made content. LMS can also be used for communication, activities, assignments, and grading. Businesses often use mobile learning for training; this can include general topics from pre-made courses or specific topics created by the company. There are a growing number of online courses that have been created by respected instructors that are offered for free and easy to access through mobile learning. Lastly, there are a variety of additional resources that are for general use or apply to more than one of the aforementioned categories. As technology progresses, many of these resources overlap and can be used in conjunction with one another.

## Authoring Software

Authoring software include *Academy of Mine, Adobe Creative Cloud, Articulate 360, Assima Atlantic Link, Composica, CourseCraft, dominKnow, Elucidat, GoMo, ISpringSolutions, LearnWorlds, Ruzuku, Softchalk, Teachable,* and *Thinkific.* The following paragraphs introduce their various features and functions, which can be fully leveraged according to different needs, requirements, tastes, and preferences of learners to enhance mobile learning experience.

Academy of Mine offers an all-in-one platform to create, sell, and market courses. Users start with a platform that can be customized and extended based on individual needs. There is a drag and drop course builder, a flexible website builder, and instructor and student dashboards. The all-in-one platform supports desktops, tablets, and mobile devices. Flexible settings allow customization. There are pre-built layouts and an integrated dashboard. Sites are set up with TLS 2.0 encryption security and translations are offered in English, Spanish, Portuguese, French, German, and more. Additional features include automated E-Commerce, certificates of completion, HTML5 and SCORM packages, marketing and analytics tools, and in-built messaging capability. Multiple choice and true-false quizzes and exams can be created, assigned, and evaluated. The integrated gradebook shows progress and grades. Academyofmine. com offers demo videos and a 30-day free trial. Prices are based on amount of monthly storage.

Adobe Creative Cloud offers the collection of Adobe desktop and mobile apps like Photoshop CC and Adobe XD CC with built-in templates and step-by-step tutorials. Various plans are offered including individual, Students and Teachers, Business, and Adobe Stock. There is an offer for a one-month free trial of Adobe Stock.

Articulate 360 can be used to create online and mobile courses easily. It is used by 78,000 organizations, 83 million learners in 151 countries. ELearning for Beginners is a free E-book that includes what eLearning is all about, how eLearning can benefit organizations and learners, step-by-step process for creating courses, how to get the right content from subject matter experts, what technology and tools to use from the toolkit, and how to design eLearning that really works. Articulate 360 offers a 60-day free trial and includes the Storyline and Rise apps and resources that are continuously updated. Templates, characters, photos, videos, and icons are included as well as live online training. Storyline is compatible with all mobile devices and offers drop buttons, dials, sliders, markers, and hotspots for interactivity, as well as, personalized interactions, videos, and simulations. Screencasts and simulations can be created and videos can be added. In addition, 25 different question types can be used to create assessments and negative scoring can be added to discourage guessing. Rise adapts courses for all devices and allows courses to be designed online using modular blocks that allow content to be arranged and provide flexibility.

Assima Atlantic Link (AAL) allows users to create interactive eLearning lessons, using three complementary products: Capture Point, Content Point, and Knowledge Point. Capture point captures screens and interactions and provides automatic narration. Capture point projects are imported into content point where eLearning is produced with media compression of sound, video, and PowerPoint files. Lastly, Knowledge point allows users to be added, edited, and deleted. Assima operates in 11 countries in Europe, North America, and Africa.

Composica is a responsive authoring platform where authors can create, collaborate, and publish eLearning. There is unlimited storage, unlimited courses, and unlimited reviewers. Published courses are HTML 5 based and will work on any platform or browser.

CourseCraft provides a way to design self-paced online courses. Lessons can include images, videos, and downloadable files. PayPal or Stripe can be connected to courses to allow for easy payment. There is also a lesson scheduler and integrate tool for blogs and websites. Forms, quizzes, and surveys can be added, and courses can be made private by adding an access code. Collaborators can be enabled to create and edit lessons. Plans are based the number of participants and uploads with monthly and transaction fees.

DominKnow is a cloud-based eLearning software for users to create courses. The authoring system is specifically designed for teams to collaborate and create interactive applications. Pricing for the system is based on the number of authors.

Elucidat is an HTML5 cloud-based authoring eLearning software. It offers engaging features such as gamification, branching, and polling. Elucidat is featured for business use; interested users must make contact for pricing.

GoMo is a cloud-based HTML5 eLearning authoring tool. Content can be created easily with responsive design that can be customized. Collaboration and distribution are simple with multi-device publishing and built-in analytics.

iSpringSolutions specializes in eLearning authoring software for PowerPoint. A toolkit can be downloaded for free and additional products are offered at various prices. Learning platforms for organizations can be created as well as eLearning courses.

LearnWorlds provides an all-in-one, create and sell online course learning platform. They offer templates for course sales pages, marketing resources for sales, interactive courses with video features, interactive buttons, and quizzes, a built-in social network for connecting with students, advanced analytics, and mobile ready, multilingual, secure, and SEO-friendly. Fees are based on the number of instructors.

Ruzuku provides a simple way to customize online courses. Ruzuku offers a 14-day free trial that includes 5 steps to your online course, Ruzuku 101: Create, Sell & Teach Your Course, 3 Steps to a Wildly Successful Online Teaching Business, and Rock Your Ruzuku.

Softchalk is an authoring program for K12 and Higher Education. It is available for individuals, teams, or enterprises. A free 30-day trial is offered and demos can be requested.

Teachable allows teachers to create online courses in a simple all-in-one platform. Membership fees are based on the number of owners and transaction fees. Unlimited video, courses, students, hosting, payment processing, student management, discussion forums, and basic quizzes, are included with all plans.

Thinkific is offered for businesses to create online courses. Features include: video and content hosting, upsell and bundle offers, single sign-on, completion certificates, payment plans, multiple instructors, quizzes, exams, and surveys, free, paid and subscription options, custom domains, affiliates, drip content, automated student emails, multi-language support, built-in landing pages, and API.

Differing in features and functions, authoring software resources reviewed here are designed to provide readers of this chapter with insightful guidelines on how to leverage the available resources to enhance mobile learning experience. Besides authoring software resources, there are also K12 resources that can also equally important for mobile learning experience enhancement.

## K12 Resources

Among the K12 resources are ABCmouse.com, Achieve LearnCast, ALEKS, BrainPOP, Carnegie Learning, ClassDojo, Desmos, Edgenuity, Edmodo, Eureka E-Learning (Zearn Math), Persona Learning, PowerSchool, and Showbie. Just like authoring software resources, K12 resources can also be fully leveraged to enhance mobile learning experience by meeting different needs, tastes and preferences of the learners.

ABCmouse.com provides a full online curriculum for kids ages 2 through 8. The step-by-step learning path offers over 9000 individual learning activities and more than 850 lessons over 10 levels. Subjects include Reading and Language Arts, Math, The World Around Us, and Art and Colors. The activities include animations, games, books, puzzles, songs, printables, and art. Progress tracking, tickets and rewards, and avatars are also available. There is a monthly subscription fee.

The mission of LearnCast is to educate learners while providing measurable results to publishers. With LearnCast, courses can be created quickly and easily using drag and drop tools for lessons, topics, quizzes, exams, and polls. Learning experience can be customized with images, videos, documents, and audio. Programming experience is not necessary. Courses can be sent to learners directly by SMS and email, posted on an app store, set up with a keyword, QR codes, or widget. LearnCast is compatible with mobile devices, laptops, or desktops. Real-time data is provided for polls, quizzes, exams, group chat discussion and comments, learner progress and grades, as well as device usage and page hits.

ALEKS is a web-based math, science, and business education program through McGraw Hill Education. It is used in K12, colleges, and universities around the world. ALEKS uses adaptive questioning to quickly and accurately determine exactly what a student knows and does not know in a course. The web-based system instructs students on the topics that they are ready to learn and reassesses prior content to ensure it is retained. The program was developed at New York University and the University of California, Irvine by a team of software engineers, mathematicians, and cognitive scientists. ALEKS uses artificial intelligence to individually and continuously assess each student. Unlike many online programs, ALEKS does not use multiple choice questions, but rather mimics what students would do with paper and pencil. Aleks.com provides a tour, research behind the program, and success stories.

BrainPOP provides cross-curricular animated movies, learning games, and interactive quizzes for grades 3+. Content includes Math, Science, Social studies, English, Engineering, Tech, Health, Art, and Music. Other features include GameUp that provides educational classroom games, Make-a-Map, a concept mapping tool, and My BrainPOP, allows teachers to keep track of learning, customize assessments, incorporate gaming into instruction, and provide feedback.

Carnegie Learning is an online math program that coaches students and adapts to their individual needs. Students work through multi-part problems and mirrors a human tutor. Mathia software is designed for grades 6-12 and provides a 1:1 learning experience with continuous formative assessment, hits, and differentiated instruction based on individual student needs. Mathia is available for Algebra I, Geometry, Algebra II and Integrated Math I, II, and III. Mika is courseware for higher education that provides a 1:1 tutoring experience.

ClassDojo connects teachers with students and parents to build classroom communities. Key features include creating a positive culture, giving students a voice, and sharing moments with parents. ClassDojo is free and works on iOS, Android, Kindle Fire, and on any computer. Parents can easily join classes and messages can be translated into over 30 languages. Students can share their learning by adding photos and videos to their digital portfolios. Teachers can create random groups and display directions.

Desmos is a math-specific website that offers classroom activities for teachers to provide feedback and offer collaboration to students. In addition to the hundreds of existing activities available, the site offers teachers free accounts to create their own activities and provides an HTML5 graphing calculator for students. Desmos also offers professional development opportunities.

Edgenuity offers core curriculum, credit recovery, intervention, test readiness, virtual learning, blended learning, summer school, ESL and special education. The courseware offers videos with direct instruction, assignments, performance tasks, and assessments for over 300 core curriculum courses for grades K-12. Students take a pre-test that is used to customize content then real-time reporting allows instructors to monitor progress.

Edmodo offers free teacher, student, and parent accounts and provides unlimited storage. Teachers can create groups, make assignments, give quizzes, and monitor student progress. Edmodo is integrated with Microsoft OneNote and Office in Google Apps for Education. Parents can receive notifications and reminders.

Eureka ELearning offers curriculum-based online educational resources for Math, Science, and English for grades K-12. The Zearn math program is designed to engage users and differentiate materials. Zearn math provides digital lessons for independent completion, lessons for small groups, lessons for whole groups, assessments, and reports.

Persona Learning is a full-service eLearning solutions provider. They combine innovative award-winning learning management system technologies with compelling custom content to enable organizations to achieve education and training goals. They provide solutions for higher education, K12, professional education, unions and apprenticeship programs, associations, government, franchises, healthcare and white label. Their on-demand services offer 24/7 support. Their idea is based on the important connection between traditional education and workforce with the need for a mix of instructor-led and self-paced capabilities for blended learning.

PowerSchool is a K12 technology platform for education. PowerSchool is used by over 32 million students in over 70 countries around the world. The platform integrates learning management, classroom collaboration, assessment, special education management, and analytics.

Showbie is an educational app that is free for students and teachers. It features an easy way to create and complete assignments and record grades. Also, there is parent sharing, collaboration, portfolios, voice notes, discussion, and annotations.

The number of K12 resources are growing faster than ever and are necessary to meet the needs of learners today. These resources allow educators to meet the needs of students but must be carefully selected and implemented. Time and money can be wasted if resources are not researched and used as intended.

## Learning Management Systems

Learning management systems include such resources as *Aduro erli bird, Blackboard, BlueVolt LMS, Desire2Learn (D2L) Brightspace, Educadium, Its Learning, Learnopia, Learnopia, Sakai, Sakai, Udutu, and WizIQ.* Among others, they can also be leveraged to enhance mobile learning experience if their features and functions can be fully utilized.

Aduro eLearning believes the best online learning management system should be personal, collaborative and intuitive. Their platform is used in a variety of industries and markets to streamline and enhance education and training practices. It can be used on PC, Mac, iOS, Android devices, Windows tablets, and more. Materials are in an HTML5 format that can be opened, viewed, and annotated in Word, PDF, or PPT. Videos, word documents, PDFs and PowerPoints can all be viewed. There are package options based on number of testers, sets of tasks, and feedback. Demos can be requested at erlibird.com.

Blackboard offers online education platforms and apps. They work to improve all aspects of education from professional training and fully online learning environments. Blackboard partners with the global education community to deliver education technology around the world. Corporate headquarters is in Washington, DC, with office locations in across North America, Asia-Pacific, Europe, Middle East, Africa, Latin America, and the Caribbean.

BlueVolt offers course development, strategic support, and integrations to turn learning into a strategic asset. It is designed for companies to align goals and provide good training. The course development includes video and animation, interactive exercises, knowledge checks, mobile optimization, multilingual translation, narration scripts and voice-overs.

D2L offers Brightspace Core for K12, Higher Education, and Enterprise. The learning management system allows users to build and manage courses for online, blended or competency-based courses. The program provides assessment options, dashboards, reports and notifications.

Educadium is an affordable cloud-hosted LMS used by nonprofits and companies to deliver self-paced training and manage eLearning courses. The EasyCampus system offers flexible discussion forums, quizzes, certificates, social networking tools, and reporting. The system is free for up to 25 users. More users can be added for a monthly fee. Free online support is provided with secure hosting and course templates.

Its Learning is a cloud-based learning management system that provided a way for school districts to put resources, strategies, lessons, and assessments in one location. The system can be used to create and manage lessons and curriculum content, view reports, and personalize instruction. The system connects student, teacher, parents, and administrators.

Learnopia is a learning management system online reseller that allows teachers to create, sell, and host online courses. Teachers create and post online courses that contain documents, videos, multiple choice assessments, and or PowerPoints. The courses are then sold for a flat rate. There are no sign-up or subscription fees.

RCampus Learning Management system offers an easy way to share materials evaluate assessments, communicate, and interact. They specialize in the iRubric comprehensive system for outcome-based assessments and are the first learning management system with built-in rubrics. The system allows for fast and easy grading and matrices provide a way to track competencies.

Sakai is an open source learning management system used in 20 countries and 20 different languages. It provides easy grading and lesson creation and features a redesigned interface and gradebook. It offers a variety of lesson tools, a PA system for alert banners and popups, enhanced assessments, functional feedback, mathematical notations, and seamless integration.

Schoology is a learning management system that connects people, content, and systems in K-12, Higher Ed, and corporate. Objective aligned assessments can be created, managed, and delivered from one location. Real-time reports are created for districts, schools, teachers, and parents. Instructional tools include course building, individualized instruction, easy grading, and tracking of student performance and engagement. Collaboration is built into schoology for students, faculty, parents, and other share-holders, with mass updates, in-platform messages, mobile notifications, and more. Collaborative spaces are prebuilt for public groups and resources. The schoology app allows grading on the go with mobile grading rubrics, annotations, and feedback. The app is compatible with Chromebooks, iOS, Android, and Kindle Fire. Data and analytics make it easy to monitor progress and quickly identify gaps in learning. There is a built-in app marketplace to allow easy integration. Schoology offers basic accounts for students, parents, and teachers. Enterprise accounts are offered for institutions.

Udutu offers a learning management system to create engaging course content designed for government and industry, retail and franchise managers, association managers, and content creators and experts. Services include course development, pre-designed courses, instructional design, graphic design, and multimedia production. Udutu offers a free trial and offers plans based on the number of users per month.

WizIQ is a virtual classroom and LMS software. Users can create and sell online courses, tutor online, create MOOCs, deliver learning on the go, and train customers and partners. WizIQ offers a free trial and monthly plans are based on storage and video quality.

Learning Management Systems can save time and provide organization but must be carefully selected. One must consider the goal of using the system, as well as the access, and ease of use. It can be difficult and time consuming to switch from one Learning Management System to another, thus, it is critical to research and carefully select before implementing.

## Employee Training

Employee training resources cover the following: *Allen Interactions, BizLibrary, Brainshark, Cegos,* and *Lessonly.* By leveraging those resources, we can also greatly enhance mobile learning experience in the context of employee training.

Allen Interactions provides custom eLearning solutions for employers to empower employees. Options include blended learning curriculums, eLearning games, mobile learning, and microlearning. They offer workshops, events, and webinars as well as learning strategy consulting.

BizLibrary is designed to engage employees as it provides training solutions to drive results. The online training library contains micro-videos on various topics with supporting materials. Topics include skills in business, HR, information technology, management and leadership, software, sales and service, and workplace safety.

Brainshark is based on four Pillars of Sales Readiness. Through Brainshark sales onboarding software organizations can accelerate training with a flipped classroom and video sales onboarding. They also specialize in solutions for continuous training.

The Cegos group is an international leader in both face-to-face and digital learning with 50 offices worldwide, 25,000 clients, and 250,000 people trained each year. They focus on developing effective learning strategies digitally. Virtual classrooms, social learning, and eLearning are options.

Lessonly provides learning software for simple training in customer service and sales. They offer a variety of templates for training guides, games, policies, announcements, handbooks, evaluations, and many more.

Employee training programs can save time and money if carefully implemented. Likewise, they can waste time and money if they are not used as intended. As with the authoring software, K12 resources, and Learning Management Systems, selection and implementation are crucial for success.

## Online Courses

Online courses include the following resources: *Coursera, Lynda.com, MOOC (edX), Open Learning Initiative, Skillshare, Skillsoft, Udacity,* and *Udemy*. They can be leveraged to enhance mobile learning greatly.

Coursera.org is an online course certificate program. Leading universities and instructors provide courses. The courses provide lectures, graded assignments, and discussion boards. For example, a course offered through Google includes five courses to prepare for jobs in IT support. It includes lecture videos and quizzes. The program is designed to be completed in approximately eight months based on eight to ten hours of coursework per week. There are 149 University partners, over 2000 courses, and over 180 specialized programs.

*Lynda.com* offers online courses through LinkedIn Learning, specializing in Higher Education, Business, and Government. There are over 1000 business courses offered and the first month is free. Lynda will make recommendations based on Linkedin profiles.

*MOOCs* (edX) or Massive Open Online Courses (MOOCs) are free online courses available for anyone. Moocs.org is an extension of edX, a non-profit, open-source learning destination educational platform that offers online courses from over 100 institutions including global colleges and universities. EdX was founded by Harvard University and the Massachusetts Institute of Technology. The Modern States Education Alliance is also part of the edX partnership and provides free online CLEP courses through the Freshman Year for Free program. The Modern States is a non-profit education alliance dedicated to college access for all. The Freshman Year for Free program aims to help students earn one year of college credit without tuition and textbook costs.

*Open Learning Initiative (OLI)* is a grant-funded program through Carnegie Mellon University. Many of the courses are free, but some do have a low cost. The courses do not provide credit or have live instructors but do offer self-assessment tools and strive to improve learning. Courses are constantly evaluated based on student performance.

*Skillshare* is an online learning community that offers over 18,000 classes in design, business, tech, and more. Basic membership is free and Premium membership is offered for a monthly fee to unlock unlimited access to all classes. There is also a team membership option.

*Skillsoft* is a cloud-based embedded learning synchronized assistant with 500 curated channels and pre-mapped competencies. It is used in 160 countries and available in 29 languages. With over 165,000 courses, videos, and books, Skillsoft is an innovative leader in online training for organizations.

*Udacity* strives to democratize education with over 160,000 students in over 190 countries enrolled. Udacity offers world-class higher education courses online. Courses are offered in Artificial Intelligence, Data Science, Programming and Development, Autonomous Systems, and Business.

*Udemy* is the world's largest online learning marketplace that offers over 65,000 courses taught by expert instructors. There is an annual subscription with a minimum user requirement.

The provided online courses are changing access to education around the world. Access to the Internet is now access to education and motivation to learn is becoming more limiting than resources to learn. Regardless of age, race, gender, location, or background, individuals have access to many of these resources. Although cost can still limit access, many resources are free.

## Additional Resources

Additional resources that can also be leveraged to enhance mobile learning experience include *EQUELLA, Google Classroom, G Suite by Google Cloud, Hippocampus, iTutorGroup, ITunesU, Kahoot!, Khan Academy, Learner.org (ANNENBERG LEARNER), MOBILIZE, Moodle, Online Educational Database Library (OEDb), OpenSesame, Peer 2 Peer University (P2PU), Pluralsight, Stoodle, Trivantis* and *VoiceThread.* The following paragraphs provide brief accounts of each of these resources.

Open EQUELLA is a Java platform for storing content in a digital repository for schools, universities, businesses, and government. EQUELLA has been used for a variety of resources. As an open course software, EQUELLA offers commercial support in North America, Asia Pacific, and EMEA.

*Google Classroom* allows instructors to create classes, invite students, assign and grade work, and send feedback. There are a variety of apps that work with Google Classroom. Instructors can integrate these into the Google Classroom.

*G Suite by Google Cloud* is a way for businesses to collaborate, access files, create projects, and manage users. It provides business email through Gmail, voice and video conferencing, secure messaging, shared calendars, and storage. Pricing is per user and is based on the amount of storage and control.

*Hippocampus* provides free academic content in the form of animations, videos, and simulations. The content is designed for middle and high school teachers and college professors to use. Content can be assigned for classwork or homework. Hippocampus is a non-profit organization sponsored by The NROC Project to provide digital content. A user guide is provided. Math subjects include Arithmetic, Algebra, Geometry, Calculus, Advanced Math, Probability, and Statistics. Natural Science subjects include Biology, Chemistry, Physics, and Earth Science. Social Science subjects include Economics, History, Government, and Sociology. Humanities subjects include English and Religion.

*iTutorGroup* specializes in English language personalized learning through online education. The online platform has teaching consultants that are always available for students. The company has over 20,000 teaching consultants in 135 countries available as the largest English language learning institute in the world.

*ITunesU* features homework hand-in, an integrated grade book, and private and class discussions. Assignments can be posted, as well as documents, web links, photos, and videos. Students can receive push notifications so that they are notified when assignments are posted and questions are answered. Students can play video and audio lessons, read books, participate in individual or class discussions, and turn in assignments. Instructors are notified when students hand in homework and can deliver lessons and grade assignments. K-12 school districts, colleges, and universities can sign up for Public Site Manager to distribute courses and educational content. ITunesU provides a best-practices guide that includes tips on descriptions, display titles, posts, assignments, discussions, diverse learning materials, my materials library, deep linking, uploading original materials, note-taking, built-in tools, grades, announcements, RSS feeds, and affiliation.

*Kahoot!* ss a free game-based learning platform designed to make learning fun, inclusive and engaging in all contexts? Kahoot! Is advertised as being flexible, simple, diverse, and engaging. It only takes a few minutes to create and will work on any device with an Internet connection. Kahoot does not require an account or log-in; participants can join as long as they have the provided game pin. Kahoot games can be used to introduce new concepts or review. The games can be played individually or as teams to build collaboration. Currently, Kahoot can be played in real-time with players in over 180 countries. Over 50 million people use Kahoot every month. Kahoot was first launch in 2013. By early 2017, they had over

1 billion cumulative participating players. Instructors can create new Kahoots as a quiz, jumble, discussion, or survey. Videos, images, and diagrams can be added. Kahoot can also be assigned. There are also hundreds of Kahoots that are all ready to use. The site provides a getting started guide, templates, tips and tricks, and video tutorials.

*Khan Academy* is a non-profit organization designed to provide a free world-class education to anyone, anywhere. Khan Academy offers a variety of online courses in math, science, history, and finance. Students can create an account and work through lessons at their own pace, watch videos, ask for hints, see math problems worked out. Khan Academy offers Learner, Teacher, and Parent accounts. Teachers can create accounts and invite students by email or a course code. Once students enroll in a course, teachers (coaches) can make assignments and track activity and progress. Class progress, as well as individual student progress, can be analyzed to identify strengths and weaknesses. There are instructional videos, lessons and interactive practice exercises. Students can see their individual progress, activity, and time. Coaches can receive weekly student highlight emails for each class and students can receive a weekly progress summary email.

*Learner.org (ANNENBERG LEARNER)* provides professional development and teacher resources for kindergarten through college. They provide interactives, lesson plans, and videos for kindergarten through college level in the Arts, Foreign Language, Literature, Language Arts, Mathematics, Science, Social Studies, and History. All resources can be searched by grade level and discipline.

MOBILIZE: Infopro Learning creates academic content for higher education. As a Fortune 500 publishing company, they led the way for eBooks, eLearning, and virtual classroom materials. Mobilize allows users to take hard copies of content and transform them into a mobile learning library. Mobisodes are short videos designed for mobile learning.

*Moodle* is an open source learning platform supported by a global community. Designed to support both teaching and learning, Moodle is used worldwide and has over 90 million users. It is the world's most widely used platform and is free with no licensing fees. Built on social constructionist pedagogy, Moodle features learner-centered tools and collaborative learning environments. The interface uses drag and drop features that are simple and easy to use. Moodle has been translated into over 120 languages. Forums, wikis, chats, and blogs can all be easily integrated into courses. Key features include a personalized dashboard, collaborative tools and activities, all-in-one calendar, file management, text editor, notifications, and a progress tracker. Administrative features include a customizable site design and layout, secure authentication and mass enrollment, bulk course creation and easy backup, user role management, and plugin management.

*Online Educational Database Library (OEDb)* is a directory for online education. There are over 8000 online college courses that are free. Colleges can be searched based on degree, content area, and rank. Featured universities include Capella, Ashford, and Liberty. There are also guides for high school students, graduate students, transfer students, and parents and guardians.

*OpenSesame* is an online eLearning course catalog. The company helps map and sync courses to any learning management system. Purchases can be made individually or in volume.

*Peer 2 Peer University (P2PU)* organizes learning circles as study groups for those who want to take online courses in-person together. Participants meet one time per week in public places like libraries or community centers. Locations and subjects can be searched or new study groups can be started on p2pu.org.

*Pluralsight* is a technology learning platform that provides courses, skill assessments, and paths for teams or individuals. Courses are offered in Software Development, IT Ops, Creative Professional,

Data Professional, Architecture and Construction, Manufacturing and Design, Business Professional, and Information and Cyber Security. Monthly personal and business plans are offered. Free courses are offered for kids.

*Stoodle* began as an interactive student network as a way for students attending different schools to collaborate. Features include a real-time interactive, collaborative virtual whiteboard, voice conferencing, text chat, permanent storage, image uploading, and drawing tools. Students can work with peers and teachers and tutors can host sessions.

*Trivantis* is an eLearning company that offers Cenario VR, Lectoria Online, Lectora Inspire, Lectora Publisher, CourseMill, ReviewLink, and Vaast – Virtual Asset and Sharing Technology.

*VoiceThread* aims to fill the gap in social presence in online learning. Students and teachers can upload pictures, videos, presentation, or documents into an online collection that looks like a slideshow. After the media is added, instructors and students will record comments about it, ask a question, critique ideas, and engage in conversation.

Many of the resources above can be used in a variety of settings and a variety of ways. Although the options are overwhelming, it goes back to Nayak's (2014) content, distribution, and delivery. The information that must be covered must be considered. This varies based on business, school, and individual needs. Content can range from a short mini lesson to courses that take years to complete. How the information is covered must be considered. The distribution options are overwhelming and must be carefully selected to ensure all users are comfortable with the system and have access. Lastly, how the information presented must be considered. Presentations can be live, pre-recorded video lessons, powerpoint presentations, texts, interactive practice, etc. The needs of learners must be considered, as well as cost and ease of use.

## REVIEWS OF MOBILE LEARNING RESEARCH

There have been numerous studies focusing on mobile learning, yet studies differ on recommendations for best practices. Mobile learning has grown rapidly and continues to transform and advance, making it difficult to define and make evaluations.

Many studies focus on best practices and effectiveness of mobile learning. More than one thousand empirical studies of online learning from 1996 through 2008 were used to evaluate evidence-based practices in a meta-analysis by Means, Toyama, Murphy, Bakia, and Jones (2009). First, they identified studies that compared online to face-to-face learning, measured learning outcomes, used rigorous research design, or provided enough information to determine an effect size. They identified 51 independent effects to be used for the meta-analysis that found students performed better online than face-to-face. The majority of the studies used in the meta-analysis were from higher education as there were very few studies available for K-12 education.

A meta-analysis of 164 studies from 2003 through 2010 to review trends in mobile learning research found the majority of studies focus on effectiveness (Wu et al., 2012). Many studies also focus on the design of mobile learning systems. Furthermore, experiments and surveys were found to be the primary methods of research. As of 2010, new technologies were replacing mobile phones and PDAs as primary devices used for mobile learning. As for studies that are cited in other research, the studies that focused on design were cited most often, but many studies focusing on effectiveness were also commonly cited. The review also found mobile learning was most frequently studied in elementary and higher education.

Research in mobile learning has expanded from small studies to large national and international studies, yet, there are few comparative studies that are evidenced based (Sharples, 2013). Studies include eLearning in classrooms, informal learning outside of the classroom, and what Sharples refers to as seamless learning for learning that connects from one location to another. Most research is based on case studies that provide observations from implementing mobile learning, but lack comparing mobile learning to traditional learning. This is often because by the time research is conducted, the technology is out of date. He cites success factors as access to technology, support for technology, integration into everyday life, and learner ownership. The challenges he states in future mobile learning are usability, the design of informal learning, and evaluation. Sharples establishes that mobile learning can be formal in a fixed setting; for example, when used in the classroom as a response system. Mobile learning can also be informal in a mobile setting; for example, outside of the classroom, when used for social networking.

Two case studies on mobile learning were examined to compare inquiry learning in formal and semi-formal environments (Jones, Scanlon, & Clough, 2013). Learner control, location, and support were analyzed to find that without teachers present, learners were able to access support through mobile devices. In a systematic review of mobile learning in preK-12 education from 2010 through 2015, Crompton, Burke, and Gregory (2017) used qualitative and quantitative analysis of mobile learning activities and found mobile learning activities were designed based on behavioral learning in 40% of studies. Over half of mobile learning research studies in the review were for science lessons. Additionally, the studies tended to focus on student learning rather than the mobile device.

The ELearning Market Trends and Forecast 2017-2021 Report (2016) by Docebo explains how the eLearning industry is complex due to the many parts, constantly changing technology, and priority changes. As the market quickly changes and grows, there are new trends and tools across the world. The report focuses on budgets for eLearning, trends in social learning, microlearning, mobile learning, MOOCs, changes in technology, such as game-based learning and gamification, and reasons for growth and changes. The report suggests lower costs and flexibility are the primary reasons for the growing eLearning markets. ELearning is constantly improving and offering increases in productivity for industries. As use expands on phones, tablets, and gadgets, distance learning and the eLearning market will continue to grow.

The report lists the United States as the leader in the adoption of eLearning technology and services with a revenue of over 23 billion, followed by Asia at almost 11 billion, Western Europe at almost 8 billion, Latin America at slightly over 2 billion, Eastern Europe at slightly over 1 billion, and the Middle East and Africa at just over half a billion. ELearning products are categorized as packaged content, service, and platforms. Packaged content creates the most revenue at slightly over 33 billion. SMAC represents social, mobile, analytics, and cloud, as the four technologies that drive business innovation (ELearning Market Trends, 2016).

There have been various studies on best practices and success factors for mobile learning. Infopro Learning, a worldwide award-winning eLearning company that provides corporate training solutions, defines the three key essentials for achieving mobile learning success: (1) Content, (2) Distribution, and (3) Delivery (Nayak, 2014). Cochrane (2012) identified critical success factors as (1) pedagogical integration of technology, (2) lecturer modeling, (3) formative feedback, and (4) appropriate mobile device and software support. Furthermore, Papanikolaou and Mavromoustakos (2006) identified the three major success factors for the development of mobile learning applications as (1) understanding characteristics, peculiarities, and constraints of mobile devices and technologies used for mobile learning,

(2) needs and requirements of learners, and (3) usability, functionality, reliability, efficiency, maintainability and probability.

In addition to best practices and success factors, cost is often a priority when implementing and studying mobile learning. The U.S. Department of Education report, *Understanding the Implications of Online Learning for Educational Productivity* (2012) identified five requirements for rigorous cost-effective studies as (1) specified design components of intervention, (2) measure of cost and outcomes, (3) comparing at least two conditions, (4) relating cost and outcome as a single ratio, and (5) controling other factors not related to study (Bakia, Shear, Toyama, & Lasseter, 2012). The report stresses that there is a lack of research for online learning models, but identified nine applications of online learning that are viewed as improving productivity: (1) broadening access, (2) engaging students in active learning, (3) individualizing and differentiating instruction, (4) personalized learning, (5) making better use of teacher and student time, (6) increasing the rate of student learning, (7) reducing school-based facilities costs (8) reducing salary costs, and (9) realizing opportunities for economies of scale.

There is a lack of research for evaluating online learning models, but cost and productivity are often a priority in all sectors. Mobile learning is becoming the new normal and accepted form of education both formally and informally in education, business, and government. As mentioned in the introduction, The U.S. Department of Education report, *Understanding the Implications of Online Learning for Educational Productivity* (2012) identified five requirements for rigorous cost-effective studies as (1) specified design components of intervention (2) measure of cost and outcomes (3) compare at least two conditions (4) relate cost and outcome as a single ratio (5) control other factors not related to study (Bakia et al., 2012). According to the report, the nine applications of online learning that are viewed as improving productivity are (1) broadening access (2) engaging students in active learning (3) individualizing and differentiating instruction (4) personalized learning (5) making better use of teacher and student time (6) increasing the rate of student learning (7) reducing school-based facilities costs (8) reducing salary costs (9) realizing opportunities for economies of scale.

In a study of 19 papers published from 2000 through 2012, Cheawjindakarn, Suwannatthachote, and Theeraroungchaisri (2012), identified five critical factors for success of online distance learning in higher education; they found institutional management, learning environment, instructional design, support services, and course evaluation should be used to implement online learning successfully.

Ozkan and Koseleer (2009) developed a six-dimension conceptual model for evaluating eLearning systems in higher education. The six dimensions include (1) system quality, (2) service quality, (3) content quality, (4) learner perspective, (5) instructor attitudes, and (6) supportive issues. The model was tested by 84 learners for validity and reliability that evaluated the quality of the system, quality of the service, perspective of the learner, quality of content, attitude of the instructor, and supported issues. The results found that all six dimensions were significant to the learners' satisfaction.

Seamless learning or the bridge between traditional education and mobile learning is characterized by ten dimensions (Wong & Looi, 2011). Through an evaluation of 54 academic papers on mobile-assisted seamless learning (MSL), a framework was developed to identify gaps in research. A seamless design for Wireless, Mobile and Ubiquitous Technology in Education (WMUTE) includes formal and informal learning, personalized and social learning, across time and location, ubiquitous knowledge access, physical and digital space, using multiple devices, smoothly transitioning between tasks, synthesis of knowledge, and a variety of pedagogical learning activities (Wong & Looi, 2011).

Chipere (2016) used a market-driven program agenda, a quality assurance system, international program standards, a cost model, course rationalization, a learning object repository, project management, document templates, and electronic workspace as a framework to implement programs for sustainable eLearning. Content, distribution, and delivery are key factors for mobile learning success (Nayak, 2014). Additionally, pedagogical integration of technology, lecturer modeling, formative feedback, and appropriate mobile device and software support are critical (Cochrane, 2012). Hwang and Wu (2014) studied the applications, impacts, and trends of mobile technology-enhanced learning in Social Science Citation Index journals from 2008 through 2012. The results were in favor of mobile learning for positively impacting student learning, interests, and motivation.

## FUTURE RESEARCH DIRECTIONS

Resources must be researched to determine the available options for specific needs. Research in this area is extremely challenges and although there are many recommendations and suggestions for best practices, all learning situations are unique. This chapter provides a starting point with a variety of resources for authoring software, K12, Learning Management Systems, Employee Training, Online Courses, and additional resources, but there are many more. Knowledge of options is often the first step for educational institutions and businesses to make choices in what will best meet their needs in mobile learning. Recommendations for mobile learning are difficult without knowing the reason for the learning and learning goal. To better determine which recommendations to follow, it is essential to consider what the mobile learning is being used for and how it is being used. One system should be implemented at a time, as it can be overwhelming to use multiple platforms simultaneously.

Critical and comprehensive reviews comparing resources are rare and difficult. Research is challenging because so much has changed so quickly with eLearning. Much of the current research on mobile learning provides examples rather than comparative studies. The examples often provide practitioner observations, advantages, and disadvantages, but lack concrete educational evaluation. In addition, it is difficult to compare more than one variable. Content, age, education, formal, informal, and cost are just a few things to consider. Many studies focus on effectiveness, but it is difficult to measure effectiveness.

There is not enough evidence based comparative studies on eLearning or mobile learning, but the market and options are quickly growing. According to eLearning Market Trends, online courses are the future of education in India. The government initiative to fund literacy in rural areas and small villages will increase demand for eLearning. In China there are increases in self-paced eLearning and many people pay for exam preparation and testing due to the competition for good jobs. In Latin America, the government and companies seek highly trained workers and ways to increase productivity, safety, and skills gaps through eLearning. In the Middle East, vocational programs and self-paced eLearning is growing (eLearning Market Trends, 2016).

The European Commission is seeking a Digital Education Action Plan as an initiative to use digital technology for learning and teaching, establish digital competences for living in the digital age, and better education with data analysis (Digital Education Action Plan, 2018). The initiative includes better use of digital technology for teaching and learning through wi-fi connections for schools, a self-reflection tool for evaluating technology use in schools, and a way to store electronic documents for qualifications. Also, digital competencies and skills include a higher education hub, science, code week, cyber security, and digital skills for girls.

Sharples, Arnedillo-Sánchez, Milrad, and Vavoula (2009) explain how it is difficult to contribute mobile learning to the theory and practice of education. Future research will be necessary as mobile learning becomes more common and cost-effective. It is important for future research to set criteria and establish ways of evaluating mobile learning effectiveness. Due to the many implementations of mobile learning, it is necessary to consider the goals of the content and reason for learning.

Regardless of the resource used, it is important to keep the educational goals in mind. In a 2014 case study, *The 8 Essentials for Mobile Learning Success in Education*, are (1) purposeful planning for mobile device use (2) leveraging content and curriculum (3) understanding the power of Internet access (4) preparing educators effectively (5) securing leadership buy-in (6) building personal learner efficacy and capacity for self-directed learning (7) measuring project results with meaningful metrics (8) creating an ecosystem that is sustainable and scalable. The report provides insight from investments to support education leaders worldwide in a quest to improve educational opportunities for students through the effective use of mobile and wireless technologies (Baker, Dede, & Evans, 2014). As more resources and options become available and others become unavailable, it can be difficult to keep up and determine what to use for course creation and mobile learning. This chapter is designed to serve as a resource for educators, industry leaders, and students to use for ideas and resources to determine how mobile learning will best meet their needs.

# REFERENCES

Andersson, A., & Grönlund, Å. (2009). A conceptual framework for eLearning in developing countries: A critical review of research challenges. *The Electronic Journal on Information Systems in Developing Countries*, *38*(1), 1–16. doi:10.1002/j.1681-4835.2009.tb00271.x

Baker, A., Dede, C., & Evans, J. (2014). *The 8 Essentials for Mobile Learning Success in Education*. Retrieved 2 May 2018 from https://www.qualcomm.com/media/documents/files/the-8-essentials-for-mobile-learning-success-in-education.pdf

Bakia, M., Shear, L., Toyama, Y., & Lasseter, A. (2012). *Understanding the Implications of Online Learning for Educational Productivity*. Washington, DC: U.S. Department of Education, Office of Educational Technology. Retrieved 8 January 2018 from https://www.sri.com/work/publications/understanding-implications-online-learning-educational-productivity

Bhuasiri, W., Xaymoungkhoun, O., Zo, H., Rho, J. J., & Ciganek, A. P. (2012). Critical success factors for e-learning in developing countries: A comparative analysis between ICT experts and faculty. *Computers & Education*, *58*(2), 843–855. doi:10.1016/j.compedu.2011.10.010

Cheawjindakarn, B., Suwannatthachote, P., & Theeraroungchaisri, A. (2012). Critical Success Factors for Online Distance Learning in Higher Education: A Review of the Literature. *Creative Education*, *03*(08), 61–66. doi:10.4236/ce.2012.38B014

Chipere, N. (2016). A framework for developing sustainable e-learning programmes. *Open Learning: The Journal of Open. Distance and E-Learning*, *32*(1), 36–55. doi:10.1080/02680513.2016.1270198

Cochrane, T. D. (2012). Critical success factors for transforming pedagogy with mobile Web 2.0. *British Journal of Educational Technology, 45*(1), 65-82. doi:10.1111/j.14678535.2012.01384.x

Crompton, H., Burke, D., & Gregory, K. H. (2017). The use of mobile learning in PK-12 education: A systematic review. *Computers & Education, 110*, 51–63. doi:10.1016/j.compedu.2017.03.013

Digital Education Action Plan. (2018). *Education and training*. European Commission. Retrieved from https://ec.europa.eu/education/initiatives/european-education-area/digital-education-action-plan_en

ELearning Market Trends and Forecast 2017-2021. (2016). *Docebo*. Retrieved 1 May 2018 from https://eclass.teicrete.gr/modules/document/file.php/TP271/Additional material/docebo-elearning-trends-report-2017.pdf

Global E-Learning Market. (2017). *Orbis Research*. Retrieved from https://www.reuters.com/brandfeatures/venture-capital/article?id=11353

Hwang, G. J., & Wu, P. H. (2014). Applications, impacts and trends of mobile technology-enhanced learning: A review of 2008-2012 publications in selected SSCI journals. *International Journal of Mobile Learning and Organisation, 8*(2), 83. doi:10.1504/IJMLO.2014.062346

Jones, A. C., Scanlon, E., & Clough, G. (2013). Mobile learning: Two case studies of supporting inquiry learning in informal and semiformal settings. *Computers & Education, 61*, 21–32. doi:10.1016/j.compedu.2012.08.008

Means, B., Toyama, Y., Murphy, R., Bakia, M., & Jones, K. (2009). *Evaluation of Evidence-Based Practices in Online Learning: A Meta-Analysis and Review of Online Learning Studies. Project Report*. Centre for Learning Technology.

Nayak, B. (2014). *3 Key Essentials to Achieving Mobile Learning Success*. Retrieved 8 May 2018 from https://cdns3.trainingindustry.com/media/17775903/essentials_of_mobile_learning_white_paper.pdf

Ozkan, S., & Koseler, R. (2009). Multi-dimensional students' evaluation of e-learning systems in the higher education context: An empirical investigation. *Computers & Education, 53*(4), 1285–1296. doi:10.1016/j.compedu.2009.06.011

Papanikolaou, K., & Mavromoustakos, S. (2006). *Critical success factors for the development of mobile learning applications*. Retrieved 8 January 2018 from https://www.researchgate.net/publication/221655699_Critical_Success_Factors_for_the_Development_of_Mobile_Learning_Applications

Sharples, M. (2013). Mobile learning: Research, practice and challenges. *Distance Education in China, 3*(5), 5–11.

Sharples, M., Arnedillo-Sánchez, I., Milrad, M., & Vavoula, G. (2009). Mobile Learning. *Technology-Enhanced Learning*, 233-249. doi:10.1007/978-1-4020-9827-7_14

Traxler, J. (2007). Defining, discussing and evaluating mobile learning. *International Review of Research in Open and Distance Learning, 8*(2). doi:10.19173/irrodl.v8i2.346

Wong, L., & Looi, C. (2011). What seams do we remove in mobile-assisted seamless learning? A critical review of the literature. *Computers & Education, 57*(4), 2364–2381. doi:10.1016/j.compedu.2011.06.007

Wu, W., Wu, Y. J., Chen, C., Kao, H., Lin, C., & Huang, S. (2012). Review of trends from mobile learning studies: A meta-analysis. *Computers & Education, 59*(2), 817–827. doi:10.1016/j.compedu.2012.03.016

## ADDITIONAL READING

Allen, M. W. (2016). *Michael Allens guide to e-learning: Building interactive, fun, and effective learning programs for any company*. Hoboken, NJ: Wiley. doi:10.1002/9781119176268

Arshavskiy, M. (2013). *Instructional design for eLearning: Essential guide to creating successful eLearning courses*. Place of publication not identified: Your eLearning World.

Christopher, D. (2015). *The succesful virtual classroom: How to design and facilitate interactive and engaging live online learning*. New York: AMACOM.

Elkins, D., & Pinder, D. (2015). *E-learning fundamentals: A practical guide*. Alexandria, VA: ADT Press.

Horton, W. (2011). *E-learning by design*. San Francisco, CA: Pfeiffer, Wiley. doi:10.1002/9781118256039

Thormann, J., & Zimmerman, I. K. (2012). *The complete step-by-step guide to designing and teaching online courses*. New York, NY: Teachers College Press.

Vai, M., & Sosulski, K. (2011). *The essential guide to online course design: A standards-based approach*. London: Routledge.

Wong, L., Milrad, M., & Specht, M. (2015). *Seamless learning in the age of mobile connectivity*. Singapore: Springer. doi:10.1007/978-981-287-113-8

# Chapter 11
# Mobile Experience in Learning Chinese:
## Review and Recommendations

**Rong Hu**
*University of Nottingham – Ningbo, China*

**Xiaoge Xu**
*University of Nottingham – Ningbo, China*

## ABSTRACT

*In the age of experience economy, students are also attaching greater importance to enlightening and entertaining experience in learning. Earlier studies on mobile learning experience have generated increasing amount of knowledge on experience-oriented learning. Through reviewing earlier studies on mobile experience in learning Chinese as a second language and recommending dimensions and directions for further studies, this chapter aims at inviting further studies on how to enhance mobile learning experience in general by applying the six stages of mobile experience, assisted with the 3M (mapping, measuring, and modeling) approach with a focus on the gap between the normative and the empirical dimensions of mobile experience in learning Chinese in the context of Confucius Institutes around the world.*

## INTRODUCTION

Learning Chinese as a second language has been prevalent around the world. Statistics (Confucius Institute Annual Report, 2017) showed that 1.7 million students and 621,000 online students from 146 countries and regions signed up for Mandarin courses with Confucius Institutes by the end of 2017. The number keeps growing. Against this backdrop, mobile has been constantly leveraged to enhance learning experience in the age of mobile devices. Use of mobile in learning Chinese as a second language has also led to accumulated studies on mobile experience in learning Chinese as a second language. With the development of social cultural approach (Comas-Quinn, 2012) and the emergence of mobile technology, more and more researchers showed their interest in improving students' language competence and experience

DOI: 10.4018/978-1-5225-7885-7.ch011

with the support of mobile technology as this advanced technology offered affordance for learners to approach target language anytime and anywhere. The first work of Mobile Assisted Language Learning (MALL) can be dated back to 1994 (Burston, 2013). Since then MALL studies increased dramatically (Hwang & Tsai, 2011; Duman, Orhon, & Gedik, 2014), coinciding with the emergence of smartphone technology (Heil, Wu, Lee, & Schmidt, 2016). Several reviews of mobile assisted second language learning revealed the research trends specifically discussed how technologies enhanced language learning experience (Hwang & Tsai, 2011; Heil et al., 2016; Persson & Nouri, 2018) .

Chinese as one of the important elements in the research field of second language acquisition attained great attention in MALL studies in recent years. Heil et al (2016) reported that following English, French, Spanish and German, Chinese ranked five as one of the top ten languages, which were learned through mobile apps. Previous studies on mobile Chinese learning investigated pedagogical design, devices and applications, learning environment, user motivation and learning strategies, specific language skills, and so on. However, the absence of a systematical review brought difficulty for scholars to draw an explicit map of what has been investigated and what lies ahead for further studies. In addition, it is hard for researchers to incorporate their studies in an international setting of mobile assisted language learning. For example, there were only two studies about mobile Chinese learning reviewed in Persson & Nouri's (2018) review of Second Language Learning with mobile technologies. Therefore, this chapter provides a critical and comprehensive review of publications of Chinese learning as a second language assisted with mobile technology. After its review of earlier studies on use of mobile technological advances in enhancing mobile learning experience, this chapter presents its identification of major components of mobile experience in learning Chinese as a second language. After locating research gaps, this chapter offers its recommendations for further studies.

## MOBILE TECHNOLOGY AND LEARNING EXPERIENCE

Since Apple launched its first generation of smartphones in 2007, mobile technological advances have been leveraged in language learning via smartphones. Learning Chinese as a second language is no exception as it has been increasingly assisted by mobile technology. The year of 2007 also witnessed the emergence of a scenario-based design(Tseng et al, 2007) and a Chinese learning support system (Chen, Chou, & Arnedillo Sánchez, 2007) supported by location-aware technology. And Student Partner System (mobile Chinese learning system focusing on collaborative learning) and CAMCLL (Context-aware Mobile Chinese Learning for foreigners) approach were introduced respectively in 2008 and 2009. (Anderson, Hwang, & Hsieh, 2008;Al-Mekhlafi, Zheng, & Hu, 2009)

In a pilot study, assisted by the framework of communicative and contextualized language learning, students took advantage of mobile assisted learning environment to learn Chinese idioms (Wong, Chin, Tan, Liu, & Gong, 2010). In addition, aided by mobile technological advances, learning Chinese characters includes game-based learning approach (Tam & Yeung, 2010; Wang et al, 2010) and correct stroke sequences of writing Chinese characters for learners to practise (Tian, Lv, Wang, Wang, Luo, Kam & Canny, 2010).

In another area of using mobile technology, Kuo et al (2011) built a Chinese learning application, HuayuNavi, based on intelligent character recognition on smartphone devices. Ever since the beginning of applying mobile technological advances in learning Chinese, researchers attached great importance to users' experience and tried to offer them better experience with easier and clearer interactive interface.

Zhang, Wang and Li (2011) developed a mobile learning system for learning Chinese pronunciation with the support of speech recognition technology and wireless communication technology. Wong, Boticki, Sun and Looi (2011) adopted DBR (design-based research) approach to learning Chinese character in a mobile computer-supported collaborative Learning environment. In particular, they presented us a completely iterative process of (re)designing and testing on Chinese game model design, effectively integrating game design into a pedagogically oriented Chinese characters learning environment. What's more, they identified the original design drawbacks, like group forming mechanism, character-forming mechanisms, and teacher solution and so on, explained the reasons and offered possible solutions by technological affordance. Another example is MicroMandarin, a contextual microlearning system to support expatriate Chinese learners to identify and acquire daily used language (Edge, Searle, Chiu, Zhao, & Landay, 2011).

By 2012, Chinese characters learning was still the dominant topic in the MALL research area. Aiming to help Filipino students to learn Chinese Hanzi and Japanese Kanji vocabularies collaboratively and creatively, Syson, Estuar and See (2012) designed a mobile learning game entitled ABKD. Feedback showed that the application made characters learning easier, more fun, and more interesting. Tam and Luo (Tam & Luo, 2012) proposed a learning object based Chinese writing system on iPhone or PAD to enhance learners' experience through interactive interface. To identify the role that mobile devices played in helping learning Chinese as a foreign language, Jing and Christine (2011) surveyed 11 students at a Midwest university in America and an interview was included. Findings suggested that collaboration and scaffolding played crucial role in mobile assisted Chinese learning.

Mobile devices have been playing an undeniably important role in learning Chinese idioms. Chunsheng and Ying (2013) gauged both short- and long-term effects of learning Chinese idioms through IPads. Their analysis showed that learners used most idioms correctly through self-generated learning activities supported by iPads and that learners reported that they enjoyed this mobile way of learning. Issues to be addressed include collaboration, peer learning and assessment and technological limitations, according to Chunsheng and Ying (2013). Wong et al (2013) further found that flexible grouping could influence collaborative patterns in mobile-assisted Chinese learning game. A new Chinese calligraphic training system based on virtual reality was also introduced (Wu et al., 2013). Further example would include use of the TRIZ theory, which originally designed for industrial design, to develop better mobile Mandarin learning system (Shih, Chen, & Li, 2013).

In light of the learning object based Chinese writing system introduced in 2012, Tam and Luo (2014) developed an Intelligent Chinese Explorer (iCExplorer) fully utilizing the smart sensors and particularly designed an intelligent algorithm to aid learners in writing Chinese characters with the correct stroke sequences. A prototype of the application was successfully built on the iOS platform for a thorough evaluation with a detailed plan. In 2014, Tam, Meng and Lu (2014) developed a Chinese character learning application for a bilingual primary school while explored experience of both teachers and designers towards enhance mobile Chinese learning. A context-based support system of mobile Chinese learning for foreigners in China was proposed in 2015 (Sun, Hou, Hu, & Al-mekhlafi, 2015). Sun et al stated that the system could support Chinese learning under the real-world communication environment while foreign learners also could experience deeper cultural implications in Chinese language in daily life.

Starting from 2016, diverse studies mobile Chinese learning were published as indicated by the following key journal articles, Chai, Wong and King (2016) developed a valid and reliable instrument MSCLQ to measure student motivation and strategies for mobile-assisted seamless Chinese language learning. Their research offered us a good model to discuss student's motivation and learning strategy from different dimensions or considering different variables, including enhance learning via learners' experience and initiatives in using technology. Lan and Lin (2016) adopted both qualitative and quantitative method to CSL oral communication enhanced by mobile seamless technology. This four-week empirical study proved that learners made fewer errors and less dependence on their first language in the mobile authentic Chinese environment. Further studies showed that mobile games or gamification approach could be utilized to enhance Chinese learning (Ying, Rawendy, & Arifin, 2016). In addition, mobile Chinese learning applications for Malaysian (Hashim et al., 2016) and for Indonesian senior high school students (Darmanto & Hermawan, 2016) were developed and evaluated. And mobile-assisted system has also been used in flipped classrooms but for native speakers' classical Chinese learning (Wang, 2016).

Several novel research topics emerged in 2017. For example, Xu and Peng (2017) studied WeChat oral feedback in teaching Chinese as a second language (CSL), the oral recordings from 13 CSL learners for one semester and feedbacks from 2 native Chinese teachers. Their study demonstrated that use of WeChat and other mobile technological advances could enhance mobile experience in Chinese. In another study, Indonesia researchers Ying, Lin and Mursitama (2017) found that a game based learning model could assist students in mastering the sentences forming in Mandarin through mobile learning. Besides, gamification and mnemonic Method (Rawendy, Ying, Arifin, & Rosalin, 2017) has been applied into design mobile mandarin learning games for 6-12 years old Indonesian students. And an interactive educational mobile apps for students learning Chinese characters writing has been proposed (Ibrahim, Kamaruddin, & Ling, 2017).

Further significant research areas in 2018 included a framework for evaluating contextualized MALL through three variables: device mobility, real world and real-life context (Cohen & Ezra, 2018), the important role of WeChat in facilitating mobile Chinese learning (Jin, 2018; Jiang & Li 2018), the integration of mobile technology in a language immersion elementary school (Eubanks, Yeh & Tseng, 2018).

## RESEARCH FOCUSES AND ELEMENTS OF MOBILE EXPERIENCE IDENTIFIED

As mobile technological advances have been widely leveraged in learning Chinese as a second language, earlier studies were largely conducted within the framework of mobile-assisted Chinese learning, focusing the following 11 major areas: (1) characters, (2) vocabulary and idioms, (3) pronunciation and oral proficiency, (4) writing, (5) grammar, (6) integrated skills, (7) feedback, (8) instructional design and design-based learning system, (9) applications and social media, (10) interaction and collaboration, and (11) motivation and influence variables (see Table 1).

As shown in Table 1, the most common research focus is on instructional design or design-based learning system, followed by Chinese characters learning assisted by mobile technology. Although differing in research focuses, earlier studies shared the same central theme, that is, how to enhance effectiveness in Chinese m-learning (Duman, Orhon, & Gedik, 2014). Earlier studies on effectiveness largely focused on m-learning system designs, in other words, project implementations (Burston, 2013, 2014; Wu et al., 2012), facilitating mobile-assisted improvement of Chinese proficiency in different contexts. One of the most frequent research topics in terms of language competence is how to use a mobile app to learn vocabulary (Burston, 2014b). Closely related to vocabulary is how to write Chinese characters. As Chinese is unique for its graphic configuration and lack of sound-script correspondence, it is particularly difficult for foreigners to learn Chinese (Li, 2017). So further focus lies in how to leverage mobile technological advances to assist learners in writing Chinese characters.

*Table 1. Research focuses in mobile Chinese learning as a second language*

| Topic | 2007 | 2008 | 2009 | 2010 | 2011 | 2012 | 2013 | 2014 | 2015 | 2016 | 2017 | 2018 | Total |
|---|---|---|---|---|---|---|---|---|---|---|---|---|---|
| Characters | | | | 2 | 1 | 2 | 2 | 1 | | | 1 | | 9 |
| Vocabulary and idioms | | | | 1 | | | 1 | | | | | | 2 |
| Pronunciation and oral proficiency | | | | | 1 | | | | | 1 | | | 2 |
| Writing | | | | | | | | | | | | 1 | 1 |
| Grammar | | | | | | | | | | | 1 | | 1 |
| Integrated skills | | | | | 1 | | | 1 | | | | | 2 |
| Feedback | | | | | | | | | | | 1 | | 1 |
| Instructional design and design-based learning system | 2 | 1 | 1 | 1 | | | | | 1 | 3 | 2 | | 11 |
| Applications and social media | | | | | | | | | | | 1 | 2 | 3 |
| Interaction and collaboration | | | | | | 1 | | | | | | | 1 |
| Motivation and influence variables | | | | | | | | | | 1 | | 1 | 2 |
| Others | | | | | 1 | | | | | | | | 1 |
| Total | 2 | 1 | 1 | 4 | 4 | 3 | 3 | 2 | 1 | 5 | 6 | 4 | 36 |

Although defined differently, mobile experience has been examined in previous studies, including the following basic elements: "fun", "interesting", "enjoyable", "facilitated learning", and so on. Although differently identified, elements of mobile experience served to enhance positive learning experience in learning Chinese in a mobile environment. Moreover, as demonstrated in previous studies, learners' Chinese language competence could be largely enhanced or facilitated by mobile technology, which could provide various and better learning experiences for learners comparing to traditional classroom teaching. For example, WeChat could land a great support to students in their learning collaboratively while receiving instant feedback from instructors or peers. And design-based learning systems or games, context-aware, seamless and authentic Chinese learning not only enabled learners to enjoy learning experience, but also encouraged them to take the initiative to enhance their Chinese learning experience.

## RESEARCH GAPS AND RECOMMENDATIONS FOR FURTHER STUDIES

As shown in the previous sections, mobile experience in learning has not been fully examined since its different elements were investigated separately in connection with use of mobile technology to assist mobile learning. Previous studies on learning Chinese as a second language via mobile technology aimed at effectiveness without considering much about learners' experience, which is probably one of the most important aspects need to be addressed. For instance, researchers attached importance on instructional design or design-based learning system, while learners' experience should be taken into account mostly to test the effectiveness of these designs or systems. Taking Chinese characters learning as another ex-

*Table 2. Mobile experience identified in earlier studies*

| Studies | Elements of Mobile Experience |
|---|---|
| Chen & Chou (2007) | Experiencing location-based materials in a context aware learning system<br>Interesting learning activities<br>Satisfied with content and services |
| Kuo et al.(2011) | Fast and friendly response<br>Easier and clearer user interface |
| Syson et al (2012) | Game-based learning experience<br>Fun, easy and interesting learning |
| Chunsheng and Ying (2013) | Enjoyment derived from mobile learning |
| Wong et al (2013) | Enjoyment generated from mobile-assisted Chinese character learning game |
| Lu et al (2014) | Both teacher and student experience<br>Usefulness and effectiveness<br>Fun, enjoyment, achievement |
| Sun et al (2015) | context-based effectiveness and efficiency |
| Heryadi & Muliamin (2016) | Gamification |
| Chai, Wong and King(2016) | seamless learning experience |
| Lan & Lin (2016) | Different experience<br>Involvement& |
| Xu & Peng (2017) | Enhancement<br>Easy feedback from Wechat-based learning<br>Relaxing and confident<br>Engagement |
| Jiang & Li (2018) | Excitement<br>Enjoyable learning |

ample, students will enjoy more to learn and practice characters if pleasant experience integrated with mobile learning rather than to be scared or even refuse to learn them.

Moreover, elements of mobile Chinese learning experience identified from the literatures seemed randomly scattered rather than being measured with a valid instrument. In addition, experience-related studies on learning Chinese as a second language via mobile technology has focused on improving learners' language skills separately, such as vocabulary, pronunciation, characters and so on, that we could think about constructing a holistic picture of learning Chinese language assisted by mobile technology by fully mobilizing various senses to enhance learners' experience.

Mobile learning has been on the rapid rise in penetration and popularity since it enables learners to learn anytime anywhere. Besides its ubiquity, it is its unique mobile learning experience that has distinguished mobile learning from the crowd as the most adaptive, blended and collaborative experience in learning. To enhance mobile learning, it is imperative for mobile learning providers to track mobile learners' changing expectations of mobile learning experience in a mobile learning app. Equally imperative it is to identify the gap between mobile learners' expectations and what mobile learning experience is actually embedded in a mobile learning app. Using ABC learning (Adaptive, Blended and Collaborative) and mobile experience indexes, we have located the venue and extent of the gap through conducting a survey of mobile learners and also a comparative analysis of indicators of mobile learning experience in selected mobile learning apps.

For further studies of mobile experience in learning Chinese, we would like to recommend the six stages of mobile experience proposed by Xu (2018), assisted with the 3M approach with a focus on the normative and empirical gap. The six stages of mobile experience include (1) enticement, (2) entertainment, (3) engagement, (4) empowerment, (5) enlightenment, and (6) enhancement (Xu, 2018). These stages in each learning or training session can be re-conceptualized and re-operationalized according to different learning settings, demographic features, learners' wants and needs, their tastes and preferences.

At the enticement stage, it is crucial to attract learners' attention to and arouse their interest in Chinese language learning. In the case of mobile learning apps, the opening screen should be designed in such a way that learners' attentions and interests are immediately aroused through leveraging attractive elements such as use of color, layout, logo, animation, video, audio or multimedia on top of the attractive presentation of Chinese language. And these elements should also be customizable so that learners can personalize them to meet their own tastes and preferences. This stage should also be constantly mapped and measured among mobile learners of different demographic features so that we can constantly adapt and adjust to the changes in enticement elements. Attention should also be paid to locate factors that have brought about these changes so that we can identify trends and patterns in terms of enticement elements related to mobile experience in learning Chinese. This is especially necessary and important for mobile learners of Chinese from different countries and cultures in the context of Confucius Institutes around the world.

Behind enticement lies another important stage, that is, entertainment. If learners are not entertained in learning Chinese, they will find it boring to learn Chinese. To entertain them in learning Chinese is one of the keys to enhance their experience in learning Chinese. Like enticement, entertainment should also be customized according to different demographic features of mobile learners. Entertainment is also culturally defined. What is entertaining to mobile learners from Russia may not be necessary entertaining to mobile learners from France.

When it comes to engagement, it is closely intertwined and interdependent with enticement and entertainment. No engagement would exist among mobile learners if they were not enticed or entertained. Engagement elements should be fully embedded in a mobile app although they can be differently conceptualized and operationalized in the context of diversified demographic features among mobile learners.

The first three stages belong to the first level of mobile experience, which focuses on the sensory intelligence used to attract, ignite and hold mobile learners' attentions to and interests in learning Chinese as well as their concentration on and enjoyment of learning Chinese. The second level of mobile experience consists of empowerment, enlightenment and enhancement, which emphasizes the advanced level of mobile experience in learning Chinese.

Once fully empowered in learning Chinese, mobile learners can choose what, when, where and how to learn in their own way. Empowerment gives mobile learners more rights to personalize their learning experience by allowing them to change the learning content, supplementary materials, language scenarios, level of difficulty, and methods of learning Chinese.

Mobile learners would not be able to be enlightened if levels of difficulty are not properly and gradually arranged and presented according to mobile learners' progress. No enlightenment would be possible in the absence of enticement, entertainment, engagement or empowerment from the learners' perspective.

The final stage of mobile experience is the stage of getting enhanced after going through the previous five stages in learning Chinese. Mobile learners will enhance their understanding, knowledge, skills and proficiency related to Chinese language at this stage, which is the highest level of mobile experience.

These six stages of mobile experience in learning Chinese should be mapped and measured on both the normative side and the empirical side so as to locate the gap between the two, resulting in a better understanding of what needs to be improved to secure better mobile experience in learning Chinese.

In investigation of mobile experience in learning Chinese, the ultimate goal is to produce a better understanding, explanation and prediction of changes and trends in mobile experience related to learning Chinese. In investigating different stages of mobile experience, it is absolutely important to locate various factors that are shaping similarities and differences in mobile experience in learning Chinese. Those factors constitute the variables to be tested for possible correlations between or among them so as to come up with a better way to model mobile experience in learning Chinese.

# REFERENCES

Al-Mekhlafi, K., Hu, X., & Zheng, Z. (2009). An Approach to Context-Aware Mobile Chinese Language Learning for Foreign Students. *Mobile Business, 2009. ICMB 2009. Eighth International Conference on*, 340-346. 10.1109/ICMB.2009.65

Anderson, T. A., Hwang, W. Y., & Hsieh, C. H. (2008). A study of a mobile collaborative learning system for Chinese language learning. *Proceedings of International Conference on Computers in Education*, 217-222.

Chai, C. S., Wong, L. H., & King, R. B. (2016). Surveying and Modeling Students' Motivation and Learning Strategies for Mobile-Assisted Seamless Chinese Language Learning. *Journal of Educational Technology & Society*, *19*(3), 170–180.

Chen, C. H., & Chou, H. W. (2007). Location-aware technology in Chinese language learning. *IADIS International Conference on Mobile Learning*.

Cohen, A., & Ezra, O. (2018). Development of a contextualised MALL research framework based on L2 Chinese empirical study. *Computer Assisted Language Learning*, 1–26. doi:10.1080/09588221.2018.1527359

Comas-Quinn, A., & Mardomingo, R. (2012). Language learning on the move: a review of mobile blogging tasks and their potential. Innovation and Leadership in English Language Teaching, 6(6), 47–65.

Confucius Institute. (2017). *Confucius Institute Annual Development Report*. Retrieved from http://www.hanban.edu.cn/report/2017.pdf

Darmanto, H. Y., & Hermawan, B. (2016). Mobile learning application to support Mandarin language learning for high school student. *Imperial Journal of Interdisciplinary Research*, 2(4), 402–407.

Edge, D., Searle, E., Chiu, K., Zhao, J., & Landay, J. A. (2011). MicroMandarin: mobile language learning in context. *Proceedings of the SIGCHI Conference on Human Factors in Computing Systems*, 3169-3178.

Eubanks, J., Yeh, H., & Tseng, H. (2018). Learning Chinese through a twenty-first century writing workshop with the integration of mobile technology in a language immersion elementary school. *Computer Assisted Language Learning*, *31*(4), 346–366. doi:10.1080/09588221.2017.1399911

Ezra & Cohen. (2018). Contextualised MALL: L2 Chinese students in target and non-target country. *Computers & Education*, *125*, 158-174.

Heil, C. R., Wu, J. S., Lee, J. J., & Schmidt, T. (2016). A Review of Mobile Language Learning Applications: Trends, Challenges, and Opportunities. *The EuroCALL Review*, *24*(2), 32–50. doi:10.4995/eurocall.2016.6402

Heryadi, Y., & Muliamin, K. (2016). Gamification of M-learning Mandarin as second language. *Game, Game Art, and Gamification (ICGGAG), 2016 1st International Conference on*, 1-4. 10.1109/ICGGAG.2016.8052645

Hwang, G.-J., & Tsai, C.-C. (2011). Research trends in mobile and ubiquitous learning: A review of publications in selected journals from 2001 to 2010. *British Journal of Educational Technology*, *42*(4), 65–70. doi:10.1111/j.1467-8535.2011.01183.x

Ibrahim, N., Kamaruddin, S., & Ling, T. (2017). Interactive educational Android mobile app for students learning Chinese characters writing. *Computer and Drone Applications (IConDA), 2017 International Conference on*, 96-101. 10.1109/ICONDA.2017.8270407

Jiang, W., & Li, W. (2018). Linking up learners of Chinese with native speakers through WeChat in an Australian tertiary CFL curriculum. *Asian-Pacific Journal of Second and Foreign Language Education*, *3*(1), 1–16. doi:10.118640862-018-0056-0

Jin, L. (2018). Digital affordances on WeChat: Learning Chinese as a second language. *Computer Assisted Language Learning*, *31*(1-2), 27–52. doi:10.1080/09588221.2017.1376687

Kuo, J. H., Huang, C. M., Liao, W. H., & Huang, C. C. (2011). HuayuNavi: a mobile Chinese learning application based on intelligent character recognition. In *International Conference on Technologies for E-Learning and Digital Entertainment* (pp. 346-354). Springer. 10.1007/978-3-642-23456-9_63

Lan, Y.-J., & Lin, Y.-T. (2016). Mobile Seamless Technology Enhanced CSL Oral Communication. *Journal of Educational Technology & Society*, *19*(3), 335–350.

Li, D. C. S. (2017). *Multilingual Hong Kong: Languages, literacies and identities (Multilingual Education 19)*. Cham: Springer. doi:10.1007/978-3-319-44195-5

Lu, J., Meng, S., & Tam, V. (2014). Learning Chinese characters via mobile technology in a primary school classroom. *Educational Media International*, *51*(3), 166–184. doi:10.1080/09523987.2014.968448

Mahamad, S., Hipani, N., Basri, S., Hashim, A., Sarlan, A., & Sulaiman, S. (2016). Development of Chinese language application in learning as a second language for Malaysian. *Computer and Information Sciences (ICCOINS), 2016 3rd International Conference on*, 596-599. 10.1109/ICCOINS.2016.7783282

Persson, V., & Nouri, J. (2018). A systematic review of second language learning with mobile technologies. *International Journal of Emerging Technologies in Learning*, *13*(2), 188–210. doi:10.3991/ijet.v13i02.8094

Rawendy, Y., Ying, Y., Arifin, Y., & Rosalin, K. (2017). Design and Development Game Chinese Language Learning with Gamification and Using Mnemonic Method. *Procedia Computer Science*, *116*, 61–67. doi:10.1016/j.procs.2017.10.009

Shih, B., Chen, C., & Li, C. (2013). The exploration of the mobile Mandarin learning system by the application of TRIZ theory. *Computer Applications in Engineering Education, 21*(2), 343–348. doi:10.1002/cae.20478

Sun, H., Hou, J., Hu, X., & Al-mekhlafi, K. (2015). A Context-based Support System of Mobile Chinese Learning for Foreigners in China. *Procedia Computer Science, 60*(1), 1396–1405. doi:10.1016/j.procs.2015.08.215

Syson, M., Estuar, M., & See, K. (2012). ABKD: Multimodal Mobile Language Game for Collaborative Learning of Chinese Hanzi and Japanese Kanji Characters. *2012 IEEE/WIC/ACM International Conferences on Web Intelligence and Intelligent Agent Technology (WI-IAT), 3,* 311-315.

Tam & Yeung. (2010). Learning to write Chinese characters with correct stroke sequences on mobile devices. *Education Technology and Computer, 2010 2nd International Conference on, 4,* 395-399.

Tam, V., & Luo, N. (2012). Exploring Chinese through learning objects and interactive interface on mobile devices. *Proceedings of IEEE International Conference on Teaching, Assessment, and Learning for Engineering,* H3C7-C9. 10.1109/TALE.2012.6360350

Tam, V., & Luo, N. (2014). An Intelligent Mobile Application to Facilitate the Exploratory and Personalized Learning of Chinese on Smartphones. *Advanced Learning Technologies (ICALT), 2014 IEEE 14th International Conference on,* 411-412. 10.1109/ICALT.2014.123

Tseng, C. C., Lu, C. H., & Hsu, W. L. (2007). A mobile environment for Chinese language learning. In *Symposium on Human Interface and the Management of Information* (pp. 485-489). Springer. 10.1007/978-3-540-73354-6_53

Wang, J., & Leland, C. H. (2012). Exploring Mobile Technologies for Learning Chinese. *Journal of the National Council of Less Commonly Taught Languages, 12,* 133–159.

Wang, Y., Ji, Y., Zhang, C., & Sun, L. (2010). An approach and implementation of Chinese character learning based on Mobile Game-Based Learning. *Network Infrastructure and Digital Content, 2010 2nd IEEE International Conference on,* 169-173. 10.1109/ICNIDC.2010.5657845

Wang, Y. H. (2016). Could a mobile-assisted learning system support flipped classrooms for classical Chinese learning? *Journal of Computer Assisted Learning, 32*(5), 391–415. doi:10.1111/jcal.12141

Wong, L., Chin, C., Tan, C., & Liu, M. (2010). Students' Personal and Social Meaning Making in a Chinese Idiom Mobile Learning Environment. *Journal of Educational Technology & Society, 13*(4), 15–26.

Wong, L. H., Boticki, I., Sun, J., & Looi, C. K. (2011). Improving the scaffolds of a mobile-assisted Chinese character forming game via a design-based research cycle. *Computers in Human Behavior, 27*(5), 1783–1793. doi:10.1016/j.chb.2011.03.005

Wong, L. H., Chin, C. K., Tan, C. L., Liu, M., & Gong, C. (2010). Students' meaning making in a mobile assisted Chinese idiom learning environment. In *Proceedings of the 9th International Conference of the Learning Sciences-Volume 1* (pp. 349-356). International Society of the Learning Sciences.

Wong, L. H., Hsu, C. K., Sun, J., & Boticki, I. (2013). How flexible grouping affects the collaborative patterns in a mobile-assisted Chinese character learning game? *Journal of Educational Technology & Society*, *16*(2), 174–187.

Wu, Yuan, Zhou, & Cai. (2013). A Mobile Chinese Calligraphic Training System Using Virtual Reality Technology. *AASRI Procedia, 5*, 200-208.

Xu, Q., & Peng, H. (2017). Investigating mobile-assisted oral feedback in teaching Chinese as a second language. *Computer Assisted Language Learning*, *30*(3-4), 173–182. doi:10.1080/09588221.2017.1297836

Yang, C., & Xie, Y. (2013). Learning Chinese idioms through iPads. *Language Learning & Technology*, *17*(2), 12–22.

Ying, Y., Lin, X., & Mursitama, T. N. (2017). Mobile learning based of Mandarin for college students: A case study of international department' sophomores. *Information & Communication Technology and System (ICTS), 2017 11th International Conference on*, 281-286.

Ying, Y., Rawendy, D., & Arifin, Y. (2016). Game education for learning Chinese language with mnemonic method. *Information Management and Technology (ICIMTech), International Conference on*, 171-175. 10.1109/ICIMTech.2016.7930324

Zhang, L., Wang, J., & Li, H. (2011). A Mobile Learning System for learning Mandarin Pronunciation. *Computer Science and Network Technology (ICCSNT), 2011 International Conference on*, *1*, 621-624. 10.1109/ICCSNT.2011.6182034

# Chapter 12
# Mobile Translation Experience:
## Current State and Future Directions

**Nancy Xiuzhi Liu**
*University of Nottingham – Ningbo, China*

**Matthew Watts**
*University of Nottingham, UK*

## ABSTRACT

*After closely examining the experiences of mobile translation in which people engage with translation on mobile platforms in the contexts of healthcare, crowdsourcing, and machine and translator training, the authors have identified a tightly intertwined relationship between mobile translation and machine translation. They have also found that the technological side is more dynamic than the user side in the case of mobile translation and machine translation, which may lead to a gradual reduction of people learning foreign languages and a possible loss of professional translators and language specialists. When it comes to contextual and textual translation, however, human translators currently outperform mobile or machine translators. Although human contribution will be determined by translation scenarios or specific translation tasks, the human-mobile/machine interaction in translation deserves further studies. It is imperative to compare mobile use and experience in human-mobile interaction related to translation in different cultures or countries so as to locate similarities and differences. Furthermore, it is also expected from the editor that further studies should focus on mapping, measuring, and modeling those identified similarities and differences.*

## INTRODUCTION

Mobile translation (MbT) is, essentially, an umbrella term which can be taken to encompass all forms of translation, both human and machine, and processes in the translation workflow carried out on a mobile device. As the number of mobile devices in the world continues to increase (GSMA Intelligence, 2018) and the underlying technology improves, it becomes increasingly possible to carry out tasks and activities on mobile devices that were traditionally performed on non-mobile, stationary devices, such as desktop computers. Translation is one of those tasks that might be carried out on mobile devices and

DOI: 10.4018/978-1-5225-7885-7.ch012

it is easy to see why – mobility is inherent in translation and its main purpose is to enable communication across different languages and cultures, which only come into contact when there is movement of people, whether physical or virtual. The form that this translation takes, however, may differ greatly depending on the context and to provide an exact definition of the term 'mobile translation' is difficult. Indeed, there is not a single, agreed upon definition for mobile translation and its definition has evolved in recent years. However, in a blogpost for the translation company Stepes, Armstrong (2016) discusses the evolution of the meaning of mobile translation into the modern idea of apps which professional translators can use to facilitate their work. Jimenez-Crespo (2016, p. 76) then draws on this to formalise three definitions as follows: "(1) the localization of apps; (2) the use of MT [machine translation] apps; and (3) the use of apps inspired by crowdsourcing workflows to carry out human translation either though post editing MT or through direct human translation". These definitions are what will be used in this chapter to reflect on different ways that users have engaged with mobile translation. As definition (1) refers to localizing (translating) apps to ensure they are ready to launch in a specific locale and does not necessarily involve using a mobile device to produce this translation, it does not specifically cover the mobile translation experience as the content could be translated using more traditional, desktop-based translation workflows. Thus, definitions (2) and (3) are of principal interest in this piece as they can be analysed in terms of users' experience in using mobile devices for translation, both from the perspective of professional translators and non-professionals. In short, mobile translation in this chapter refers to translation, human or machine, that is performed through the use of mobile devices or realized through apps operated on mobile systems.

Numerous mobile apps have been developed to facilitate communication between speakers of different languages and the list of such apps is inexhaustible with the technology constantly evolving and new apps constantly being developed. Although we may think of MbT technologies as rather new, they can be traced back further than the modern systems available on smartphones to electronic pocket translators. Furthermore, in 1993, the German government, with collaborators around the world, initiated a huge project called Verbmobil, a speaker-independent and bidirectional speech-to-speech translation system for spontaneous dialogue in mobile situations between three languages (German, English and Japanese) (Wahlster, 2013). However, it is with the rise of smartphones that MbT has proliferated and become democratised, with increasing numbers of people using the technology. Indeed, our modern understanding of MbT apps and apps more generally only emerged in the late 2000s with the development of app stores that enabled easy installation on a device without a wired connection. Taking China as an example, according to a report by the Sootoo Institute (2018), a Chinese research institute specialized in analysing big data on the Internet, MbT users have increased substantially in recent years, especially since 2015, as shown in Figure 1. By 2019, the number of MbT users is projected to reach 382 million.

According to the report, mobile translation has benefited from the development of other technologies, such as Virtual Reality, Artificial Intelligence, Computer Vision and Neural Machine Translation (NMT) among others, allowing both the accuracy and speed of MbT to be improved to a much higher level. The report categorizes MbT into the two groups of travelling translation and language learning. Apps with the function of travelling translation can provide instant translation in over 100 languages through screen tapping, picture taking or sentence recording, while those designed for language learning can function between 2 to 12 languages, mostly with English as the main language by providing key word translation in a couple of other languages. The report also provides the updated number of downloads for a variety of MbT apps as of 17th April 2018 as shown in Figure 2.

*Figure 1. Mobile translation users and the rate of increase in China (Translated and adapted from Sootoo Report, 2018)*

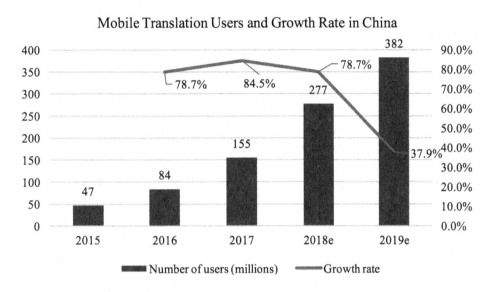

*Figure 2. Number of MbT app downloads (Translated and adapted from Sootoo Report, 2018)*

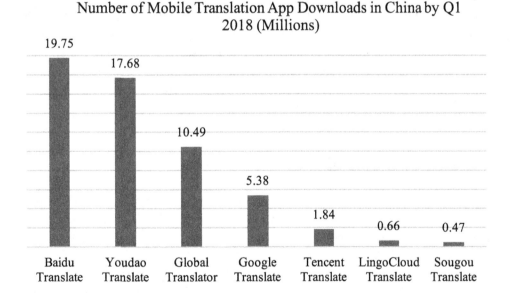

Baidu Translate and Youdao Translate are the top two apps downloaded with 19.75 and 17.68 million downloads respectively, while Google Translate falls somewhere in the middle with 5.38 million. Furthermore, the mobile app WeChat, developed by Tencent Company and released in January 2011, is the most popular social network mobile app in China and increasingly around the world. It provides users with many multimedia communication functions, including instant text translation and scanning for translation and serves different countries, languages, operating systems and network formats. According to Qi and Mei (2016, p. 387), it took only 15 months for WeChat to reach 100 million users, compared to 56 months for Facebook and 30 months for WhatsApp. In addition, the number of WeChat users in China now exceeds 500 million and the number of overseas users exceeded 100 million by the end of 2014. Although these figures are for China, a country known for its willingness to embrace technological developments, it is likely that this mirrors the scenario in other countries as the trend of MbT is on the rise with unstoppable momentum due to the benefits and opportunities it provides users. However, it is likely that Google Translate is far more popular in other countries as it only became available in China in 2017 (Sootoo Report, 2018) and Google products are much more limited in China than other countries. Scarce data on MbT, in terms of the number of users and increased over time, is available in other countries. In the UK, for example, Ofcom conducts yearly surveys to understand how citizens are interacting with and using technologies, including smartphones. The survey ask respondents whether they use apps that fall within certain categories (e.g., social media, news, games), but translation or even language learning/services are not in the categories, so figures regarding MbT are unavailable. It is, thus, worth highlighting here that further research and data gathering on MbT is necessary in other countries in order to produce comparable data and for a greater understanding of how the number of users are increasing and how exactly people are using such technologies.

Despite the fact that Translation Studies (TS) is undergoing a technological turn (Cronin, 2010; O'Hagan, 2012, 2016b), little attention has been given to MbT within current research, indicating that it is an exciting, emerging area that provides new forms of translatioOn. Given the large numbers of people engaging with MbT, it thus seems timely and necessary for TS to begin exploring this phenomenon. As such, this chapter aims to provide a review of the mobile translation experience by focussing on four specific areas in which users may engage with MbT: machine translation, crowdsourcing, health care and translator training. It reflects on the experiences of people engaging with translation on mobile devices in both professional and non-professional contexts through presenting and discussing previous research conducted on the MbT experience in these areas. The chapter concludes by discussing and highlighting the importance of potential new areas and questions for research around the mobile translation experience and how TS could engage with this area.

## MACHINE TRANSLATION AND MOBILE TRANSLATION

Machine Translation is translation carried out by automatic means, translation in which the conversion from language A to language B (the translation act) is carried out not by a human, but by a computer. Although the first MT system was developed in 1954, it was only during the late 1990s that systems started to become readily available online and accessible by anyone, thanks in large part to the accessibility and spread of the Internet. Indeed, one of the most famous online MT systems, Google Translate, only launched its desktop-based website in 2006 (Och, 2006), with the app version for mobile devices launching much later on, firstly on the Android operating system in 2010 (Verma, 2011). So, whilst MT systems have existed for over 60 years, new ways of accessing MT have emerged and the technology has become increasingly democratised and used on a large scale. As of June 2018, Google Translate has over 500,000,000 installs according to the Google Play Store (this only includes Android devices), which indicates that large numbers of people are engaging with this form of mobile and machine translation. In addition to this new way of non-professional translators engaging with translation, mobile translation presents new opportunities for translation professionals, enabling them to change the way they work and carry out tasks on the move. Mobile translation is a new experience for users as, unlike other aids for travellers such as phrasebooks, it allows the user to determine and input the content to be translated, thus perhaps being more useful and more user-centric. Indeed, more recently, MbT technologies have also removed the need for the user to speak the phrase thanks to the integration of text-to-speech technology, thus in theory providing more seamless and smoother experiences.

Furthermore, recent years have seen the emergence of a plethora of smart devices in addition to smartphones – smart lighting, smart fridges, smart washing machines, smartwatches, to name but a few. In 2017, a PwC report found that "eighty-one percent of US internet users are aware of the concept of smart devices" and "roughly one in every four (26%) US internet users currently owns a smart home product or device" (Bothun & Lieberman, 2017, p. 5/6). A key feature of smart devices is their ability to connect to smartphones and enable further new experiences, as is the case with smart headphones. For example, Pilot Translator, developed by Waverly Labs, and Pixel Buds, developed by Google, are earphones which can pair with mobile devices and enable users to receive spoken translations directly into their ears, thus facilitating conversations with other people with whom they do not share a language. However, reviews for such devices are not entirely positive, with James Temperton (2017) of Wired magazine stating that "you don't actually need the Pixel Buds to do this [translation], they're an entirely pointless accessory" and Leo Mirani (2017) of 1843 magazine also providing a negative review due to the impractical nature of the Pixel Buds. Although it is impossible to determine how successful these technologies will be, it is to be expected that when they are still in their initial stages there will be difficulties and that it is only likely to be technophiles who embrace them. Furthermore, their success is somewhat limited by the underlying MT systems which actually produce the translation as, even if the technology itself works perfectly, if the translation is of a poor quality the users will have a negative experience. However, to date, there are no studies exploring how users of mobile devices might use them to engage with MT or their experience of MT on mobile devices. Indeed, as O'Brien (2017, p. 313) highlights, "research into the usability and acceptability of MT by end users is still in its nascent stages" and she is here referring to all forms of MT, not just MT on mobile devices, which is itself an even more specific area of study.

In the context of professional translation and translators, mobile translation is also a new, developing area. Currently, translators still work primarily at a desktop workstation, using CAT tools which are only available in desktop versions, although there has been some initial work to investigate the possibility of completing translation jobs on mobile devices. For instance, the ADAPT centre at Dublin City University developed the Kanjingo post-editing app, initially available through a browser and then developed into a native iOS app, in order to investigate the viability of translating and post-editing machine translation on mobile devices (Torres-Hostench, Moorkens, O'Brien, & Vreeke, 2017, p. 139). The researchers have carried out three rounds of testing which provide insight into the way that users experience such an app. O'Brien, Moorkens and Vreeke (2014) describe the first round of testing in which five users from different backgrounds (including one professional translator) were asked to evaluate the app by Think Aloud Protocol and were quite positive about the app in general. Their main concerns were practical, relating to problems with punctuation and auto-capitalization, retention of work if the user stops half way through, insufficient help available and input problems due to the speed of typing on a keyboard on a mobile device. Moorkens, O'Brien and Vreeke (2016) describe the subsequent round of testing based on improvements made after the first round, with 13 people completing a feedback survey. In this round, users were again positive about the app, with most respondents liking the app's interface and some of the issues, such as problems with accented characters and spellcheck could have been dealt with in a controlled-testing environment. Users generally stated they preferred desktop applications for post-editing due to ease of use, but the app would be useful in situations where desktop software is unavailable. They also provided several suggestions for improvements to the app, such as clicking words to see other possible translations or editing segments after they have been submitted and the ability to view the finished translation at a level longer than the sentence level. Torres-Hostench et al. (2017) then describe another round of user testing in which five participants used the app while five observers watched them testing the app. The testers highlighted problems with the ergonomics of inputting text via a small keyboard on a smartphone device, indicating that this would be a major limitation for using the app for professional purposes, but that it might be more useful for not-for-profit projects. They also reviewed using voice input for the app and found that it was more useful when re-translating longer sentences and discussed the usability of the app suggesting two major improvements - changing the location of the accept button and allowing a user to edit anywhere within a tile. The researchers themselves call for further research and testing to explore different uses of the app and this is something the authors of this piece echo.

The use of MT apps by non-professional translators represents a large, growing area, which suggests that users are happy to engage with MT on mobile devices to aid their communication. Further study is needed to explore their experience of this form of MT, to understand the positives and negatives and possibilities and limitations. In addition, MT apps and apps for post-editing MT are still new and may represent a growing area of interest, both from a research perspective and a practical perspective for translators, particularly with the increasing hybridity of devices capable of being used in both a mobile and desktop form. Kanjingo is one example of an app designed to take elements of the translation process traditionally completed on desktop devices and enable them to be completed on a mobile device and might be described as a pioneer in this area. What is encouraging is the willingness of users to engage with app versions of translation tools and as smartphone usage becomes increasingly commonplace it is thus likely that people will be more willing to try out new activities and experiences which traditionally have not been carried out on a mobile device.

## CROWDSOURCING

In 2006, Jeff Howe and Mark Robinson, editors at Wired magazine, coined the term crowdsourcing, explaining that "simply defined, crowdsourcing represents the act of a company or institution taking a function once performed by employees and outsourcing it to an undefined (and generally large) network of people in the form of an open call" (Howe, 2006). Crowdsourcing may refer to either paid or unpaid work and has grown on a large scale thanks to the communicative possibilities of the Internet, with small work packages easily distributed to large numbers of users almost instantaneously. Although it was possible to crowdsource work prior to the Internet by phone or post, it is the Internet, thanks to its communicative power, which has enabled this phenomenon to grow on a large scale as tasks can easily and speedily be split up and disseminated to workers. For translation, this model of working has facilitated the rise of new forms of translation, for which a variety of terms exists, such as community translation, collaborative translation and volunteer translation, often used interchangeably (O'Hagan, 2016a, p. 940). They all essentially refer to the translation task being broken down into smaller chunks and being carried out by large numbers of people on the Internet – in some cases they are paid for the work, in others they are unpaid, and in some cases it is professional translators who do the work and in others it is anyone who volunteers.

In recent years, the percentage of website traffic generated through mobile phones worldwide has risen rapidly, from just 0.7% in 2009 to 52.2% in 2018, with the percentage even higher in certain areas such as Asia (65.1%) and Africa (59.5%) (Statistia, 2018). The statistics indicate that people are increasingly using their mobile phones to access the Internet, thanks to the spread of 3G/4G networks and the normalisation of plans which include a certain quota of data each month. As such, it is understandable that the crowdsourcing model, which relies on small chunks of work which lend themselves well to the physical limitations of the smaller screen of mobile devices, is being explored for mobile devices and translation. Users can simply take the device out of their pocket, select the relevant app or go to the relevant webpage in the browser, and complete as many translation tasks as they wish. In addition, crowdsourcing on mobile devices enables users to complete work on the go, whenever is convenient for them, perhaps whilst travelling or during breaks at work. For example, Duolingo is a popular language-learning platform that was developed in response to the question "how can we get a 100 million people translating the web into every major language for free?" (Von Ahn, 2011). Although Duolingo is available on both desktop and mobile devices, over 80% of traffic comes from mobile apps (Amazon, n.d.), indicating that it is a clear example of mobile translation. Firstly, Duolingo teaches its users a language and then "as learners progress, they will be invited to translate simple sentences at first before eventually being asked to translate sentences from live web pages" (Cronin, 2017, p. 95). Indeed, Duolingo did manage to produce usable translations with this approach as it collaborated with CNN and Buzzfeed to translate their news articles into other languages, with some users doing the translating and others checking the translation. However, the success of this initiative is unclear given that Duolingo decided to retire their immersion tool, the name given to this translation service, in January 2017 citing lack of resource as the reason for their decision.

Furthermore, crowdsourcing via mobile devices represents a commercial opportunity, potentially through the monetisation of people's spare time, just as apps such as Kanjingo seek to enable users to use their spare time productively for voluntary work. One such company using the crowdsourcing model for their workflow is Stepes. Stepes, established in 2015 by CSOFT International, a localization company, functions based on a crowdsourcing model and launched its mobile app, employing a chat interface, in December 2015 (Lomas, 2015). On its website (https://www.stepes.com/mobile-translation/), the company states that

Stepes unlocks convenient and accurate human translation services not yet witnessed from a mobile device. Previously, translation tools were desktop-based. However, with Stepes' patent pending mobile translation technology, translators utilize the advanced mobile application to confidently translate and earn income from virtually anywhere at any time.

It also describes itself as "The World's First Uber Translation Service" as it uses a network of over 100,000 translators to provide "translation of short texts in minutes", that is, it operates by sending short snippets to translators via the chat interface and they then input the translation into the chat. The premise of the app is that there are 3.6 billion people around the world who speak two or more languages who could potentially translate during their free time and earn money for doing so (Collins, 2017). In an interview with a translator who translates via Stepes, the translator was positive about Stepes, citing its "flexibility" and "ease of use" (Adams, 2017).

However, Stepes is still a relatively new company and this way of translating is still rather new and innovative; it remains to be seen what effect and level of disruption apps such as these may have on the translation industry and the uptake of this model of working by translators. In addition, the idea that being bilingual is the only requirement for a person to be a translator or perform translation may be an erroneous one (Kolawole, 2012) and does also raise concerns over the quality and professionalism of such translations. Nevertheless, in the era of ephemeral, user-generated content, perhaps translation of a non-professional quality is all that is needed in certain circumstances if all that is required is a fast translation, which will not have a long shelf life and which will not be used in demanding, professional circumstances. Stepes does indeed offer Twitter and Facebook translation services for users, which will translate their social media content. Whilst this might generally be considered the circumstances in which MT would be useful, this is not always true given the brevity and informal nature of communication on the Internet. Crowdsourcing which uses humans to translate may produce better quality translations of such content as they are more likely to be able to take into account any slang or extra-textual features such as emoticons or images that will accompany the translation. Furthermore, this form of translation is not without precedent and has become widespread in the modern era on the Internet, with large companies such as Facebook and Twitter sourcing translations from their user community rather than professional translators (see Littau, 2016; O'Hagan, 2016a). What is clear is that this new translation workflow is developing and is becoming increasingly prominent thanks to the spread of MbT and is something which will require further study in the future to understand its implications on translation, the translation industry and translators. Crowdsourced translation seems to lend itself well to translation of short content, designed for digital consumption and it is understandable that it may become increasingly popular through mobile devices as their physical form also suits short content due to their size and content designed for digital consumption.

## HEALTH CARE

This section will discuss the use of MT apps in healthcare settings. According to Buijink, Visser, and Marshall (2013):

*There were 10,000 apps available in the 'medical section' of Apple's 'App store' and over 3,000 on Google's 'Play store'. Since these platforms facilitate development and distribution of mobile applications by clinicians and other developers, rapid proliferation of the market will likely continue. (p. 90)*

However, with the potential growth in the usage of such apps, peer-reviewed content, appropriate regulatory measures and professional involvement should be in place in order to reduce potential risks. Regarding the usage of mobile translation apps specifically, research carried out thus far concerning use by medical workers and patients in healthcare settings is still quite limited. It is believed that patients with a different cultural background and language than their health nursing staff are more likely to be disadvantaged in their access to the health system (Gerrish, Chau, Sobowale, & Birks, 2004). Relevant studies (e.g., Karliner, Jacobs, Chen, & Mutha, 2007) report positive benefits of professional interpreters on communication (errors and comprehension), utilization, clinical outcomes and satisfaction with care, although some medical workers tend to use family and friends more. Mobile electronic translation tools are ubiquitously available, mainly hassle-free, and provide quick translations for users, which seem attractive at first glance. Several MbT apps are already available for mobile devices, such as MediBabble Translator by NiteFloat, Inc, or Universal Doctor Speaker by Universal Projects and Tools SL (Albrecht, Behrends, Matthies, Von Jan, & Schmeer, 2013).

In the sphere of mobile translation experience, Oladosu and Emuoyibofarhe's (2012) study highlights the use of mobile communication technology for distant communication in Nigeria, in Yoruba (one of its three major languages). Such technology helps to provide healthcare services to the disadvantaged such as semi-literates (those who cannot communicate in an official language) and those in a rural setting without physical contact with the doctors. In order to achieve a better healthcare delivery system, a doctor–patient chat application was developed using various components such as a mobile client terminal, the Internet and a server. The model was simulated using JAVA 2 Micro Edition (J2ME) and the JAVA 2 Enterprise Edition (J2EE). The tool developed in this work has been tested and usability assessment conducted among medical practitioners and a number of clients.

Adopting MT technology, the approach employed in this work is known as example-based translation in three stages of the language translation process: source language (SL) word/expression analysis, SL to target language (TL) sentential structure translation and TL generation. Some semi-structured message templates are programmed into the mobile user's device from which a word or phrase can be selected. Their work uses the mobile programming technology provided by JAVA to design and simulate an instant message exchanger and language mapping that allows patients to reach out to physicians and have access to healthcare services. It addresses the observed shortcomings and implements an e-health system that enables mobile ubiquitous service delivery. This work introduces a novel paradigm of real-time, online, indigenous-official language mapping thus enabling illiterate patients to communicate with highly educated medical personnel. In their survey of both medical experts and patients, the user experience was very satisfactory. Medical experts consisting of doctors, nurses and pharmacists rated it above 80% in its relevance to mobile health care provisioning. In general, users were satisfied with having tools that can assist medical practitioners in doing their job and patients in accessing medical services anywhere and at any time. "It also gives the patient the opportunity of having remote access to various specialists and it allows for language flexibility, through the use of the language translator. If adequately used, it will help reduce long queues in the hospitals, as it is a real-time and instant information exchange via mobile devices" (p. 155). Therefore, it has the potential to be implemented in more indigenous languages in Nigeria and Africa.

Albrecht et al. (2013) conducted a study in Hannover Medial School to examine nurses' opinions of a specific mobile medical translation app called xprompt. The application was chosen based on the highly positive reviews it received from healthcare professionals. According to the researchers, "the purpose of xprompt was simply to provide additional means for alleviating communication problems between the nursing staff and non-German speaking patients" (p. 2). It can be helpful in many different settings as it contains a large phrase set (800 phrases, currently available in 23 languages), covering nursing care as well as daily life communications. The application usage is simple. The phrases are provided in tailored menus for the nursing staff and the patients, grouped according to the situation in which they might be used, and it is quick and simple to navigate to the desired content. Via simple point and touch actions, selected phrases are translated into the target language. This study was conducted as part of a wider project called 'iPads in Nursing', examining more generally how such mobile technology could be integrated into a nurse's workflow. The app used is not an MT app as such, as users are not able to freely input text or speech to be translated. Instead, they use the menu to navigate to a certain situation and select the phrase and the target language and the device can then show or speak this translation to the patient or the user can enter a word and search for a phrase for translation. It is, therefore, more akin to an electronic phrasebook rather than a translator or MT. The results showed an obvious discrepancy between the expert assessments of xprompt, stated in the user comments on the App Store and the actual usefulness attributed to it by the participants. The above appears to indicate that ultimately a distinction must be made between the individual use of xprompt and its use in the context of nursing care. For integrating it into the daily routine of inpatient care, detailed instructions with respect to the application's use must be provided. It can be assumed that individual users who installed xprompt on their own initiative had clear expectations about the program. They were searching the App Store for solutions to a specific problem they had encountered, that is, an "always available mobile medical translation." The results regarding the deployment of xprompt showed that when introducing new technologies it is especially important to adequately train the nursing staff and adapt the training according to their job requirements.

Similarly, Villalobos, Lynch, DeBlieck, and Summers (2017) explore the use of the Canopy Translation App in the interactions in mental health care between English and Spanish. This kind of interaction involves a significant exchange of information to identify symptoms and formulate a working diagnosis. The inability of patients to communicate due to language limitations may lead to situations that will be detrimental, such as receiving the wrong diagnosis or treatment due to inadequate translation from English to Spanish. The Canopy Translation App from Canopy Innovations Incorporated was developed in an effort to improve communication with patients with limited English proficiency. It has a wide array of medical phrases commonly utilized in clinical settings available and organized into categories. The medical phrases available can be translated into 15 languages, including Spanish. The methods used in their study were mixed: firstly, they assessed the participants' (medical residents, registered nurses, and psychiatric nursing assistants) speaking and listening proficiency in Spanish; they then provided a tutorial and online resources showing them how to use the Canopy Translation App. Participants used the translation app during their assessments of Spanish-speaking patients. After 2 weeks, participants were asked to complete the System Usability Scale (SUS)-1 test questionnaire, including the demographics and usage questionnaire. Post-usage was measured after a period of 6 weeks, by completing the SUS-2 posttest questionnaire, which included a question asking the total number of times the app was used followed by semi-structured interviews. According to the SUS scores, participants found the Canopy Translation App useful during their interactions with Spanish-speaking patients, but also found some limitations with the app. According to Villalobos et al. (2017):

The main drawback discussed was compromising rapport. Participants expressed that a translation app might hinder the therapeutic relationship between patients and health care professionals. However, participants consider using a mobile translation app more practical than other means, such as contacting translators. Participants continued using the translation app and encouraged other staff members to download it; they believe Canopy should be considered for other settings to assess both the physical and mental needs of patients not proficient with English. (p. 377)

The findings emphasize the importance of exploring new resources and technologies to ameliorate language barriers.

Chang, Thyer, Hayne, and Katz (2014) have carried out studies of mobile interpreting services in healthcare settings as an adjunct to human service. In Australia, because of its large migrant population with variable fluency in English, interpreting services help ensure that healthcare services are delivered appropriately to these populations. However, the use of professional interpreters in hospitals is expensive and there are also issues with service availability and convenience. They set out to examine how mobile technology played a role as an adjuvant option of improving communication with the patient during the care of a non-English speaking patient with a complex past medical history. In one of their case studies, they used a professional interpreter in a Persian patient's history taking and consent obtaining. However, some issues arose such as timing inconveniences and high costs during their everyday ward rounds, new findings, plan updates among others. They decided to try Google Translate on a smartphone as it allows two-way interpretation between Persian and English. The team typed or spoke English into Google Translate, which translated it into written Persian words that could be either read from the screen or spoken aloud by the device using the embedded speech engine and vice versa. The use of Google Translate on a smartphone is both convenient and inexpensive where patient review and ward rounds can be done spontaneously, and simple medical procedures can be performed conveniently while nursing staff could describe the prescribed medications to the patient and respond better to patient requests. Even kitchen staff made use of this technology to discuss food options appropriate to the patient's religion and needs. Furthermore, the patient was more satisfied as most of his questions were answered. Their study has found that mobile devices containing software with translating abilities have promising potential to improve communication between patients and hospital staff, particularly as they are highly convenient and inexpensive. However, they also found that there are concerns about the accuracy of the interpretation done with such software and more research needs to be carried out to support or allay these concerns. Similarly, using free, online MT services for translating confidential information or data may raise issues relating to data protection. For now, clinically important and medicolegal related interpretation should be undertaken by professional interpreters, whereas less crucial tasks may be performed with the help of interpreting software on mobile devices.

## TRANSLATOR TRAINING

Mobile experiences in relation to translator training fall mainly into three categories, namely the technology, language learning and translation per se. In terms of the technological side, the focus is mainly on the development of various apps aimed at facilitating translation, while in language learning the focus is mostly on improving the learning experience through translation. As for the translation experience, the main focus of this chapter, studies mainly concern how mobile devices are used in translation classrooms. Mobiles apps are more often used for language learning (e.g., C.-K. Chang & Hsu, 2011; Godwin-Jones, 2011; Groves & Mundt, 2015; Mundt & Groves, 2016) than in translator training. Over the last ten years, such apps have been used widely in vocabulary practice, quiz delivery, live tutoring, emailing lesson content and delivery, to name a few examples (Godwin-Jones, 2011).

For example, Chang and Hsu's (2011) research introduced mobile devices into an intensive reading course and allowed functions that are usually found only in the language laboratory to be easily and flexibly utilized in the general classroom. They integrated a computer-assisted language-learning (CALL) system with an instant translation, annotation and multiusers shared mode that can support a synchronously intensive reading course in the normal classroom. Their experiments in analysing the usage of the system, including the attitude and satisfaction of users, show that the system enhances and improves the reading comprehension of English as a foreign language (EFL) readers, in small groups of two to four in particular. The researchers were encouraged to find that many participants agree that the developed technology tool was useful and interesting in their language learning. They were willing to conduct their task (reading) through the CALL system, which helped them remain focused on their tasks. As a result, the comprehension outcomes of EFL readers were likely to be improved by using the CALL system. There are also studies verifying that mobile technologies play a positive role in improving writing skills and vocabulary leaning (e.g., Agca & Özdemir, 2013; Lee & Kim, 2013). Rahimi and Miri's (2014) investigation on the impact of mobile dictionary use on language learning finds that EFL learners who used the mobile dictionary to learn English improved their language ability more than those who used the printed dictionary. Furthermore, they also find that mobile phones play a vital role in extending learning out of the classroom anywhere anytime. Similarly, Lilley and Hardman (2017) examine mobile dictionary use in an advanced EFL class in South Africa with the findings both contested and contradictory in that mobile dictionary use is "changing the object from 'understanding' to 'translation', as well as changing labour from 'collaborative' to 'autonomous'" (p. 143).

Bahri and Mahadi (2016) investigate the usage of mobile devices in the translation classroom through a mixed-method approach. They gathered qualitative data through a focus group/interview with 4 translator trainers and used these findings to inform the design of a quantitative survey. They found that generally instructors encourage students to use mobile devices to complete tasks, assignments, perform searches or discuss relevant content on social media, but discourage and found it less useful using such devices for tasks not related to the lesson or learning content. In this study, the mobile devices under consideration included laptops, tablets and smartphones and a preference was expressed for laptops over other kinds of mobile devices. Most of the tasks described in the study are not specific to translation students and are employed by a wide range of students. However, they found that more advanced students were encouraged to search terminology databases using their mobile devices and it was highlighted by one focus group participant that students rarely carry paper dictionaries anymore and instead rely on digital dictionaries accessible through mobile devices. The authors recognize the limited nature of their study, in that it was only sent out to instructors for one language pair (English ⇔ Persian) and that the findings may not be generalizable and also call for further research to verify their findings. The findings highlight the potential of mobile devices to have a positive impact on the translation classroom activities and underline the need for their systematic integration into the translation curriculum as part of the tools contributing to the development of technological competence.

Arnáiz-Uzquiza and Álvarez-Álvarez (2016) use a survey to explore the use of technology in the learning process for 280 undergraduate translation and interpreting students across 13 different Spanish universities. Although the survey questioned respondents about their usage of various devices, including desktop computers as might be used in lab-based sessions, it focusses specifically on the usage of mobile devices such as laptops, smartphones, tablets/iPads and smartwatches. They found that 77% of respondents used a laptop in the classroom and 55% used a smartphone, whilst only 10% used a tablet/iPad. They also found that device usage increased as students progressed through their course, with first years using devices the least and final years using devices the most. Similarly, perceptions of advantages and disadvantages varied over the years, with "limited access to information" being the main disadvantage for first years, whereas for fourth years it was the fact that using mobile apps was distracting and did not create an ideal learning environment. They acknowledge the relatively small sample size of the study and the fact that not all institutions in Spain were represented in the findings. Similarly, they call for further research to investigate if there would be any differences in the findings if undergraduates and postgraduates were compared. They report that smartphones were used for communication, social media and managing work, whilst laptops were used for taking notes and doing work, accessing the VLE, carrying out searches, checking email and carrying out homework and classroom tasks. In addition to the limitations recognised by the authors, the study also does not go into depth on how people actually use these apps or specifically explore how they are used for translation or interpreting activities rather than usage that students of any discipline would make of them. In addition to increasing the sample size, it would be useful to gather more qualitative data regarding students' insights into their experience of using mobile devices for translation activities, exploring these other issues and also to replicate this study in other countries to enable international comparison.

Besides the aforementioned three areas related to training, there are also sporadic studies concerning MbT such as Zapata's (2016) study on interactive translation dictation (ITD). He provides a general overview of interactive translation dictation (ITD), an emerging translation technique that involves interacting with multimodal voice-and-touch enabled devices such as touch-screen computers, tablets and smartphones. This could be described as a more modern version of the old-fashioned dictating mode used by senior members of staff dictating to secretaries or writers dictating their work. ITD integrates new techniques and technologies into the translation sector and Zapata provides a brief description of a recent experiment investigating the potential and challenges of ITD and outlines avenues for future work. The extraordinary aspect about Zapata's paper is the fact that no physical keyboard was used in its preparation. He predicts that multimodal interaction and ubiquitous, mobile and cloud computing appear to be promising avenues for translation technology research. He concludes that "experiments that explore voice, touch and stylus input (and even other emerging interaction modes such as gaze, gesture and brain input) will play a crucial role in the design and development of new user-friendly tools and devices that are adapted to translators' needs and to the changing reality of the industry in the twenty-first century" (p. 71-72).

While the presence of mobile electronic devices in the classroom has posed real challenges to instructors, a growing number of teachers believe they should seize the chance to improve the quality of instruction. The aspects that need to be emphasized are that we cannot afford to lose the opportunity of technological upgrading in this context while users' differentiated needs cannot be overlooked, just as observed by Godwin-Jones (2011) that, as personal devices, smartphones are ideal for individualized informal learning. Therefore, users can determine which apps to acquire and how to use them. Godwin-Jones (2011) further highlights that, "as language educators, we should encourage and assist the learner autonomy this enables and provide means for learners to combine formal and informal learning" and "as mobile devices become even more powerful and versatile, we are likely to see more users make them their primary, perhaps their sole computing devices. This is not a trend language educators can ignore" (p. 8).

## CONCLUSION AND FUTURE RESEARCH DIRECTIONS

This chapter has explored the experiences of mobile translation by discussing specific instances in which people engage with translation on mobile platforms, examining the relationship between MT and MbT, MbT and crowdsourcing, MbT and health care, as well as MbT and translator training. It is clear that there is a close, intertwined relationship between MbT and MT, largely because mobile devices greatly increase accessibility to MT and, as such, MT is the principle type of translation conducted on mobile devices. In conclusion, firstly, this study has found that the technological side is more dynamic than the user side. While the tech specialists are keeping their heads to the grindstone in order to overcome the language barriers in communication, the users or customers, language and translator trainers in particular, are reluctant to embrace the technology with various concerns, such as that it does not help to improve trainees' language proficiency for example. That may account for the fact that relatively few scholars have investigated translation technology in the workplace, and also that research on translators' interaction with translation tools and how this affects their minds and work processes are rather scarce (Christensen, Flanagan, & Schjoldager, 2017).

Secondly, there is the undertone, although less openly expressed by professional trainers and translators, that if the technology becomes so readily accessible and easy to use, as the tech side strives to achieve, it might ultimately lead to less people learning foreign languages and further deprive people of their livelihood as professional translators and language specialists. However, the future of the translation profession may not be so bleak after all. For one thing, machines are unlikely to take over all translation processes, at least not in the foreseeable future; for another, more automated translation processes do not necessarily mean that there will be *no* need for human translators.

Thirdly, there can be no doubt that human translators and MT systems will still need to complement each other, as humans currently outperform computers in tasks that involve contextual and textual interpretation, but the degree of human involvement will depend on the translation scenario, including "the purpose, value and shelf-life of the content" (Way, 2013, p. 2). In the context of health care, for example, complicated communications between patients and health workers have to rely on humans for the purposes of accuracy and safety, while MbT serves as a convenient tool for daily or routine communications between the two to ameliorate language barriers (Villalobos et al, 2017).

Fourthly, regarding specific translation tasks, there might be a growing demarcation line between MT and human translation. MbT, particularly MT apps, may prove to be more handy and popular in the context of carrying out more of a facilitator's role such as during travelling or other less sophisticated communications. But, in other areas such as post-editing or literary translation, where semantic disambiguation, contextual knowledge and genre expertise are required, human translation will still be irreplaceable (Christensen et al., 2017). So, the future of translation will witness a collaborative scenario where MT and humans complement each other.

Overall, the tendency of mobile translation development is unstoppable with its numerous, easily accessible apps, particularly machine translation apps. Increasingly, these apps provide people with more choice around translation, allowing them to input text in writing, orally using a device's microphone, or even visually by using a device's camera and text recognition technology. MbT is becoming an anywhere, anytime translating tool for people who need different languages in various contexts, thanks largely to the growing ubiquity of the internet, with Wi-Fi and mobile data becoming increasingly used and widespread. Furthermore, users now have the possibility to use MT apps without an internet connection thanks to downloadable language packs and the advent of neural networks on mobile devices increasing the quality of MT even without an internet connection. MbT, and particularly MT apps, is becoming a life-changing tool for people who speak no other languages while traveling in other countries. It could be seen as empowering more people than ever to become translators for themselves, who are always on the go, and redefining our understanding of and relationship with translation and what it is to be a translator.

In terms of future research directions, the authors would like to highlight mobile translation as a research priority. Whilst people are experiencing MbT on a daily basis, such as Syrian refugees using the GT app to aid their daily communication in the UK as described by Vollmer (2017), our understanding of these experiences is limited and research into how people are engaging with and using translation on mobile devices is still in its nascent stages. Similarly, MbT has the potential to disrupt and revolutionize the tourism industry and change the way we experience travel by providing instant translations on our mobile devices. More work is needed to understand how, why, when and where users are experiencing MbT, to better explore the positives, negatives, opportunities and limitations of this new experience.

In addition, further research would enable us to discover the effects that MbT is having on people's perception of translation and translation itself on the ground. Indeed, MT apps represent engagement with a new form of translation where the agent becomes a machine, usable by anyone, rather than a human translator. It opens the field for investigation not just of the phenomenon of MbT, but the effects it is having on translators, translation, perceptions of translation and society more generally. This could include larger-scale studies to investigate further the way in which users engage with translation on mobile devices, studies to understand the differences and challenges of using mobile devices for translation or studies to examine the differences between the experiences of professional and non-professional translators.

As technology updates continuously, the growing hybridity between desktop and mobile devices will inevitably lead to more people engaging with MbT as the devices improve and become capable of allowing translators to carry out their job on them with the same efficiency and ergonomics as their traditional desktop workstations. Furthermore, coupled with continual improvements to Internet access, this hybridity may contribute to overcoming some perennial drawbacks with mobile devices such as the small screen size, the possible consequential harm to eyesight, awkward text entry, slow network connectivity, and limited storage, among others.

The authors believe that research into the user experience of MbT should seek to consider a variety of perspectives (e.g., translators, clients, translator trainers, students and language service providers) and some key questions to consider are: how, if at all, does MbT differ from other forms of translation? What are people's experiences of using and engaging with MbT? Why are people engaging with MbT? What impedes people from engaging with MbT? Do people understand the limitations of MbT and MT apps? In each case, it would be useful to apply and aim to answer the following basic questions:

- What content is being translated? Is it a specific text type? Is it limited to shorter texts? Is it "high value" texts? Is it texts that will be externally facing or potentially used by customers?
- Who is commissioning the translation? Who is performing the translation (e.g., professional translators or non-professional translators)?
- How are people translating using the app? Do they simply type the translation straight away in the app's interface? Do they dictate the translation? Do they use a variety of resources (e.g., websites, online dictionaries, forums)?
- Where are people carrying out these translations? Is it when they are travelling? Is it at a desk?
- When are they carrying them out? Is it during work hours? Is it after work hours to earn additional income?
- Why are people commissioning the translation in this way? Why use this method over a more traditional method, such as contacting a language service provider? What are the benefits of this method?

Furthermore, as the ethics of Artificial Intelligence (AI) is of increasing interest more broadly (see, for example, The Malicious Use of Artificial Intelligence: Forecasting, Prevention, and Mitigation by Brundage et al. 2018), research should be conducted on the ethics of technologies employing AI and machine learning in the context of translation. Indeed, some key areas that need to be considered for MbT and MT more widely are questions regarding the ethics, risks and data protection concerns related with the technology. These are areas that have also received little attention to date, but will be of growing importance as the number of users grows. For example, who is responsible for any translation errors that occur through using MT and MbT apps, especially if it is not a human conducting the translation? Is the use of MT apps by language or translation students considered plagiarism or cheating? Are policies in place, at institutional and national levels, to ensure that there are clear guidelines around the use of MT and MbT technologies in education? Do users of MbT, particularly MT apps, understand the implications for their data and data protection? For instance, in health care, using Google Translate or similar apps may lead to issues concerning doctor-patient confidentiality and the inadvertent sharing of data beyond those who should have access in the health care environment. These are topical issues in broader contexts, but also of great pertinence for translation as we seek to understand the evolving nature of the risks and ethics of translation in a world in which it is not only a human agent, professional translator or not, who can perform a translation. By researching these areas, translators, linguists and translation studies scholars will be able to proactively engage with the technology developers rather than be in a reactive position to the technology as has often been the case.

Finally, research into human interaction with MT and MbT is of the essence, given that it is becoming increasingly used in real-life situations. To highlight a further example, at the Bo'ao Asian Forums in 2018 in China, the organizers decided not to use human interpreters, but to use AI-powered translation instead. The system translated the Belt and Road Strategy in a word-for-word manner, producing translations of such a poor quality that they were actually humorous for the audience. From this event, it can be interpreted that MbT is still in its infancy and far from achieving the sophisticated level of comprehension of the cultural-specific elements or subtle subtexts in communication under different contexts as humans are able to do. However, the technology is constantly improving and MbT systems, powered by MT, are increasingly able to translate complicated expressions of meaning rather than just basic communication (Sootoo Report, 2018). As more and more people are engaging with translation, particularly MT, on mobile devices, humankind's dream of overcoming the plurality of languages caused by the Tower of Babel may not seem so unrealistic. It is, nevertheless, unlikely to be realised any time soon and is likely only to be realised for a small number of the world's languages. Although this technology is designed to facilitate communication, it may potentially have the opposite effect or at least change the nature of communication, where our communication is increasingly mediated via machines. We could potentially see less people learning languages and so less instances of human-human communication or our ability to have human-human communication in languages other than our mother tongue may be hindered. As Cronin (2013, p. 3) suggests, it is the "dual dangers of terminal pessimism and besotted optimism" that must be avoided, as the technological situation of reality will, in actuality, find itself somewhere between the two ends of this spectrum.

# REFERENCES

Adams, M. (2017). *An Interview with a Stepes Translator*. Retrieved June 13, 2018, from https://blog. stepes.com/an-interview-with-a-stepes-translator/

Agca, R. K., & Özdemir, S. (2013). Foreign Language Vocabulary Learning with Mobile Technologies. *Procedia: Social and Behavioral Sciences*, *83*, 781–785. doi:10.1016/j.sbspro.2013.06.147

Albrecht, U.-V., Behrends, M., Matthies, H. K., Von Jan, U., & Schmeer, R. (2013). Usage of multilingual mobile translation applications in clinical settings. *JMIR mHealth and uHealth*, *1*(1), e4. doi:10.2196/ mhealth.2268 PMID:25100677

Amazon. (n.d.). *Duolingo Case Study*. Retrieved June 24, 2018, from https://web.archive.org/ web/20170530040432/https://www.amazon.com/p/feature/x4et6o3v69rc8rd

Armstrong, T. (2016). *The Evolution of Mobile Translation*. Retrieved May 3, 2018, from https://blog. stepes.com/the-evolution-of-mobile-translation/

Arnáiz-Uzquiza, V., & Álvarez-Álvarez, S. (2016). El uso de dispositivos y aplicaciones móviles en el aula de traducción: Perspectiva de los estudiantes. *Revista Tradumàtica: tecnologies de la traducció, 14*, 100-111.

Bahri, H., & Mahadi, T. S. T. (2016). The Application of Mobile Devices in the Translation Classroom. *Advances in Language and Literary Studies, 7*(6), 237–242. doi:10.7575/aiac.alls.v.7n.6p.237

Bothun, D., & Lieberman, M. (2017). *Smart home, seamless life: Unlocking a culture of convenience.* Academic Press.

Brundage, M., Avin, S., Clark, J., Toner, H., Eckersley, P., Garfinkel, B., … Amodei, D. (2018). *The Malicious Use of Artificial Intelligence : Forecasting, Prevention, and Mitigation.* Retrieved June 15, 2018, from https://maliciousaireport.com

Buijink, A., Visser, B. J., & Marshall, L. (2013). Medical apps for smartphones: Lack of evidence undermines quality and safety. *Evidence-Based Medicine, 18*(3), 90–92. doi:10.1136/eb-2012-100885 PMID:22923708

Chang, C.-K., & Hsu, C.-K. (2011). A mobile-assisted synchronously collaborative translation–annotation system for English as a foreign language (EFL) reading comprehension. *Computer Assisted Language Learning, 24*(2), 155–180. doi:10.1080/09588221.2010.536952

Chang, D. T. S., Thyer, I. A., Hayne, D., & Katz, D. J. (2014). Using mobile technology to overcome language barriers in medicine. *Annals of the Royal College of Surgeons of England, 96*(6), e23–e25. doi:10.1308/003588414X13946184903685 PMID:25198966

Christensen, T. P., Flanagan, M., & Schjoldager, A. (2017). Mapping Translation Technology Research in Translation Studies. An Introduction to the Thematic Section. *HERMES-Journal of Language and Communication in Business,* (56), 7-20.

Collins, R. (2017, February). A Glimpse into the Future of Work. *Huffpost.* Retrieved from https://www.huffingtonpost.com/entry/a-glimpse-into-the-future-of-work_us_5893effee4b061551b3dfd33

Cronin, M. (2010). The Translation Crowd. *Revista Tradumàtica: tecnologies de la traducció, 8*, 1–7. Retrieved from http://revistes.uab.cat/tradumatica/article/view/100/pdf_15

Cronin, M. (2013). *Translation in the Digital Age.* Abingdon, UK: Routledge.

Cronin, M. (2017). Response by Cronin to Translation and the materialities of communication. *Translation Studies, 10*(1), 92–96. doi:10.1080/14781700.2016.1243287

Gerrish, K., Chau, R., Sobowale, A., & Birks, E. (2004). Bridging the language barrier: The use of interpreters in primary care nursing. *Health & Social Care in the Community, 12*(5), 407–413. doi:10.1111/j.1365-2524.2004.00510.x PMID:15373819

Godwin-Jones, R. (2011). Mobile apps for language learning. *Language Learning & Technology, 15*(2), 2–11.

Groves, M., & Mundt, K. (2015). Friend or foe? Google Translate in language for academic purposes. *English for Specific Purposes, 37*, 112–121. doi:10.1016/j.esp.2014.09.001

GSMA Intelligence. (2018). *Global Data*. Retrieved January 15, 2018, from https://www.gsmaintelligence.com/

Holmes, J. S. (1994). The Name and Nature of Translation Studies. In *Translated! Papers on Literary Translation and Translation Studies* (2nd ed.). Amsterdam: Rodopi.

Howe, J. (2006). *Crowdsourcing: A Definition*. Retrieved May 10, 2018, from http://crowdsourcing.typepad.com/cs/2006/06/crowdsourcing_a.html

Jimenez-Crespo, M. A. (2016). Mobile apps and translation crowdsourcing: The next frontier in the evolution of translation. *Revista Tradumàtica: tecnologies de la traducció*, *14*, 75–84. doi:10.5565/rev/tradumatica.167

Karliner, L. S., Jacobs, E. A., Chen, A. H., & Mutha, S. (2007). Do professional interpreters improve clinical care for patients with limited English proficiency? A systematic review of the literature. *Health Services Research*, *42*(2), 727–754. doi:10.1111/j.1475-6773.2006.00629.x PMID:17362215

Kolawole, S. O. (2012). Is every bilingual a translator? *Translation Journal*, *16*(2). Retrieved from http://translationjournal.net/journal/60bilingual

Lee, K. J., & Kim, J. E. (2013). A Mobile-based Learning Tool to Improve Writing Skills of Efl Learners. *Procedia: Social and Behavioral Sciences*, *106*, 112–119. doi:10.1016/j.sbspro.2013.12.014

Lilley, W., & Hardman, J. (2017). "You focus, I'm talking": A CHAT analysis of mobile dictionary use in an advanced EFL class. *Africa Education Review*, *14*(1), 120–138. doi:10.1080/18146627.2016.1224592

Littau, K. (2016). Translation's Histories and Digital Futures. *International Journal of Communication*, *10*, 907–928. Retrieved from http://ijoc.org/index.php/ijoc/article/view/3508

Lomas, N. (2015). *Stepes Is A Bet That A Chat App Can Mobilize Crowdsourced Translation*. Retrieved June 12, 2018, from https://techcrunch.com/2015/12/17/stepes-is-a-bet-that-a-chat-app-can-mobilize-crowdsourced-translation/?ncid=rss

Mirani, L. (2017, December 12). No, Google's Pixel Buds won't change the world. *1843 Magazine*. Retrieved from https://www.1843magazine.com/technology/the-daily/no-googles-pixel-buds-wont-change-the-world

Moorkens, J., O'Brien, S., & Vreeke, J. (2016). Developing and testing Kanjingo: A mobile app for post-editing. *Revista Tradumàtica: tecnologies de la traducció*, *14*, 58–66.

Mundt, K., & Groves, M. (2016). A double-edged sword: The merits and the policy implications of Google Translate in higher education. *European Journal of Higher Education*, *6*(4), 387–401. doi:10.1080/21568235.2016.1172248

O'Brien, S. (2017). Machine Translation and Cognition. In J. W. Schwieter & A. Ferreira (Eds.), *The Handbook of Translation and Cognition* (pp. 311–331). Hoboken, NJ: John Wiley & Sons Inc.; doi:10.1002/9781119241485.ch17

O'Brien, S., Moorkens, J., & Vreeke, J. (2014). Kanjingo: A Mobile App for Post-Editing. In M. Tadic, P. Koehn, J. Roturier, & A. Way (Eds.), *Proceedings of the 17th Annual Conference of the European Association for Machine Translation (EAMT 2014)* (pp. 137–141). Dubrovnik, Croatia: EAMT.

O'Hagan, M. (2012). The impact of new technologies on translation studies. In C. Millán & F. Bartrina (Eds.), *The Routledge Handbook of Translation Studies* (pp. 503–518). Routledge. doi:10.4324/9780203102893

O'Hagan, M. (2016a). Massively Open Translation: Unpacking the Relationship Between Technology and Translation in the 21st Century. *International Journal of Communication*, *10*, 929–946. Retrieved from http://ijoc.org/index.php/ijoc/article/view/3507/1572

O'Hagan, M. (2016b). Response by O'Hagan to "Translation and the materialities of communication." *Translation Studies*, *9*(3), 322–326. doi:10.1080/14781700.2016.1170628

Och, F. (2006). *Statistical machine translation live*. Retrieved November 8, 2017, from https://research.googleblog.com/2006/04/statistical-machine-translation-live.html

Oladosu, J. B., & Emuoyibofarhe, J. O. (2012). A Yoruba—English Language Translator for Doctor—Patient Mobile Chat Application. *International Journal of Computers and Applications*, *34*(3), 149–156. doi:10.2316/Journal.202.2012.3.202-3079

Prasad, R., Natarajan, P., Stallard, D., Saleem, S., Ananthakrishnan, S., Tsakalidis, S., & Challenner, A. (2013). BBN TransTalk: Robust multilingual two-way speech-to-speech translation for mobile platforms. *Computer Speech & Language*, *27*(2), 475–491. doi:10.1016/j.csl.2011.10.003

Qi, Y., & Mei, W. (2016). Examining the Role of WeChat in Advertising. In X. Xiaoge (Ed.), *Handbook of Research on Human Social Interaction in the Age of Mobile Devices* (pp. 386–405). Hershey, PA: IGI Global.

Rahimi, M., & Miri, S. S. (2014). The Impact of Mobile Dictionary Use on Language Learning. *Procedia: Social and Behavioral Sciences*, *98*, 1469–1474. doi:10.1016/j.sbspro.2014.03.567

Report, S. (2018). Retrieved June 12, 2018, from http://www.sootoo.com/content/675436.shtml

Statistia. (2018). *Percentage of all global web pages served to mobile phones from 2009 to 2018*. Retrieved June 12, 2018, from https://www.statista.com/statistics/241462/global-mobile-phone-website-traffic-share/

Temperton, J. (2017). Google's Pixel Buds aren't just bad, they're utterly pointless. *Wired*. Retrieved from http://www.wired.co.uk/article/pixel-buds-review-google

Torres-Hostench, O., Moorkens, J., O'Brien, S., & Vreeke, J. (2017). Testing interaction with a mobile MT postediting app. *Translation and Interpreting*, *9*(2), 138–150. doi:10.12807/ti.109202.2017.a09

Verma, A. (2011). *A new look for Google Translate for Android*. Retrieved October 25, 2017, from https://googleblog.blogspot.co.uk/2011/01/new-look-for-google-translate-for.html

Villalobos, O., Lynch, S., DeBlieck, C., & Summers, L. (2017). Utilization of a Mobile App to Assess Psychiatric Patients With Limited English Proficiency. *Hispanic Journal of Behavioral Sciences*, *39*(3), 369–380. doi:10.1177/0739986317707490

Vollmer, S. (2017). Syrian newcomers and their digital literacy practices. *Language Issues: The ESOL Journal, 28*(2), 66–72.

Von Ahn, L. (2011). *Massive-scale online collaboration*. Retrieved May 10, 2018, from https://www.ted.com/talks/luis_von_ahn_massive_scale_online_collaboration#t-529477

Wahlster, W. (2013). *Verbmobil: foundations of speech-to-speech translation: Springer Science & Business Media*. Berlin: Springer.

Way, A. (2013). Traditional and Emerging Use-Cases for Machine Translation. Proceedings of Translating and the Computer, 35.

Zapata, P. (2016). Translating On the Go? Investigating the Potential of Multimodal Mobile Devices for Interactive Translation Dictation. *Revista Tradumàtica: tecnologies de la traducció*, (14), 66-74.

# Chapter 13
# Contemporary Mobile Experience Among Bottom of Pyramid

**Pooja Sehgal Tabeck**
*Amity University, India*

**Anurupa B. Singh**
*Amity University, India*

## ABSTRACT

*Mobile has proven to a most successful tool for bottom of pyramid markets as it is the most affordable means to generate utilities for mobile users of lower income strata of pyramid and to provide them with quick and low-cost access to information, government systems, business opportunities, access to education, and health. This chapter presents the landscape of contemporary experience through mobile phones in the field of economic value creation, social value, and health among the bottom of the pyramid in India.*

## INTRODUCTION

Bottom of Pyramid markets is lowest strata of economies. Constraints for survival here are not only money rather the need for basic necessities of life. The lowest strata always attract many politicians (U.S. President Franklin D. Roosevelt on April 7, 1932 in a Radio address), several academicians (Prahalad & Hurt, 2002) and marketers to not only address the issues rather implement the innovations.

United nation development programme reported that there are 700 million people around the world living under extreme poverty and their income is as low as $ 1.90 per day. Priority has always been given to "poverty reduction in World "as the most important one out of seventeen different goals set by them. India, China, Nigeria, Indonesia and South Africa have been identified as top five Bottom of Pyramid markets.

Bottom of Pyramid includes small farmers, slum dwellers, informal sector workers, unskilled /semi-skilled workers, migrants (IFC). The population of worlds' poorest is distributed amongst the middle

DOI: 10.4018/978-1-5225-7885-7.ch013

*Table 1. Key Facts: Bottom of pyramid*

---

- About million people still live below the US $ 1.90
- An additional 800 million people are close to the poverty threshold, social, economic and environmental shocks could push them back in poverty
- In 108 countries, covering a total population of 5.4 million
- 1.6 billion are MPI*-Poor
- 85% of the MPI poor live in rural areas
- Access to different opportunities particularly by the poor and most vulnerable, is the key to reduce poverty.

---

MPI*= Multidimensional Poverty Index (developed by Oxford Poverty and Human Development Initiative and UNDP)
Source: United National Development Goals

and low income developing countries which includes 95 percent of South Asia's population, 68 percent of the Middle East and North Africa's population, and 27 percent of Latin America's population.

UNDP also defined poverty (bottom of pyramid) as multidimensional in demographical aspect where in the segment has not only income constrained rather there is lack of other resources like education, sanitation, water, food, health and living standards.

Prahalad and Hurt (2002) had taken reference of Millennium Development Goals by United Nations and Defined Bottom of Pyramid as,

*The real source of market promise is not the wealthy few in the developing world, or even the emerging middle-income consumers: It is the billions of aspiring poor who are joining the market economy for the first time.*

*This is a time for MNCs to look at globalization strategies through a new lens of inclusive capitalism. For companies with the resources and persistence to compete at the bottom of the world economic pyramid, the prospective rewards include growth, profits, and incalculable contributions to humankind.*

Socio Economic Caste and Census 2011 defined BoP as people living in slums are 60.90 million and categorized as urban poor. SECC had taken several question in record and on the basis of source of generation of income, defined urban poor as household whose main source of income is as following:

- Rag-picking, domestic work, Street vendor/cobbler/hawker/other service provider working on streets;
- Construction worker/ plumber/ mason/ labor/ painter/ welder/ security guard/coolie and other head-load worker;
- Sweeper/ sanitation worker / gardener; Home-based worker/ artisan/handicrafts worker / tailor;
- Transport worker/ driver/ conductor/helper to drivers and conductors/cart puller/ rickshaw puller;
- Shop worker/ assistant/ peon in small establishment/ helper/ delivery assistant / attendant/ waiter;
- Electrician/ mechanic/ assembler/repair worker; Washer-man;
- Other work/Non-work; Non-work (Pension/ Rent/ Interest, etc.).

In pursuit to improve quality and reduce constraints of BoP markets, several product and services have been launched in past by both Private and Government sectors.

Penetration of mobile phones is much faster in Bottom of Pyramid markets then other infrastructural facilities like electricity, road and health services. Hence to achieve better and cost effective results for most ignorant strata of population, many services in developing countries and emerging markets has been piggybacked on mobile technology.

As per the report of World Bank 10% increase in penetration of mobile phones in any country results in increased per capita GDP by .081% to 1.38%.

It has been observed that Mobile telephony is one of the most successful scaled innovation for Bottom of pyramid Markets (Foster& Heeks, 2013). According to Telecom Regulatory authority of India, Tele density in 2017 in India reached to 93.5% which includes wireless and wire line subscriptions. (TRAI, 2017).The mobile telephony penetrated not only fastest but also deepest in India (Jack& Suri, 2011).

## LITERATURE REVIEW

Bottom of pyramid markets are full of potentials. Still there are many needs which have not been addressed, including hygiene, health and education.

Marketers can adopt the models where they can take poor as producers or partners not only to eradicate poverty but also to change their behavior (Cross, J., & Street, A. 2009).The major problem is to understand the needs of poor by MNCs where they were offering low quality products to BOP due to lack of affordability (Barki, E., & Parente, J. 2010), hence value creation cannot be achieved.

Bottom of Pyramid customers desire and are able to pay for quality products tailored to their needs, which increase their income, improve health and increase their standard of living.

In response, firms need to develop new products specific to the demands and conditions of this low-income population, despite the barrier of poverty (Nakata & Weidner 2011). There is a need to launch new services/products in the areas of that BOP innovations need to be introduced in the field of healthcare, housing, food and water (Mendoza and Thelen 2008).

Any innovation for BOP should enhance productivity of the individual in the terms of income previous studies supported that mobile phones have positive impact on social capital and subjective well being .(Chan, M.,2015).

The Bottom of the Pyramid offers the private companies an opportunity to serve poor with the help of right combination of scale technology, price, sustainability, and usability (Shyle, 2011) Many research studied support the fact that penetration of mobile phones are one of the fastest in bottom of pyramid (Joshi, A. 2009) and many services have been delivered in past using mobile channels. Technologies which have empowered resource constrained poor have permanent and significant impact on them which include mobile phones and internet technologies. (Trujillo, et al (2010). Any product to get adoption in Bottom of Pyramid markets require affordability, availability, accessibility, and awareness (Goyal, S.,et al 2014), mobile phones and its applications fulfilled all the criteria and accepted all over world .

Government of different countries also utilized the mobile phone channel to disseminate their information and policies among poor in countries like china (Liu, Y. et. al 2014) and India. Poverty can also be reduced if marginal community will accept any of mobile system based innovations or applications (Rahman, S. A, et. al, 2017).M-pesa in kenya and b-Kash in bangladesh has been considered as most successful project which brought the changes in marginal community after adoption of mobile based applications.

Value Creation through Mobile Phone Applications: - Mobile phones are most affordable means to generate several utilities for lower income strata of pyramid. They are able to provide millions of people around the world quick and low cost access to information, government systems, business opportunities, access to education and health which in past were not accessible to them. (Qiang, C.Z.,et al,2012). Adoption of mobile phones are one of the fastest accepted innovation among poor due to its 24 hour availability in a day, perception to connect with people at distance without any geographical barrier and communication from both the ends.(Takavarasha Jr, S.,et al,2018). To uplift the lives of economically week strata following values can be created through mobile phones among bottom of pyramid:-

1. Economic Value creation
2. Social Value Creation
3. Health Value Creation

Economic Value Creation through Mobile Phones: Mobile phone industry created direct and indirect jobs in different facets of society, specifically for Bottom of Pyramid Markets. Direct jobs include small mobile phone sellers, mobile phone repairing services and pre-paid mobile phone charge. Indirect jobs include call center jobs, jobs and customer search with help of mobile phones. Bottom of Pyramid strata use mobile phones to reduce different types of transaction cost once they mastered over "how to use'. (Cáceres, R. B., & Fernández-Ardèvol, M. 2012).

Different mobile based apps have been launched by companies to support employability of this stratum of society and eradication of poverty. Autonomous micro entrepreneurs reported to increase productivity of their businesses by using mobile phones through business expansion, employment search, transaction cost, financial interaction, (Bhavnani, A., et al, 2008), information dissemination (Cecchini, S.,& Scott, C.2003).

Business Expansion: Informal business sectors are backbone of developing economies and provide employment to poor (Johnson, S. C., & Thakur, D. 2015). Use of mobile phones have increased business opportunities for poor in informal sector. Small vegetable vendors, street vendors (Mramba, N., et al,2015)hawkers, semi-skilled workers like plumbers, washer man, food vendors expanded their business with the help of mobile device. Mobile phone helped them to connect with major chunk of customers and provide better services. Food and vegetable vendors have used mobile device very effectively for home delivery and customized their services according to customer; need, results not only in efficiency rather increased incomes.

Employment Generation: Mobile phones has successfully generated indirect and direct jobs in different area and eradicated poverty among lower strata. Direct job generation has been done in different tangent related to phone (Aker, J. C., et al, 2010) and service providers. Most of BoP dependent upon prepaid mobile services, hence vendors are required to recharge their talk time and disseminate information pertaining new promotion offers and most suitable plans .Mobile service providers to provide these services trust only on someone who belongs to this strata and able to communicate in their native dialect, which has given opportunity of employment to BoP for not only recharge the talk time rather to sell SIM (subscriber identity module) cards also. Same has applied for selling of affordable mobile handsets.

Indirect employment generation includes call center jobs, jobs and customer search with help of mobile phones. Many telecom operators open their call centers in rural areas to support poor rural youth (Majumder, S.et. al, 2014) and to provide services in vernacular languages. HDFC bank, Ruralshores are the few names those who had started call centers in rural India

Financial Interaction: Financial Transaction and their cost had always put hindrance for BOP in different strata be it rural poor or migrants workers living in urban areas.6 billion people out of Worlds 7 billion population carrying mobile phones, but only 2 million have their bank accounts.(Gupta S.,2013). India has a population of 1.2 billion with 900 million mobile phones but only 250 million bank accounts. Due to unavailability of bank branches in accessible area, poor need to travel distant branches, for low transactions, which increase their transaction cost, hence major chunk of the poor population opted out from formal banking system (Mas, I.2011).

Many mobile service providers seek this opportunity, and launched customized financial products (apps) for poor. MPESA is Kenya is one of the most successful cases of mobile money that allowed poor to make financial transaction using their mobile phones. (Jack, W., et al, 2011). Other similar Mobile money services include MTN in Uganda (Ssonko, G. W., 2010), Vodacom in Tanzania. Realini, C., & Mehta, K. (2015) described about different mobile based financial products for bottom of pyramid customers in various countries i.e. Globe Gcash in phillipines . Dahabshiil in Kenya and bKash in Bangladesh

Later mobile money options have been change with other nomenclature mobile wallets i.e. paytm, mobiwik, phonepe etc.in India and these mobile wallets also successfully adopted by BoP.

Majority of poor need assistance to carry their financial transactions due to lack of required skills and confidence, different mobile money app and products has given financial independence to Bottom of Pyramid (BoP) customers and reduced their transactions cost which usually includes travel time and leakages due to corruption.

Information Dissemination: Low-cost infrastructure is required to disseminate information among the bottom of pyramid related to their business/job, health, finances, politics, new technologies (Aker, J. C.et al, 2010). Mobile phones successfully addressed different information related needs of consumers those who otherwise dependent upon more traditional sources like newspapers, radio, television, personal cost to travel to obtain information and different social groups. Phones also able to connect similar interest groups i.e. artisans and small farmers to markets, which further help them to get business and reduce poverty.

Mobile Applications for Artisans: In India to help artisans Govt. and entrepreneurs launched different mobile applications which helped artisans to now customer's changing need and customize products accordingly. It became major facilitator to eradicate poverty from ignorant strata.

## A CASE OF GOCOOP

The size of handloom and handcraft market in India is about $ 4 million, which provides employment to nearly 250 million people around the country. These 250 million people belong to unorganized sector (Banik, S. 2017), which further deteriorate their condition in absence of quality of work life (Dhingra, V., et al, 2017) and right enumeration of their work. Many times they also face problems like irregular order, irregular supplies of raw material, irregular payment and rejections of orders (Mohapatra, S.,et al,2011) results in irregular incomes which keep them to remain in BoP strata. For artisans marketing channel also filled with several middlemen which often lead to lower realization of their product. To help them to sustain in business an application has been launched which has been named as GoCoop.

GoCoop works with more than 10 states of India and connected with artisans and provide them platform to bridge with national and international buyers. To widen the customer base application founder has also hired the team of designers which inform artisans about changing taste and preferences of customers.

Gocoop team works at regional cluster in different regions of the country to create awareness about computers, e-platforms and mobile applications among artisans. Once artisan joined their products are listed online and they were connected directly to customers.

Mobile Applications for Small Farmers:-Awareness about right price, information access to climate updates, plant protection, cultivation best practices, market distribution practices, demand and logistics (Mittal, S.,et al,2010). availability of quality inputs for farming, were the major obstacles pertaining to farmingin different agriculture based economies. Through right, affordable effective and timely communication mobile phone reduced several transaction costs related to agriculture and farming. (De Silva, H. et al, 2008)

Digital technologies for farmers (Seth, A. et al, 2017) transformed agriculture practices in many developing countries which have maximum farmers in bottom of pyramid segments . It has been proved that adoption of different technology based application results in better farming, yield and better market price. Government of India also acknowledged ubiquitous presence of mobile phones and launched Kisan Suvidha(farmer facilitation) mobile application .The application is available in five different languages and provide information related to weather, dealers, market price, plant protection, agro advisory and Kisan Call Centre. The application has designed as single window concept which enabled farmers to receive all relevant information through single interface.

## A CASE OF M-KILMO

Kenya is developing country where 70 percent of population is living in rural areas and majority of them are dependent upon farming. The full potential of small farm holding were not realized and used by owners, they were dependent upon traditional farming practices which had been handed down from generation to generation.

To uplift the standard of Kenyan farmers agricultural mobile value added service products (Agri VAS), the Development Fund, supported by the Rockefeller Foundation launched M-kilmo a mobile based helpline for Kenyan farmers. The service was launched in several local languages and allowed farmers to call anytime between &am to 11 pm.

M-Kilmo provided information in different facets related to farmers namely horticulture, aquaculture, climatology, pests, plant disease, animal husbandry, agricultural engineering, market advice, environmental and veterinary issues.

Kenya is a tribal country, with more than 40 tribes and different languages spoken. Official government language of Kenya is English and Swahili. While all marketing materials are developed in Swahili for better outreach, 'M-Kilmo' was also marketed on vernacular radio-stations and local events.

Mobile Applications for Semi-skilled workers:-Mobile played ubiquitous role to assist semi-skilled workers of BoP. Saral Rozgar is an initiative of Tech Mahindra group to provide opportunities to blue collar job seekers from low income segment. Saral Rozgar has partnered with CMS and SREI sahaj to provide opportunities even low income segments of deep rural areas. Five million job seekers are registered with Saral Rojgar and saral has reduced the gap by providing 1.5 million jobs. Likewise The Kabariwala is an app which helps people to collect junk and scrap from their houses. The Kabariwala is platform which provided business to many rag pickers indirectly and increases their income.

Social Value Creation through Mobile Phones: - Social value is a utility acquired by association with one's own social groups. Social value can be generated through social interaction, entertainment value, experiential learning and knowledge or skill enhancement among BOP consumers. In many studies researchers found that mobile phones also creates symbolic value creation for poor which includes fashion and improved social status. (Rashid, A. T., & Elder, L. 2009). Mobile phones facilitates both formal and informal communication in the changing environment which lead towards making social networks and personal security (Galperin, H., & Mariscal, J., 2007) hence create value for customers.

Mobile phones contributed not only to reduce poverty rather increased and expanded social interaction (Sife, A. S et al, 2010) at low cost. It also increased poor's ability to deal with emergencies which includes medical emergencies, emergencies related to robbery (specifically to inform police).

Literacy rate is one of the key measurements of socio-economic development of any country. Illiteracy became one of the key hindrances to eradicate poverty in many Asian and African countries. In India literacy rate is 69.3%(census, 2011), country has 270 million people living below poverty line and 272 million are illiterates.35% of world's illiterate population is living in India. The case of education get worst with unavailability of infrastructure, teachers, amenities and non-conducive learning environments, geographical remoteness and access challenges and other traditional material required (Datta, D. et al. 2010). We in this chapter argue that education and skill enhancement for BoP can be achieved through mobile telephony in developing country. Education can easily bridge the gap between poor and non-poor. Mobile technology can deliver education in more affordable, accessible and effective way than traditional learning. (Oluwatobi, S.et al, 2015).

## A CASE OF HUL KAN KHAJURA TESAN (EAR WORM RADIO CHANNEL)

Media dark villages (no access to any media i.e. print or electronic) of Hinterland in India created challenges in front of many multinationals. These villages consist of 200 million people, which is a huge chunk of potential customers. Although reach of traditional media is major hurdle but every household has at least one mobile phone. Money constraints put these mobile users to use innovative way of missed call which is a communication virtually without any cost (Sivapragasam, N.,2008).

To reach these media dark villages FMCG giant in India Unilever limited launched its innovative mobile based radio channel Kan Khajura Tesan with missed call feature. Missed call feature has been perceived by users that they are receiving the content without paying any cost. Kan Khajura tesan is demand based entertainment channel, so subscribers need not to wait for particular day or time to listen the content, rather whenever wherever they would have been get free time, were need to give a missed call and able to listen desired content. Content which is available on Kan Khajura Tesan includes music, jokes, educational content, current affairs etc. followed by unilever limited brand promotion. Success story of Kan Khajura Tesan includes 50 million subscribers and 1billion advertisement impressions. Kan Khajura station not only entertain target audience rather brand awareness of unilever products like close-up, wheel, ponds also increased in target region.

Health Value Creation through Mobile Phones: - Resource poor environment, constrained of income and education made BoP as most vulnerable strata for multiple diseases and poor health. Mobile phones have potential to solve problems related to health in BoP because of its unique characteristics like reach, customized solutions and location based services. (Akter, S., et al, 2010). 30% of Indian each year fall below poverty line due to healthcare expenses which can be reduced through prevention and affordable medical care.

According to report by PWC in India it has been stated that India ranked second in among developing countries studied on maturity for mHealth adoption. mhealth is viewed as the delivery of healthcare services through mobile phones. mhealth applications can manage to deliver quality services due to its timeliness services and improved coverage area. (Garai, A., 2011). Many researchers opined their views that mobile health applications (Nieroda, M.,et al, 2015) can be seen as transformatory factor which can change BoP engagement towards health.

To increase health awareness, to support health of mother and child, to reduce risk of cardiovascular diseases several apps which will deliver cost effective and reliable (Misra, R., & Srivastava, S. 2016) health services are need of the hour.

Several life and health insurance related technologies leveraged the mhealth services among poor, so that they will better expose to facilities.

## A CASE OF POSHAN (IAP HEALTH PHONE PROGRAMME)

Every fifth child of world lives in India and out of this every second child suffers from malnutrition which causes different types of diseases. To prevent malnutrition and to make women aware about malnutrition of their child and themselves, Poshan a mobile based initiative has been launched in India specifically for people who are living in areas without accessibility of medical practices. Poshan is an initiative which tried to reaped benefits of easy accessibility and availability of mobile phones in the hands of poor. Main objective of poshan was to reach 6 million girls and woman of India till year 2018.

IAP health phone programme was a public private partnership of Health phone with Indian pediatric association, UNICEF, Vodafone and Govt. of India. To overcome the challenges related to language barrier, Poshan series has been launched in 18 different Indian languages and distributed through poshan mobile application in different languages and preloaded micro sd cards.

The very second challenge which impediment success of any programme is its reach to right audience, poshan equipped the human health enablers like Anganwadi workers (mother and childcare centers in India for poor), ASHAs (belongs to rural community, an interface between community and public health systems) and ANMs(Auxiliary nurse midwife, a village level female health worker) to share knowledge with target audience. To penetrate the information among target audience Vodafone supported the cause with free talk time (if subscriber will watch 4 poshan videos they will receive 10 Indian rupee free talk time).

### Theoretical Model of Value Creation Through Mobile Applications

Bottom of Pyramid markets are most deprived strata of any country's population, which is not only deprived by income rather health and social perspective also. Authors of this chapter have seen ray of hope in mobile applications through which up-liftment of this segment can be done. Researchers have suggested different mobile applications in various fields for the mentioned deprivations.

Challenges Related to Mobile Phones Experience among Bottom of Pyramid: The major challenge is related to language barrier, in most of the mobile handsets numbers and keypad is in English while BOP has constraint related to literacy. The second major constraint is availability of electricity for limited duration of time, hence one mobile set which serves multiple people and multiple needs (the handset which is with chief wage earner during work serve the purpose to connect with others and entertain-

*Figure 1. Value creation through mobile applications among bottom of pyramid*
*Source: Compilation by Authors*

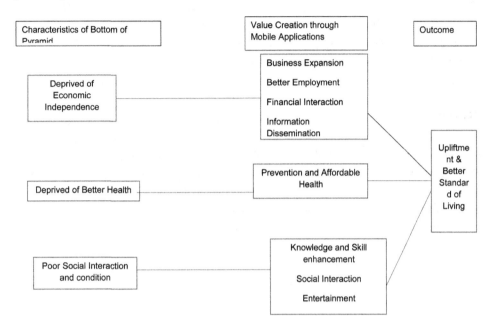

ment, same phone works as source of entertainment for kids when he reached home) require charging several times in a day.

Illiteracy: Illiteracy is a major challenge for adoption of mobile phones and its usage. When combined with digital illiteracy it creates a powerful impact. Digital illiteracy has been ranked highest by providers and consumers as a hindrance for mobile phone usage. Although developers tried to solve these challenges through voice-based services and initiated to develop skills to use mobile phones.

Language Barriers: Most of the content available on internet and mobile phones are in the English language (Banerjee, S., et al, 2014). in a country like India where total official languages are 22, is a major constraint for mobile phone adoption. Mobile and app developers are suggested to reap the benefit of this gap by launching phones and app with native language accessibility. Information provided through mobile phones should be easy to understand and stakeholders are advised to avoid any technical words while dissemination of information.

Week Physical Infrastructure: In the less developed countries access to infrastructure and electricity is major cause of inequality (Dugoua, E., et al., 2014). Electricity deprived poor in India still dependent upon kerosene oil for lighting. The lack of accessibility to electricity and use of more traditional resources widened the gap between poor and non- poor. (Ranjan, R., et al, 2017) Unavailability of electricity or availability for shorter duration of time has become main challenge for mobile experience also. The device which requires charging battery to operate can only be done with the help of electricity; the poor infrastructure of developing nations does not allow users to do so.

Network Connectivity issues: Total tele density in India is 82.7% while internet and broad band density is only 28.76%(TRAI, 2016) results in poor network accessibility, in deep rural areas, conflict zones, slums where most of BOP strata live. Poor signal strength and connectivity lead toward frustration among users. Greater and better accessibility definitely will facilitate more participation of ignored population in e governance, generate new employment opportunities and increase contemporary experience with technology.

Security: There are two major challenges related to security of mobile phones i.e. physical security and cyber security. Physical security issues are related to handset theft and vandalism while cyber security includes malfunction by viruses, theft of data, spam etc.

## CONCLUSION

Mobile phones have enormous potential to provide contemporary experience in different fields to poor, a small device also helps to improve standard of living of poor. Through this chapter we tried to touch different facets of mobile experiences like economic, social and health value creation in different BoP segments. Need of the hour for different stakeholders i.e. government, service providers, handset manufacturers are to overcome different infrastructural and personal challenges which impediment success of different mobile based experiences.

## REFERENCES

Aker, J. C., & Mbiti, I. M. (2010). Mobile phones and economic development in Africa. *The Journal of Economic Perspectives*, *24*(3), 207–232. doi:10.1257/jep.24.3.207

Akter, S., & Ray, P. (2010). mHealth-an ultimate platform to serve the unserved. *Yearbook of Medical Informatics*, *2010*, 94–100. PMID:20938579

Banerjee, S., Mandal, K. S., & Dey, P. (2014, April). A Study on the Permeation and Scope of ICT Intervention at the Indian Rural Primary School Level. CSEDU, (2), 363-370.

Banik, S. (2017). *A Study on Financial Analysis of Rural Artisans in India: Issues and Challenges*. Academic Press.

Barki, E., & Parente, J. (2010). Consumer Behaviour of the Base of the Pyramid Market in Brazil. *Greener Management International*, (56).

Bhavnani, A., Chiu, R. W. W., Janakiram, S., Silarszky, P., & Bhatia, D. (2008). *The role of mobile phones in sustainable rural poverty reduction*. Academic Press.

Cáceres, R. B., & Fernández-Ardèvol, M. (2012). Mobile phone use among market traders at fairs in rural Peru. *Information Technologies & International Development, 8*(3), 35.

Cecchini, S., & Scott, C. (2003). Can information and communications technology applications contribute to poverty reduction? Lessons from rural India. *Information Technology for Development*, *10*(2), 73–84. doi:10.1002/itdj.1590100203

Chan, M. (2015). Mobile phones and the good life: Examining the relationships among mobile use, social capital and subjective well-being. *New Media & Society*, *17*(1), 96–113. doi:10.1177/1461444813516836

Cross, J., & Street, A. (2009). Anthropology at the bottom of the pyramid. *Anthropology Today*, *25*(4), 4–9. doi:10.1111/j.1467-8322.2009.00675.x

Datta, D., & Mitra, S. (2010). *M-learning: mobile-enabled educational technology*. Innovating.

De Silva, H., & Ratnadiwakara, D. (2008). *Using ICT to reduce transaction costs in agriculture through better communication: A case-study from Sri Lanka*. LIRNEasia, Colombo, Sri Lanka.

Dhingra, V., Mudgal, R. K., & Dhingra, M. (2017). Safe and Healthy Work Environment: A Study of Artisans of Indian Metalware Handicraft Industry. *Management and Labour Studies*, *42*(2), 152–166. doi:10.1177/0258042X17714071

Dugoua, E., & Urpelainen, J. (2014). Relative deprivation and energy poverty: When does unequal access to electricity cause dissatisfaction? *International Journal of Energy Research*, *38*(13), 1727–1740. doi:10.1002/er.3200

Foster, C., & Heeks, R. (2014). Nurturing user–producer interaction: Inclusive innovation flows in a low-income mobile phone market. *Innovation and Development*, *4*(2), 221–237. doi:10.1080/215793 0X.2014.921353

Galperin, H., & Mariscal, J. (2007). *Poverty and mobile telephony in Latin America and the Caribbean. Dialogo Regional sobre Sociedad de la Informacion (DIRSI)*. IDRC.

Garai, A. (2011). *Role of mHealth in rural health in India and opportunities for collaboration*. Indira Gandhi National Open University.

Goyal, S., Sergi, B. S., & Kapoor, A. (2014). Understanding the key characteristics of an embedded business model for the base of the pyramid markets. *Economia e Sociologia (Evora, Portugal)*, *7*(4), 26.

Gupta, S. (2013). The mobile banking and payment revolution. *European Finance Review*, *2*, 3–6.

Jack, W., & Suri, T. (2011). *Mobile money: The economics of M-PESA (No. w16721)*. National Bureau of Economic Research. doi:10.3386/w16721

Johnson, S. C., & Thakur, D. (2015). Mobile phone ecosystems and the informal sector in developing countries–cases from Jamaica. *The Electronic Journal on Information Systems in Developing Countries*, *66*(1), 1–22. doi:10.1002/j.1681-4835.2015.tb00476.x

Joshi, A. (2009, October). Mobile phones and economic sustainability: perspectives from India. In *Proceedings of the First international conference on Expressive Interactions for Sustainability and Empowerment* (pp. 2-2). British Computer Society.

Kaur, H., Lechman, E., & Marszk, A. (Eds.). (2017). *Catalyzing development through ICT adoption: the developing world experience*. Springer. doi:10.1007/978-3-319-56523-1

Liu, Y., Li, H., Kostakos, V., Goncalves, J., Hosio, S., & Hu, F. (2014). An empirical investigation of mobile government adoption in rural China: A case study in Zhejiang province. *Government Information Quarterly*, *31*(3), 432–442. doi:10.1016/j.giq.2014.02.008

Londhe, B. R., Radhakrishnan, S., & Divekar, B. R. (2014). Socio economic impact of mobile phones on the bottom of pyramid population-A pilot study. *Procedia Economics and Finance*, *11*, 620–625. doi:10.1016/S2212-5671(14)00227-5

Majumder, S., & Sharma, R. P. (2014). Indian ITES Industry Going Rural: The Road Ahead. *Journal of Business and Economic Policy*.

Mas, I. (2011). Why are banks so scarce in developing countries? A regulatory and infrastructure perspective. *Critical Review*, *23*(1-2), 135–145. doi:10.1080/08913811.2011.574476

Mendoza, R. U., & Thelen, N. (2008). Innovations to make markets more inclusive for the poor. *Development Policy Review*, *26*(4), 427–458. doi:10.1111/j.1467-7679.2008.00417.x

Misra, R., & Srivastava, S. (2016). M-education in India: An effort to improve educational outcomes with a special emphasis on Ananya Bihar. *On the Horizon*, *24*(2), 153–165.

Mittal, S., Gandhi, S., & Tripathi, G. (2010). *Socio-economic impact of mobile phones on Indian agriculture*. New Delhi: Indian Council for Research on International Economic Relations.

Mohapatra, S., & Dash, M. (2011). Problems Associated with Artisans in Making of Handicrafts in Orissa, India. *Management Review: An International Journal*, *6*(1), 56–64.

Mramba, N., Apiola, M., Sutinen, E., Haule, M., Klomsri, T., & Msami, P. (2015, June). Empowering street vendors through technology: An explorative study in Dar es Salaam, Tanzania. In *Engineering, Technology and Innovation/International Technology Management Conference (ICE/ITMC), 2015 IEEE International Conference on* (pp. 1-9). IEEE.

Nakata, C., & Weidner, K. (2012). Enhancing new product adoption at the base of the pyramid: A contextualized model. *Journal of Product Innovation Management*, *29*(1), 21–32. doi:10.1111/j.1540-5885.2011.00876.x

Nieroda, M., Keeling, K., & Keeling, D. (2015). Acceptance of Mobile Apps for Health Self-management: Regulatory Fit Perspective. UMAP Workshops.

Oluwatobi, S., & Olurinola, O. I. (2015). *Mobile learning in Africa: strategy for educating the poor*. Academic Press.

Prahalad, C. K., & Hart, S. L. (2002). The fortune at the bottom of the pyramid. *Strategy and Business*, *26*, 54–67.

Qiang, C. Z., Kuek, S. C., Dymond, A., & Esselaar, S. (2012). *Mobile applications for agriculture and rural development*. Academic Press.

Rahman, S. A., Taghizadeh, S. K., Ramayah, T., & Alam, M. M. D. (2017). Technology acceptance among micro-entrepreneurs in marginalized social strata: The case of social innovation in Bangladesh. *Technological Forecasting and Social Change*, *118*, 236–245. doi:10.1016/j.techfore.2017.01.027

Ranjan, R., & Singh, S. (2017). *Energy deprivation of Indian households: evidence from NSSO data*. Academic Press.

Rashid, A. T., & Elder, L. (2009). Mobile Phones and Development: An Analysis of IDRC-Supported Projects. *The Electronic Journal on Information Systems in Developing Countries, 36*(1), 1–16. doi:10.1002/j.1681-4835.2009.tb00249.x

Realini, C., & Mehta, K. (2015). *Financial Inclusion at the Bottom of the Pyramid*. FriesenPress.

Seth, A., & Ganguly, K. (2017). Digital technologies transforming Indian agriculture. *The Global Innovation Index*, 105-111.

Shyle, I. (2011). Global crisis and its effects in the developed and emergent countries-"The bottom of the pyramid" as an innovation resource. *EMAJ: Emerging Markets Journal, 1*(2), 48–58. doi:10.5195/EMAJ.2011.12

Sife, A. S., Kiondo, E., & Lyimo-Macha, J. G. (2010). Contribution of mobile phones to rural livelihoods and poverty reduction in Morogoro region, Tanzania. *The Electronic Journal on Information Systems in Developing Countries, 42*(1), 1–15. doi:10.1002/j.1681-4835.2010.tb00299.x

Sivapragasam, N. (2008). *Hit me with a missed call: The use of missed calls at the bottom of the pyramid*. Academic Press.

Ssonko, G. W. (2010). *The role of mobile money services in enhancing financial inclusion in Uganda*. Bank of Uganda.

Takavarasha, S. Jr., & Adams, C. (Eds.). (2018). *Affordability Issues Surrounding the Use of ICT for Development and Poverty Reduction*. IGI Global. doi:10.4018/978-1-5225-3179-1

Trujillo, C. A., Barrios, A., Camacho, S. M., & Rosa, J. A. (2010). Low socioeconomic class and consumer complexity expectations for new product technology. *Journal of Business Research, 63*(6), 538–547. doi:10.1016/j.jbusres.2009.05.010

# Chapter 14
# Mobile Experience in War and Conflict Reporting

**Shixin Ivy Zhang**
*University of Nottingham – Ningbo, China*

## ABSTRACT

*Defining mobile experience as a process of usage, affordance, roles, and impacts of mobile phones during wars and conflicts, the author has located similarities and differences in mobile experience in war and conflict reporting among professional journalists, citizen journalists, governments, militaries, rebels, NGOs, activists, and communities. After exploring the changes and trends related to mobile experience in war and conflict reporting, the author also offered specific dimensions and directions for further studies.*

## INTRODUCTION

As a subset of journalism, war and conflict reporting has attracted scholarly attention as it undergoes major changes with the emergence of digital media (Kristensen & Mortensen, 2013). Among the digital media, mobile phone has become an indispensable multimedia device for producing and disseminating information as well as interacting and networking with the wider world in times of crisis. Mobile experience can be defined as both a process and an outcome of a user's interaction with a product, a service, a content or their combination in an interactive, personalized, immersive, and mobile context (Xu, 2017). In this chapter, mobile experience is defined as a process of usage, affordance, roles and impacts of mobile phones during wars and conflicts. The term of mobile phone will be mainly used. Other terms such as mobile media, smart phones, satellite phones, cell phones, camera phones will also be used interchangeably to refer to telephony devices 'on the go' in particular contexts.

Much as Achilles was impervious to war, mobile phones survive and thrive in conflict-stressed environment (Best, 2011, p.15). In similar veins, mobile phones are essential and significant in contemporary war and conflict reporting due to their 'ubiquity and wearable multimedia functionality' (Gilboa et al, 2016). Mobile phones have become the most ubiquitous communication device worldwide with worldwide users penetration rate of 67% in 2019 (Poblet, 2011; statistica.com). In addition, with the convergence of digital media technologies, the mobile has developed into a handy multimedia computer

DOI: 10.4018/978-1-5225-7885-7.ch014

that incorporates cameras, video cameras, audio recording devices, GPS, Wi-Fi, and many other functionalities. The mobile phone is unique in being a wearable rather than portable device (Gilboa et al, 2016). However other scholars such as Reading (2009) believe the mobile phone is more portable than wearable. In her book *Mobile Witnessing*, Reading (2009, p.64) claims that the mobile phone is rare in being 'a wearable rather than portable device'. It is carried with us and on us, often in handbags or pockets, on an everyday basis. No matter whether it is wearable or portable, the mobile phone's 'ubiquity', 'mobility' and 'multimedia functionalities' features make it indispensable in war and conflict reporting.

War and conflict news is no longer produced exclusively by trained journalists, but involves different actors ranging from citizens, bloggers, NGOs to diaspora communities (Meyer, Baden & Frère, 2017). In times of crisis, creative solutions appeared beyond the 'normal' use of cell phones and new media to draw attention, create interactions, and obtain information (Malka, Ariel and Avidar, 2015). Thus different actors' usage and affordance of mobile phones as well as the mobile phone's roles and impacts in contemporary wars and conflicts will be discussed next, starting with professional journalists.

## MOBILE EXPERIENCE OF PROFESSIONAL JOURNALISTS

The mobile phone has become an indispensable part of journalists' work. Journalists have widely used mobile phones as professional tools in war and conflict reporting. This process includes live reporting (or video streaming) from the field (MoJo), the convergence of mobile phone and social media (UGC, storify), the use of mobile photo apps in war photography, and the use of mobile phones in drone journalism.

Mobile communication technologies allowed journalists to report live with ease from anywhere in the world. With mobile technologies, foreign correspondents also gained greater independence from military or political control. Matherson and Allan (2009, pp. 62-63) noted that "satellite phone and videophone stand at the heart of major changes in war reporting and in journalism's relations with combatants and governments". Afghanistan war was even described as 'the first videophone war'. In late 2001, as the Taliban forces retreated, reporters equipped with portable digital satellite phones crossed from Pakistan, with sixty soon based in Jalalabad near the border. From there, *the Independent*'s Richard Lloyd Parry was able to report the destruction of village of Kamo Ado and the deaths of an estimated 115 people. Parry called the satellite phone "the most essential and closely guarded item…without which the most astonishing news story is useless" (Matherson and Allan, 2009, p.63). Portable communications also empowered the independent foreign correspondents. By 2007, freelancer Vaughan Smith who accompanied the British Grenadier Guards on deployment in Helmand province, Afghanistan could travel with an Apple MacBook Pro laptop, a compact satellite phone and two portable video cameras (ibid).

Mobile journalism (MoJo), meaning integrating mobile technology within the workflow of professional journalists (Mills et al, 2012), emerged and developed fast in recent years. It is especially useful in war journalism. Mills and his colleagues (2012) pointed out, if truly embedded within newsroom systems, MoJo could become a streamlined, efficient, and effective process. In war and conflict zones, smartphones could function as 'a converged newsroom in their own right' (Mills et al, 2012). These researchers conducted four international smartphone-centric case studies using a beta-stage editorial commissioning platform and smartphone. In one case, they examined the Nokia N900 as a multimedia content collection device in Afghanistan. The trial results indicated that the smartphone's Wi-Fi capabilities and camera enabled the reporter to fact-check on the go and take publishable images. The lightweight nature of the phone and its mobility was also praised. But the ergonomic design of the keyboard was criticized. In addition, content management systems in newsrooms shall be adaptable to the mobile phones. The researchers concluded, mobile phones should act as a communications and content generation hub that allows users to not just collect content, but to edit, publish, and then form a dialogue while on the move and working remotely from the newsroom. The largest barrier seems to be the tension between the development of the technology workflow and its lack of integration with editorial processes (ibid).

In comparison, Westlund (2013)'s study focused on journalists and defined MoJos as journalists who use mobile devices in their news reporting. He argued that mobile devices have enhanced the possibilities for journalists to work and report from the field. Internet connectivity and advanced search functionality, and many intelligent and easily accessible apps, have provided journalists with new and powerful tools for reporting news. Rather than using the mobile only for traditional reporting, journalists have also used applications for live video streaming (e.g. Bambuser) and live blogging (e.g. Disqus) (Westlund, 2013). Moreover, since the mobile device is personal and easy to use, legacy news media can use the mobile's functions of personalization and positioning to provide instance and in-depth news in everyday life (ibid).

Mobile phones have provided new and alternative news sources by integrating with social media platforms and the UGC (user generated content). They have enhanced the interactions between professional journalists and audiences (Hoskins and O'Loughlin, 2010). Mobile technologies have impacted on newsgathering with news organizations now using camera phone images generated by the public (Reading, 2009). In the social media era, the divide between elite sources and netizens is disappearing (Mare, 2014). Journalists stress the value of audience visuals as historical documents, which can help journalists to reconstruct the chain of events (Papadopoulos & Mervi, 2013). Social media platforms are being incorporated into news production processes and they are complementing traditional news-sourcing strategies. Story ideas are "crowd-sourced" on social media platforms before professional journalists use tools such as Storify to cleanse and validate them (ibid). Mainstream media have made space for UGC production within newsrooms, for instance, BBC created a UGC Hub (Bivens, 2008), Yahoo! launched youwitnessnews to accommodate camera phone generated images with a section specifically on news from Iraq. BBC and Sky have online services dedicated to user generated images largely taken using camera phones (Reading, 2009). Associated Press, CNN, Al Jazeera and other news agencies offer mobile apps displaying content created by citizen journalists (Westlund, 2013). In addition, media organizations created new roles, such as social media editors and community managers, to collate online content. Such restructuring has allowed them to search and verify eyewitness accounts online and introduce new online formats (Bennett, 2013).

However, such increasing reliance on citizen imagery also posed challenges to the journalists who are renegotiating the conventional model of objectivity in favor of the model of transparency (Papadopoulos & Mervi, 2013). Bivens (2008) argued, along with the rising 'power of the people' to produce the so called 'Exposuregates', news organizations are facing issues of public accountability for their use of forged material and the rise in e-mail and blog-based campaigns in response to specific news coverage. Scholars have criticized the big legacy media organizations' use or exploitation of citizens' footages. Using a case study of mobile phone footage of the Iranian woman Neda Agha Soltan who was killed in Iran in June 2009, Mortensen (2011) examined the ethical dilemmas of the western news media's use of citizen photojournalism as a unique and headline grabbing source. She pointed out that there is a general lack of editorial procedures for accommodating these new sources. Lorenzo-Dus and Bryan (2011) examined the role of mobile media in British broadcasters' news coverage of the 2005 London bombings. Their analysis also revealed that journalists selected and used uniform, repetitive and sanitized mobile media footage (UGC) in their live news reporting. Citizens who generated the mobile imagery were backgrounded or marginalized and they were given no attributions. While examining news coverage of the capture and death of Muammar Gaddafi in October 2011, Kristensen and Mortensen (2013) found that one hour after the story of Gaddafi's capture breaks, a picture taken with a mobile camera that shows the now deceased Muammar Gaddafi was sent out by AFP. The image was soon validated by AFP, Reuters and the *Guardian*, and distributed globally. However while the text states that an amateur took the photo with a mobile camera, the caption credits the photo to AFP. AFP later stated in the press release that the photo was taken in Sirte by a rebel fighter using a mobile phone. AFP photographer Philippe Desmazes took a photo of the mobile's screen a few minutes later and transmit the picture. Thus illustrations may originate from amateur videos but are credited to news agencies (ibid).

Meanwhile, professional journalists cannot be replaced, and their expertise is still significant in war reporting. Otto and Meyer (2012) argued, most conflicts occur in the most deprived areas where mobile phones are the most advanced technology used. While alternative sources have opened up avenues for reporting, one still needs knowledge and expertise of journalists permanently based in a country or region to use these sources and provide context. Thus, the availability of 'new' sources of information does not compensate for individual journalistic expertise. Both taken together represent an added value (Otto and Meyer, 2012).

For war (photo)journalists, mobile phone cameras are especially important in war photography. Schwalbe (2006) noted that Iraq war was the first U.S. war photographed by embedded journalists. Satellite phones, high-quality digital cameras, and Internet access enabled instant dissemination of battlefield images. These technologies allowed journalists to feed stories and images quickly back to their newsrooms, and freed news teams from being tethered to a large satellite uplink (Schwalbe, 2006; Quinn, 2006). Photojournalists tend to be early adopters of mobile technologies. For instance, photojournalists in Austria carried camera-enabled cell phones as well as standard digital cameras since 2002. On arrival at a news event, they sent MMS (multimedia messaging system) images to online editors via the cell phone and then telephoned the newsroom to dictate two or three sentences about what happened (Quinn, 2006). Alper (2014) examined embedded photojournalists who visually documented the experience of US soldiers in Afghanistan using mobile photo application Hipstamatic. He pointed out that compared to traditional camera, a smartphone has many benefits on the front lines: smaller size, lighter weight, less additional equipment, easier battery charging, rugged, traps less dust, and more efficient workflow when pushing images to a site such as Tumblr. For the embedded photojournalist, one of the affordances of the iPhone is that it enables a level of intimacy between the photojournalist and the soldier. "The soldiers often take photos of each other with their phones, so they were more comfortable than if I had my regular camera," said a war photographer in Afghanistan (ibid) These embedded photographers not only identify with the troops they are following, but adopt the storytelling tools favored by troops to reflect that reality to the outside world (ibid). Alper (2014) also argued, the existing discourse about the use of mobile apps overlooks an important ethical issue: the implications of non-soldiers mimicking the smartphone-equipped US soldier.

In recent years, drone journalism starts to emerge and develop with the unmanned aerial vehicle (UAV). Mobile phone is used to operate and control UAV. Gynnild (2014) noted, traditional eyewitness reporting on warfare, the very act of being there, often involves considerable risk taking for reporters. Drone journalism will reduce the journalists' risk taking especially when covering wars and conflicts. He predicted that drone journalism – what he called 'robot witnessing'- will replace or supplement visual news coverage on the ground with aerial views. In the future, the drone journalist may arrive on the scene of a breaking story using the UAVs in the news coverage (ibid). Mobile phones are critical elements in practicing drone journalism. Corcoran (2014) argued, rather than flying larger, complex drones that require a stand-alone radio control network, a cheaper simpler craft shall be operated by smart phone and on-board Wi-Fi. "Even the commercial mobile phone network can be used to control this, so it would be difficult to target in on a mobile phone, if there are thousands of those about the place. So it depends on that communications network you are using, if one stands out, that's easy to detect that's going to be a military communications device or its going to be something else, then that can easily be detected but if you blend into the background it's a little bit more difficult," said Swinsburg (Corcoran, 2014). Meanwhile there are restrictions on practicing drone journalism in war zones. Drone newsgathering was more suited to assignments located in a relatively static location. A small drone, flying low and slow, may not survive for long in wars and conflicts before being shot down. Media coverage of conflicts such as Syria's civil war, may present additional problems. In 2012, thirty-three Syrian and international media were killed covering the conflict. The Committee to Protect Journalists (CPJ) reported that at least two journalists' deaths may be attributed to the interception and location of their satellite phone transmissions by Syrian regime forces (ibid).

## MOBILE EXPERIENCE OF CITIZEN JOURNALISTS

The focus of this section is on mobile-enabled citizen journalism or participatory journalism featured by the blurring boundary of amateur and professional journalists. Westlund (2013) noted, the mobile device has opened new spaces for citizen journalism. Mobile-enabled citizen journalism involves facilitating two-way communications between people who have traditionally been considered producers and users of media (ibid). Citizen war reporters used mobile camera-phone as personal witness device to record the unfolding crisis and conflicts. The current debate is revolving about the citizen journalism, protest journalism, mobile witnessing, and ethical issues, which will be discussed in more details next.

Today anyone who has a mobile phone can be a citizen reporter. Richardson (2017) noted, anyone who owns a mobile device that is equipped with a camera is a 'dormant, potential journalist ready for activation'. These citizen journalists contribute to and act as new sources in war journalism. Elite and non-elite sources as well as professionals and amateurs interplay. Pantti (2013) observed that digital communication technologies have become an inseparable part of a conflict as extended spaces of appearance and political struggle, the so called 'new visibility'. In particular, mobile phone videos, shot in the midst of conflict in areas less accessible to professional photographers, are dominated by nonprofessional footage. In contemporary wars, the citizen is not a bystander but central to the conflict. Civilians are potential citizen soldiers in ideological war. In this context, "individualized digital technologies such as the cell phone or personal computer become tools deployed in the war" (Matherson and Allan, 2009, p.95). Kristensen and Mortensen (2013) also pointed out, on account of their proximity to unfolding events, amateur sources often break the news by means of raw and fragmented bits of visual and verbal information. Elite sources rarely possess the same exclusive access to information from war zones, but are instead brought in to comment on, validate and grant legitimacy to amateur sources.

The convergence of mobile phone and social media is widely used to produce protest journalism. *Time* magazine named the protester its person of the year noting how the self-immolation of a Tunisian street-vendor gained global political currency in part because images taken by smart phones quickly made their way onto the Internet, where blogs and news reports appended meaning onto his actions (Creech, 2015). Creech (2015) used the 2011 Arab Spring protests as a case study to argue that events and individuals captured by mobile phones are granted broader meaning and visibility through journalism's meaning-making practices. Throughout the Middle East and North Africa, new media and blogger are now quasi-synonyms for protest and protester. Still, the images and tweets do not gain political consequence without journalistic practices (ibid). However, although this black "oppositional gaze" is afforded by the ubiquity of mobile devices and social networking sites, most of the academic literature are limited to case studies of amateur news reports during the 2011 Arab Spring revolts or the Occupy Wall Street demonstrations (Richardson, 2017).

Contemporary war reporting carries a long tradition of individual reporters bearing witness to the brutality of war (Matherson and Allan, 2009, p.92). Media witnessing, as a centuries-old practice, has experienced a renaissance period amid the proliferation of mobile devices. The proliferation of mobile devices lowered the entry barriers for citizen participation (Richardson, 2017). Anyone with a smartphone and a Wi-Fi connection could create and disseminate multimedia storytelling throughout the network (ibid). The media witnessing, also called digital witnessing or mobile witnessing, can be done by both citizens and professional media. It can shape shared conceptions of history and society (Belair-Gagnon, 2015). Mobile witnessing established new ways of recording events and afforded individuals networking power. Mobile phone images are part of the global news reporting and public witnessing of events in the 'war on terror' (Reading, 2009). The mobile camera-phone, as 'a wearable digital mobile prosthetic' that is connected to global digital networks, has enabled, if not produced, the now-ubiquitous performance of photography as a standard response to crisis (Papadopoulos, 2014). In the age of mobile camera-mediated mass self-publication or citizen camera-witnessing, individuals have generated important eyewitness imagery for global audiences, e.g. the September 11 attacks (2001), the London bombings (2005), and the Arab uprisings (2011–2012).

However there are moral and ethical issues with citizen journalists, the so called 'accidental photojournalists' (Allan, 2014). On the one hand, the legitimacy of citizen photo-reportage was morally problematic. "You're not a journalist just because you have your smartphone in your pocket and can take pictures of someone who has just had their leg blown off and their life shattered," someone commented on Facebook (ibid). Other criticisms include the callousness of individuals too busy taking images of victims to lend assistance, the prospect of media celebrity proving impossible to resist, and the depictions of carnage and panic fulfilling the perpetrators' desire for notoriety or inviting 'copy-cat' responses. On the other hand, citizen photojournalism was defended by those who argue that amateurs create and share 'unfiltered' news as alternative to professionals' self-censorship and template-centred coverage (ibid). The issues are also about public trust. In face of de-professionalization of news industry, both photojournalists and cellphone wielding citizens shape what we see and remember 'as a shared experience in the fullness of time.' (ibid) Meanwhile terrorists may exploit the bystanders with camera phones and the mainstream media to serve their own interests. For instance, regarding the violence occurred in Woolwich, London on 22 May 2013, a front-page picture showed Adebolajo gesturing with a blood-soaked hand while holding a stained knife and meat cleaver in his other hand. "It is this echo chamber of horror, set up by the media, public figures and government, that does much of terrorism's job for it", political commentator Simon Jenkins (2013) contended. "It converts mere crimes into significant acts. It turns criminals into heroes in the eyes of their admirers. It takes violence and graces it with the terms of a political debate." The danger is that "the terrorist might sometimes win." (Allan, 2014)

Carrying mobile phones can also be dangerous in the conflict zones. Harkin (2013) wrote: when the uprising came along, pointing camera phones or going equipped with new media became a more dangerous proposition. In April 2011, after the demonstrators were gunned down in the huge demonstration at the clock tower, Syria's official news agency SANA brought crowds armed with camera phones to the main square to show that life was returning to normal. It was premature – a sniper saw the camera phone snappers and instinctively opened fire (Harkin, 2013).

In sum, there are similarities and differences in terms of mobile experiences between professional and citizen journalists. As Table 1 indicates, the similarities lie in the fact that the boundaries between professional journalists and citizen reporters have blurred. Professionals and citizen journalists interplay and interact while social media platforms and citizen reporters act as new sources. Both professionals and citizen journalists integrate mobile phones with social media and both face ethical issues. The differences are that professional journalists use mobile phones as professional tools in war reporting such as MoJo, war photography and drone journalism whereas citizen journalists use mobile camera-phone as personal witness device to record the ongoing conflicts, the so-called mobile witnessing. Professional journalists and citizen reporters face different moral and ethical issues.

## MOBILE EXPERIENCE OF GOVERNMENTS, MILITARIES, AND REBELS

This section illustrates the control, censorship and surveillance of governments and militaries over mobile networks and flow of information during war/conflict times. The use of mobile phones by soldiers and rebels for different purposes will also be discussed.

The Web 2.0 technologies has turned the Internet into a digital war zone, the so called 'cyber conflict', 'web wars', or 'another war zone'. In today's 'global matrix of war', digital technologies are used by both state and non-state actors, blurring the line between military and entertainment and between control and resistance (Kuntsman, 2010). Against this backdrop, Kuntsman (2010) coined a new term 'cybertouch' referring to "ways in which past and current events can touch us through the monitors of our computers and mobile phones, whether by creating an immediate emotional response or by leading to long-lasting changes in the ways we remember and experience war and conflicts." The digital battlefield thus feels real.

*Table 1. Comparison of mobile experiences between professional and citizen journalists in war and conflict reporting.*

|  | **Similarities** | **Differences** |
|---|---|---|
| Professional journalists | • The distinctions between professionals and amateurs, journalists and citizens have blurred. <br> • Mobile phones have opened new sources for war reporting. <br> • Mobile phone and social media have converged <br> • Both face moral and ethical issues | Professional journalists use mobile phones as professional tools in war and conflict reporting, including <br> • MoJo <br> • Mobile war photography <br> • Drone journalism <br> • News organizations face issues of public accountability for their use of forged material and exploitation of citizens' footages. |
| Citizen journalists |  | Citizen journalists used mobile camera-phone as personal witness device to record the unfolding crisis and conflicts, the mobile witnessing, including: <br> • Citizen journalism <br> • Protest journalism <br> • Mobile witnessing <br> • The legitimacy of citizen photo-reportage was morally problematic. Terrorists may exploit the bystanders with camera phones. |

During wars and conflicts, governments and militaries attempted to control the online flow of information through blocking internet connections, censorship and surveillance. Reading (2009) argued, many governments are developing personal and mobile messages to populations through m-government (mobile-government). Mobile phones can and are providing governments with individual tracking devices via mobile data. Meier and Leaning (2009) noted that regimes are becoming increasingly savvy in their ability to control and monitor communication. For instance, in Timor-Leste, the government maintains a monopoly over the mobile phone network and has little incentive to add coverage beyond the capital city from which the state-owned company derives the majority of the profit. In Burma, the military junta closely controls the sale of SIM cards and prices of mobile phones. Shehabat (2013) also noted that in response to the social media activism, the Syrian government switched off mobile phones and Internet connections in Darra and Homs to hinder communication and the dissemination of news about the revolution. However, proxy modems, satellite phones and international mobile phones SIM cards were smuggled in from neighboring countries. Smart phones and the 3G wireless internet access became tools so significant for the revolutionaries that the Syrian government banned the use and import of iPhones into the country. Moreover, Papadopoulos and Pantti (2013) observed that Assad's security forces have proven adept at using Internet and mobile phone surveillance to track down dissidents. The government's cyber-army also took to impeding the online flow of information, particularly images of government violence toward protesters. In Sudan, Mancini (2013) stated that government censored and controlled cell phone and Internet services under the National Telecommunication Corporation (NTC). A number of media outlets were closed down for publishing content contrary to the views of the government. The government also required telecom networks to disconnect any mobile prepaid subscribers who do not provide personal information. The government also occasionally requested mobile network providers to cut off connections in certain areas for national security.

Meanwhile the new media have challenged governmental and military control over the flow of information, leading to adaptations in military tactics and new forms of intelligence gathering and surveillance (Meyer, Baden & Frère, 2017). During the Israel-Palestine conflict, phones, particularly cell phones, played an instrumental role in the Gaza offensive (Kuntsman & Stein, 2010). Kuntsman and Stein (2010) observed, Israeli soldiers were asked to surrender their cell phones before entering the Strip in order to prevent security breaches in the form of illicit calls, tweets or photographs. The IDF (Israeli Defense Force) employed electronic signal jamming to stop Palestinian fighters from communicating with each other or detonating roadside bombs. The best-known use of phones was the "knock on the roof" of Gazan houses, whereby Arabic-speaking military personnel would ring residents with live or recorded warnings that their house was marked for demolition. These phone calls were defended by the IDF and the Western mainstream media as evidence of Israel's morality in times of mortal peril. Meanwhile, the use of cell phones by Gazans was deemed evidence of terrorist activity. One soldier testified: "If I detect a lookout, someone holding binoculars or a cell phone, he's an accomplice…. If he stands on a roof holding a cell phone, that's suspect." (Kuntsman & Stein, 2010). Furthering to the soldiers' use of mobile phones, Pinder et. al, (2009) studied perceptions of the coverage of the Iraq War among British service personnel deployed during the 2003 invasion of Iraq. They find that mobile phone use has been widespread. The e-mail, Internet, cellular phones, and television conferencing has reduced the isolation felt by deployed personnel. This increasing telecommunication has been described as a mixed blessing: while some individuals feel reassured by communication with home, others feel distracted or disempowered by domestic problems (ibid).

Regarding the rebels and insurgent forces' use of mobile phones, scholars have studied the linkage between cellular coverage and violence. Pierskalla and Hollenbach (2013) examined Africa and found that cell phone coverage facilitated insurgent activity and increased violence. They argued that the distribution of cell phones helped the coordination of actions especially during asymmetric insurgent warfare. Cell phone allowed insurgent commanders to better plan, coordinate and implement operations. By contrast, Shapiro and Weidmann (2015)'s study about attacks against Coalition and Iraqi government forces showed that mobile communications reduced insurgent violence in Iraq. Meanwhile terrorist networks and organizations used mobile phone networks as a propaganda medium to circulate personalized messages globally through m-terror (mobile-terror). For instance, Al-Qaeda's media wing, a-Sahab, announced in 2008 that it was reissuing video recordings in mobile formats (Reading, 2009). Taliban used SMS to narrowcast propaganda in Afghanistan (Dafoe & Lyall, 2015). All these scenes from the contemporary battlefield, including Syrian rebels using Google Maps to correct mortar fire, illustrate the democratization of ICTs and the role of modern technology in changing the dynamics of conflict (ibid).

In sum, as Table 2 shows, on the one hand, state and non-state actors use digital technologies for their own interests, blurring the line between control and resistance. On the other hand, governments attempted to control the online flow of information via m-government, censorship and surveillance; militaries used new tactics and forms of intelligence gathering; rebels and terrorist organizations used mobile phone networks to coordinate actions, to propagate personalized messages and to implement m-terror.

## MOBILE EXPERIENCE OF NGOS, ACTIVISTS, AND COMMUNITIES

The rise of social media and new communication technology has helped to amplify the voice of less well-resourced local and semi-local NGOs (Meyer, Baden & Frère, 2017). NGOs collaborate or train activists about tactics, methods and usage of ICTs in time of crisis or violence. For instance, Mancini (2013) noted that UNDP supported a mobile-technology project for preventing election-related violence during the 2011 presidential election by a network of NGOs. This project sought to dispel various provocative rumors. European social movements advised activists how to use "ghost servers" in order to confuse the online monitoring of the government; 'Global Voices' from Tunisia and the 'Egyptian Initiative for Personal Rights' created a digital guide about how to use mobile phones and Twitter to share information about arrested activists. NGOs from the United States trained activists with a focus on video reporting and media-skills. The human rights organization 'Witness' taught them about camera operations and the use of audio recording devices, and the Kenyan NGO 'Ushahidi' built their online capabilities for reporting securely with mobile phones and building online content around it (Mancini,

*Table 2. Comparison of mobile experiences among governments, militaries and rebels during wartime*

| | Similarities | Differences |
|---|---|---|
| Governments | Digital technologies are used by both state and non-state actors, blurring the line between control and resistance | Governments control the online flow of information through m-government, blocking internet connections, censorship and surveillance. |
| Militaries | | Militaries use new tactics and new forms of intelligence gathering such as electronic signal jamming, 'knock on the roof'. |
| Rebels | | Rebels and terrorist organizations use mobile phones to coordinate actions, propagate and narrowcast messages, and practice m-terror (mobile-terror). |

2013). Meier and Leaning (2009) argued that after radios, mobile phones are most prevalent technology used to communicate in the developing world. The use of SMS is widespread during the post-election violence in Kenya. Toolkits such as Frontline SMS enable NGOs to run SMS campaigns directly from a computer. TXTmob, an open source program, also facilitates SMS broadcasting. These tools could be used to facilitate mass communication in crisis zones. Livingston (2011) also argued that mobiles are not just for making telephone calls, but data transmissions are more important for the networking and collective actions. For instance, the objective of MobileActive is to assist the "effectiveness of NGOs around the world who recognize that the 4.5 billion mobile phones provide unprecedented opportunities for organizing, communications, and service and information delivery" (Livingston, 2011). FrontlineSMS freely distributes a software program that enables users to send and receive text messages with large groups of people through mobile phones. Networks are the key that identify problems, monitor conditions, and implement solutions (ibid).

For activists, mobile phones are critical counter-government tools and part of their social media strategies in times of crisis. Harkin (2013) observed that after the Syrian regime shut down the internet for several days in Syria in November 2012, many activists and armed rebels continued to use Skype using smuggled satellite phones and dial-up modems. "How the government used its weapons against the revolution", a 27-year-old activist told *The New York Times*, "that is how activists use Skype." (ibid) Papadopoulos (2014) argued that mobile phones have created incessant media exposure and a 'perpetual crisis-awareness'. Such awareness involves the dissemination and normalization of protocols for how citizens and activists are expected to behave at the scene of a breaking crisis event: they turn their cameraphones into a personal recording device in order to make their eyewitness experience more evidential. The Arab uprisings of 2011–2012 also revealed how tech-savvy activists today routinely employ their smartphones as a key tool of insurgency and opposition to state power. For example, in Syria, as Assad's restricted foreign media, opposition activists swiftly turned their mobile phones into apparatuses of dissent and mobilization, providing the outside world with video testimony to state violence. It gives force to the audiovisual device of the mobile camera. International news media, in turn, included the raw feeds in their coverage, thus enabling activists and protesters on the ground to put their case across to international audiences and governments (Papadopoulos, 2014). However based on his personal experience, Harkin (2013) argued that many media activists' enthusiasm for the virtual weapon of new media had given way to a thirst for the real thing because the Syrian regime may have preferred all this Facebooking and Youtubing: a young man carrying a laptop or mobile phone can't also be aiming a gun. Meanwhile Pantti (2013) pointed out that activists have made abundant visual source material but there is little empirical evidence of how the conflict has been made visible in the international media, or how Syrian activists' images figure in this mediated visibility.

For civilians and communities, mobile applications have played an important role in citizens' lives. The emergence of smartphones (smart mobile phones) that combines functions of cell phones with PCs has upgraded user experience of regular mobile phones (Malka, Ariel and Avidar, 2015). Malka, Ariel and Avidar (2015) examined the roles WhatsApp played in the lives of Israeli citizens during military operation Protective Edge between Israel and Hamas in July 2014 and named the Protective Edge operation the first WhatsApp war. Based on a survey of 500 Israeli citizens, they found that the mobile application has become a significant factor for its community of users. WhatsApp has contributed to the management of 'normal life during wartime' through disseminating news updates, sending humorous, satirical and critical messages, checking on the welfare of relatives, recruiting volunteers, and organizing activities. They also found that there was a significant difference in the frequency and variety of use in accordance with the proximity of the place of residence to the firing zones. The more citizens were exposed to injury, the more central WhatsApp was in their lives (Malka, Ariel and Avidar, 2015). As a new communicative tool, WhatsApp and other applications facilitate communication in a conflict that are community-oriented as well as individually-oriented. WhatsApp helped users to 'stay in the know' rather than promoting in-depth understanding of the situation (ibid).

Similarly, Schejter and Cohen (2013) studied the role of the mobile phone during crises in Israeli society. Using real-time data on mobile phone usage provided by Cellcom, they revealed mobile usage by civilians in Israel during The Second Lebanon War in 2006 and the Operation Cast Lead in 2008/9. They concluded that the perception of the mobile is not only what it is in our lives, but also what it has the potential of becoming. In times of crisis, the mobile has become an important interpersonal communication device. When war and terror occur, the mobile is particularly useful due to its basic feature – mobility – that is, the ability to use it and reach people almost everywhere and at any time. This remains its most important quality despite the mobile's many value-added features.

In the United States, Bracken and his colleagues (2005) highlighted the importance of cell phones during the September 11 attacks. They argued that the timing of the event and the time people learned about the terrorist attacks influenced the selection of communication channels. Those who learned about the attacks earlier were more likely to contact others on their cell phone or via e-mail. Cell phones afforded them with immediacy and a feeling of closeness. The speed with which feedback could be provided also seemed to play a factor (Bracken et al., 2005). Katz and Rice (2002) examined the meaning of cell phones and how ordinary people used cell phones during the 9/11 terrorist attacks. They proposed that the mobile phone was effective and important for personal emergency communication. People used mobile phones to transmit both information and affection to their families and friends as well as alert authorities. For many, "letting others know they were loved by their special someone was their highest priority." (Katz and Rice, 2002) People without hearing were also able to use their mobile technology for text-based communication to reassure friends in the midst of the horror. The authors also argued, technical characteristics of media and boundaries across media are not particularly salient to people who have pressing personal and social communication needs (ibid).

In other parts of the world such as Sudan and South Sudan, Mancini (2012) observed that mobile technology is more prominent in poor countries. The Internet is largely an urban phenomenon. Mobile phones and interactions on popular websites played a role on the community level in fostering group action toward fight or flight. But these technologies were predominantly used to help mobilize violent mobs, issue threats to the opposing community, and propagate conflict. Meier and Leaning (2009) also argued that the lack of information communication infrastructure and basic utility infrastructure in developing countries limited the widespread use of ICTs, thus restricts the communication from the 'big world' to the areas in crisis, and vice versa.

In sum, the rise of social media and new communication technology has amplified the voice and facilitated communications of NGOs, activists and communities. While NGOs mainly provided trainings about tactics, methods and usage of ICTs as well as organize mobile SMS based campaigns, activists used mobile phones as counter-government tools and part of their social media strategies. For citizens and communities, mobile applications play an important role in interpersonal communication and normalizing citizens' lives during wartime.

*Table 3. Comparison of mobile experiences among NGOs, activists and communities in times of crisis*

| | Similarities | Differences |
|---|---|---|
| NGOs | The rise of social media and new communication technology has amplified the voice and facilitated communications of NGOs, activists and communities | NGOs provide trainings about tactics, methods and usage of ICTs; organize mobile SMS campaigns/movements |
| Activists | | Activists use mobile phones as counter-government tools and part of their social media strategies; 'virtual weapon of new media'. |
| Communities | | Mobile applications are important in interpersonal communication and normalizing citizens' lives during wartime. |

## TRENDS AND FUTURE RESEARCH DIRECTIONS

Mobile media has become a game changer to news reporting and it makes a moving target involving transformations to both the technology and usage patterns of mobile devices (Westlund, 2014). This review chapter reveals the instrumental role that mobile media played during wars and conflict. The mobile, as an important communication device, has changed the dynamics of conflict and has impacted on the conflict process. Mobile phones have become tools and virtual weapons in the war. In the digital war zone, both state and non-state actors use digital technologies to protect and champion their interests. The diving line between professionals and amateurs, between journalists and citizens, between military and entertainment, and between control and resistance has blurred. Mobile media have enabled the government and militaries to adopt m-governance, new military tactics, and new forms of information control and surveillance. Meanwhile mobile phones and social media have facilitated professional journalists' work in the field, enabled journalists gain greater independence from military and political control, enhanced the interactions between professional journalists and audiences, brought in 'new visibility' to the conflicts, empowered citizen reporters, and amplified voices of NGOs and activists who used new media as counter-government tools. Mobile communication can either increase or reduce insurgent violence in different conflict areas. Mobile phones and mobile applications also play an important role in citizens' lives, facilitating communication in community and helping normalize people's life during wartime.

There are similarities and differences in terms of mobile usage patterns and affordance in war reporting. The similarities lie in the use of mobile phones converged personalized communication, multi-media and networking functionalities. State and non-state actors use or control the use of mobile phones as mass self-communication tools. Mobile phone's basic characters such as mobility, making phone calls, cameras, sending SMS and MMS as well as other advanced functionalities such as audio-video recording, Internet access, Wi-Fi, convergence with social media platforms, Apps, personalization and positioning enable users to connect, to network, to inform, to produce, to transmit information and emotions, to mobilize, and to have conversations when wars and conflicts occur. Mobile phones have become an indispensable part of life and work for all sides who have developed and adopted their own mobile strategies.

Despite the similarities, different actors used mobiles for different purposes. They focused on different aspects or characters of mobile phones and encountered different issues. Evidence suggests that satellite phones, mobile phones' Wi-Fi capabilities, camera, mobility and lightweight are most important for professional journalists to do live reporting in war zones. Mobile phones have opened new sources for news outlets and made the news workflow more efficient. But news organizations still need to adapt their content management systems to the mobile phones as well as setting up editorial procedures to avoid exploitation of citizens' footages. Iphone and its photo applications also enabled intimacy between embedded photojournalist the soldier. For mobile-enabled citizen journalists, they mainly used mobile camera-phone as personal witness device to record the unfolding crisis and conflicts, the so-called mobile witnessing. Governments and militaries tried to control the online flow of information using the mobile phones' individual tracking functionalities. In addition to the conventional internet disconnection, censorship and surveillance, governments and militaries adapted new tactics and forms of surveillance, and practiced m-government. Similarly, terrorist groups also used mobile networks to circulate personalized messages globally, the so-called m-terror. In response to the governments' information control, NGOs and activists used mobile phones as counter-government tools and adopted social media strategies. Mobile based data transmissions in the form of audiovisual recording and text messaging (SMS, MMS) are essential for the networking and collective actions. For civilians and communities, mobile phones (cell phones) afforded immediacy and closeness in times of crisis. Mobile applications such as WhatsApp keep users 'stay in the know'.

In a word, the mobile phone's 'incessant media exposure' (Papadopoulos, 2014) and networking power has made it indispensable in contemporary wars and conflicts. State and non-state actors all adopted mobile strategies and tactics as well as developing their discourses and paradigms. Previous researches have shed lights on the usage, roles and implications of mobile communication technologies in time of crisis. Moral and ethical issues in relation to the use of mobile phones have also been addressed. Previous literatures have revealed a mixture of technological, political, economic, social and cultural factors that have shaped mobile experience in war reporting. However, most studies focused upon the political implications of mobile media and the empowering of citizen reporters. In terms of geographical locations and events, the existing literatures mainly addressed the US, UK, Africa and the Middle East as well as the 9/11 terrorist attacks, London bombing, the Arab Spring, Iraq War, Syrian War, and Israeli-Palestinian conflict. Other wars, conflicts and crisis in other parts of the world such as Asia and Latin America are missing. More comparative and cross-cultural studies on mobile experience in times of crisis are needed.

In the future, research agendas for studying the mobile experience in war journalism may include exploring the convergence of mobile media with the latest computing technologies such as AI (Artificial Intelligence), VR (Virtual Reality) and AR (Augmented Reality); the use and affordance of mobile media in practicing drone journalism; and most importantly, the institutional and non-institutional usage patterns of different mobile applications in the current and future conflicts. Empirical and comparative analysis in different types of wars and conflicts and in different countries and regions shall be conducted. An integrated theoretical framework can be developed and applied in researches combing mobile studies with digital war journalism. ANT (Actor Network Theory) is highly recommended to trace and describe the innovations, associations and interactions among human and non-human actors. Mobile experience varies among professional and citizen journalists, governments, militaries, rebels, NGOs, and communities. Similarities and differences are shaped by different factors, actors, and networks. ANT can be leveraged to explain and predict the mobile experience in contemporary war reporting.

Based on the literature review and observations, the following research questions can be further explored: with the fast development of mobile technologies, what are the recent changes in the usage patterns and affordance of mobile phones in wars and conflicts? What are the complexities and special features of human-technology interactions in times of crisis? What are the actions and reactions of mobile media users in different wars and/or countries? How does the mobile media influence the war/peace journalism in the digital and social media era? How to embed and apply the AI, VR and AR technologies in mobile phones for the purpose of war reporting? How do the mobile users perceive and interact with the newly converged technologies or functionalities? What are users' likes and dislikes towards the mobile-based war reporting in different cultural contexts? How to use ANT to identify, describe, explain and predict mobile experience in war reporting? What are the moral and ethical issues in relation to the mobile experience in war reporting?

Regarding research methods, both quantitative and qualitative research approaches can be employed ranging from online and offline questionnaire surveys, conventional and digital ethnographic research, in-depth interviews, focus group discussions, case studies, participant or non-participant observations, as well as laboratory experiments.

# REFERENCES

Allan, S. (2014). Witnessing in crisis: Photo reportage of terror attacks in Boston and London. *Media. War & Conflict*, *7*(2), 133–151. doi:10.1177/1750635214531110

Belair-Gagnon, V. (2015). Book review: Mette Mortensen, Journalism and Eyewitness Images: Digital Media, Participation, and Conflict, Routledge: London, 2015, 182. Media, War & Conflict, 8(3), 384–386.

Bennett, D. (2013). Exploring the impact of an evolving war and terror blogosphere on traditional media coverage of conflict. *Media. War & Conflict*, *6*(1), 37–53. doi:10.1177/1750635212469907

Bivens, R. K. (2008). The internet, mobile phones and blogging. *Journalism Practice*, *2*(1), 113–129. doi:10.1080/17512780701768568

Bracken, C. C., Jeffres, L. W., Neuendorf, K. A., Kopfman, J., & Moulla, F. (2005). How cosmopolites react to messages: America under attack. *Communication Research Reports*, *22*(1), 47–58. doi:10.1080/0882409052000343516

Creech, B. (2015). Disciplines of truth: The 'Arab Spring', American journalistic practice, and the production of public knowledge. *Journalism*, *16*(8), 1010–1026. doi:10.1177/1464884914550971

Dafoe, A., & Lyall, J. (2015). From cell phones to conflict? Reflections on the emerging ICT–political conflict research agenda. *Journal of Peace Research*, *52*(3), 401–413. doi:10.1177/0022343314563653

Gynnild, A. (2014). The robot eye witness. *Digital Journalism*, *2*(3), 334–343. doi:10.1080/21670811.2014.883184

Harkin, J. (2013). *Good media, bad politics? New media and the Syrian conflict. Reuters Institute Fellowship Paper*. University of Oxford.

Kristensen, N. N., & Mortensen, M. (2013). Amateur sources breaking the news, metasources authorizing the news of Gaddif's death. *Digital Journalism*, *1*(3), 352–367. doi:10.1080/21670811.2013.790610

Kuntsman, A. (2010). Online Memories, Digital Conflicts and the Cybertouch of War, Digital Icons: Studies in Russian. *Eurasian and Central European New Media*, *4*, 1-12.

Kuntsman, A., & Stein, R. L. (2010). *Another War Zone Social Media in the Israeli-Palestinian Conflict*. Retrieved from http://www.merip.org/mero/interventions/another-war-zone

Livingston, S. (2011). The CNN effect reconsidered (again): Problematizing ICT and global governance in the CNN effect research agenda. *Media, War & Conflict, 4*(1), 20–36. doi:10.1177/1750635210396127

Lorenzo-Dus, N., & Bryan, A. (2011). Recontextualizing Participatory Journalists' Mobile Media in British Television News: A Case Study of the Live Coverage and Commemorations of the 2005 London Bombings. *Discourse & Communication, 5*(1), 23–40. doi:10.1177/1750481310390164

Malka, V., Ariel, Y., & Avidar, R. (2015). Fighting, worrying and sharing: Operation 'Protective Edge' as the first WhatsApp war. *Media, War & Conflict, 8*(3), 329–344. doi:10.1177/1750635215611610

Mancini, F. (Ed.). (2013). *New technology and the prevention of violence and conflict. April 2013*. New York: International Peace Institute. Retrieved from https://ssrn.com/abstract=2902494

Mare, A. (2014). New Media Technologies and Internal Newsroom Creativity in Mozambique. *Digital Journalism, 2*(1), 12–28. doi:10.1080/21670811.2013.850196

Matherson, D., & Allan, S. (2009). *Digital war reporting*. Cambridge, UK: Polity Press.

Meier, P., & Leaning, J. (2009). *Applying Technology to Crisis Mapping and Early Warning in Humanitarian Settings*. Working Paper Series.

Meyer, C. O., Baden, C. N., & Frère, M. (2017). Navigating the complexities of media roles in conflict: The INFOCORE approach. *Media, War & Conflict*, 1–19.

Mills, J., Paul, E., Omer, R., & Heli, V. (2012). MoJo in Action: The Use of Mobiles in Conflict, Community, and Cross-platform Journalism. *Continuum, 26*(5), 669–683. doi:10.1080/10304312.2012.706457

Mortensen, M. (2011). When citizen photojournalism sets the news agenda: Neda Agha Soltan as a Web 2.0 icon of post-election unrest in Iran. *Global Media and Communication, 7*(1), 4–16. doi:10.1177/1742766510397936

Otto, F., & Meyer, C. O. (2012). Missing the story? Changes in foreign news reporting and their implications for conflict prevention. *Media, War & Conflict, 5*(3), 205–221. doi:10.1177/1750635212458621

Pantti, M. (2013). Seeing and not seeing the Syrian crisis: New visibility and the visual framing of the Syrian conflict in seven newspapers and their online editions. *Journal Journalism, Media and Cultural Studies*. Retrieved from cf.ac.uk/jomec/jomecjournal/4-november2013/Pantti_Syria.pdf

Papadopoulos, K. A. (2014). Citizen camera-witnessing: Embodied political dissent in the age of 'mediated mass self-communication'. *New Media & Society, 16*(5), 753–769. doi:10.1177/1461444813489863

Papadopoulos, K. A., & Pantti, M. (2013). The Media Work of Syrian Diaspora Activists: Brokering Between the Protest and Mainstream Media. *International Journal of Communication, 7*, 2185–2206.

Papadopoulos, K. A., & Pantti, M. (2013). Re-imagining crisis reporting: Professional ideology of journalists and citizen eyewitness images. *Journalism, 14*(7), 960–977. doi:10.1177/1464884913479055

Pierskalla, J. H. & Hollenbach, F. M. (2013). Technology and Collective Action: The Effect of Cell Phone Coverage on Political Violence in Africa. *American Political Science Review, 107*(2).

Pinder, R. J., Murphy, D., Hatch, S. L., Iversen, A., Dandeker, C., & Wessely, S. (2009). A Mixed Methods Analysis of the Perceptions of the Media by members of the British Forces during the Iraq War. *Armed Forces and Society*, *36*(1), 131–152. doi:10.1177/0095327X08330818

Quinn, S. (2006). War Reporting and the Technologies of Convergence. In R. D. Berenger (Ed.), *Cybermedia go to war: role of converging media during and after the 2003 Iraq War*. Washington, DC: Marquette Books. Retrieved from http://marquettebooks.com/images/CyberMediaNP.pdf

Reading, A. (2009). Mobile Witnessing: Ethics and the Camera Phone in the 'War on Terror'. *Globalizations*, *6*(1), 61–76. doi:10.1080/14747730802692435

Richardson, A. V. (2017). Bearing Witness While Black. *Digital Journalism*, *5*(6), 673–698. doi:10.1080/21670811.2016.1193818

Schejter, A. M., & Cohen, A. A. (2013). Mobile phone usage as an indicator of solidarity: Israelis at war in 2006 and 2009. *Mobile Media & Communication*, *1*(2), 174–195. doi:10.1177/2050157913476706

Schwalbe, C. B. (2016). Changing faces: the first five weeks of the Iraq war. In R. D. Berenger (Ed.), *Cybermedia go to war: role of converging media during and after the 2003 Iraq War*. Washington, DC: Marquette Books. Retrieved from http://marquettebooks.com/images/CyberMediaNP.pdf

Shapiro, J., & Weidmann, N. B. (2015). Is the phone mightier than the sword? Cell phones and insurgent violence in Iraq. *International Organization*, *69*(2), 247–274. doi:10.1017/S0020818314000423

Shehabat, A. (2013). *The social media cyber-war: the unfolding events in the Syrian revolution 2011* (Australian Edition). Global Media Journal.

Westlund, O. (2013). Mobile news. *Digital Journalism*, *1*(1), 6–26. doi:10.1080/21670811.2012.740273

Xu, X. (2017). Comparing Mobile Experience. In Handbook of Human-Computer Interaction. Hoboken, NJ: Wiley. doi:10.1002/9781118976005

# Chapter 15
# Mobile News Experience:
## Comparing *New York Times* and *The Guardian*

**Wendi Li**
*The University of Melbourne, Australia*

**Xiaoge Xu**
*University of Nottingham – Ningbo, China*

## ABSTRACT

*Mobile has become a mainstream medium for news consumption on the go. To cater to the growing demand for mobile news, traditional news providers have switched from "mobile too" to "mobile first" strategies. To enhance mobile news communication, it is imperative for mobile news providers to stay abreast of mobile news consumers' changing expectations of mobile news experience in a news app. It is equally imperative to identify the gap between news consumers' expectations and what mobile news experience is embedded in a mobile news app. Using a mobile experience index, the authors of this chapter have located the venue and extent of the gap through conducting a survey of mobile news app users and a comparative analysis of indicators of mobile news experience in selected news apps.*

## INTRODUCTION

As the most popular and powerful medium for communication, mobile has been increasingly and widely used as a tool for journalists to gather and disseminate news and as a platform for consumers to seek and follow news. Consequently, going mobile has been prioritized by more and more news providers. Mobile first has become a strategy for most news organizations. On top of mobile news sites, news apps have also been growing in popularity among mobile users. Its popularity has led to more and more focus on how to enhance mobile news experience in designing and delivering news through apps. It has also brought about a strong desire for enhanced mobile news experience among mobile users. There is seemingly, however, a gap between what mobile news experience mobile users expect to have and what

DOI: 10.4018/978-1-5225-7885-7.ch015

mobile news experience has been offered by news providers. To narrow the gap is ultimately important in enhancing mobile news experience. To that end, we need to be on the constant alert as what changes have taken place, what challenges are confronting us and what chances lie ahead in terms of the gap in mobile news experience between news providers and consumers.

## EARLIER STUDIES ON MOBILE NEWS EXPERIENCE

Among earlier studies on news apps (excluding news aggregator apps), we selected studies on mobile news experience expected and offered. The goal of this section is to identify what has been done and what remains unsolved in terms of news apps with a special focus on the gap the normative and the empirical in mobile news experience.

Mobile news experience can be reflected in personalized content recommendations by leveraging their special and specific interests, desires and preferences inferred from their locations, Facebook and/ or Twitter feeds and in-app actions (Kazai, Yusof & Clarke, 2016) or the access history (Kiritoshi & Ma, 2015). It may also be embedded in customization and enhancement of engaging experience through leveraging ubiquitous geolocation metadata (Oppegaard & Rabby, 2016). Thirdly, mobile news experience may also be part of a habitual personalization process through recognizing kinds of news reading behavior and adapting its display and interaction methods (Constantinides, 2015).

Mobile news experience can fall under two themes (repurposing and customizing) and four categories (human-led repurposing, technology-led repurposing, human-led customization, and technology-led customization (Westlund, 2013). In the case of podcasting apps, aggregated from a fragmented media environment, listeners can form a stronger relationship with content producers through "increasing sonic interactivity, encouraging ubiquitous listening, curating and packaging podcasts as visual media, and emphasizing social features that allow users to share podcasts with each other" (Morris & Patterson, 2015).

In their investigation of the circumstances, in which users want or do not want tailor-made news, Groot Kormelink and Costera Meijer (2014) found that users have limited interest in personalizing or participating in news. Different news app users may have different expectations. Per Constantinides, Dowell, Johnson and Malacria (2015), news app users can largely be grouped into three major categories: (1) trackers, (2) reviewers and (3) dippers. Due to their different motivations, news app users may have different expectations and experiences too in terms of frequency, daily reading time, browsing strategy, reading style, and location (Constantinides, Dowell, Johnson & Malacria 2015). In experiencing, mobile news, the role of social media can't be belittled as they can help mobile users filter news per their prioritized needs and tastes for news in the context of information overload (Pentina & Tarafdar, 2014). In their longitudinal study of personalization of mobile news sites/apps, Thurman and Schifferes (2012) found that mobile users were reluctant to engage with complex forms of active personalization and a significant growth of user social-network based 'social collaborative filtering' as a form of passive personalization.

The results of our brief review of previous studies on mobile news experience have located a research problem, that is, few studies have been conducted to investigate the gap between what mobile news consumers expect to have and what they actually get in terms of mobile news experience in news apps.

## CONCEPTUALIZATION OF MOBILE NEWS EXPERIENCE

Our gap investigation was equipped with a 3M approach (mapping, measuring and modeling) proposed by the first author of this chapter. Mobile news experience can be mapped, measured and modeled through its six different stages (enticement, entertainment, engagement, empowerment, enlightenment and enhancement) developed by the first author of this chapter on the normative level, the empirical level and the gap between the two (Xu, 2018).

In terms of enticement, which is the first stage of mobile news experience, we look into the following three fundamental dimensions of mobile news experience: (a) appealing interface, (b) easy navigating, and (c) interest-arousing (Xu, 2018). Appealing interface refers to simple, easy and attractive interface, which is designed to attract and maintain mobile news consumers' attention and interest. Easy navigation means a simple and easy device-specific navigation mechanism, allowing mobile news consumers to navigate easily from one place to another on the screen. Interest-arousing, as the third dimension of enticement, is to arouse mobile news consumers' interest in reading or viewing a story on the homepage of a news app.

Mobile news experience at the entertainment stage can be located in the following areas: (a) fun, (b) pleasure, and (c) satisfaction (Xu, 2018). By fun, we refer to what mobile news consumption is amusing, entertaining and enjoyable while pleasure means sensual gratification. Satisfaction refers to fulfillment of mobile news consumers' wishes, expectations or needs.

In terms of engagement, which is the third stage, we map mobile news experience through three dimensions proposed by Xu (2018): (a) searching, (b) interacting, and (c) sharing. To search before, in the middle of or even after consuming a news story in an app is to keep mobile news consumers engaged throughout the whole process of consuming a mobile news story. The engagement also involves interacting with others via social media or online chatting or interacting mechanisms such as chatrooms, blogging, emailing, and instant messengers. To share with what mobile news consumers have consumed also enables them to stay engaged during their consumption.

Mobile news experience can also be located at the empowerment stage by looking into the following three dimensions designed by Xu (2018): (a) selecting (content, font size, layout, color, background, theme, region, media, edition, and language), (b) commenting (correcting, rating and feedback), and (c) producing (tweeting, blogging, writing, editing, publishing, and broadcasting). Mobile news experience at this stage largely refers to the extent mobile news consumers are empowered to select, comment and produce what they would like via mobile and the corresponding outcomes.

Enlightenment-specific mobile news experience can be located in the following aspects: (a) awareness, (b) understanding, and (c) consciousness (Xu, 2018). To consume a news story in an app will enable mobile news consumers to be enlightened in one way or another through enhancing their awareness, understanding and consciousness respectively of events or issues in a news story.

Ultimately, mobile news experience can be mapped through the following ways: (a) knowledge, (b) skills, and (c) abilities (Xu, 2018). The ultimate goal of enriching mobile news experience is to enhance mobile news consumers' interaction with news stories in an app through enhancing their knowledge, skills and abilities respectively.

# OPERATIONALIZATION OF MOBILE NEWS EXPERIENCE

To measure mobile news experience, previous studies have employed different methods. Among them are (a) use of metrics to describe and measure it, (b) using stories to provide its rich details, (c) using usability tests to gauge it, and (d) using a hierarchy of user needs to measure its different layers. For example, in using metrics, some of earlier studies employed a number of metrics to describe and measure user experience (Hassenzahl 2003; Jordan 2000; Norman 2004). The story approach can be illustrated by other studies, which provided stories of how users experience a product or service (Forlizzi & Ford, 2000) or still other studies, which provided in-depth views and insights or social and cultural factors located in narratives or storytelling in the hope of capturing and interpreting user experience in a holistic and constructionist way by listening to users' stories of experience with technology (McCarthy & Wright 2005).

The usability approach can be exemplified in other earlier studies, where scholars focused on either "the ability to complete some functional or goal-directed task within a reasonable time" or "the degree, to which a product is desirable or serves a need beyond the traditional functional objective" (Logan, 1994). Some of earlier usability studies also examined the following new dimensions: (a) "efficiency, affect, control, and learnability, and helpfulness" (Kirakowski & Corbett, 1993), (b) "performance and image/impression" (Han, Yun, Kwahk & Hong, 2001), and (c) "ease of use, helpfulness, affective aspect, minimal memory load, and efficiency" (Ryu & Smith-Jackson, 2006).

Finally, the hierarchy approach can be illustrated by a study which investigated a hierarchy of user needs: safety, functionality, usability, and a pleasurable experience (Jordan, 2000).

To measure the normative level of mobile news experience, we investigate what components and levels of mobile news experience to be expected by a mobile user through a combination of surveys, focus groups, and interviews. To measure the empirical level, we gauge both components and level of mobile experience to be actually experienced by a user by using the same combination of surveys, focus groups, and interviews. Beyond that, however, we can also conduct comparative content analyses of what elements of mobile experience are actually embedded in apps.

To standardize the quantification of both normative and empirical mobile news experience, we use percentage to measure each of the six components of mobile experience. Both normative and empirical mobile experience can also be measured on both micro and macro levels. By micro level, we refer to the level of mobile news experience at each stage of mobile news experience while macro level is meant to be about the total level of all six stages of mobile news experience (Xu, 2018).

On the micro level, each component of mobile news experience can be measured in three ways: (a) the normative measurement of each component of mobile experience, (b) the empirical measurement of each component of mobile experience, and (c) the gap measurement between the normative and empirical level of each component of mobile experience.

At the enticement stage, a scale of 1 to 5 is assigned to each of the three components: appealing interface, easy navigation and interest-arousing, making 15 points for the entire enticement dimension of mobile experience. In both normative and empirical dimension of enticement, 1-5 refers to the low range of enticement, 6-10 means the medium range, and 11-15 is the high range. And the gap between the normative and the empirical can be gauged by using this formula: the normative enticement 15 - the empirical enticement (15), with three major categories of gap: 1-5 (the narrowest gap), 6-10 (the medium gap), and 11-15 (the widest gap).

At the entertainment stage, a scale of 1 to 5 is assigned to each of the three components: fun, pleasure and satisfaction, making 15 points for the entertainment dimension of mobile experience. In both normative and empirical dimension of entertainment, 1-5 refers to the low range of entertainment, 6-10 means the medium range, and 10-15 is the high range. And the gap between the normative and the empirical can be gauged by using this formula: the normative entertainment (15) - the empirical entertainment (15), with three major categories of gap: 1-5 (the narrowest gap), 6-10% (the medium gap), and 11-15 (the widest gap).

At the engagement stage, a scale of 1-5 is assigned to each of the three components: searching, interacting and sharing, making 15 for the engagement dimension of mobile experience. In both normative and empirical dimension of engagement, 1-5 refers to the low range of engagement, 6-10 means the medium range, and 11-15 is the high range. And the gap between the normative and the empirical can be gauged by using this formula: the normative engagement (15) - the empirical engagement (15), with three major categories of gap: 1-5 (the narrowest gap), 6-10 (the medium gap), and 11-15 (the widest gap).

At the empowerment stage, a scale of 15 is assigned to each of the three components of enlightenment experience: selecting, commenting and producing, making 15 for the empowerment dimension of mobile experience. In both normative and empirical dimension of empowerment, 1-5 refers to the low range of empowerment, 6-10 means the medium range, and 6-15 is the high range. And the gap between the normative and the empirical can be gauged by using this formula: the normative empowerment (15) - the empirical empowerment (15), with three major categories of gap: 1-5 (the narrowest gap), 6-10 (the medium gap), and 11-15 (the widest gap).

At the enlightenment stage, a scale of 1-5 is assigned to each of the three components of enlightenment experience: awareness, understanding and knowledge, making 15 for the enlightenment dimension of mobile experience. In both normative and empirical dimension of empowerment, 1-5 refers to the low range of enlightenment, 6-10 means the medium range, and 11-15 is the high range. And the gap between the normative and the empirical can be gauged by using this formula: the normative enlightenment (15) - the empirical enlightenment (15), with three major categories of gap: 1-5 (the narrowest gap), 6-10 (the medium gap), and 10-15 (the widest gap).

At the enhancement stage, a scale of 1-5 is assigned to each of the three components of enhancement experience: knowledge, skills and abilities, making 15 for the enhancement dimension of mobile experience. In both normative and empirical dimension of enhancement, 1-5 refers to the low range of enhancement, 6-10 means the medium range, and 11-15 is the high range. And the gap between the normative and the empirical can be gauged by using this formula: the normative enhancement (15) - the empirical enhancement (15), with three major categories of gap: 1-5 (the narrowest gap), 6-10 (the medium gap), and 11-15 (the widest gap).

On the macro level, mobile experience can be measured in three ways: (a) measuring the total normative mobile experience, (b) measuring the total empirical mobile experience, and (c) measuring the gap between the normative and empirical measurement of the total mobile experience (Xu, 2018). The mobile experience gap can be gauged by using the following formula: the gap = the total normative mobile experience (90) – the total empirical mobile experience (90). The gap may fall under any of the three categories: the narrowest gap (1-30), the medium gap (31-60) and the widest gap (61-90) (Xu, 2018).

# MOBILE NEWS EXPERIENCE: SURVEY AND CONTENT ANALYSIS

Guided by the above-stated conceptualization and operationalization of mobile news experience, we investigated and compared mobile news experience between *New York Times* and *The Guardian*. As the most popular and powerful medium for communication, mobile has been increasingly and widely used as a tool for journalists to gather and disseminate news and as a platform for consumers to seek and follow news. Consequently, going mobile has been prioritized by more and more news providers. Mobile first has become a strategy for most news organizations. On top of mobile news sites, news apps have also been growing in popularity among mobile users. Its popularity has led to more and more focus on how to enhance mobile news experience in designing and delivering news through apps. It has also brought about a strong desire for enhanced mobile news experience among mobile users. There is seemingly, however, a gap between what mobile news experience mobile users expect to have and what mobile news experience has been offered by news providers. To narrow the gap is ultimately important in enhancing mobile news experience. To that end, we need to be on the constant alert as what changes have taken place, what challenges are confronting us and what chances lie ahead in terms of the gap in mobile news experience between news providers and consumers.

To investigate and compare news apps of *New York Times* and *The Guardian*, the following five research questions were addressed in the following paragraphs:

**RQ1:** To what extent did New York Times enhance mobile news experience in its app?

**RQ2:** To what extent did The Guardian enhance mobile news experience in its app?

**RQ3:** Is there a big difference between New York Times and The Guardian in offering mobile news experience?

**RQ4:** To what extent did mobile news users expect to have from a mobile news app?

**RQ5:** Is there a big difference between mobile news providers and consumers in mobile news experience?

To address these questions, a comparative content analysis and a survey were conducted. The comparative content analysis was conducted to analyze *New York Times* news app and *The Guardian* news app by using a mobile news experience index (Xu, 2018) The index consists of six stages: 1. Enticement, 2. Entertainment, 3. Engagement, 4. Empowerment, 5. Enlightenment, and 6. Enhancement. And each stage consists of three indicators.

Enticement is measured by appealing interface, easy navigation and interest-arousing while entertainment is gauged in terms of fun, pleasure and satisfaction. Engagement is measured on three fronts, that is, searching, interacting, and sharing. Empowerment lies in the following three areas: selecting, commenting and producing. Enlightenment is gauged by awareness, understanding and consciousness while enhancement is measured by knowledge, skills and abilities. Each indicator is measured on a range from 1 through 5 scale. The maximum score for each part is 15 marks while the total is 90 marks. At each individual stage level, a mark range of 1-5 would be considered as low level of mobile news experience, 6-10 as medium and 11-15 as high. Putting all six stages together, any marks ranging from 1 through 30 would be considered low level of mobile news experience. Anywhere between 31 and 60 would be considered as medium. The high level would fall anywhere between 61 and 90.

In measuring mobile news experience, the homepage of each of the two selected news apps was investigated. The homepage investigation focused on three dimensions of the first stage of mobile news experience, that is, appealing interface, easy navigation, and interest-arousing. After the homepage investigation, a major story of the day was picked up for an investigation for the remaining five stages: 1. Entertainment, 2. Engagement, 3. Empowerment, 4. Enlightenment, and 5. Enhancement.

The survey was conducted online using Google Forms from 7th Sep to 8th September 2016. A purposive sampling method was used to select 35 students of media and communication studies at The University of Melbourne in Australia. The same codebook was used in coding *New York Times* and *The Guardian* apps to ask the respondents to rate the importance of each indicator of each stage of mobile news experience per their respective perspectives and preferences.

## FINDINGS AND DISCUSSION

**RQ1:** To what extent did *New York Times* enhance mobile news experience in its app?

On its homepage, New York Times scored 10 out of 15 marks on enticement. In terms of its story page, New York Times scored 9 out of 15 on entertainment, 10 out of 15 on engagement, 9 out of 15 on empowerment, 11 out of 15 on enlightenment, and 10 out of 15 on enhancement. The total level of mobile news experience was 59 out of the total 90 marks. As seen from these figures, overall, New York Times can be placed at the higher end of the medium range in terms of level of overall mobile news experience.

**RQ2:** To what extent did *The Guardian* enhance mobile news experience in its app?

The Guardian scored 12 out of the total 15 marks on the homepage in terms of enticement. Its story page scored 10 out of 15 on entertainment, engagement, empowerment, and enhancement respectively while 11 on enlightenment. The grand total score was 63. Judging by these figures, The Guardian can be placed at the lower end of the high range.

**RQ3:** Is there a big difference between *New York Times* and *The Guardian* in offering mobile news experience?

Placed against each other for comparison, there is no big difference between *New York Times* and *The Guardian* in either six individual stages of mobile news experience or their respective total level of mobile news experience. The difference between the two is only four marks apart with *New York Times* at the high end of the medium range while *The Guardian* at the lower end of the high range. Although they can be placed above average in terms of overall mobile news experience, both *New York Times* and *The Guardian* still have some room for improvement, especially in the areas of enlightenment and enhancement.

**RQ4:** To what extent did mobile news users expect to have from a mobile news app?

To divide the 5-point scale into two parts, those below 3 (with 3 included) would be considered low expectation while whose above 3 would be viewed as high expectation. Measured by this division, while a large portion of the participants placed "easy navigation" (71.5%) and "interesting content" (82.9%) respectively within the high range of expectation, slightly more than half of the participants considered "appealing interface" (65.7%) as important. It shows that enticement as the first stage of mobile news experience matters to most mobile news users. Within this same stage, however, participants gave greater importance to "easy navigation" and "interesting content" than "appealing interface". When asked about the importance of entertainment, only about half of participants put "fun" (51.4%), "pleasure" (48.6%) and "satisfaction" (57.2%) respectively into the high expectation range. When it comes to engagement, less than half of the participants selected "searching" (34.3%) and "sharing" (48.5%) respectively as important while only slightly more than half of the participants viewed "interacting" (57.1%) as important. In measuring empowerment, far less than half of the participants attached importance to selecting (48.6%), commenting (40%), and producing (37.2%) respectively. In terms of enlightenment, the percentage of the participants went up, who viewed "awareness" (62.9%), "understanding" (74.3%) and "consciousness" (65.7%) respectively as important. In measuring enhancement, however, there is a difference even within the same stage. While most the participants viewed "knowledge" (80%) as very important, only slightly more or less than half of the participants considered "skills" (60%) and "abilities" (48.5%) respectively as very important.

**RQ5:** Is there a big difference between mobile news providers and consumers in mobile news experience?

The combined results of the content analysis and the survey suggest that there is no big difference between mobile news providers and consumers in enticement and enlightenment. When it comes to entertainment, engagement, empowerment, and enhancement, however, there is a big difference with the only exception of the "knowledge" dimension of enhancement. Although mobile news providers seemed to be very keen on providing more entertainment, engagement, empowerment, and enhancement, news consumers were not so keen on some of these aspects, such as "fun" (51.4%), "pleasure" (48.6%), "satisfaction" (57.2%), selecting (48.6%), commenting (40%), and producing (37.2%), skills (60%), and abilities (48.5%).

As leading news providers in USA and UK respectively, New York Times and The Guardian performed decently well in providing their respective news consumers with a decent amount of mobile experience on major indicators of the mobile news experience index in an increasingly fragmented media environment where news media, old and new, are fiercely competing for news consumers on many fronts. Conventionally defined, New York Times and The Guardian remain the mainstream news media in USA and UK respectively. In the age of the Internet and especially when citizen media and mobile media have become more and more popular and powerful than those used to be claimed as mainstream or leading news media, New York Times and The Guardian cannot afford to rest on their conventional comfort zone. It is undeniable that more and more news consumers have switched to mobile devices for seeking and following news on the go instead of holding on to the dying print or broadcast news media. To compete with new news media such as facebook, twitter, and Instagram, they have changed their strategies from "mobile too" to "mobile first". While some news providers have even gone so far as to adopting the "mobile only" strategy to stay ahead of others, the majority have tried every possible way to entail as much as possible enticement, entertainment, engagement, empowerment, enlightenment and enhancement to stop losing their readers to new news media.

Mobile news consumers, however, although they do expect much mobile news experience, may not be aware of what is available and what can be leveraged in terms of mobile news features that may enable them to enhance their mobile news experience. Even if they are aware of those features, they have their preferences and prioritizations. To most of mobile news users, although enticement and enlightenment

*Table 1. Comparing New York Times and The Guardian: Results*

| Stages | Indicators | Measurements | The Guardian | New York Times |
|---|---|---|---|---|
| Enticed | Appealing Interface | 1 least ..... 5 most | 4 | 3 |
| | Easy Navigation | 1 least ..... 5 most | 4 | 4 |
| | Interest-Arousing | 1 least ..... 5 most | 4 | 3 |
| **Sub-Total** | | | 12 | 10 |
| Entertained | Fun | 1 least ..... 5 most | 3 | 3 |
| | Pleasure | 1 least ..... 5 most | 3 | 3 |
| | Satisfaction | 1 least ..... 5 most | 4 | 3 |
| | **Sub-Total** | | 10 | 9 |
| Engaged | Searching | 1 least ..... 5 most | 3 | 3 |
| | Interacting | 1 least ..... 5 most | 3 | 3 |
| | Sharing | 1 least ..... 5 most | 4 | 4 |
| | **Sub-Total** | | 10 | 10 |
| Empowered | Selecting | 1 least ..... 5 most | 3 | 3 |
| | Commenting | 1 least ..... 5 most | 4 | 3 |
| | Producing | 1 least ..... 5 most | 3 | 3 |
| | **Sub-Total** | | 10 | 9 |
| Enlightened | Awareness | 1 least ..... 5 most | 4 | 3 |
| | Understanding | 1 least ..... 5 most | 4 | 4 |
| | Consciousness | 1 least ..... 5 most | 3 | 4 |
| | **Sub-Total** | | 11 | 11 |
| Enhanced | Knowledge | 1 least ..... 5 most | 4 | 4 |
| | Skills | 1 least ..... 5 most | 3 | 3 |
| | Abilities | 1 least ..... 5 most | 3 | 3 |
| | **Sub-Total** | | 10 | 10 |
| **Grand Total** | | | 63 | 59 |

seem to be more important than entertainment, engagement, empowerment, and enhancement. While attaching importance to different stages, they do not give equal importance to specific indicators of those stages of mobile news experience.

While the results of this preliminary investigation seem to suggest that mobile news consumers have not reached beyond meeting their fundamental needs to be enticed and enlightened when it comes to news consumption. The findings also seem to suggest that mobile news experience remains largely sequential, which is, however, different from what Maslow's (1970a, 1970b) hierarchy of needs or even the hierarchy of desires which emphasizes the strength of layered desires (Turner, 2003). Using the six-stage approach in measuring mobile news experience, as shown by the results of the content analysis and the survey, we could conclude that the index can work in a way to provide us with a more accurate picture of how mobile news experience works. It works largely in the sequence of importance instead of exposure since the sequence of importance can change without necessarily following the sequence of exposure. In other words, mobile users may attach greater importance to one aspect of one stage of mobile experience, as indicated by the results of this comparative examination.

In the age of mobile connection, conversation and contribution, although content remains fundamentally important, what is more vital is how to enhance mobile experience. In the case of mobile news communication, how to cultivate mobile news experience at different stages of mobile news communication has become paramount to the overall enhancement of mobile news communication. Equally important is to narrow the gap between the embedded and the experienced in terms of mobile news experience. To that end, mobile news providers keep abreast of changes in needs, tastes, and preferences for mobile news experience and spare no efforts to enhance mobile news experience in their news apps.

Our preliminary investigation was restricted in that we did not examine native news apps, in other words, news apps without being affiliated with any traditional news media such as *New York Times* and *The Guardian*. For future investigation, we would like to suggest a comprehensive and comparative study of news apps of both native mobile news media and digitized news media as well as a wider range of users of news apps.

## REFERENCES

Constantinides, M. (2015, April). Apps with habits: Adaptive Interfaces for News Apps. In *Proceedings of the 33rd Annual ACM Conference Extended Abstracts on Human Factors in Computing Systems* (pp. 191-194). ACM. 10.1145/2702613.2702622

Constantinides, M., Dowell, J., Johnson, D., & Malacria, S. (2015). *Exploring mobile news reading interactions for news app personalization. 17th International Conference on Human-Computer Interaction with Mobile Devices and Services*, Copenhagen, Denmark.

Forlizzi, J., & Ford, S. 2000. Building Blocks of Experience: An Early Framework for Interaction Designers. In *Proceedings of Designing Interactive Systems Conference (DIS 2000)* (pp. 419–23). Brooklyn, NY: Association for Computing Machinery. 10.1145/347642.347800

Groot Kormelink, T., & Costera Meijer, I. (2014). Tailor-Made News: Meeting the demands of news users on mobile and social media. *Journalism Studies*, *15*(5), 632–641. doi:10.1080/1461670X.2014.894367

Han, S. H., Yun, M. H., Kwahk, J., & Hong, S. W. (2001). Usability of Consumer Electronic Products. *International Journal of Industrial Ergonomics*, *28*(3-4), 143–151. doi:10.1016/S0169-8141(01)00025-7

Hassenzahl, M. (2003). The thing and I: Understanding the relationship between user and product. In M. A. Blythe, A. F. Monk, K. Overbeeke, & P. C. Wright (Eds.), *Funology: From usability to enjoyment* (pp. 31–42). Kluwer Academic Publisher. doi:10.1007/1-4020-2967-5_4

Jordan, P. W. (2000). *Designing Pleasurable Products*. New York: Taylor and Francis. doi:10.4324/9780203305683

Kazai, G., Yusof, I., & Clarke, D. (2016, July). Personalised News and Blog Recommendations based on User Location, Facebook and Twitter User Profiling. In *Proceedings of the 39th International ACM SIGIR Conference on Research and Development in Information Retrieval* (pp. 1129-1132). ACM. 10.1145/2911451.2911464

Kirakowski, J., & Corbett, M. (1993). SUMI: The Software Usability Measurement Inventory. *British Journal of Educational Technology, 24*(3), 210–212. doi:10.1111/j.1467-8535.1993.tb00076.x

Kiritoshi, K., & Ma, Q. (2015). A Diversity-Seeking Mobile News App Based on Difference Analysis of News Articles. In *Database and Expert Systems Applications 26th International Conference, DEXA 2015* (pp. 73-81). Academic Press. 10.1007/978-3-319-22852-5_7

Logan, R. J. (1994). Behavioral and Emotional Usability: Thomson Consumer Electronics. In M. E. Wiklund (Ed.), *Usability in Practice* (pp. 59–82). New York: AP Professional.

Mahlke, S. (2005). Understanding Users' Experience of Interaction. In *Proceedings of the 2005 Annual Conference on European Association of Cognitive Ergonomics* (pp. 251–4). Chania, Greece: Academic Press.

McCarthy, J., & Wright, P. C. (2005). Putting 'Felt-Life" at the Centre of Human-Computer Interaction (HCI). *Cognition Technology and Work, 7*(4), 262–271. doi:10.100710111-005-0011-y

Morris, J. W., & Patterson, E. (2015). Podcasting and its apps: Software, sound, and the interfaces of digital audio. *Journal of Radio & Audio Media, 22*(2), 220–230. doi:10.1080/19376529.2015.1083374

Naftali, M., & Findlater, L. (2014, October). Accessibility in context: understanding the truly mobile experience of smartphone users with motor impairments. In *Proceedings of the 16th international ACM SIGACCESS conference on Computers & accessibility* (pp. 209–216). ACM. doi:10.1145/2661334.2661372

Norman, D. (1999). *The Invisible Computer*. Cambridge, MA: MIT Press.

Norman, D. (2004). *Emotional Design: Why We Love (or Hate) Everyday Things*. New York: Basic Books. doi:10.1145/985600.966013

Oppegaard, B., & Rabby, M. K. (2016). Proximity: Revealing new mobile meanings of a traditional news concept. *Digital Journalism, 4*(5), 621–638. doi:10.1080/21670811.2015.1063075

Pentina, I., & Tarafdar, M. (2014). From "information" to "knowing": Exploring the role of social media in contemporary news consumption. *Computers in Human Behavior, 35*, 211–223. doi:10.1016/j.chb.2014.02.045

Ryu, Y. S., & Smith-Jackson, T. L. (2006). Usability Questionnaire Items for Mobile Products and Content Validity. *Proceedings of the Human Factors and Ergonomics Society 50th Annual Meeting*.

Thurman, N., & Schifferes, S. (2012). The future of personalization at news websites: Lessons from a longitudinal study. *Journalism Studies, 13*(5-6), 775–790. doi:10.1080/1461670X.2012.664341

Turner, R. (2003). *Desire and the Self*. Retrieved from http://robin.bilkent.edu.tr/desire/desire3.pdf

Westlund, O. (2013). Mobile news: A review and model of journalism in an age of mobile media. *Digital Journalism, 1*(1), 6–26. doi:10.1080/21670811.2012.740273

Xu, X. (2018). Comparing mobile experience. In K. Norman & K. Kirakowski (Eds.), *Wiley Handbook of Human-Computer Interaction Set* (Vol. 1, pp. 225–238). Wiley.

# Chapter 16
# Smart Tourist Experiences:
## Impacts of Smartphones on Leisure Travels

**Natalia Menezes**
*University of Aveiro, Portugal*

**Belem Barbosa**
*University of Aveiro, Portugal*

**Carolina Barrios Laborda**
*Universidad Tecnológica de Bolivar, Colombia*

**Dayana R Pinzón Callejas**
*Universidad Tecnológica de Bolivar, Colombia*

## ABSTRACT

*After a comprehensive review on mobile tourism experience, the authors have identified the benefits and impacts of mobile use to tourists and their experiences. Besides locating similarities and differences in using mobile for tourism, the authors have confirmed that mobile empowers tourists to get more from their vacations and to have more flexible planning, resulting in satisfaction and accomplishment. This chapter enlightens tourism operators, among other stakeholders, on the opportunities for contextualized mobile advertising, which would attract and convert tourists into potential customers.*

## INTRODUCTION

Technology is acknowledged by managers, scholars, and consumers alike for its unavoidable impact on the tourism sector in recent years. Its influence is clear when observing tourists' behavior and experience, and the consequent adaptation of tourism operators that reinvent the use of the internet to better serve, attract, and satisfy prospects and customers. Tourists are increasingly technology savvy, they are connected, and they naturally use technology to improve their tourism experience, gaining in ubiquity, personalization, flexibility, and unlimited access to information. Therefore, they may be seen as smart and empowered tourists that require the delivery of complete, relevant and timely information, during their

DOI: 10.4018/978-1-5225-7885-7.ch016

travels (Dickinson, Ghali, Cherrett, Speed, Davies & Norgate, 2014). In this context, several authors (e.g., Gretzel & Jamal, 2009; Hannam, Butler & Paris, 2014; Lamsfus, Wang, Alzua-Sorzabal & Xiang, 2015; Liang, Schuckert, Law & Masiero, 2017) have discussed mobile tourism and suggest that there is a new class of tourists that construct personalized traveling experiences relying mainly on their smartphones.

Further, smartphones influence tourists' decision processes before and during the trip, playing an essential role as facilitators of the tourism experience, not only for communication purposes but also for information search, entertainment and purchase. In fact, tourists use smartphones for different utilitarian and entertainment functions, and its usage depends on several factors, including patterns of use in their daily life (Wang, Xiang & Fesenmaier, 2016).

Based on the most relevant contributions in the literature on mobile tourism and the results of a qualitative exploratory study comprising 20 semi-structured interviews, this chapter addresses the impact of smartphones on tourist consumer behavior during leisure trips. The research objectives include understanding the impact of using smartphones on leisure travel experiences, explaining differences in the patterns of use during leisure travel and everyday life, and providing relevant cues for tourism sector managers so that they may include them strategically in destination management in order to enhance tourists' travel experience.

## BACKGROUND

Recent research has shown that Information and Communications Technology (ICT), especially the Internet, has substantially transformed travel behavior (Gretzel, Fesenmaier & O'Leary, 2006; Buhalis & Law, 2008; Xiang, Gretzel & Fesenmaier, 2009). However, the impact of ICT is considered more significant when travelers are equipped with mobile technologies (Xiang & Gretzel, 2010; Wang, Park & Fesenmaier, 2012). Among the existing mobile devices, the smartphone has been one of the most popular, especially in travel, particularly because of its ease of transport. According to Statista (2018), the number of smartphone users in 2020 will be around 2.87 billion. In a 2017 World Travel & Council (WTTC) Report, Rob Torres, Managing Director of Google, claims that in the next four years 1 more billion people will be coming online, mostly on a mobile device. Today, it is estimated that around two-thirds of adults worldwide own a smartphone (Zenith, 2017). With these growing numbers, organizations offering hospitality related products and services must learn to manage new technologies, regardless of company size. Unlike tablets and laptops, smartphones fit in hands and pockets, making it easy to use while traveling. Its small size and familiarity of use make the smartphone a more intimate and personalized device, used both in different contexts of the user's life (Miller, 2012). High market penetration reflects the great technological evolution that smartphones have had since its inception and the multiple functions that this device exerts. The first smartphone was introduced in the market by IBM in 1993. Since then, these devices have evolved over the years from a simple mobile phone to make calls and perform simple activities to a handheld computer with its own operating system, running several apps at a time for communication, information and entertainment purposes (Miller, 2012). Indeed, smartphones owners use their handsets for a variety of activities, such as to access the internet, to access e-mails, take pictures, to use and download applications, and for social networking (Watson, McCarthy & Rowley, 2013). But devices are better exploited through what is known as the Internet of Things (IoT), which enables communication vessels between the physical and the digital world, making virtually any object smart and identifiable. According to Intel (2018), the number of connected devices (IoT) will grow from

2 billion in 2006 to 200 billion in 2020, which will represent on average 26 smart objects per person. These technological advances are echoed in the frequent use of this device in everyday life and, consequently, in other contexts like travel. The impact of mobile technology on tourism can be measured by the increasing number of studies that intend to summarize and review the literature on the subject, as is the case of Law, Chan & Wang (2018) or Kim & Law (2015).

## TOURISTS' MOBILE EXPERIENCE

Mobile devices, especially smartphones are among the technological innovations that have impacted tourism the most and have directly influenced the tourism experience (Hyun, Lee, & Hu, 2009; Wang et al., 2012). Researchers point out that the use of smartphones in the experiential process of travel has modified the travel patterns and behavior before, during, and after the touristic experience, increasingly merging all these stages through the process of search and use of information (Wang, Xiang & Fesenmaier, 2014; Grün, Werthner, Pröll, Retschitzegger & Schwinger, 2008). The stages can be explained as follow (Law, Chan & Wang, 2018):

- **Pre-Trip:** Tourists may plan less knowing they will have access to information onsite relying in their mobile devices.
- **During the Trip:** Tourists feel more empowered and are more connected and participative. They have more valuable and interactive experiences through the use of mobile devices. Mobile devices allow them to be informed, to communicate easily with family and friends, to do work-related tasks, to have more fun and share their experiences in the social media.
- **Post-Trip:** Mobile devices are paramount for storing memories that are shared through social media.

The general objective of smart technologies is to evaluate the environment and facilitate the processes so that they can be carried out in a more intelligent, efficient, effective and useful way (Neuhofer, Buhalis & Ladkin, 2015). Given the impact of these technologies on the travel experience, they are being studied more deeply in the literature. This new concept of using mobile devices to access tourism content is called mobile tourism or m-tourism, a term created by Brown and Chalmers in 2003. Tourists use these mobile services for a variety of functions, but mainly to perform planning on a travel route, that is, seeking information to make decisions during the trip itself. This information is largely about transportation, accommodation reservations, flight reservations, and other services directly related to travel and that are context sensitive (Goh, Ang & Lee, 2010).

The great differential of the use of smartphones during the tourism experience is the capacity that the tourist has of accessing information in different contexts and refining it according to his/her objectives. To achieve this, smart mobile systems need to meet three technological requirements: information aggregation, ubiquitous mobile connectedness and real-time synchronization of information (Neuhofer, Buhalis & Ladkin, 2015). Information will only positively influence the tourism experience if it is adequate to the contextual awareness of the place, time and activity in which the tourist is inserted and to the profile of each tourist (Ferdiana & Hantono, 2014). However, mobile technology should be able to provide contextual triggers, give personalized and proactive recommendations that have the power to influence tourist behavior and improve experiences (Tussyadiah & Wang, 2016).

Improving contextual awareness is pointed out in the study by Dickinson et al. (2014) as one of the three areas of development in relation to tourism trips. The second factor is the greater use of sharing capabilities and the internet of things to revolutionize the organization of travel within social networks. Finally, the third area of growth is data mining techniques, which will be able to integrate personal data and objectives to reveal new opportunities to the tourist, especially given the wide variety of travel resources available. All these trends prove to be a challenge for tourism managers who want to offer a travel experience through smartphones as communication tools. Researchers suggest that organizations in charge of destination marketing should integrate marketing programs in a wide variety of channels, adopting new business models that take advantage of m-tourism (Kim & Law, 2015). To understand the impact of using smartphones on travel and giving them a more strategic use, it is important to distinguish the unique characteristics of mobile communication and capitalize on them. According to Siau, Lim and Shen (2001), the mobile commerce features are:

- **Ubiquity:** Through mobile devices connected to the internet, customers can be impacted anywhere and anytime. The great advantage for customers is the possibility of obtaining information at convenience, regardless of location, on a real-time basis.
- **Personalization:** Despite the large amount of information available on the internet, users search for information relevant to the context in which they are inserted for which information regarding the experiences must be collected. Therefore, tailored services may be offered which are preferred for long-term successful commercial relationships. Smartphones are a good fit for this feature since they are normally used by a sole individual.
- **Flexibility:** The inherently portable feature of mobile devices enables owners to use it in the course of other activities, allowing its frequent use.
- **Dissemination:** It is the ability to simultaneously reach a wide number of users within a specific geographic region, thus being more efficient.

The great challenge of delivering sophisticated information to tourists is only possible thanks to these characteristics of mobile communication. Tourists recognize this enhanced capability of their mobile devices and increasingly want information that is space-time relevant (Dickinson, et al., 2014). Other attributes like convenience (the right information at the right moment and the right time), localization (through GPS systems in mobile devices) and accessibility (receiving information in a timely manner) make mobile commerce extremely attractive. All these particularities of mobile devices influence the way tourists use smartphones before, during and after the tourism experience.

Given the great impact that mobile technologies have on tourism experiences, it is important to understand more deeply how smartphone´s characteristics have an effect on usage during travel. Tourist behavior is directly associated to the usage of devices in everyday life. The different uses of mobile devices in daily life are in turn directly associated with the use and capability of smartphones to allow tourists to communicate, use their time and enjoy their travel (Wang et al., 2016). The study by Wang et al. (2016) proposed a model that covers the use of smartphones in daily life and during travel. In daily life, the authors summarize use in six functions: communication, social activity, information acquisition, information search, entertainment and facilitation. These functions bring some changes in users' daily lives, such as greater communication with family and friends or exploration of new applications, among others. These behaviors or feelings have "spill over" effects on travel, so that the six functions mentioned are maintained. Moreover, the use of smartphones during travel enhances the travel experience.

There are however, tasks that are specifically carried out by smartphone owners exclusively for travel purposes. Smartphones are used for planning and sharing experiences through social media during travel, leading to less planning prior to the trip and obtaining more gratification during the travel experience. Feelings are also affected by the use of smartphone during travel. Tourists who use smartphones during the trip feel different: feel more connected, more informed, more secure, more confident, more flexible and more comfortable (Wang et al., 2014; 2016). These changes in tourist behavior generate impacts for managers of tourism destinations and must be considered when thinking about marketing strategies that must be in line with tourists' motivations of mobile device usage.

Understanding the motivations for using mobile devices is the starting point for understanding modern tourist behavior. The study by Kim, Park & Morrison (2008) proposes two external variables - technological experience and travel experience - and two influential determinants - perceived ease and perceived utility of use - as decisive factors for adopting mobile devices in the context of tourism. The study confirmed that recurrent travelers are positively influenced to use them on other trips, use them more frequently and perceive them as necessary. Regarding usage of technology, perceived usefulness had a stronger impact than perceived ease of use on the perceived need for mobile devices. Davis (1989) defines perceived usefulness as the degree to which a person believes that using a particular system will enhance performance, while perceived ease of use is the degree to which a person believes that using a particular system will be effortless. For the tourist, the use of mobile devices should be convenient, effective and productive while traveling.

Eriksson & Strandvik (2009) identify factors that may influence tourist´s use of mobile services (such as apps) during their travel experience: perceived mobile value, perceived service value, perceived (financial) risk, ease of use, the influence of other people (social influence) and the unique characteristics of the tourist context (which relate to the tourist experience and the type of trip). Among these factors, the perceived utility (mobile and service value) was also the aspect that most influenced the use of mobile services by tourists. Both studies point out that service utility is paramount and the main factor influencing the value perspective for the tourist during their experience. This information is of great relevance for hospitality managers and organizations that intend to offer new travel-related applications.

Recent research addresses other factors that motivate the use/reuse of mobile devices during travel as summarized by Law, Chan & Wang (2018). They recap their findings into five factors, being utilitarian reasons (mentioned above) the most important one. The other factors are hedonic, dispositional, behavioral and environmental. Hedonic factors are the intrinsic motivations (Kim, Kim, Kim & Kim, 2016) for both tourists that in their everyday life enjoy and do not enjoy the use of smartphones. Travel preferences, destination image, social acceptance and emotional attachment are other reasons classified in this factor. As dispositional factors Law, Chan & Wang (2018) classify the elements related to personal traits of the tourist. These include self-confidence on technology usage, innovativeness (this characteristic usually drives usage), privacy concerns, trust and involvement with service providers, among others. Behavioral factors are related to consumer behavior affecting usage. Everyday habits such as expense patterns, smartphone usage and technological experience, on one hand, and frequency of travel, on the other, have an effect on the use of smartphones during trips. Finally, environmental factors such as service environment may affect the use and consumption of mobile technology according to the authors.

Wang et al. (2014) explored extrinsic and intrinsic motivations for smartphone use. The extrinsic motivations are related to the results achieved. In the case of smartphone use on travel, the three main goals are social entertainment, staying connected to the world and get food information while traveling. The main intrinsic motivation is the habit of everyday usage replicated in their trips. The authors also explore cognitive benefits that motivate usage, in particular the ease and convenience of using the mobile devices while traveling. This point has a direct relation with two other categories that motivate tourists to take their smartphone for a trip: experience of daily use and usage of the device in previous trips. The better the tourist´s satisfaction, the more likely she/he will use it in future trips (Wang et al., 2014).

The relationship of continuity of use of the smartphone in later trips was confirmed in the study of Tussyadiah & Wang (2016). For tourists, smartphones are considered companions of travel that help them in the decision-making process resulting in a more pleasant touristic experience. Smartphones provide the opportunity to travelers to access online information anytime and anywhere. In view of these motivational factors, it is imperative that tourism managers are aware of motivations as well as perceived benefits of smartphone usage in order to be able to offer a pleasant experience and when difficulties arise be capable of offering worthy solutions.

## SMARTPHONE'S BENEFITS FOR TOURISTS

The continuity of use of a tool has a direct relationship with the benefits perceived by the owner when using it. The more a tourist uses mobile technologies during a positive experience, the more likely she/he will use it in their next destination (Wang et al., 2014). Mobile tourism scholars have carried out research to identify the benefits that smartphone use can bring throughout the tourism experience process. We will focus on the benefits that have a direct impact during the travel period.

The study by Wang et al. (2014) covers the influence of smartphone usage during the tourism experience. They point out that the main effects of mobile device use are connectivity, the ability to navigate through geolocation, flexibility of travel planning and, the possibility to carry out online transactions. The use of the smartphone as a tool for communication, entertainment, facilitation, and information search results in changes in tourist behavior and perceived benefits of a good experience during the trip. We can point out five main benefits in the literature:

- **The Tourist Feels Better Informed:** The uninterrupted provision of up-to-date information tailored to the profile of tourists is directly related to the intention to use mobile device services. Therefore, the service used during the trip should be perceived as useful and satisfy the users in their search for relevant and contextual information (Kim, Ahn & Chung, 2013; Tussyadiah & Wang, 2016). This factor has an impact on the following two aspects.
- **The Tourist Feels More Confident:** Travelers who use smartphones throughout the travel process feel safer and more confident because they are aware that they can access more information from reliable sources throughout the journey (Wang et al., 2014). Getting more information before and during the stay makes the tourist feel less lost in the destination and provide them more possibilities to choose what to do during the trip (Jansson, 2007).
- **The Tourist Feels More Connected:** Another advantage is the connectivity offered by the smartphone. Through the internet, tourists can share their tourism experience through social media instantly and receive feedback and social reward immediately (Jansson, 2007). This makes them feel more comfortable and connected with their daily life (Wang et al., 2014). They, therefore, seek to establish and maintain this copresence in order to feel integrated, even if they are physically distant. For this reason, copresence is an integral part of the tourism experience (White & White, 2007).

- **The Tourist Is Better Located:** The location offer that the smartphone offers during the tourist experience has several impacts for the traveler. The study by Tussyadiah & Zach (2012) points out that geolocation-based technology helps people to have more geographical knowledge of places and help with exploration while traveling. Consequently, it makes the tourism experience more meaningful and authentic, as it allows tourists to be better informed and make more assertive decisions during their trip. The use of geolocation-based technology also impacts on the sensorial and emotional dimensions of the tourism experience, creating a greater value for the tourist, which generates a stronger emotional attachment to the destination.
- **The Tourist Can Plan With More Flexibility:** The connectivity factor, along with mobility, makes travel planning more objective and flexible. If the itinerary previously proposed does not please the tourist, he/she has the ability to modify it more easily through the use of a mobile device with access to the internet (Wang et al., 2014). Routing planning has a direct impact on pre-trip planning, while the tourism experience may be built during the journey, in accordance to the context and needs that arise.

All these benefits together maximize the value of the tourism experience, proving that the use of smartphone brings benefits to the tourist before, during and after the trip (Wang, et al., 2014). In order to bring all these benefits to the tourist, travel applications need to evolve in three directions: firstly by investing in the contextual awareness of the applications by delivering information so that tourists have a better perception of space-time of the destination, offering more confidence through ease of navigation; secondly, improving information-sharing capabilities through social networks, which have the potential to revolutionize travel organization, allowing independence and flexibility (to plan on demand); thirdly, developing information-seeking techniques integrated with personal data and thus providing relevant information and enabling tourists to make better informed travel decisions according to their profile (Dickinson et al., 2014).

Despite the mentioned benefits, there are still some barriers to smartphone usage pointed out in the literature, factors that hinder or obstruct its performance during the trip. These factors are related to difficulty to access information on mobile devices, including screen dimensions, limited access to WiFi and battery chargers during the trips, and roaming costs that could be associated to some of the tourists' destinations (Lamsfus et al., 2015). The need to feel disconnected from their daily lives during leisure (Dickinson, Hibbert, & Filimonau, 2016; White & White, 2007) stands out as a barrier to smartphone use. Hence, the use of smartphones during the trip are also influenced by external and technological factors, making the use of smartphones during the tourism experience a complex phenomenon worth exploring.

## METHOD

Based on the contributions presented in the literature review and considering the objectives defined, the authors performed an exploratory qualitative research in order to further understand how the use of smartphones affects tourist experiences. The study was conducted in Portugal. Semi-structured interviews are particularly useful for obtaining insights of individual views and experiences, and thus are frequently used in consumer behavior studies. The interview outline comprised two sections: smart tourism experience and a comparison with everyday smartphone usage.

Principles for ethical research were included guaranteeing that the participation is voluntary, confidential and anonymous. Complete information prior to data collection was provided so participants could give their informed consent. Interviews were audio-recorded and then transcribed.

The study population comprised Portuguese adults that owned a smartphone and that had at least one tourism experience in the prior 12 months. Participants' selection was performed to identify tourists with diverse sociodemographic profiles. The sampling method was non-probabilistic, as researchers used their professional, community, and personal networks to identify potential participants. Twenty individuals accepted to participate in the study, and their profiles are presented in Table 1. Data was collected between October and November 2015. Data saturation was reached on the fifteenth interview.

Qualitative content analysis was performed following the recommendations by Bardin (2008) and Ritchie et al. (2013): initial identification of themes and concepts, followed by coding, labeling, and classification. The categories identified include benefits of using the smartphone in everyday life, types of smartphone usage during tourism activities, perceived smartphone benefits during the travel experience, and positive and negative impacts of smartphones during leisure travel. The authors used NVIVO qualitative data analysis software for organizing and coding content categories and themes.

*Table 1. Sample characteristics*

| Interviewee | Gender | Age | Education Level | Occupation |
|---|---|---|---|---|
| Interviewee 1 | Female | 29 | Master's degree | Researcher |
| Interviewee 2 | Female | 32 | Master's degree | Journalist |
| Interviewee 3 | Female | 35 | PhD | Teacher |
| Interviewee 4 | Male | 54 | PhD | Computer engineer |
| Interviewee 5 | Female | 19 | Undergraduate | Student |
| Interviewee 6 | Female | 23 | Undergraduate | Student |
| Interviewee 7 | Male | 25 | Graduate | Musician |
| Interviewee 8 | Male | 27 | Master's degree | Researcher |
| Interviewee 9 | Female | 33 | Master's degree | Journalist |
| Interviewee 10 | Female | 29 | Graduate | Manager |
| Interviewee 11 | Female | 27 | Graduate | Teacher |
| Interviewee 12 | Female | 29 | Graduate | Architect |
| Interviewee 13 | Male | 27 | Undergraduate | Machine operator |
| Interviewee 14 | Male | 23 | Master's degree | Biomedic |
| Interviewee 15 | Female | 22 | Undergraduate | Clerk |
| Interviewee 16 | Male | 32 | Graduate | Pilot |
| Interviewee 17 | Male | 27 | Graduate | Computer engineer |
| Interviewee 18 | Male | 29 | Master's degree | Agronomist |
| Interviewee 19 | Female | 27 | Master's degree | Food engineer |
| Interviewee 20 | Male | 35 | Graduate | Businessman |

# RESULTS

Overall, intensity and diversity of smartphone usage during travels is expected to depend on usage in everyday life. Our results corroborate this in the sense that people that use smartphones frequently in their daily routines, are particularly familiar with the applications and functionalities that may be useful during travel. Still, the next sections will evidence notorious differences, both in the type of functionalities and the usage intensity during travel. The impact of tourism experience and consumer behavior will also be highlighted.

## Smartphones as Travel Facilitators: Information Search, Location, Entertainment, and Communication

This study demonstrates that information search is one of the main uses of smartphones during travel. Besides the functional benefits, it enhances the tourism experience, as stressed by one of the participants "For me the smartphone is a very interesting tool! It makes traveling much more interesting because it provides information when I need it" (Interviewee 4). Interviewees often mentioned contextualized information search such as geographical location, search for restaurants and hotels, and detailed information about the destination´s touristic activities. This contextualization made participants feel like they get the most out of their travel, particularly considering time constraints: "Because you save time, you can use it doing more activities that fit your traveling goals" (Interviewee 9). These results are similar to extant literature, which highlight the gains that information access through smartphones present for the tourist experience (Wang et al., 2012; Wang et al., 2014; Tussyadiah & Zach, 2012).

As highlighted in the literature, information access also affects travel planning, making it more flexible, as Interviewee 15 explains: "Now I'm more impulsive, because oh, I'll take care of that later, I don't need to prepare so much. I guess I used to prepare travels more back then." There is increased flexibility for choosing places to visit, where to stay and where to eat, meaning that tourists are more open to last call opportunities and use apps to find the best last-minute deals. Collected data therefore show the importance of providing accurate, relevant, and contextualized information to tourists for their immediate decisions during travels, keeping in mind information functionality. This result is similar to the ones obtained by Wang et al. (2012), who recommended that tourists be provided relevant and contextualized information. Hence, information access makes traveling more efficient and less planned. In fact, with the help of smartphones, tourists participating in this study stated that they changed their plans constantly, in order to meet their needs and to comply with real-time offers they found online. Consequently, during travel tourists make more consumer decisions, making them particularly sensitive to proposals and online alternatives available as an option to recommendations from tourism operators such as travel agencies that used to plan their trips in detail in the past.

Location tools also provide increased feeling of safety and autonomy. The participants in this study mentioned preferring apps that they are more familiar with, and thus imply no extra effort for them. In fact, all the reported apps were highlighted for their usefulness, thus stressing the importance of the relationship between perceived utility and satisfaction of using mobile devices in the tourism context as pointed out in the literature (e.g. Kim et al., 2008; Kim et al., 2013, Tan, 2016). For example, Interviewee 4 concludes:

*As a general rule, whenever I had a smartphone and had the opportunity to use it, it was always interesting to have it. Today's apps are becoming more and more sophisticated and allow a range of things that were not possible before. Smartphones have always been useful.*

The entertainment activities that stood out were taking photos and sharing them on social network sites such as Facebook and Instagram. In fact, the real-time sharing of experience is one of the key changes that smartphone use enables. Wang et al. (2012) refer to this as "sharing of happiness", meaning tourists seek immediate social feedback and reinforcement. This feature also keeps tourists close to their family and friends, as for instance Interviewee 19 explained:

*[Last December] I travelled alone. So it's a bit dumb, but when you're alone and you have your cell phone, it seems you're not alone. It's like, you're in the restaurant alone, you end up connecting to the Wi-Fi and stay there, you know? Chatting and getting distracted more, using things, it keeps you close.*

Hence, these results show that interacting with family and friends through the ease of accessibility and ubiquity of smartphones is part of tourists' travel experience, a characteristic already present in previous literature (e.g., White & White, 2007; Hannan, Butler & Paris, 2014). Participants' narratives also confirmed the importance of smartphones' entertainment features for dealing with waiting time, making trips less monotonous, as noted by Hyun et al. (2009). Interviewee 11 provides a clear example:

*My flight departed at six in the morning and I got at the airport before midnight the previous day (...) So, poor me without a smartphone, because I picked it up, I accessed the Wi-Fi, and I communicated, I searched, I browsed...*

Finally, participants communicate during travel mostly through social network sites and messaging applications, especially when they travel abroad, in order to avoid communication costs.

## Everyday vs. Tourism Usage of Smartphones

Literature points to the relationship between smartphone usage during travel, in previous travels, and in everyday life (Wang et al, 2014). The present study shows that usage intensity might be lower during travels, namely because of the additional cost of internet access when traveling abroad, the need for evasion and escapism, and the focus on the actual tourism experience. Interviewee 1 shared her experience:

*When I'm on leisure trips there is little time for procrastination. (...) But what happens is that if I don't have Wi-Fi, I will not be in the Internet for superfluous things. (...) So, I use it only when I need it.*

Thus, both intensity and type of usage changes during travel, which has a more utilitarian nature and less communication or social interaction outside the tourism experience. Interviewee 17 added: "During the trip I'll focus more on being a tourist. I'm not responding to messages, I'll just use [the smartphone] for specific things like a map to know where I'm going to. So, during vacations I use it less." Interviewee 5 explained that the main reason of usage is the quest for a greater immersion on the travel context:

*I have the smartphone with me all the time, but I'm much more interested in getting to know the city, it's all very new. So, if I get stuck in the cell phone a lot, I'll end up not seeing the surroundings.*

Hence, such behavior seems particularly adherent in the cases when tourists are visiting a new destination. In fact, participants who stressed the decline in smartphone usage during travel, were the ones that most value the escape from daily lives provided by vacations. This quest for detachment was also found by White and White (2007) and Dickinson et al. (2016), as a need of some tourists to keep their distance from their daily lives and to disconnect from the people that are not traveling with them.

The participants in this study who recognized using the smartphone as intensely as in everyday life stressed that the types of activities of usage change during tourism travels. These activities are mostly related to search and location contextualized activities, booking and shopping, managing travel, and sharing experiences with friends and family, as mentioned earlier.

## IMPACT ON TOURISM EXPERIENCE

The benefits of using the smartphone during travel identified by the participants in this study include access to information, saving money, greater use of travel, autonomy, more flexible planning, and safety, as already evident in the previous sections. These results corroborate the findings in the literature, namely of Kim et al. (2013) who assert that satisfaction of the user has a strong relationship with information factors, as it unfolds and impacts on various perceptions about the travel experience. For instance, Interviewee 8 explained that smartphone usage had an important impact on his overall satisfaction during travel. Participants in general recognized that they increasingly seek immediate and relevant tourist information in order to maximize their experience and gain in terms of flexibility and autonomy. In fact, with their smartphones, tourists can make better consumer decisions by planning while traveling. Wang et al (2014) also emphasized this en-route planning which is exemplified in this study by Interviewee 16:

*It was different because I felt I was making the most of the site, I had to have a lot more work before going on vacation to get to see all I've seen. Because I do not like going somewhere and thinking that I did not visit the best sites. The smartphone allows me to plan on the moment or a day in advance and allows me to be more flexible.*

Thus, aside from overall satisfaction, this study also evidenced the sense of accomplishment as one of the impacts of the use of smartphones during tourism experience. Figure 1 synthetizes the results of the present study by proposing a conceptualization based on the main findings.

Hence, our results demonstrate that tourism experience is associated to a decreased smartphone usage for communication and procrastination activities as compared to everyday life, while tourists use the mobile device for other activities directly related to managing their tourism experience (e.g., information search, geo-localization, online purchases) and sharing the experience. Smartphone usage is leveraged by a set of benefits that were acknowledged by tourists and are identified in the proposed conceptual framework as the gains associated to usage. Benefits such as saving time and money, enjoying more flexibility in planning the trip, being more autonomous and feeling safe are combined with

*Figure 1. Proposed conceptual framework*

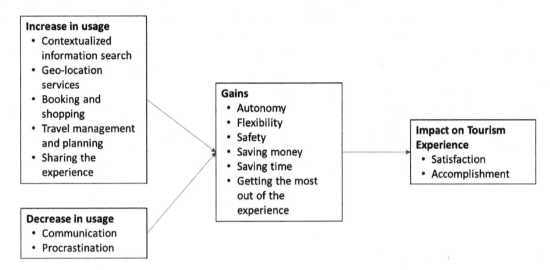

the feeling expressed by tourists that smartphones help them get the most out of the tourism experience. Consequently, the perceived gains of using the smartphone will have a positive impact on the tourism experience, in the form of both satisfaction and feeling of accomplishment.

## SOLUTIONS AND RECOMMENDATIONS

By using smartphones, tourists become more proactive, autonomous and well informed, having a greater control over their experience and being able to make better decisions according to their needs and preferences. Consequently, the smart tourist depends on technology and reliable information to design her experience in an autonomous and flexible way. This obviously creates further challenges and opportunities for tourism operators. Practitioners ought to consider this target segment by anticipating and satisfying their needs and wants, using mobile marketing strategies to attract, convert, and please smart tourists. In order to do that, practitioners need to develop ubiquitous information services and focus on creating value and developing relationships with tourists. Geolocation data can help improve tourism management effectiveness by identifying tourists' patterns and preferences, and by presenting tourists with customized and real-time offers. Overall, the smartphone helps tourists accomplish their traveling goals and increases their overall satisfaction and sense of accomplishment, namely by helping them get the most out of their experience in a flexible and autonomous way.

Amid so many benefits offered to tourists, it is imperative that organizations in the tourism sector adopt strategic mobile technologies that attract and retain tourists, improve destination awareness, satisfy travel experience, and create loyalty to the destination (Kim et al., 2008; Hyun et al., 2009). At the end, the ultimate goal of using smart technologies in the tourism sector is to improve experiences, generating added value and increasing competitiveness (Neuhofer, 2012).

Indeed, benefits perceived by tourists using mobile devices provide opportunities for organizations related to the tourism sector to develop innovative services that meet this new market demand (Goh, Ang & Lee, 2010). One of the important steps to attain this, is through the involvement of tourism destination managers, for instance in online social communities, or by monitoring and participating in social media to develop mobile services that add value to tourists and tourism itself (Bader et al, 2012). Understanding the context in which travel decisions are made has become vital for destination managers to seek effective decision support and advertising targeting strategies as there is an emerging class of tourists that rely heavily on information technology, namely mobile technology, which gives the possibility of building a connected travel experience (Lamsfus et al., 2015).

In other words, the tourism manager must be aware of the complexity of a networked tourism experience, which includes the tourist being at a destination while his entire online contacts are sharing his/her experience from the distance. This dual-channel communication makes the tourism experience more collaborative and dynamic (Dickinson et al., 2014), and can be a valuable source of information for tourism managers that want to include mobile devices strategically in destination management, especially because they can be great allies in improving the travel experience.

## FUTURE RESEARCH DIRECTIONS

Considering that this study has an exploratory and descriptive nature, it is necessary that future research further explores the role of smartphones in tourists' behavior. Some interesting topics include attitudes and behaviors regarding contextualized advertising, m-commerce activities, the impact of online word-of-mouth in en-route tourists' decisions and the effect of planning before traveling.

Despite the interesting findings, this study reflects the experiences of a limited number of tourists; hence similar approaches should be applied to different samples and populations in order to validate results. Business travels were beyond the scope of this study. Still, considering that smartphones are particularly relevant in that context too, the authors recommend that future research replicates this study to that population, including comparing the results among tourists in either business or leisure trips. Other variables that should be explored in future research include the role of tourist profile (e.g., frequency of traveling, tourist preferences), the characteristics of trips (e.g., duration, type of destination), as well as the relationship with satisfaction and post-consumption behavior, including online word-of-mouth communication.

## CONCLUSION

This chapter provides a comprehensive review on extant literature on mobile tourism experience, highlighting the benefits that smartphones provide to tourists and the overall impact on their experience, namely on their consumer behavior, from planning and searching to buying and giving feedback. Moreover, by exploring the relationship with smartphone usage in everyday life, this chapter also contributes with important insights on similarities and differences that were evidenced by a group of 20 tourists that shared their views and experiences during interviews.

This study shows that the use of smartphone impacts tourists' experience even before traveling. In fact, and according extant literature (e.g., Lamsfus et al., 2015; MacKay & Vogt, 2012; Oh, Lehto, & Park, 2009; Wang, et al., 2012), smartphones help make planning more flexible and personalized. Tourists use smartphones for activities directly related to the context of the trip, namely for searching information, making purchases, and also for geographic localization, as pointed by extant literature (e.g., Lamsfus et al., 2015; MacKay & Vogt 2012; Wang, Xiang & Fesenmaier, 2016). This study also provides additional empirical support for contributions in the literature that stress the motivation for using the smartphone during tourism activities dependent on the perceived benefits, including feeling more autonomous and safer, well informed and confident, as previously stressed by Wang et al. (2012, 2014). This study found clear differences between smartphone daily usage and during tourism, confirming that tourists use smartphones less, particularly for communication and leisure, as pointed out by Wang et al. (2016).

Overall, this study demonstrates that the smartphone empowers the tourist and has a positive impact on tourism experience as a result of its utilitarian benefits (e.g., information search, location and itineraries, mobile shopping, managing tickets and other documents, translating foreign languages). Entertainment activities pertaining to smartphone usage in tourism context include taking and sharing photos and spending waiting time leisurely. In fact, the results presented evidence that the smartphone enables tourists to get more from their vacations and to have more flexible planning. Clear consequences of smartphone usage during tourism travels are satisfaction and accomplishment.

Obviously, it is essential for tourism operators to provide timely and accurate information, corroborating contributions in the literature that stress that providing contextualized and reliable information may revolutionize tourists' experiences before, during, and after the trip (Dickinson et al., 2014; Ferdiana & Hantono, 2014; Lane et al., 2010; Wang et al., 2014). Moreover, there is a need for accessible, affordable and good quality Internet connections, as well as entertainment and utilitarian apps that enable offline usage.

For tourism operators, companies and managers, this chapter evidences very interesting opportunities, including developing contextualized and timely offers targeted at tourists that are searching information, making for instance mobile advertising particularly interesting to attract and convert potential customers. Considering the dynamism and relevance of smart tourists, this is a topic that is expected to continue to deserve attention and effort of practitioners and academics in the future.

# REFERENCES

Ashton, K. (2009). That "Internet of Things" Thing. *RFID Journal*, *22*(7), 97–114.

Bader, A., Baldauf, M., Leinert, S., Fleck, M., & Liebrich, A. (2012). Mobile Tourism Services and Technology Acceptance in a Mature Domestic Tourism Market: The Case of Switzerland. In M. Fuchs, F. Ricci, & L. Cantoni (Eds.), *Information and Communication Technologies in Tourism 2012* (pp. 296–307). Vienna: Springer. doi:10.1007/978-3-7091-1142-0_26

Bardin, L. (2008). *Análise de conteúdo*. Edições 70.

Brown, B., & Chalmers, M. (2003). Tourism and mobile technology. In *Proceedings of the eighth conference on European Conference on Computer Supported Cooperative Work* (pp 335-354). Helsinki: Kluwer Academic Publishers.

Buhalis, D., & Law, R. (2008). Progress in information technology and tourism management: 20 years on and 10 years after the internet- The state of eTourism research. *Tourism Management*, *29*(4), 609–623. doi:10.1016/j.tourman.2008.01.005

Davis, F. D. (1989). Perceived usefulness, perceived ease of use, and user acceptance of information technology. *Management Information Systems Quarterly*, *13*(3), 319–340. doi:10.2307/249008

Dickinson, J. E., Ghali, K., Cherrett, T., Speed, C., Davies, N., & Norgate, S. (2014). Tourism and the smartphone app: Capabilities, emerging practice and scope in the travel domain. *Current Issues in Tourism*, *17*(1), 84–101. doi:10.1080/13683500.2012.718323

Dickinson, J. E., Hibbert, J. F., & Filimonau, V. (2016). Mobile technology and the tourist experience: (Dis)connection at the campsite. *Tourism Management*, *57*, 193–201. doi:10.1016/j.tourman.2016.06.005

Eriksson, N., & Strandvik, P. (2009). Possible determinants affecting the use of mobile tourism services. In *International Conference on E-Business and Telecommunications* (pp. 61-73). Springer. 10.1007/978-3-642-05197-5_4

Ferdiana, R., & Hantono, B. S. (2014). *Mobile tourism services model: A contextual tourism experience using mobile services.* 2014 6th International Conference on Information Technology and Electrical Engineering (ICITEE), Yogyakarta.

Goh, D. H., Ang, R. P., & Lee, C. K. (2010). Determining Services for the Mobile Tourist. *Journal of Computer Information Systems*, *51*(1), 31–40.

Gretzel, U., Fesenmaier, D. R., & O'Leary, J. T. (2006). The transformation of consumer behaviour. In D. Buhalis & C. Costa (Eds.), *Tourism business frontiers: Consumers, products and industry* (pp. 9–18). Oxford, UK: Elsevier. doi:10.1016/B978-0-7506-6377-9.50009-2

Gretzel, U., & Jamal, T. (2009). Conceptualizing the creative tourist class: Technology, mobility and tourism experiences. *Tourism Analysis*, *14*(4), 471–482. doi:10.3727/108354209X12596287114219

Grün, C., Werthner, H., Pröll, B., Retschitzegger, W., & Schwinger, W. (2008). Assisting tourists on the move-an evaluation of mobile tourist guides. *7th International Conference on Mobile Business*, 171–180. 10.1109/ICMB.2008.28

Hannam, K., Butler, G., & Paris, C. (2014). Developments and Key Issues in Tourism Mobilities. *Annals of Tourism Research*, *44*, 171–185. doi:10.1016/j.annals.2013.09.010

Hyun, M. Y., Lee, S., & Hu, C. (2009). Mobile-mediated virtual experience in tourism: Concept, typology and applications. *Journal of Vacation Marketing*, *15*(2), 149–164. doi:10.1177/1356766708100904

Intel. (2018). *A guide to the internet of things: Infographic.* Retrieved from https://www.intel.com/content/www/us/en/internet-of-things/infographics/guide-to-iot.html

Jansson, A. (2007). A sense of tourism: New media and the dialectic of encapsulation/ decapsulation. *Tourist Studies*, *7*(1), 5–24. doi:10.1177/1468797607079799

Kim, D., Park, J., & Morrison, A. M. (2008). A Model of Traveler Acceptance of Mobile Technology. *International Journal of Tourism Research*, *10*(5), 393–407. doi:10.1002/jtr.669

Kim, H., & Law, R. (2015). Smartphones in Tourism and Hospitality Marketing: A Literature Review. *Journal of Travel & Tourism Marketing*, *32*(6), 692–711. doi:10.1080/10548408.2014.943458

Kim, J., Ahn, K., & Chung, N. (2013). Examining the factors affecting perceived enjoyment and usage intention of ubiquitous tour information services: A service quality perspective. *Asia Pacific Journal of Tourism Research, 18*(6), 598–617. doi:10.1080/10941665.2012.695282

Kim, M. J., Kim, W. G., Kim, J. M., & Kim, C. (2016). Does knowledge matter to seniors' usage of mobile devices? Focusing on motivation and attachment. *International Journal of Contemporary Hospitality Management, 28*(8), 1702–1727. doi:10.1108/IJCHM-01-2015-0031

Lamsfus, C., Wang, D., Alzua-Sorzabal, A., & Xiang, Z. (2015). Going mobile: Defining context for on-the-go travelers. *Journal of Travel Research, 54*(6), 691–701. doi:10.1177/0047287514538839

Lane, N. D., Miluzzo, E., Lu, H., Peebles, D., Choudhury, T., & Campbell, A. T. (2010). A survey of mobile phone sensing. *IEEE Communications Magazine, 48*(9), 140–150. doi:10.1109/MCOM.2010.5560598

Law, R., Chan, I. C. C., & Wang, L. (2018). A comprehensive review of mobile technology use in hospitality and tourism. *Journal of Hospitality Marketing & Management, 27*(6), 626–648. doi:10.1080/19368623.2018.1423251

Liang, S., Schuckert, M., Law, R., & Masiero, L. (2017). The relevance of mobile tourism and information technology: An analysis of recent trends and future research directions. *Journal of Travel & Tourism Marketing, 34*(6), 732–748. doi:10.1080/10548408.2016.1218403

MacKay, K., & Vogt, C. (2012). Information technology in everyday and vacation contexts. *Annals of Tourism Research, 39*(3), 1380–1401. doi:10.1016/j.annals.2012.02.001

Miller, G. (2012). The Smartphone Psychology Manifesto. *Perspectives on Psychological Science, 7*(3), 221–237. doi:10.1177/1745691612441215 PMID:26168460

Neuhofer, B. (2012). An analysis of the perceived value of touristic location based services. In M. Fuchs, F. Ricci, & L. Cantoni (Eds.), *Information and Communication Technologies in Tourism 2012*. Vienna: Springer. doi:10.1007/978-3-7091-1142-0_8

Neuhofer, B., Buhalis, D., & Ladkin, A. (2015). Smart technologies for personalized experiences: A case study in the hospitality domain. *Electronic Markets, 25*(3), 243–254. doi:10.100712525-015-0182-1

Oh, S., Lehto, X. Y., & Park, J. (2009). Travelers' intent to use mobile technologies as a function of effort and performance expectancy. *Journal of Hospitality Marketing & Management, 18*(8), 765–781. doi:10.1080/19368620903235795

Ritchie, J., Lewis, J., Nicholls, C. M., & Ormston, R. (Eds.). (2013). *Qualitative research practice: A guide for social science students and researchers*. London: Sage.

Siau, K., Lim, E., & Shen, Z. (2001). Mobile commerce: Promises, challenges, and research agenda. *Journal of Database Management, 12*(3), 4–13. doi:10.4018/jdm.2001070101

Statista. (2018). *Number of smartphone users worldwide from 2014 to 2010 (in billions)*. Retrieved from https://www.statista.com/statistics/330695/number-of-smartphone-users-worldwide/

Ström, R., Vendel, M., & Bredican, J. (2014). Mobile marketing: A literature review on its value for consumers and retailers. *Journal of Retailing and Consumer Services, 21*(6), 1001–1012. doi:10.1016/j. jretconser.2013.12.003

Tan, W.-K. (2016). The relationship between smartphone usage, tourist experience and trip satisfaction in the context of a nature-based destination. *Telematics and Informatics, 34*(2), 614–627. doi:10.1016/j. tele.2016.10.004

Tussyadiah, I. P., & Wang, D. (2016). Tourists' Attitudes toward Proactive Smartphone Systems. *Journal of Travel Research, 55*(4), 493–508. doi:10.1177/0047287514563168

Tussyadiah, I. P., & Zach, F. J. (2012). The role of geo-based technology in place experiences. *Annals of Tourism Research, 39*(2), 780–800. doi:10.1016/j.annals.2011.10.003

Wang, D., Park, S., & Fesenmaier, D. R. (2012). The Role of Smartphones in Mediating the Touristic Experience. *Journal of Travel Research, 51*(4), 371–387. doi:10.1177/0047287511426341

Wang, D., Xiang, Z., & Fesenmaier, D. R. (2014). Adapting to the mobile world: A model of smartphone use. *Annals of Tourism Research, 48*, 11–26. doi:10.1016/j.annals.2014.04.008

Wang, D., Xiang, Z., & Fesenmaier, D. R. (2016). Smartphone Use in Everyday Life and Travel. *Journal of Travel Research, 55*(1), 52–63. doi:10.1177/0047287514535847

Watson, C., McCarthy, J., & Rowley, J. (2013). Consumer attitudes towards mobile marketing in the smart phone era. *International Journal of Information Management, 33*(5), 840–849. doi:10.1016/j. ijinfomgt.2013.06.004

White, N. R., & White, P. B. (2007). Home and away: Tourists in a Connected World. *Annals of Tourism Research, 34*(1), 88–104. doi:10.1016/j.annals.2006.07.001

World Travel & Tourism Council. (2017). *Travel & Tourism. Global Economic Impact & Issues 2017*. London: WTTC.

Xiang, Z., & Gretzel, U. (2010). Role of social media in online travel information search. *Tourism Management, 31*(2), 179–188. doi:10.1016/j.tourman.2009.02.016

Xiang, Z., Gretzel, U., & Fesenmaier, D. R. (2009). Semantic Representation of Tourism on the Internet. *Journal of Travel Research, 47*(4), 440–453. doi:10.1177/0047287508326650

Zenith. (2017). *Smartphone penetration to reach 66% in 2018*. Retrieved from https://www.zenithmedia. com/smartphone-penetration-reach-66-2018/

# Conclusion

In this ever-changing world, the only unchanged thing is to change constantly. It is specially so in this mobile world, where everything and everyone are wirelessly wired. With mobile technological advances being constantly introduced, mobile is ever changing its use and corresponding experience as different uses of mobile will naturally result in different mobile experiences. When mobile was first introduced, it was used for making phone calls only. Soon after that, with new features and functions being introduced, it was used for texting, taking photos, and video recording. With the advent of smartphone features, functions, and social media, mobile has become even more powerful and popular than ever, becoming the most popular and powerful medium in the world.

Equipped with ever changing features and functions, mobile enables users to produce, provide, advertise, market, sell, or consume content anytime anywhere, be it a love story, a poem, a bottle of wine, a brand, an image, a video, an audio, or a multimedia product. Interactive, integrated and immersive, mobile also enables users to experience a content, product or service in a media-rich environment for any purpose, ranging from governing to shopping. Across time and space, mobile has been widely used in different areas, including mobile journalism, mobile advertising, mobile marketing, mobile public relations, mobile social communications, mobile dating, mobile parenting, mobile government, mobile health, mobile learning, mobile banking, mobile shopping, mobile commerce, mobile entertainment, mobile creativity, and mobile business.

Both mobile use and mobile experience have been widely examined, leading to increasingly accumulated body of knowledge. Previous studies, however, have largely focused their investigations in country-specific mobile use and experience in different areas. Few studies have been conducted in a comparative fashion to compare mobile use and experience in different countries. Another need for comparison is that fact that previous studies have been published differently in different languages, there has also emerged a pressing need for different language comparative studies. Through comparative studies, we will be able to locate similarities and differences at both the macro and the micro levels. Furthermore, we can also identify similarities and differences at the normative and the empirical levels as well as the gap between two.

Since different countries around the world have become so interdependent and interconnected in so many ways, whatever we share or differ in will affect our interaction with each other in the mobile space, where any event or issue in one place will affect those in other places in one way or another across time and space, generating ripple implications socially, culturally, economically, and politically (Xu, 2018). To achieve a holistic picture and a better understanding of the mobile world, it is necessary to conduct a comparative investigation of mobile use and experience in different communities, countries or cultures to enhance our global mobile communication literacy and competence (Xu, 2018).

The essence of comparative studies is to validate, revise, triangulate, and glocalize research results, as explained fully by the authors The Handbook of Comparative Communication Research in the following: (1) to validate or revise country-specific interpretations, (2) to avoid country-specific overgeneralizations, (3) to challenge country-specific paradigms, (4) to contextualize country-specific understanding, (5) to establish global scholarship, and (6) to obtain locally-applicable knowledge and experience (Esser & Hanitzsch, 2012, pp. 4-5).

Being interdisciplinary, comparative and applied, dimensions and directions of further studies of mobile use and experience may be guided by the most important questions identified by scholars include (a) how humans interact via mobile devices and the consequent implications (suggested by James E. Katz), (b) how to theorize mobile media and communication, (proposed by Harmeet Sawhney), and (c) the nature of mobile media and communication (suggested by Michele H. Jackson) (Craig, 2007).

Guided by these central themes and following a comparative approach, it is recommended that focuses should be placed on comparing mobile use and experience in different areas in different countries in the presence of a basis for comparison. Furthermore, a 3M framework was proposed to use, that is, to map, to measure and to model. To map is to locate where different components of mobile use and experience are located. To measure is to gauge different levels of different components of mobile use and experience. To model is to describe, explain and predict a pattern in mobile use and experience respectively (Xu, 2018).

In terms of research methods, further studies of mobile use and experience should involve a mix of different research methods. For instance, it may be more accurate and reliable to use a mix of surveys, focus groups, in-depth interviews, content analysis and/or experiments in comparing similarities and differences in mobile use and experience around the world in order to provide more accurate and reliable findings through triangulating and or complementing the results of studies using different research methods.

## FURTHER STUDIES OF MOBILE USE: DIMENSIONS AND DIRECTIONS

Mobile use has been examined in terms of its informational, relational (Lee, Kwak, Campbell, & Ling, 2014), extractive, or immersive use (Humphreys, Von Pape & Karnowski, 2013). It has also been investigated in terms of different demographic features, resulting in differences in age group, gender, physical conditions, or occupation (e.g., Ventola, 2014; Organista-Sandoval & Serrano-Santoyo, 2014; Brandenburg, Worrall, Rodriguez & Copland, 2013; Hashim, Tan & Rashid, 2015; Lee & Kim, 2014).

Further studies of mobile use and its impacts should also include the following areas: (a) mobile forms, (b) mobile formats, (c) mobile features, and (d) mobile functions, which are closely related to mobile use and its impacts on actors and activities. They can also be investigated and compared by using the 3M approach using mixed method research. For illustration, the following paragraphs investigate mobile features:

### Further Studies of Mobile Features

Mobile features, by definition, refer to those features that mobile devices are equipped with. Some features are related to hardware, such as camera, video, audio, multi-touch gestures, voice sensor, clock, alarm, timer, reminder, search, calculator, calendar, color touch screen, gyroscope, accelerometer, GPS navigation, sensors, high-speed data transfer, a knowledge repository, analysis resources, multiple device access, augmented reality, barcode scanner, WAP browser, HTML browser, flicking, tapping, pinching, stretching, Infrared, NFC, GPS, GPRS, FM, Bluetooth, MMS, personal hotspot, Java, and location tracking. Other features are related to software: mobile instant messengers such as WeChat, Skype, Facebook messengers, visualization, audio production apps, video production apps, writing apps, editing apps, data analysis apps, social networking apps, and to name a few. Still other features are network-related, such as EDGE, WCDMA, WLAN, HSAPA, 3G, 4G, 5G, Wi-Fi, context-aware mobile cloud services, connectivity, mobility, compatibility, and interoperability.

As mobile has different operating systems, such as iOS, Android, Windows, BlackBerry OS, or FireFox OS, it can be equipped with different mobile features accordingly. They may share some fundamental features while differ in other features. Essentially, however, mobile features of all kinds are leveraged in one way or another to enhance mobile performance for different purposes.

As the cornerstone of mobile, mobile features have been playing a significant role in mobile communications of all kinds, ranging from mobile government to mobile parenting. Therefore, it is imperative to compare how mobile features have been leveraged by mobile users. And it is equally imperative to compare how mobile features have been leveraged in different countries or cultures since this world is increasingly interdependent and inter-influenced.

For comparing mobile features in an effective way, it is crucial to design a better way of categorizing them. Since they can be categorized differently, mobile features should be properly grouped according to different purposes of comparing mobile features.

In an earlier study, mobile features were classified into three major categories: hardware, software and communication (Flora, Wang, & Chande, 2014). By hardware features, Flora, Wang and Chande meant to refer to less power, input mechanism, screen size and form factor, start-up time, physical parameters, device fragmentation while by software features, they referred to user experience, user interface, integration with other apps, action feedback, error notification, application focus, experienced resource, convenience, responsiveness, personalization, localization, readability, encryption, expire sessions, request validity period, prevent repeat request. In their definition, communication features referred to network connectivity. Although Flora, Wang and Chande did a neat and nice job in grouping different features into three categories, their categorization, however, has overemphasized the technological side of mobile features, missing out other important features of the mobile-human interaction side.

Mobile features are used to perform different functions. Some features are used to enhance device performance while others are leveraged to produce content. Some are employed to entertain users while others are for empowering users. Therefore, for better comparison, mobile features can be grouped as follows: (a) device enhancement, (b) user entertainment, (c) content production, and (d) user empowerment.

Specifically, device enhancement features include location-awareness, multi-touch gestures, gyroscope, accelerometer, and wireless communication capability, sensors, high-speed data transfer, a knowledge repository, analysis resources, multiple device access, augmented reality, near field communication (NFC), personal hotspot, context-aware mobile cloud services, location tracking, connectivity, compatibility, interoperability, mobility, and usability. These enhancement features have been examined in isolation or different combinations (e.g., Head & Ziolkowski, 2012; Ho, 2012; Qi & Gani, 2012; Wu, Wu, Chen, Kao, Lin & Huang, 2012; Aharony, 2013; Chang, Wu & Hsu, 2013; Gohil, Modi & Patel, 2013; Jambulingam, 2013; Kim, Park, Lim & Kim, 2013; Madden, Lenhart, Cortesi & Gasser, 2013; Pagoto, Schneider, Jojic, DeBiasse & Mann, 2013; Oliveira, Noguez, Costa, Barbosa & Prado, 2013; Plaza, Demarzo, Herrera-Mercadal & García-Campayo, 2013; Ramanathan, Swendeman, Comulada,

Estrin & Rotheram-Borus, 2013; Riikonen, Smura, Kivi & Töyli, 2013; Shapiro-Mathews & Barton, 2013; Fortunati & Taipale, 2014; O'bannon & Thomas, 2014; Pielot, De Oliveira, Kwak & Oliver, 2014; Sendra, Granell, Lloret & Rodrigues, 2014; Ciampa, 2014; Suominen, Hyrynsalmi & Knuutila, 2014; Wei & Lu, 2014; Ben-Zeev, Schueller, Begale, Duffecy, Kane & Mohr, 2015; Zhao & Balagué, 2015; Yang, Li, Jin, Zeng, Wu & Vasilakos, 2015; Espada, Díaz, Crespo, Martínez, G-Bustelo & Lovelle, 2015). Although these features may vary in different operating systems, fundamentally, they do the same job, that is, to enhance mobile performance for different purposes.

One of the major functions of mobile is to entertain mobile users anytime anywhere. Instead of turning to TV in our living rooms for a movie or a show, we just take out our mobile devices to watch it. If we want to listen to music, we do not have to use a separate music player. If we want to enjoy singing, we do not have to go to a KTV lounge anymore. Instead, we just take our mobile to sing away a few hours with friends or family across time and space. We have also found ourselves ending up playing games via our mobile devices with friends or family instead of switching on a computer or a game console. Even for gambling, we do not have to fly or drive to a casino anymore, instead we can gamble via mobile. In brief, mobile devices are also be equipped with such user entertainment features (see Kim, Kim & Wachter, 2013 and Yang & Kim, 2012).) as games, music (Jambulingam & Sorooshian, 2013), play games for learning (Lu, Chang, Huang & Ching-Wen, 2014), and gamifications (Eng & Lee (2013).

Mobile is used largely to produce content of all kinds for different purposes, ranging from texting to video production. Among production features are SMS, email (Jambulingam & Sorooshian, 2013), story-telling apps (Bonsignore, Quinn, Druin & Bederson, 2013), visualization of invisible information (Schall, Zollmann, & Reitmayr, 2013), and productivity software (Wu, Wu, Chen, Kao, Lin, & Huang, 2012). Among countless different production apps, we can choose what we need and like to make a movie, a photo album, an audio clip or a video clip. We can also use them to write, edit and publish articles, blogs, and reports right from our mobile devices without switching on our desktop or laptop computers. Production features of our mobile devices have been playing a pivotal role in our mobile activities.

Mobile users can further be empowered to enhance their communications for different purposes by leveraging such user empowerment features as authenticity, collaboration, personalization, crowd-powered data collection, cross-space data mining, data analysis, and interactivity (e.g. Kearney, Schuck, Burden & Aubusson, 2012; Lee, Moon, Kim & Mun, 2015; Guo, Yu, Zhou & Zhang, 2014; Kearney & Maher, 2013; Gao, Bai, Tsai & Uehara, 2014).

As part of the essential components of mobile, mobile features have been examined in many different dimensions. In comparing mobile features, it is crucially necessary to map what has been done in the past to locate similarities and differences in leveraging mobile features. For illustration, we reviewed the most recent studies that have been published since 2012. Using key words "mobile features", our search on Google Scholar generated 150 research articles, including some important conference papers that have been cited by other scholars in their respective studies.

Out of 150 studies we reviewed, only seven studies were comparative in nature. In their study, Fortunati and Taipale (2014) compared the advanced use of mobile phones in five European countries (Italy, France, Germany, Spain and UK). They found that substantial differences existed in the advanced use of mobile phone and its predictors in these countries. Another finding was that only about one third of the studied mobile features were exploited. And the extensive under-utilization of mobile features, they found, the mobile phone as a tool of social labor was efficiently exploited by the minority. Their further finding was that limited use of advanced features resulted in the new patterns of social stratification (Fortunati & Taipale, 2014).

In another comparative study, Lu, Chang, Huang and Ching-Wen (2014) investigated use of two game features (context-awareness and story generation) really influenced students' attitudes towards using such educational mobile role-playing games in Canada and Taiwan. They found that the story generated in CAM-RPG positively influenced users' attitude towards game use and increased users' perceived game usefulness, especially for the male students. They also found that natural language processing, location-awareness, multiple input forms, social networking, and student modeling could provide students with effective and efficient mobile learning experiences. Their further findings showed that context-awareness features of the game might be less important for the game-play although it did not affect their attitude towards using the game and that subject selection, such as learning environment, selected learning topic, and learning materials, would be an important issue in order to make users aware of the advantages of a context-aware mobile educational game (Lu, Chang, Huang & Ching-Wen, 2014).

In still another comparative study, Chong, Chan and Ooi (2012) investigated the factors that would predict consumer intention to adopt m-Commerce in Malaysia and China. They found that new variables such as trust, social influence, variety of services, and cost would predict consumer decisions to adopt m-Commerce in both Malaysia and China. Malaysian consumers, however, they found, were more concerned than their Chinese counterparts with the variety of services offered by m-Commerce. But both Malaysian and Chinese consumers, they further found, were price conscious in m-Commerce (Chong, Chan and Ooi (2012).

In comparing users' attitudes toward the use of mobile devices in second and foreign language learning in higher education in China and Sweden, Viberg and Grönlund (2013) found that respondents' attitudes toward mobile learning were very positive with individualization being most positive (83%) followed by collaboration (74%), and authenticity (73%). Hofstede's factors of cultural dimensions (power distance index, individualism versus collectivism, masculinity versus femininity, uncertainty avoidance index, long term orientation versus short term normative orientation, and indulgence versus restraint) were found to be unable to explain the differences in mobile-assisted language learning (MALL) attitudes. Among the personal factors, however, gender could explain the differences in students' attitudes toward MALL (Viberg and Grönlund, 2013).

In their comparative investigation of use of the web browser, e-mail, applications such as Facebook or Google Maps, extractive usage, and immersive use in USA and Germany, Humphreys, Von Papeand Karnowski (2013) found no siginificant difference between mobile devices and non-mobile devices. Besides linguistic differences in calling mobile Internet, they did find any other significant difference in using mobile Internet. In both countries, according to their findings, the context very much shaped the mode through which people used the mobile Internet and having only a mobile phone at hand could foster an extractive use and make an immersive Internet use less likely (Humphreys, Von Pape & Karnowski, 2013).

To compare mobile and desktop, Güler, Kılıç and Çavuş (2014) investigated difficulties in instructional design processes. They found difficulties experienced in instructional design processes for mobile devices and desktop computers tended to be similar in developing learning content. And they also found that difficulties in internal design and production and front-end analysis were significantly different in terms of the Internet connection in personal mobile devices. Further differences were also located in their study that external design and development difficulties, rolling-out difficulties and total scores were significantly different with regard to levels of Internet experience (Güler, Kılıç & Çavuş, 2014).

In comparing product and service brand categories, Kim, Lin and Sung (2013) investigated their similarities and differences through addressing the following research questions: RQ1: To what extent do branded apps incorporate vividness, novelty, motivation, control, customization, feedback, and multiplatforming? What are the differences and similarities between product and service brand categories? RQ2: Do branded app types (informational versus experiential apps) differ according product and service brand categories? RQ3: What message strategies are used most frequently in branded apps? What differences and similarities exist between product and service brand categories? RQ4: To what extent do branded apps offer consumers brand-related content, and what types of content are frequently offered?

And among other findings, Kim, Lin and Sung (2013) found that use of each engagement attribute (vividness, novelty, motivation, control, customization, feedback, and multiplatforming) varied by brand category and that animation and background sound were more frequently used in product brand apps than service brand apps. Further difference was also found in that informational message strategies were more frequently employed by service branded apps than those of product brands and just the opposite for transformational message strategies.

Judging by the seven comparative studies, it is obvious that only limited number of mobile features have been compared only in a small number of countries being compared: Malaysia and China, Canada and Taiwan, China and Sweden, USA and Germany, and Italy, France, Germany, Spain and UK.

Although non-comparative in nature, the majority of the most recent studies provided us with clues and inspirations for further comparative studies. For instance, based on the research questions addressed and/or hypotheses tested in the most recent studies, the following areas may inspire future scholars in their comparative studies of mobile features: (a) most important features being used (Jambulingam & Sorooshian, 2013), (b) perceptions of usefulness of mobile features, (Mayfield, Ohara, & O'Sullivan, 2013; O'bannon, & Thomas, 2014), (c) diffusion patterns of mobile features (Riikonen, Smura, Kivi & Töyli, 2013), (d) effects of mobile features (Ho, 2012; Glackin, Rodenhiser, & Herzog, 2014), (e) motivations of using mobile features (Teodoro, Ozturk, Naaman, Mason & Lindqvist, 2014, February; Ciampa, 2014), (f) use of mobile features in providing personalized sightseeing tours (Anacleto, Figueiredo, Almeida, & Novais, 2014), (g) likes and dislikes of mobile features (Grindrod, Li, & Gates, 2014), (h) users' readiness to accept marketing (Persaud, & Azhar, 2012), (i) facilitators and barriers of using mobile features (Dale Storie MLIS, 2014), (j) implementation of behavior change techniques in apps (Yang, Maher, J& Conroy, 2015), (k) interactive design and gamification of ebooks for mobile and contextual learning (Bidarra, Figueiredo, & Natálio, 2015), (l) mobile security (La Polla, Martinelli & Sgandurra, 2013; Jain & Shanbhag, 2012), (m) mobile augmented reality (Pendit, Zaibon & Bakar, 2014; Ke & Hsu, 2015), (n) context-aware mobile features (Mizouni, Matar, Al Mahmoud, Alzahmi, & Salah, 2014).

In future comparative studies of mobile features, we may wish to compare perceptions of usefulness of mobile features. And closely related to perception comparison is reality check, that is, to compare in reality to what extent and what type of mobile features are actually embedded in apps and used by mobile users. Similar investigations can be conducted to compare motivations, diffusion patterns, areas, and effects of using mobile features. And topics to be compared include likes and dislikes of mobile features, facilitators and barriers of using mobile features, mobile privacy, safety, and security, use of mobile augmented reality, context-aware mobile features. And all these comparative studies can be conducted among different demographic (age, gender, race, education, sexual orientation, marriage status, occupation, income, level of mobile savviness) groups within one country or culture and between countries or cultures.

To get a holistic picture of what mobile features have been fully leveraged as well as similarities and differences in leveraging mobile features, there should be more countries and more features to be compared due to changes in mobile technologies, user, behavior, needs, tastes, and preferences. For instance, more studies should be conducted to examine how countries would fare in leveraging mobile features, how mobile applications would fare in leveraging mobile features, how different types of mobile features would fare in terms of being leveraged, and how factors would work together in shaping similarities and differences in leveraging mobile features.

In terms of country-specific comparison, we would like to suggest that mobile features should be compared between developed countries, between developing countries, between underdeveloped countries, or between developed and developing or underdeveloped countries.

As far as mobile applications are concerned, it is also desirable to compare them in order to locate similarities and differences between mobile apps in leveraging mobile features. For instance, are there any similarities or differences between mobile health apps and mobile banking apps in leveraging mobile features? Similar questions could be addressed between mobile learning and mobile teaching, between mobile dating and mobile romancing, between mobile public relations and mobile branding, between mobile advertising and mobile journalism, or between mobile shopping and mobile entertainment.

In comparing different types of mobile features, it is important to locate similarities and differences in leveraging mobile features. For instance, do user entertainment features differ from user empowerment features in terms of how and to what extent they have been leveraged. Are production features different from device features in terms of how and to what extent they have been leveraged?

To locate shaping factors, comparison should also be made to identify different factors that have worked in isolation or combination in shaping similarities and differences in leveraging mobile features.

To compare mobile features in the future, we would like to propose a 3M framework: to map, to measure and to model. To map mobile features is designed to locate where mobile features are leveraged. To measure mobile features is to gauge the levels of leveraging mobile features in different areas. To model mobile features is ultimately meant to come up with a model to describe, explain and predict use and effects of mobile features.

It is not exaggerated to say by now that mobile is everything and everything is mobile since mobile has penetrated in not only every country's population but also every single occupation. Against such a backdrop, we need to map where mobile features are being leveraged.

Mobile features can be mapped under the following four categories: (a) device enhancement, (b) user entertainment, (c) content production, and (d) user empowerment in the following three domains: (a) countries, (b) mobile devices and (c) apps. To map mobile features in countries, we may take into account the developed, developing and underdeveloped countries in our comparison. In terms of mobile devices, it is necessary to identify what mobile features are leveraged in smartphones and non-smart phones. As far as apps are concerned, it is equally necessary to locate what mobile features are being leveraged in different apps, ranging from mobile government to mobile gambling.

It may be advisable to employ social network analysis to map mobile feature in different countries, mobile devices and apps. With the results of social network analysis, we will be able to have a holistic picture of both dimensions and frequencies of four types of mobile features being leveraged in three domains.

To measure mobile features is to gauge the number, frequency and degree of (a) device enhancement, (b) user entertainment, (c) content production, and (d) user empowerment in three domains: (a) countries, (b) devices, and (c) apps by using one standardized codebook.

In terms of measurement, the number of leveraging mobile features can be measured on a 5-point scale with 1 being the minimum while 5 being the maximum in terms of number of mobile features being leveraged under each mobile feature category. And the same 5-point scale is also applied in measuring the frequency with 1 being the least frequent while 5 being the most frequent. And the degree of leveraging mobile feature can be measured on a 5-point scale with 1 being the least use while 5 being the heaviest use.

Use of mobile features can be influenced and shaped by different factors. Among them are political, economic, cultural and social factors in a country. These factors can be called external factors. Politically, if a country is democratic, open and free, then its mobile users will tend to use more mobile features to seek and follow information flow within and outside the country than those in less democratic, open and free countries. Economically, if a country is more developed, its mobile users tend to afford to use of more cost-incurring mobile features than those in less or under developed countries. Culturally, if a country is more open, aggressive, and globalized, then its mobile users may use more mobile features than those in less open, aggressive or globalized countries. Socially, if a country is more tolerant and welcome to new things and/or ideas, then its mobile users may use more mobile features than those in less tolerant and welcome to new things and/or ideas.

On top of these factors, there are other factors. For instance, demographics of mobile users including age, gender, education, income, sexual orientation, marital status, race, occupation and level of mobile savviness can play a very important role in influencing and shaping the number and frequency of leveraging mobile features. Besides demographics, motivations, expectations, experience, and perceptions may also shape use of mobile features. These factors can be called internal factors.

Among those factors, which factor or factors play a bigger part in shaping use of mobile features? And how does it or do they shape use of mobile features? Is there any pattern being formed among any shaping factors? Answers to these questions and more will generate a model for describing, explaining and predicting changes in use of mobile features.

In order to get the answers to all the questions regarding use of mobile features, mixed research methods, in other words, mixing quantitative methods with qualitative methods, should be employed instead of purely quantitative or qualitative methods. Different research methods used can triangulate the results of each research method so that more reliable and accurate results can be generated.

Specifically, a nation-wide survey can be conducted in different countries to collect data regarding mobile users' motivations, expectations, experience, preferences, and perceptions in relation to mobile features. Those data, together with those demographic data generated from the nation-wide survey, can be compared from country to country. As for the external factors, country-specific data can be obtained from the authoritative sources such as UN, World Bank, and UNESCO.

## Further Studies of Mobile Journalism

As another example of dimensions and directions of further studies of mobile use, mobile journalism can be investigated and compared as follows:

Mobile has been leveraged by professional and freelancing journalists in news reporting and writing, revolutionizing the way news is reported, designed, delivered, and consumed around the world. Mobile has been redefining news, journalists, and journalism, restructuring and reshaping the news industry. We are not just passive recipients of news as we used to be in the age of mass media and communication. Instead, through social media, especially mobile social media, we can follow news and news cam follow us anytime anywhere. Mobile journalism has been practiced by not only professional journalists but also non-professional journalists or simply every citizen who would like to report or share news. It is an open concept where everyone can be a journalist, be it a reporter, an editor or an anchorperson. In the age of mobile journalism, there is no fixed deadline for news as it can be reported anytime anywhere in the mobile space. There is no more clear media division like in the old days when we had radio news, TV news or newspaper news separately. Instead, we have multimedia news or news designed in different media but delivered to the same medium, that is, mobile. Mobile journalism is not just a gear but a game changer (Bivens, 2008) in journalism and the news industry.

Mobile journalism has been examined in the past by many scholars around the world, generating an increasing amount of scholarship. A quick review of Google-Scholar-curated academic publications on mobile journalism has spotted a general focus on largely any of the technology-specific, country-specific, citizen-specific, professional-specific, method-specific, time-specific topics.

Among previous studies, for instance, some have investigated (a) mobile news as a model (Mills, Egglestone, Rashid & Väätäjä, 2012; Scolari, Aguado & Feijóo, 2012; Westlund, 2013; Wolf & Hohlfeld, 2012), (b) user experience (Bethell, 2010; Väätäjä, 2008; Väätäjä, Koponen, & Roto, 2009; Koponen & Väätäjä, 2009; Väätäjä, 2010, November; Wigelius & Väätäjä, 2009; Väätäjä, Männistö, Vainio & Jokela, 2009; Väätäjä & Roto, 2010), and (c) skills (Powers, 2012; Wenger, Owens, & Thompson, 2014).

Others have examined (a) tools, (Jokela, Väätäjä & Koponen, 2009), (b) impact (Martyn, 2009; Clarke & Bromley, 2012), and (c) perceptions, needs and challenges (Väätäjä & Egglestone, 2012).

Still others have looked into (a) immediacy and openness (Mudhai, 2011), (b) delivery (Fidalgo, 2009), (c) location-based design (Bjørnestad, Tessem & Nyre, 2011; Goggin, Martin & Dwyer, 2015) and augmented reality design (Pavlik & Bridges, 2013), (d) production and consumption (Aguado & Martínez, 2008; Westlund, 2014), (e) participation (Verclas & Mechael, 2008;), and (f) empowerment (Berger, 2011).

For further studies, as Westlund (2012) proposed, we should use "mixed approaches and methods, preferably aiming for cross-cultural comparisons rather than national studies, and time-series rather than cross-sectional studies" (p. 22). Moreover, he also suggested inclusion of both production and consumption of mobile news for future research.

Through comparing mobile journalism being practiced and researched around the world, we will be able to identify universal features and particular characteristics of mobile journalism in different countries or cultures as well as to locate changes and patterns in mobile journalism globally or locally.

Specifically, mobile news designing and delivering should also be compared since these are the two major areas in mobile journalism.

Obviously different from the traditional news media and communication in designing news, mobile journalism has its own unique ways of designing and packaging news for mobile consumption on the go. In order to grab fragmented attention on the go in competition with traditional news media, mobile news designing has become increasingly essential and indispensable. By leveraging mobile features available, mobile news designing is expected to entice, entertain, engage, empower, and enlighten mobile users on the go anywhere anytime via any device.

Similarities and differences have been generated in designing mobile news in different countries but yet they have not been well examined and compared. Few longitudinal or comparative studies have been conducted to map and measure those similarities and differences systematically. Therefore, it is highly urgent to come up with a systematic way to compare and keep track of changes and patterns in similarities and differences in designing mobile news.

Mobile journalism can also differ in delivery. And delivery can be highly customized and personalized according to mobile users' needs, tastes and preferences in terms of mode, time, venue, or channel to deliver mobile news.

Mobile news can be delivered through subscriptions, micro-payment or free of charge with ads. It can also be delivered through email to our inbox directly according to our needs, tastes and preferences for news. In terms of time, mobile news can be further delivered to a specific time of our choice. It can also be personalized in light of our chosen venue or channel.

Delivery can also be done in a cross-media fashion. In other words, one news story can be designed and packaged differently using different media and it can be delivered across media to us so that we can decide when to read, view or listen to news according to different situations we happen to find ourselves in or different moods or needs for different media. For instance, after knocking off in a car while driving, we would prefer for a news story to be delivered via radio so that we can listen to news if we have used our eyes on different screens for too long and it is about the time for us to rest our eyes. After they reach home, we would prefer to watch news stories delivered in video format, which can be scheduled to be watched over dinner time with our families.

Mobile news can also be delivered according to types of phones we use. If we use feature phones, we would prefer to have news delivered to us in text or at least low graphic format if some photos are still expected to go along with stories. If we use smart phones, we would definitely like to have multimedia stories to be delivered to our devices. And we may even want ourselves to be totally immersed in a mix reality, where we would experience news stories in a totally different way, which would allow us to experience news in a virtual, real and mixed fashion.

Based on the results of previous studies and what should be examined, mobile journalism in different countries can be compared in many different areas. In this chapter, we would like to select the following areas for comparison: (a) roles, (b) features, (c) experience, (d) impacts, (e) issues, and (f) factors.

As mobile journalists can play different roles, ranging from informing to entertaining, it is crucial to compare what roles mobile journalists want to play, what roles are actually played and the gap between the two. Among the normative roles, there may exist different emphases on the same set of roles to be played by mobile journalists. And it is also necessary to run the whole set of comparative studies in the case of freelancing mobile journalists and mobile news consumers. What roles do freelancing mobile journalists want to play? What roles do mobile news consumers want mobile journalists to play? What roles do they actually play most in reality in the eyes of freelancing mobile journalists and mobile news consumers respectively?

Another area to be compared lies in features of mobile journalism. There are many different features, ranging from ubiquitous news reporting to unique news consumption. Due to political, social, economic and cultural differences, the same features can be treated differently or emphasized differently. Some features can be universal as they are treated more or less the same around the world. Other features can be country-specific or particular. Still other features can be hybrid, meaning a combination of universal and particular features in practicing mobile journalism.

Features can be generally reflected in design and delivery of mobile news. In terms of design, mobile journalism can be very different from that of traditional news media. In the age of mass media and communications, news was designed for radio, TV, newspapers or magazines, featuring medium-specific design and presentation. In mobile journalism, news is designed for entice, entertain, engage, empower, and enlighten mobile users on the go anywhere anytime across time and space barrier within fragmented time and space. Different from traditional news delivery, mobile news is delivered through not only apps and mobile-friendly sites but also various mobile social media. Its delivery is not only push, pull or hybrid but also tailored and customized to the full according to different needs, tastes and preferences by leveraging mobile features and functions.

A comparison should also be conducted to investigate what specific features mobile users, freelancing mobile journalists and mobile journalists respectively would like to have in mobile journalism, what they actually have as well as the gap between the two.

In the mobile space, although content remains essential, what is indispensable is experience. News is everywhere these days. If you do not get it through one medium, you will surely get it through another medium. If you are not being followed by news via one medium, you will end up being followed via another medium. News is everywhere. In this ubiquitous news environment, what has become more important is not just news itself, but rather how it is being experienced, which makes a huge difference in terms of impression and impact. Experiencing news on the go is very different from the conventional way of experiencing news on radio, TV, newspapers, or news magazines. And it can also be very different in different countries. Hence it is essentially important to compare mobile news experience to locate similarities and differences. Mobile news experience can be compared at six different stages of mobile news consumption: (a) enticement, (b) entertainment, (c) engagement, (d) empowerment, (e) enlightenment, and (f) enhancement. And the comparison should be three-fold, meaning we should compare the normative experience, the empirical experience and the gap between the two.

Ever since its inception, mobile journalism has been influencing and reshaping the way news is reported, designed, delivered, consumed, and leveraged. Furthermore, it has also incurred great changes in different countries socially, economically, culturally, and politically. Therefore, it is another major area for comparative studies of mobile journalism.

Issues in mobile journalism can be similar or different in different countries. It should also be part our comparative investigation to locate similar or different issues in different countries so that we will have a better understanding what is universal and what is particular when it comes to issues in mobile journalism.

The last but not the least area is the question why mobile journalism is practiced differently in different countries. To map and measure those differences is not the end of the story. To take one step further, those identified differences should be well described, explained and predicted by locating different factors.

To collect data on those three areas of mobile journalism, it is our suggestion that surveys, focus groups and interviews should be conducted in isolation or combination. On top of these research methods, we also recommend a cross-country content analysis of mobile news apps or mobile sites to gather related data for comparison. Different methods, qualitative or quantitative, can be used employed to triangulate the results of different comparisons using different methods.

Among the identified roles of mobile journalism are (a) to inform, (b) to entertain, (c) to educate, (d) to share, (e) to monitor, (f) to collaborate, (g) to investigate, and (h) to change. But these roles may not be supposed to be played universally. Instead, mobile journalists may play different roles in different countries. Even within the same country, different mobile journalists from different news media organizations may also play different roles. Furthermore, amateur mobile journalists may also play different roles. The differences may also exist in their orientations and prioritizations. In other words, although they may play a set of similar roles, but they may prioritize them differently.

These roles can be compared first on the normative level. In other words, we should compare what roles mobile journalists and freelancing mobile journalists want to play and what roles they have actually played. Furthermore, we also need to compare what roles mobile users want mobile journalists to play and what roles they have actually played in the eyes of mobile users. The comparison can be conducted through a survey by asking mobile journalists, freelancing mobile journalists and mobile users to rate the importance of the identified roles on a Likert scale: 1 for the least important while 5 for the most important.

After being compared on the normative level, these roles can be further compared on the empirical level. That means that we should compare what roles mobile journalists and freelancing mobile journalists have actually played in a country. And mobile users should also be asked to rate the level of actual performance of these roles on a Likert scale: 1 for the least played while 5 for the most played.

After comparing both normative and empirical roles respectively, we can place the results of the two comparisons in perspective so that we can locate the various gaps between the normative and empirical roles. The gaps can also be measured on a Likert scale: 1 for the narrowest gap while 5 for the widest gap.

Features of mobile journalism can differ in terms of availability, accessibility and usability from one device to another, from one person to another, from one country to another although features can be universally the same due to the same set of mobile technological advances available to everyone in the world. Features can also be highly different when it comes to how they are leveraged in reporting, writing, designing and delivering news anytime anywhere via any device.

The available features of mobile journalism can be broadly divided into the following categories: (a) immediacy, (b) openness, (c) interactivity, (d) multimedia, (e) immersion, (f) ubiquity, (g) geo-tagging, (h) gamification, (i) location-based services, and (j) mixed reality.

Immediacy is a feature that distinguishes mobile journalism from its conventional counterparts. News reporting in the case of mobile journalism can be immediate, or specifically speaking, not only instantaneously but also simultaneously. No traditional news media can have that level immediacy. With a mobile device, anyone can report and share anything via social media sites or apps immediately. It is even more important and crucial in terms of live coverage of a general election (Mudhai, 2011).

Another outstanding feature of mobile journalism is openness (Mudhai, 2011). Closely related to this feature is the civic engagement in mobile journalism (Lasica & Firestone, 2008). And it is also the main pitch of what open journalists have been passionately advocating around the world (Aitamurto, 2013; Rusbridger, 2012).

As one of the prominent features of mobile journalism, interactivity can mean different things to different journalists or different news media. It may also differ in terms of how interactivity should be and is actually leveraged in mobile journalism. Interactivity can also mean co-creation of content on mobile between mobile journalists and mobile news consumers, in which a news story is unfolding involving user participation in contributing new information, views and new elements to be added to enrich the news story.

Multimedia is another feature, which can differ among mobile journalists, mobile news providers and mobile news consumers. It may refer to photo, image, audio or video galleries separately to some while to others it may also refer to an integrated use of different media elements within a story.

As an emerging feature, immersion has become more and more aware of and used in designing and delivering news for mobile devices. Specifically speaking, immersion can also be different in not only definition but also how it has been actually leveraged. Gamification is an example of how to immerse a mobile user in mobile news production and consumption. And use of mixed reality is another.

As one of the outstanding features of mobile journalism, ubiquity is of ultimate importance in reporting, designing and delivering news for mobile users who can be connected and who are in need of news anytime anywhere on the go. So to what extent has this feature been leveraged by mobile journalists or mobile users? The answer can also be different in different countries.

Mobile news, co-created by professional or amateur journalists, can be geo-tagged in that geo-sensitive-or-specific information or elements can be clearly marked, stored, embedded in a news story so that it can be called up anytime anywhere via any device for in-depth and further reference, understanding or update. In terms of what to be geo-tagged and how it should be different on the individual level according to different individual needs, tastes and preferences. It can also be very different on the corporate level according to the nature, type, target audience, and venue of news media.

Gamification, by definition, refers to use of gaming elements to enhance mobile experience in the context of mobile news communication.

Location-based services are becoming increasingly leveraged in mobile journalism to enhance mobile news communication (Bjørnestad, Tessem & Nyre, 2011; Goggin, Martin & Dwyer, 2015; Watkins, Hjorth & Koskinen, 2012; Väätäjä & Egglestone, 2012).

Although not often or rarely leveraged in designing and delivering mobile news, mixed-reality has presented itself as a high promising turf (Väätäjä & Männistö, 2010) to enhance mobile news communication anytime anywhere via any device.

The ten key features of mobile journalism should first be compared among mobile journalists, freelancing mobile journalists and mobile users to measure the level of importance being paid by mobile journalists, freelancing mobile journalists and mobile news consumers to each of these ten features by using a Likert scale with 1 for the least important while 5 for the most important.

Besides the normative side, these ten key features should also be compared on the empirical level, meaning that the level of each of these ten features can be compared among mobile journalists, freelancing mobile journalists and mobile news consumers with 1 for the least leveraged while 5 for the most.

As far as the gap between the normative and the empirical features of mobile journalism are concerned, naturally once the results of both normative and empirical comparisons are placed in perspective, we will obtain a holistic picture of the gap among mobile journalists, freelancing mobile journalists and mobile news consumers.

In the mobile news industry, although content remains essential, what is indispensable and important, however, is experience. News is everywhere these days. If you do not get it in a medium, you will surely get it in another. If you are not being followed by news in a medium, you will end up being followed in another. In this ubiquitous news environment, what has become more important is not just news itself, but rather how it is being experienced, which makes a huge difference in terms of impression and impact. Experiencing news on the go is very different from the conventional way of experiencing news on radio, TV, newspapers, or news magazines. And it can also be very different in different countries. Hence it is essentially important to compare mobile news experience to locate similarities and differences.

Mobile news experience can be compared at different stages of mobile news consumption, which normally starts with how mobile users are enticed to news within the first few seconds by three key elements at the enticement stage (a) appealing interface, (b) easy navigation, and (c) interest-arousing. Using a Likert scale with 1 for the least enticed while 5 for the most enticed, mobile news experience at this stage can be compared through measuring its normative and empirical dimensions as well as the gap between the two with 1 for the narrowest gap while 5 for the widest gap.

At the entertainment stage, mobile news experience can be compared through measuring three key elements: (a) fun, (b) pleasure, and (c) satisfaction. Using a Likert scale with 1 for the least entertained while 5 for the most entertained, mobile experience at this stage can be compared through measuring its normative and empirical dimensions as well as the gap between the two with 1 for the narrowest gap while 5 for the widest gap.

In the case of engagement, mobile news experience can be compared through measuring three key elements: (a) searching, (b) interacting, and (c) sharing on a Likert scale with 1 for the least engaged while 5 for the most engaged. Mobile experience at this stage can be compared through measuring its normative and empirical dimensions as well as the gap between the two with 1 for the narrowest gap while 5 for the widest gap.

At the empowerment stage, mobile news experience can be compared through measuring three key elements: (a) selecting, (b) commenting, and (c) producing. Using a Likert scale with 1 for the least empowered while 5 for the most empowered, mobile experience at this stage can be compared through measuring its normative and empirical dimensions as well as the gap between the two with 1 for the narrowest gap while 5 for the widest gap.

At the enlightenment stage, mobile news experience can be compared through measuring three key elements: (a) awareness, (b) understanding, and (c) consciousness, using a Likert scale with 1 for the least enlightened while 5 for the most enlightened on both normative and empirical levels. The gap between the two can be compared with 1 for the narrowest gap while 5 for the widest gap.

Mobile news experience at the enhancement stage can be compared through measuring three key elements: (a) knowledge, (b) skills, and (c) abilities on both normative and empirical levels, using a Likert scale with 1 for the least enhanced while 5 for the most enhanced. The gap between the normative and empirical enhancement can also be compared with 1 for the narrowest gap while 5 for the widest gap.

Mobile news services and mobile apps have been reshaping news communication and newsroom operations, making news reportable and available besides what Westlund (2012) described as accessible anytime, anywhere, and through any device. News writing, reporting, designing and delivering have all become personalized on the go way beyond what Goggin (2006) had described as radically personalized news-gathering. These are just examples of the industry-specific impacts of mobile journalism.

The impacts of mobile journalism are also social, cultural, economic, political, and individual since mobile journalism has been changing the way we vote in elections of all kinds, the way we understand our world, the way we interact with each other, the way we receive, process and share information and knowledge locally, nationally or internationally.

To compare the impacts of mobile journalism, we have to come up with an overall comparative framework. The one we would like to suggest is to cover impacts on the societal, industry-specific and individual levels. On the societal level, we would like to propose to focus on the social, economic, political and cultural impacts of mobile journalism. And those impacts should be measured on both normative and empirical levels. And the gap between the two should also be measured. On the industry-specific level, we include the impacts on news writing, reporting, designing, delivering, and functions. On the individual level, our proposed focus would be placed on the impacts of mobile journalism on the individual news needs, tastes, preferences, and experience. And those impacts should be measured on both normative and empirical levels. And the gap between the two should also be measured. It is our belief that this three-fold framework would be more accurate in capturing and comparing similarities and differences regarding the impacts of mobile journalism in different countries.

Among the identified issues in mobile journalism are infrastructure, safety, mobile capacity (Wigelius & Väätäjä, 2009), privacy, (Quinn, 2011), digital literacy, functionality, usability (Watkins, Hjorth & Koskinen, 2012), connectivity cost and reliability (Bethell, 2010), citizenship, identiy and local public sphere (Berger, 2011), and user experience (Väänänen-Vainio-Mattila & Väätäjä, 2008).

Issues can differ from country to country due to different political, social, economic and cultural backgrounds and conditions. For a cross-country comparison of issues in mobile journalism, we can compare those identified issues falling into the following categories: (a) credibility, (b) accuracy, (c) authority, (d) reliability, (e) depth, (f) privacy (g) ethics, (h) copyrights, (i) safety, (j) capacity, and (k) quality. And those issues can be compared on both normative and empirical levels.

Generally speaking, political, social, economic, cultural and technological factors may influence or shape the way mobile journalism operates in different countries. They constitute the overall environment. In this chapter, however, we would like to focus on those factors that have direct influence on mobile journalism. Specifically, mobile journalism can be more directly influenced by obstacles to access, limits on content, and violations of user rights.

The above paragraphs handled mobile features to illustrate how different dimensions of mobile use can be further investigated and compared. Using the same 3M approach and focusing on the gap between normative and empirical dimensions of mobile use related to mobile forms, formats, features, and functions may yield better and accurate findings regarding the impacts of mobile use. Mobile use can also be investigated and compared among different social media, native digital media or immigrant digital media. Mobile use may also be compared between or among immigrants, migrants, the marginalized, the handicapped, and the digital divides. Furthermore, integrated with mobile use are artificial intelligence, big data, deep machine learning, augmented reality, virtual reality, and mixed reality. These constitute further dimensions and directions of further studies of mobile use.

## FURTHER STUDIES OF MOBILE EXPERIENCE: DIMENSIONS AND DIRECTIONS

As experience economy is taking the centerstage and the world is increasingly dominated by mobile media and communication, mobile experience has become crucially important. Interconnected and interdependent with mobile use, mobile experience exists in all mobile activities. Besides the general conceptualization and operationalization briefly introduced in the first chapter of this volume, mobile experience has many dimensions and directions for further studies, including, among others, mobile experience in entertainment, gaming, tourism, leisure, journalism, healthcare, parenting, dating, government, advertising, marketing, and education. In this section, as a way to show how further studies of mobile experience may be conducted in what specific dimensions and directions, we propose a new approach to explain and predict mobile experience and recommend a new framework on how to maximize mobile experience to secure sustainable development.

## A NEW APPROACH TO EXPLANATION AND PREDICATION OF MOBILE EXPERIENCE

Factors influencing mobile experience in general that have been identified by earlier studies include location, social context, mobility, battery, application interface design, application performance, phone features, routine, and lifestyle needs (Ickin, Wac, Fiedler, Janowski, Hong, & Dey, 2012); emotions and memories (Kujala & Miron-Shatz, 2013), ease of use, awareness, security, usefulness, availability, and accessibility (Sarmento & Patrício, 2012); mobile site optimization (Djamasbi, McAuliffe, Gomez, Kardzhaliyski, Liu, & Oglesby, 2014); motor impairments and accessibility (Naftali and Findlater,

2014); gender differences such as responsive mobile design for female while dedicated mobile design for male (Silverstein, 2014); trust and privacy (Krontiris, Langheinrich and Shilton, 2014); load speed, site format, calculated download speed, social media presence, and app presence (Silverstein, 2013); personalization of content, structural navigation and representation (Chua, Wan, Chang, and Yi, 2014); and device capabilities and settings, network performance (Patro, Rayanchu, Griepentrog, Ma, and Banerjee, 2013), values, emotions, expectations, prior experiences, physical characteristics, personality, motivation, skills, usefulness, reputation, adaptivity (Arhippainen & Tähti, 2003), and attitude, social influence, media influence, perceived mobility, and perceived monetary value (Hong, Thong, Moon, Tam, 2008). Those factors can be generally described as internal factors, as they are closely related to all dimensions of mobile, that is, users, devices, apps, and networks.

Besides the internal factors, there are other factors that may shape differences in mobile experience. They can be called external factors. They may refer to social structure (degree of Homogeneity, extent to which egalitarian, communication patterns with the outside world, gender, ethnicity, generational cultures, religion, education provision and the support for literacy, and language), temporal structures of daily life (subjective experience of time, societal time use structures and cultural expectations about time), values (openness to technological innovation, the degree to which societies are individualistic or group-oriented, other culture values), communication (communication patterns and expectations and low and high context cultures), and material cultures (special considerations, national differences in housing characteristics, and artefacts) (Thomas, Haddon, Gilligan, Heinzmann & de Gournay, 2005). Other external factors may refer to social factors (time pressure, pressure of success and fail, etc.), cultural factors (habits, norms, religion, etc.), context of use (time, place, accompanying persons, temperature, etc.) (Arhippainen & Marika Tähti, 2003).

The above identified factors, internal and external, may shape mobile experience in one way or another. And most of earlier studies focused on factors that have shaped empirical experience, leaving the normative mobile experience and the gap between the normative and empirical mobile experience largely untouched.

Even within the examination of the empirical mobile experience, earlier studies did not investigate whether factors such as connection, curiosity, consumption, competitiveness, and creativity, which may have a role to play in shaping mobile experience. By connection, we refer to mobile connection, which constitutes the foundation of our mobile world, without which everything mobile would be impossible. Curiosity can enable mobile users to stay interested in exploring everything mobile. The more curious mobile users are, the more eager they are to experience everything mobile. Consumption means the overall consumption of mobile users in a country. The more mobile users consume on mobile, the more they would be exposed to mobile experience. Competitiveness refers to the level of the overall competitiveness of mobile users in a country. The more competitive mobile users are, the more they would be exposed to mobile experience. By creativity, we refer to the level of mobile users being creative in a nation. The more creative mobile users are, the more eager they would be to experience everything mobile experience (Xu, 2018).

After identifying these shaping factors, it is crucial to test different associations or correlations between or among those variables so as to design a model to describe, explain and predict mobile experience. The five factors are combined to constitute the environment that can shape different stages of mobile experience. The level of each of the five factors also constitutes the individual level of the environmental conduciveness. Combined, the five levels represent the level of the total environmental conduciveness.

As the overall the environmental conduciveness may be correlated with the overall mobile experience, it is also necessary to test the individual association between different variables at different stages of mobile experience. All these possible associations or correlations should be tested to determine their different inter-connections, interactions, and inter-influences, which can serve as a model to describe, explain and predict mobile experience in a country or in different countries around the world.

Guided by the results of testing all the different associations or correlations among different variables, one further important area not to be ignored is to compare how mobile experience can be enhanced in different countries or cultures. For effective and better comparison, we propose three key areas: (a) how all indicators of mobile experience can be enhanced in apps or mobile sites, (b) how mobile experience can be enhanced by leveraging users' changing tastes, desires and preferences, (c) how mobile experience can be enhanced by narrowing the gap between the expected and actual mobile experience, and (d) how mobile experience can be enhanced through general education on mobile experience enhancement.

Specifically, we should compare how all dimensions of mobile experience are properly embedded in an app or a mobile site so that its users can choose, customize and prioritize them according to their own desires, tastes and preferences. The second key area is to compare how to leverage all changes in mobile users' desires, tastes and preferences for mobile experience by combining different methods such as surveys, focus groups, interviews, and participant observations. The third area is to compare how the gap is narrowed between the normative and the empirical mobile experience. The results of the gap measurement should also be compared and triangulated to identify areas for enhancement. Further education, consulting and services on mobile experience can be provided to mobile professionals and general publics in different countries. A general education on mobile experience should be offered to mobile users at different levels through different means. Mobile experience won't be fully enhanced before mobile users and mobile content providers fully understand its changing forms, features and functions as well as how it can be leveraged to enhance mobile communication.

## MAXIMIZING MOBILE EXPERIENCE FOR SUSTAINABLE DEVELOPMENT

Other dimensions and directions for further studies of mobile experience include how to maximize mobile experience to secure sustainable development in the context of experience economy. It is especially important and crucial for developing and underdeveloped countries, where mobile may be the only widely and easily affordable and accessible medium which can be fully leveraged to secure and enhance sustainable development.

In September 2015, the United Nations approved 17 sustainable development goals to be obtained by 2030. They are (1) no poverty, (2) zero hunger, (3) good health and well-being, (4) quality education, (5) gender equality, (6) clear water and sanitation, (7) affordable and clean energy, (8) decent work and economic growth, (9) industry, innovation and infrastructure, (10) reduced inequalities, (11) sustainable cities and communities, (12) responsible consumption and production, (13) climate action, (14) live below water, (15) life on land, (16) peace, justice and strong institutions, and (17) partnerships for the goals (Division for Sustainable Development Goals, 2015). To achieve each of these 17 goals, mobile can be leveraged to assist, facilitate and expedite global joint efforts as mobile for development has become a global tagline and also a common practice worldwide, especially in the developing and underdeveloped countries. For instance, GSMA, representing "the interests of mobile operators worldwide, uniting more than 750 operators with over 350 companies in the broader mobile ecosystem, including handset and device makers, software companies, equipment providers and internet companies, as well as organisations in adjacent industry sectors," (GSMA, n.d.) has put together a global team within its organization to "identify opportunities and deliver innovations with socio-economic impact in financial services, health, agriculture, digital identity, energy, water, sanitation, disaster resilience and gender equality" (GSMA Mobile for Development, n.d.).

As the only and single technology cutting across geographies, cultures and income levels, mobile is expected to bring about tremendous socio-economic growth to the emerging markets, as demonstrated by one of the examples provided by GSMA's Mobile for Development team, whose work has impacted more than 30 millions lives across 49 countries (GSMA Mobile for Development, n.d.). Another example is that GSMA Connected Society program is working to connect 1.09 billion people around the world who lack mobile broadband coverage and still another example is that "34 mobile operators worldwide have made 49 commitments to reduce the gender gap in their mobile money or mobile internet customer base" (GSMA Mobile for Development, n.d.).

According to GSMA Mobile for Development (n.d.), "over 1.5 million women and their families reached with life-saving health and nutrition information while 12.5 million smallholder farmers reached with mobile agricultural services to improve their crop yields and income". And GSMA Mobile for Development also reports that "1.6 million mobile-enabled, pay-as-you-go solar home systems installed" and that "the GSMA Digital Identity program is working with partners to develop and scale innovative identity solutions for the people who lack identification".

Having made such achievements in narrowing the digital or mobile gap between the haves and the have-nots, what seems to be more imperative is to maximize mobile experience behind each use of mobile to secure sustainable development goals. Undeniably, mobile use is a crucial step for development. To secure sustainable development, however, mobile use alone won't be effective or sufficient. Any emphasis of mobile use over its experience will result in reduced effectiveness and efficiency in leveraging mobile for development. It is recommended that both use and experience in leveraging mobile for development should be examined as an interconnected and interdependent whole.

It is also recommended that the six stages of mobile experience and the 3M approach with a focus on the gap between the normative and the empirical proposed by Xu (2018) can be applied in examining how mobile experience can be fully maximized to secure each of the 17 sustainable development goals. The six stages of mobile experience are (1) enticement, (2) entertainment, (3) engagement, (4) empowerment, (5) enlightenment, and (6) enhancement (Xu, 2018). Each of the generic stage of mobile experience can be specifically re-conceptualized and re-operationalized according to each of the 17 sustainable development goals.

For instance, the first goal is to eliminate poverty. To facilitate the global efforts to obtain that goal, mobile can be fully leveraged in national, regional and global coordination and collaboration in enabling mobile users in developing and underdeveloped countries to create or co-create opportunities for social and economic development. Resources and efforts can also be globally allocated through use of mobile. In light of this understanding, the first stage of mobile experience is to entice mobile users to join the efforts to eliminate poverty globally. The specific dimensions of enticement should be conceptualized and also operationalized so that it can gauged. For example, enticement may include being enticed to part of mobile inclusion and literacy, which is a must in leveraging mobile for development. The same procedure can be applied to the other five stages of mobile experience.

Once all the six stages of mobile experience are re-conceptualized and re-operationalized according to each of the 17 sustainable development goals, the 3M approach should also be applied, which consists of mapping, measuring and modeling (Xu, 2018). To map is to locate where each dimension of mobile experience lies while to measure is to gauge the level, extent or strength of each dimension of mobile experience. To model is to test the association or correlation between or among identified variables that shape mobile experience. The gap focus of the 3M approach is to locate and gauge the disparity between what is expected and what actually is in terms of mobile experience. The identification and measurement of the gap can also aid and facilitate global efforts to maximize mobile experience for securing sustainable development.

The proposed framework of the six stages of mobile experience, accompanied by the 3M approach with a special focus on the normative and empirical gap, can also be applied to examine how mobile experience can be maximized in facilitating global efforts to obtain the other 16 sustainable development goals.

# REFERENCES

Aguado, J. M., & Martínez, I. J. (2008). Massmediatizing mobile phones: Contents development, professional convergence and consumption practices. *New Media and Innovative Technologies: Industry and Society*, 211-239.

Aharony, N. (2013, April). Librarians' attitudes towards mobile services. In Aslib Proceedings (Vol. 65, No. 4, pp. 358-375). Emerald Group Publishing Limited.

Aitamurto, T. (2013). Balancing between open and closed: Co-creation in magazine journalism. *Digital Journalism*, *1*(2), 229–251. doi:10.1080/21670811.2012.750150

Anacleto, R., Figueiredo, L., Almeida, A., & Novais, P. (2014). Mobile application to provide personalized sightseeing tours. *Journal of Network and Computer Applications*, *41*, 56–64. doi:10.1016/j.jnca.2013.10.005

Arhippainen, L., & Tähti, M. (2003, December). Empirical evaluation of user experience in two adaptive mobile application prototypes. In *MUM 2003. Proceedings of the 2nd International Conference on Mobile and Ubiquitous Multimedia* (No. 011, pp. 27-34). Linköping University Electronic Press.

Ben-Zeev, D., Schueller, S. M., Begale, M., Duffecy, J., Kane, J. M., & Mohr, D. C. (2015). Strategies for mHealth research: Lessons from 3 mobile intervention studies. *Administration and Policy in Mental Health*, *42*(2), 157–167. doi:10.100710488-014-0556-2 PMID:24824311

Berger, G. (2011). Empowering the youth as citizen journalists: A South African experience. *Journalism*, *12*(6), 708–726. doi:10.1177/1464884911405466

Bethell, P. (2010). Journalism students' experience of mobile phone technology: implications for journalism education. *Asia Pacific Media Educator*, (20), 103.

Bidarra, J., Figueiredo, M., & Natálio, C. (2015). Interactive design and gamification of ebooks for mobile and contextual learning. *International Journal of Interactive Mobile Technologies*, 24-32.

Bivens, R. K. (2008). The Internet, mobile phones and blogging. *Journalism Practice*, *2*(1), 113–129. doi:10.1080/17512780701768568

Bjørnestad, S., Tessem, B., & Nyre, L. (2011, February). Design and evaluation of a location-based mobile news reader. In *New Technologies, Mobility and Security (NTMS), 2011 4th IFIP International Conference on* (pp. 1-4). IEEE. 10.1109/NTMS.2011.5720634

Bonsignore, E., Quinn, A. J., Druin, A., & Bederson, B. B. (2013). Sharing stories "in the wild": A mobile storytelling case study using StoryKit. *ACM Transactions on Computer-Human Interaction, 20*(3), 18. doi:10.1145/2491500.2491506

Brandenburg, C., Worrall, L., Rodriguez, A. D., & Copland, D. (2013). Mobile computing technology and aphasia: An integrated review of accessibility and potential uses. *Aphasiology, 27*(4), 444–461. doi:10.1080/02687038.2013.772293

Chang, H. Y., Wu, H. K., & Hsu, Y. S. (2013). Integrating a mobile augmented reality activity to contextualize student learning of a socioscientific issue. *British Journal of Educational Technology, 44*(3), E95–E99. doi:10.1111/j.1467-8535.2012.01379.x

Chong, A. Y. L., Chan, F. T., & Ooi, K. B. (2012). Predicting consumer decisions to adopt mobile commerce: Cross country empirical examination between China and Malaysia. *Decision Support Systems, 53*(1), 34–43. doi:10.1016/j.dss.2011.12.001

Chua, W. Y., Wan, M. P. H., Chang, K., & Yi, W. (2014). *Improving Mobile Applications Usage Experience of Novice Users through User-Acclimatized Interaction: A Case Study*. Academic Press.

Ciampa, K. (2014). Learning in a mobile age: An investigation of student motivation. *Journal of Computer Assisted Learning, 30*(1), 82–96. doi:10.1111/jcal.12036

Clarke, J., & Bromley, M. (Eds.). (2012). *International news in the digital age: East-west perceptions of a new world order* (Vol. 4). Routledge. doi:10.4324/9780203804674

Craig, R. T. (2007). Issue Forum Introduction: Mobile Media and Communication: What are the Important Questions? *Communication Monographs, 74*(3), 386–388. doi:10.1080/03637750701543501

Dale Storie, M. L. I. S. (2014). Mobile devices in medicine: A survey of how medical students, residents, and faculty use smartphones and other mobile devices to find information*. *Journal of the Medical Library Association: JMLA, 102*(1), 22–30. doi:10.3163/1536-5050.102.1.006 PMID:24415916

Dinh, H. T., Lee, C., Niyato, D., & Wang, P. (2013). A survey of mobile cloud computing: Architecture, applications, and approaches. *Wireless Communications and Mobile Computing, 13*(18), 1587–1611. doi:10.1002/wcm.1203

Division for Sustainable Development Goals. (2015). *17 Sustainable Development Goals*. Retrieved from https://sustainabledevelopment.un.org/sdgs

Djamasbi, S., McAuliffe, D., Gomez, W., Kardzhaliyski, G., Liu, W., & Oglesby, F. (2014, June). Designing for success: Creating business value with mobile user experience (UX). In *International conference on HCI in Business* (pp. 299-306). Springer. 10.1007/978-3-319-07293-7_29

Eng, D. S., & Lee, J. M. (2013). The promise and peril of mobile health applications for diabetes and endocrinology. *Pediatric Diabetes, 14*(4), 231–238. doi:10.1111/pedi.12034 PMID:23627878

Espada, J. P., Díaz, V. G., Crespo, R. G., Martínez, O. S., G-Bustelo, B. C. P., & Lovelle, J. M. C. (2015). Using extended web technologies to develop Bluetooth multi-platform mobile applications for interact with smart things. *Information Fusion*, *21*, 30–41. doi:10.1016/j.inffus.2013.04.008

Esser, F., & Hanitzsch, T. (2012). *The Handbook of Comparative Communication Research*. New York: Routledge.

Fidalgo, A. (2009). PUSHED NEWS: When the news comes to the cellphone. *Brazilian Journalism Research*, *5*(2), 113–124. doi:10.25200/BJR.v5n2.2009.214

Fiordelli, M., Diviani, N., & Schulz, P. J. (2013). Mapping mHealth research: A decade of evolution. *Journal of Medical Internet Research*, *15*(5), e95. doi:10.2196/jmir.2430 PMID:23697600

Flora, H. K., Wang, X., & Chande, S. V. (2014). An investigation on the characteristics of mobile applications: A survey study. *International Journal of Information Technology and Computer Science*, *6*(11), 21–27. doi:10.5815/ijitcs.2014.11.03

Fortunati, L., & Taipale, S. (2014). The advanced use of mobile phones in five European countries. *The British Journal of Sociology*, *65*(2), 317–337. doi:10.1111/1468-4446.12075 PMID:24697752

Gao, J., Bai, X., Tsai, W. T., & Uehara, T. (2014). Mobile application testing: A tutorial. *Computer*, *47*(2), 46–55. doi:10.1109/MC.2013.445

Gascon, H., Uellenbeck, S., Wolf, C., & Rieck, K. (2014, March). *Continuous Authentication on Mobile Devices by Analysis of Typing Motion Behavior*. Sicherheit.

Glackin, B. C., Rodenhiser, R. W., & Herzog, B. (2014). A library and the disciplines: A collaborative project assessing the impact of eBooks and mobile devices on student learning. *Journal of Academic Librarianship*, *40*(3), 299–306. doi:10.1016/j.acalib.2014.04.007

Goggin, G., Martin, F., & Dwyer, T. (2015). Locative news: Mobile media, place informatics, and digital news. *Journalism Studies*, *16*(1), 41–59. doi:10.1080/1461670X.2014.890329

Gohil, A., Modi, H., & Patel, S. K. (2013, March). 5G technology of mobile communication: A survey. In *Intelligent Systems and Signal Processing (ISSP), 2013 International Conference on* (pp. 288-292). IEEE.

Grindrod, K. A., Li, M., & Gates, A. (2014). Evaluating user perceptions of mobile medication management applications with older adults: A usability study. *JMIR mHealth and uHealth*, *2*(1), e11. doi:10.2196/mhealth.3048 PMID:25099993

GSMA. (n.d.). *About Us*. Retrieved from https://www.gsma.com/aboutus/

GSMA Mobile for Development. (n.d.). *Overview*. Retrieved from https://www.gsma.com/mobilefordevelopment/

Güler, Ç., Kılıç, E., & Çavuş, H. (2014). A comparison of difficulties in instructional design processes: Mobile vs. desktop. *Computers in Human Behavior*, *39*, 128–135. doi:10.1016/j.chb.2014.07.008

Guo, B., Yu, Z., Zhou, X., & Zhang, D. (2014, March). From participatory sensing to mobile crowd sensing. In *Pervasive Computing and Communications Workshops (PERCOM Workshops), 2014 IEEE International Conference on* (pp. 593-598). IEEE. 10.1109/PerComW.2014.6815273

Hashim, K. F., Tan, F. B., & Rashid, A. (2015). Adult learners' intention to adopt mobile learning: A motivational perspective. *British Journal of Educational Technology*, *46*(2), 381–390. doi:10.1111/bjet.12148

Head, M., & Ziolkowski, N. (2012). Understanding student attitudes of mobile phone features: Rethinking adoption through conjoint, cluster and SEM analyses. *Computers in Human Behavior*, *28*(6), 2331–2339. doi:10.1016/j.chb.2012.07.003

Ho, S. Y. (2012). The effects of location personalization on individuals' intention to use mobile services. *Decision Support Systems*, *53*(4), 802–812. doi:10.1016/j.dss.2012.05.012

Hong, S. J., Thong, J. Y., Moon, J. Y., & Tam, K. Y. (2008). Understanding the behavior of mobile data services consumers. *Information Systems Frontiers*, *10*(4), 431–445. doi:10.100710796-008-9096-1

Hu, X., Chu, T. H., Leung, V., Ngai, E. C. H., Kruchten, P., & Chan, H. C. (2014). A survey on mobile social networks: Applications, platforms, system architectures, and future research directions. *IEEE Communications Surveys and Tutorials*, *17*(3), 1557–1581. doi:10.1109/COMST.2014.2371813

Humphreys, L., Von Pape, T., & Karnowski, V. (2013). Evolving mobile media: Uses and conceptualizations of the mobile internet. *Journal of Computer-Mediated Communication*, *18*(4), 491–507. doi:10.1111/jcc4.12019

Ickin, S., Wac, K., Fiedler, M., Janowski, L., Hong, J. H., & Dey, A. K. (2012). Factors influencing quality of experience of commonly used mobile applications. *IEEE Communications Magazine*, *50*(4), 48–56.

Jabeur, N., Zeadally, S., & Sayed, B. (2013). Mobile social networking applications. *Communications of the ACM*, *56*(3), 71–79. doi:10.1145/2428556.2428573

Jain, A. K., & Shanbhag, D. (2012). Addressing security and privacy risks in mobile applications. *IT Professional*, *14*(5), 28–33. doi:10.1109/MITP.2012.72

Jambulingam, M. (2013). Behavioural intention to adopt mobile technology among tertiary students. *World Applied Sciences Journal*, *22*(9), 1262–1271.

Jambulingam, M., & Sorooshian, S. (2013). Usage of mobile features among undergraduates and mobile learning. *Current Research Journal of Social Sciences*, *5*(4), 130–133.

Jokela, T., Väätäjä, H., & Koponen, T. (2009, September). Mobile Journalist Toolkit: a field study on producing news articles with a mobile device. In *Proceedings of the 13th International MindTrek Conference: Everyday Life in the Ubiquitous Era* (pp. 45-52). ACM. 10.1145/1621841.1621851

Jordan, P. W. (2000). *Designing Pleasurable Products*. New York: Taylor and Francis.

Ke, F., & Hsu, Y. C. (2015). Mobile augmented-reality artifact creation as a component of mobile computer-supported collaborative learning. *The Internet and Higher Education*, *26*, 33–41. doi:10.1016/j.iheduc.2015.04.003

Kearney, M., & Maher, D. (2013). Mobile learning in math teacher education: Using iPads to support pre-service teachers' professional development. *Australian Educational Computing*, *27*(3), 76–84.

Kearney, M., Schuck, S., Burden, K., & Aubusson, P. (2012). Viewing mobile learning from a pedagogical perspective. *Research in Learning Technology*, 20.

Kharrazi, H., Chisholm, R., VanNasdale, D., & Thompson, B. (2012). Mobile personal health records: An evaluation of features and functionality. *International Journal of Medical Informatics*, *81*(9), 579–593. doi:10.1016/j.ijmedinf.2012.04.007 PMID:22809779

Kim, C., Park, T., Lim, H., & Kim, H. (2013). On-site construction management using mobile computing technology. *Automation in Construction*, *35*, 415–423. doi:10.1016/j.autcon.2013.05.027

Kim, E., Lin, J. S., & Sung, Y. (2013). To app or not to app: Engaging consumers via branded mobile apps. *Journal of Interactive Advertising*, *13*(1), 53–65. doi:10.1080/15252019.2013.782780

Kim, Y. H., Kim, D. J., & Wachter, K. (2013). A study of mobile user engagement (MoEN): Engagement motivations, perceived value, satisfaction, and continued engagement intention. *Decision Support Systems*, *56*, 361–370. doi:10.1016/j.dss.2013.07.002

Koponen, T., & Väätäjä, H. (2009, September). Early adopters' experiences of using mobile multimedia phones in news journalism. In *European Conference on Cognitive Ergonomics: Designing beyond the Product---Understanding Activity and User Experience in Ubiquitous Environments* (p. 2). VTT Technical Research Centre of Finland.

Krontiris, I., Langheinrich, M., & Shilton, K. (2014). Trust and privacy in mobile experience sharing: Future challenges and avenues for research. *Communications Magazine, IEEE*, *52*(8), 50–55. doi:10.1109/MCOM.2014.6871669

Kujala, S., & Miron-Shatz, T. (2013, April). Emotions, experiences and usability in real-life mobile phone use. In *Proceedings of the SIGCHI Conference on Human Factors in Computing Systems* (pp. 1061-1070). ACM. 10.1145/2470654.2466135

La Polla, M., Martinelli, F., & Sgandurra, D. (2013). A survey on security for mobile devices. *IEEE Communications Surveys and Tutorials*, *15*(1), 446–471. doi:10.1109/SURV.2012.013012.00028

Lasica, J. D., & Firestone, C. M. (2008). *Civic Engagement on the Move: How mobile media can serve the public good*. Washington, DC: The Aspen Institute.

Lee, D., Moon, J., Kim, Y. J., & Mun, Y. Y. (2015). Antecedents and consequences of mobile phone usability: Linking simplicity and interactivity to satisfaction, trust, and brand loyalty. *Information & Management*, *52*(3), 295–304. doi:10.1016/j.im.2014.12.001

Lee, H., Kwak, N., Campbell, S. W., & Ling, R. (2014). Mobile communication and political participation in South Korea: Examining the intersections between informational and relational uses. *Computers in Human Behavior*, *38*, 85–92. doi:10.1016/j.chb.2014.05.017

Lee, J. H., & Kim, J. (2014). Socio-demographic gaps in mobile use, causes, and consequences: A multi-group analysis of the mobile divide model. *Information Communication and Society*, *17*(8), 917–936. doi:10.1080/1369118X.2013.860182

Lu, C., Chang, M., Huang, E., & Ching-Wen, C. (2014). Context-Aware Mobile Role Playing Game for Learning-A Case of Canada and Taiwan. *Journal of Educational Technology & Society*, *17*(2), 101.

Martyn, P. H. (2009). The Mojo in the third millennium: Is multimedia journalism affecting the news we see? *Journalism Practice, 3*(2), 196–215. doi:10.1080/17512780802681264

Mayfield, C. H., Ohara, P. T., & O'Sullivan, P. S. (2013). Perceptions of a mobile technology on learning strategies in the anatomy laboratory. *Anatomical Sciences Education, 6*(2), 81–89. doi:10.1002/ase.1307 PMID:22927203

Mills, J., Egglestone, P., Rashid, O., & Väätäjä, H. (2012). MoJo in action: The use of mobiles in conflict, community, and cross-platform journalism. *Continuum, 26*(5), 669–683. doi:10.1080/10304312.2012.706457

Mizouni, R., Matar, M. A., Al Mahmoud, Z., Alzahmi, S., & Salah, A. (2014). A framework for context-aware self-adaptive mobile applications SPL. *Expert Systems with Applications, 41*(16), 7549–7564. doi:10.1016/j.eswa.2014.05.049

Mudhai, O. F. (2011). Immediacy and openness in a digital Africa: Networked-convergent journalisms in Kenya. *Journalism, 12*(6), 674–691. doi:10.1177/1464884911405470

Naftali, M., & Findlater, L. (2014, October). Accessibility in context: understanding the truly mobile experience of smartphone users with motor impairments. In *Proceedings of the 16th international ACM SIGACCESS conference on Computers & accessibility* (pp. 209-216). ACM. 10.1145/2661334.2661372

Nicholas, J., Larsen, M. E., Proudfoot, J., & Christensen, H. (2015). Mobile apps for bipolar disorder: A systematic review of features and content quality. *Journal of Medical Internet Research, 17*(8), e198. doi:10.2196/jmir.4581 PMID:26283290

Noulas, A., Scellato, S., Lathia, N., & Mascolo, C. (2012, December). Mining user mobility features for next place prediction in location-based services. In *Data mining (ICDM), 2012 IEEE 12th international conference on* (pp. 1038-1043). IEEE. 10.1109/ICDM.2012.113

O'bannon, B. W., & Thomas, K. (2014). Teacher perceptions of using mobile phones in the classroom: Age matters! *Computers & Education, 74*, 15–25. doi:10.1016/j.compedu.2014.01.006

Oliveira, R. R., Noguez, F. C., Costa, C. A., Barbosa, J. L., & Prado, M. P. (2013). SWTRACK: An intelligent model for cargo tracking based on off-the-shelf mobile devices. *Expert Systems with Applications, 40*(6), 2023–2031. doi:10.1016/j.eswa.2012.10.021

Organista-Sandoval, J., & Serrano-Santoyo, A. (2014). Appropriation and educational uses of mobile phones by students and teachers at a public university in Mexico. *Creative Education.*

Pagoto, S., Schneider, K., Jojic, M., DeBiasse, M., & Mann, D. (2013). Evidence-based strategies in weight-loss mobile apps. *American Journal of Preventive Medicine, 45*(5), 576–582. doi:10.1016/j.amepre.2013.04.025 PMID:24139770

Patro, A., Rayanchu, S., Griepentrog, M., Ma, Y., & Banerjee, S. (2013, December). Capturing mobile experience in the wild: a tale of two apps. In *Proceedings of the ninth ACM conference on Emerging networking experiments and technologies* (pp. 199-210). ACM. 10.1145/2535372.2535391

Pavlik, J. V., & Bridges, F. (2013). The emergence of augmented reality (AR) as a storytelling medium in journalism. *Journalism & Communication Monographs, 15*(1), 4–59. doi:10.1177/1522637912470819

Pendit, U. C., Zaibon, S. B., & Bakar, J. A. A. (2014). Mobile Augmented Reality for Enjoyable Informal Learning in Cultural Heritage Site. *International Journal of Computers and Applications*, *92*(14).

Persaud, A., & Azhar, I. (2012). Innovative mobile marketing via smartphones: Are consumers ready? *Marketing Intelligence & Planning*, *30*(4), 418–443. doi:10.1108/02634501211231883

Pielot, M., De Oliveira, R., Kwak, H., & Oliver, N. (2014, April). Didn't you see my message?: predicting attentiveness to mobile instant messages. In *Proceedings of the 32nd annual ACM conference on Human factors in computing systems* (pp. 3319-3328). ACM. 10.1145/2556288.2556973

Pine, B. J., Pine, J., & Gilmore, J. H. (1999). *The experience economy: work is theatre & every business a stage*. Harvard Business Press.

Plaza, I., Demarzo, M. M. P., Herrera-Mercadal, P., & García-Campayo, J. (2013). Mindfulness-based mobile applications: Literature review and analysis of current features. *JMIR mHealth and uHealth*, *1*(2), e24. doi:10.2196/mhealth.2733 PMID:25099314

Powers, E. (2012). Learning to do it all: When it comes time to fill one of those precious hiring slots, news outlets are looking for journalists with a wide array of skills. *American Journalism Review*, *34*(1), 10–14.

Qi, H., & Gani, A. (2012, May). Research on mobile cloud computing: Review, trend and perspectives. In *Digital Information and Communication Technology and it's Applications (DICTAP), 2012 Second International Conference on* (pp. 195-202). IEEE.

Quinn, S. (2011). *Mojo-mobile journalism in the Asian Region*. KAS.

Ramanathan, N., Swendeman, D., Comulada, W. S., Estrin, D., & Rotheram-Borus, M. J. (2013). Identifying preferences for mobile health applications for self-monitoring and self-management: Focus group findings from HIV-positive persons and young mothers. *International Journal of Medical Informatics*, *82*(4), e38–e46. doi:10.1016/j.ijmedinf.2012.05.009 PMID:22704234

Riikonen, A., Smura, T., Kivi, A., & Töyli, J. (2013). Diffusion of mobile handset features: Analysis of turning points and stages. *Telecommunications Policy*, *37*(6), 563–572. doi:10.1016/j.telpol.2012.07.011

Rusbridger, A. (2012, February 29). Alan Rusbridger on open journalism at the Guardian:'Journalists are not the only experts in the world'–video. *The Guardian*.

Sarmento, T., & Patrício, L. (2012). Mobile Service Experience-a quantitative study. *AMAServsig2012*.

Schall, G., Zollmann, S., & Reitmayr, G. (2013). Smart Vidente: Advances in mobile augmented reality for interactive visualization of underground infrastructure. *Personal and Ubiquitous Computing*, *17*(7), 1533–1549. doi:10.100700779-012-0599-x

Scolari, C. A., Aguado, J. M., & Feijóo, C. (2012). Mobile Media: Towards a Definition and Taxonomy of Contents and Applications. *International Journal of Interactive Mobile Technologies*, *6*(2).

Sendra, S., Granell, E., Lloret, J., & Rodrigues, J. J. (2014). Smart collaborative mobile system for taking care of disabled and elderly people. *Mobile Networks and Applications*, *19*(3), 287–302. doi:10.100711036-013-0445-z

Shapiro-Mathews, E., & Barton, A. J. (2013). Using the patient engagement framework to develop an institutional mobile health strategy. *Clinical Nurse Specialist CNS, 27*(5), 221–223. doi:10.1097/ NUR.0b013e3182a0b9e2 PMID:23942098

Silverstein, M. (2013, September). *Introducing the Mobile Experience Scorecard.* Retrieved from http:// www.thesearchagents.com/2013/09/introducing-the-mobile-experience-scorecard/

Silverstein, M. (2014). *Female-Oriented Sites More Optimized for Mobile?* Retrieved from http://www. thesearchagents.com/2014/03/female-oriented-sites-more-optimized-for-mobile/

Sims, F., Williams, M. A., & Elliot, S. (2007, July). Understanding the Mobile Experience Economy: A key to richer more effective M-Business Technologies, Models and Strategies. In *Management of Mobile Business, 2007. ICMB 2007. International Conference on the* (pp. 12-12). IEEE.

Singh, V. K., Freeman, L., Lepri, B., & Pentland, A. S. (2013, September). Predicting spending behavior using socio-mobile features. In *Social Computing (SocialCom), 2013 International Conference on* (pp. 174-179). IEEE. 10.1109/SocialCom.2013.33

Su, Z., Xu, Q., Zhu, H., & Wang, Y. (2015). A novel design for content delivery over software defined mobile social networks. *IEEE Network, 29*(4), 62–67. doi:10.1109/MNET.2015.7166192

Suominen, A., Hyrynsalmi, S., & Knuutila, T. (2014). Young mobile users: Radical and individual–Not. *Telematics and Informatics, 31*(2), 266–281. doi:10.1016/j.tele.2013.08.003

Teodoro, R., Ozturk, P., Naaman, M., Mason, W., & Lindqvist, J. (2014, February). The motivations and experiences of the on-demand mobile workforce. In *Proceedings of the 17th ACM conference on Computer supported cooperative work & social computing* (pp. 236-247). ACM. 10.1145/2531602.2531680

Thomas, F., Haddon, L., Gilligan, R., Heinzmann, P., & de Gournay, C. (2005). *Cultural factors shaping the experience of ICTs: An exploratory review. In International collaborative research. Cross-cultural differences and cultures of research* (pp. 13–50). Brussels: COST.

Väänänen-Vainio-Mattila, K., & Väätäjä, H. (2008, September). *Towards a life cycle framework of mobile service user experience. In 2nd MIUX Workshop at MobileHCI*, Amsterdam, The Netherlands.

Väätäjä, H. (2008, September). Factors affecting user experience in mobile systems and services. In *Proceedings of the 10th international conference on Human computer interaction with mobile devices and services* (pp. 551-551). ACM.

Väätäjä, H. (2010, November). User experience of smart phones in mobile journalism: early findings on influence of professional role. In *Proceedings of the 22nd Conference of the Computer-Human Interaction Special Interest Group of Australia on Computer-Human Interaction* (pp. 1-4). ACM.

Väätäjä, H. (2010, November). User experience evaluation criteria for mobile news making technology: findings from a case study. In *Proceedings of the 22nd Conference of the Computer-Human Interaction Special Interest Group of Australia on Computer-Human Interaction* (pp. 152-159). ACM.

Väätäjä, H., & Egglestone, P. (2012, February). Briefing news reporting with mobile assignments: perceptions, needs and challenges. In *Proceedings of the ACM 2012 conference on Computer Supported Cooperative Work* (pp. 485-494). ACM.

Väätäjä, H., Koponen, T., & Roto, V. (2009, September). Developing practical tools for user experience evaluation: a case from mobile news journalism. In *European Conference on Cognitive Ergonomics: Designing beyond the Product---Understanding Activity and User Experience in Ubiquitous Environments* (p. 23). VTT Technical Research Centre of Finland.

Väätäjä, H., Männistö, A., Vainio, T., & Jokela, T. (2009). Understanding user experience to support learning for mobile journalist's work. *The Evolution of Mobile Teaching and Learning*, 177-210.

Väätäjä, H., & Männistö, A. A. (2010, October). Bottlenecks, usability issues and development needs in creating and delivering news videos with smartphones. In *Proceedings of the 3rd workshop on Mobile video delivery* (pp. 45-50). ACM. 10.1145/1878022.1878034

Väätäjä, H., & Roto, V. (2010, April). Mobile questionnaires for user experience evaluation. In CHI' 10 Extended Abstracts on Human Factors in Computing Systems (pp. 3361-3366). ACM.

van Velsen, L., Beaujean, D. J., & van Gemert-Pijnen, J. E. (2013). Why mobile health app overload drives us crazy, and how to restore the sanity. *BMC Medical Informatics and Decision Making, 13*(1), 1. doi:10.1186/1472-6947-13-23 PMID:23399513

Ventola, C. L. (2014). Mobile devices and apps for health care professionals: Uses and benefits. *P&T, 39*(5), 356. PMID:24883008

Verclas, K., & Mechael, P. (2008). *A mobile voice: The use of mobile phones in citizen media. Mobile-Active. Org, Pact*. USAID.

Viberg, O., & Grönlund, Å. (2013). Cross-cultural analysis of users' attitudes toward the use of mobile devices in second and foreign language learning in higher education: A case from Sweden and China. *Computers & Education, 69*, 169–180. doi:10.1016/j.compedu.2013.07.014

Watkins, J., Hjorth, L., & Koskinen, I. (2012). Wising up: Revising mobile media in an age of smartphones. *Continuum, 26*(5), 665–668. doi:10.1080/10304312.2012.706456

Wei, P. S., & Lu, H. P. (2014). Why do people play mobile social games? An examination of network externalities and of uses and gratifications. *Internet Research, 24*(3), 313–331. doi:10.1108/IntR-04-2013-0082

Wenger, D., Owens, L., & Thompson, P. (2014). Help Wanted Mobile Journalism Skills Required by Top US News Companies. *Electronic News, 8*(2), 138–149. doi:10.1177/1931243114546807

Westlund, O. (2013). Mobile news: A review and model of journalism in an age of mobile media. *Digital Journalism, 1*(1), 6–26. doi:10.1080/21670811.2012.740273

Westlund, O. (2014). The production and consumption of news in an age of mobile media. *The Routledge companion to mobile media*, 135-145.

Wigelius, H., & Väätäjä, H. (2009). Dimensions of context affecting user experience in mobile work. In *Human-Computer Interaction–INTERACT 2009* (pp. 604–617). Springer Berlin Heidelberg. doi:10.1007/978-3-642-03658-3_65

Wolf, C., & Hohlfeld, R. (2012). Revolution in Journalism? In Images in Mobile Communication (pp. 81-99). VS Verlag für Sozialwissenschaften. doi:10.1007/978-3-531-93190-6_5

Wu, W. H., Wu, Y. C. J., Chen, C. Y., Kao, H. Y., Lin, C. H., & Huang, S. H. (2012). Review of trends from mobile learning studies: A meta-analysis. *Computers & Education*, *59*(2), 817–827. doi:10.1016/j.compedu.2012.03.016

Xu, X. (2018). Comparing mobile experience. In K. Norman & K. Kirakowski (Eds.), *Wiley Handbook of Human-Computer Interaction Set* (Vol. 1, pp. 225–238). Wiley.

Yang, C. H., Maher, J. P., & Conroy, D. E. (2015). Implementation of behavior change techniques in mobile applications for physical activity. *American Journal of Preventive Medicine*, *48*(4), 452–455. doi:10.1016/j.amepre.2014.10.010 PMID:25576494

Yang, K., & Kim, H. Y. (2012). Mobile shopping motivation: An application of multiple discriminant analysis. *International Journal of Retail & Distribution Management*, *40*(10), 778–789. doi:10.1108/09590551211263182

Yang, M., Li, Y., Jin, D., Zeng, L., Wu, X., & Vasilakos, A. V. (2015). Software-defined and virtualized future mobile and wireless networks: A survey. *Mobile Networks and Applications*, *20*(1), 4–18. doi:10.100711036-014-0533-8

Zhao, Z., & Balagué, C. (2015). Designing branded mobile apps: Fundamentals and recommendations. *Business Horizons*, *58*(3), 305–315. doi:10.1016/j.bushor.2015.01.004

# Compilation of References

Abraham, R. (2007). Mobile phones and economic development: Evidence from the ashing industry in India. *Information Technologies and International Development, 4*(1), 5–17. doi:10.1162/itid.2007.4.1.5

Adams, M. (2017). *An Interview with a Stepes Translator.* Retrieved June 13, 2018, from https://blog.stepes.com/an-interview-with-a-stepes-translator/

Agca, R. K., & Özdemir, S. (2013). Foreign Language Vocabulary Learning with Mobile Technologies. *Procedia: Social and Behavioral Sciences, 83*, 781–785. doi:10.1016/j.sbspro.2013.06.147

Aguado, J. M., & Martínez, I. J. (2008). Massmediatizing mobile phones: Contents development, professional convergence and consumption practices. *New Media and Innovative Technologies: Industry and Society*, 211-239.

Aguado, J. M., Feijóo, C., & Martínez, I. J. (2015). *Emerging perspectives on the mobile content evolution.* IGI Global.

Aharony, N. (2013, April). Librarians' attitudes towards mobile services. In Aslib Proceedings (Vol. 65, No. 4, pp. 358-375). Emerald Group Publishing Limited.

Ahn, H., Wijaya, M. E., & Esmero, B. C. (2014). A Systemic Smartphone Usage Pattern Analysis : Focusing on Smartphone Addiction Issue. *International Journal of Multimedia and Ubiquitous Engineering, 9*(6), 9–14. doi:10.14257/ijmue.2014.9.6.02

Ahuvia, A. C., & Adelman, M. B. (1992). Formal intermediaries in the marriage market: A typology and review. *Journal of Marriage and the Family, 54*(2), 452–463. doi:10.2307/353076

Aitamurto, T. (2013). Balancing between open and closed: Co-creation in magazine journalism. *Digital Journalism, 1*(2), 229–251. doi:10.1080/21670811.2012.750150

Ajzen, I. (1985). From intentions to actions: A theory of planned behavior. In J. Kuhl & J. Beckman (Eds.), *Action-control: From cognition to behavior* (pp. 11–39). Heidelberg, Germany: Springer. doi:10.1007/978-3-642-69746-3_2

Akamai. (2017). *Akamai's State of the Internet Report.* Retrieved from https://www.akamai.com/uk/en/about/our-thinking/state-of-the-internet-report/global-state-of-the-internet-connectivity-reports.jsp

Akca, E., & Kaya, B. (2016). The different approaches to digital divide in the concept of gender equality and it's dimensions. *Intermedia International, 3*(5). Available from: http://dergipark.gov.tr/download/article-file/399689

Aker, J. C. (2008). *Does digital divide or provide? The impact of mobile phones on grain markets in Niger (Working paper No. 154).* Berkeley, CA: Department of Agriculture and Resource Economics, University of California. Retrieved at http://are.berkeley.edu/aker/cell.pdf

Aker, J. C., & Mbiti, I. M. (2010). Mobile phones and economic development in Africa. *The Journal of Economic Perspectives, 24*(3), 207–232. doi:10.1257/jep.24.3.207

Akter, S., & Ray, P. (2010). mHealth-an ultimate platform to serve the unserved. *Yearbook of Medical Informatics, 2010,* 94–100. PMID:20938579

Albrecht, U.-V., Behrends, M., Matthies, H. K., Von Jan, U., & Schmeer, R. (2013). Usage of multilingual mobile translation applications in clinical settings. *JMIR mHealth and uHealth, 1*(1), e4. doi:10.2196/mhealth.2268 PMID:25100677

Albury, K., Burgess, J., Light, B., Race, K., & Wilken, R. (2017). Data cultures of mobile dating and hook-up apps: Emerging issues for critical social science research. *Big Data & Society, 4*(2), 2053951717720950. doi:10.1177/2053951717720950

Al-Emran, M., Elsherif, H., & Shaalan, K. (2016). Investigating attitudes towards the use of mobile learning in higher education. *Computers in Human Behaviors [Online], 56,* 93–102. doi:10.1016/j.chb.2015.11.033

Alhassan, A. A., Alqadhib, E. M., Taha, N. W., Alahmari, R. A., Salam, M., & Almutairi, A. F. (2018). The relationship between addiction to smartphone usage and depression among adults: A cross sectional study. *BMC Psychiatry, 18*(1), 148. doi:10.118612888-018-1745-4 PMID:29801442

Al-Kandari, A., Melkote, S. R., & Sharif, A. (2016). Needs and Motives of Instagram Users that Predict Self-disclosure Use: A Case Study of Young Adults in Kuwait. *Journal of Creative Communications.*

Allan, S. (2014). Witnessing in crisis: Photo reportage of terror attacks in Boston and London. *Media. War & Conflict, 7*(2), 133–151. doi:10.1177/1750635214531110

Al-Mekhlafi, K., Hu, X., & Zheng, Z. (2009). An Approach to Context-Aware Mobile Chinese Language Learning for Foreign Students. *Mobile Business, 2009. ICMB 2009. Eighth International Conference on,* 340-346. 10.1109/ICMB.2009.65

Almog, R., & Kaplan, D. (2015). The nerd and his discontent: The seduction community and the logic of the game as a geeky solution to the challenges of young masculinity. *Men and Masculinities.* doi:10.1177/1097184X15613831

Amazon. (n.d.). *Duolingo Case Study.* Retrieved June 24, 2018, from https://web.archive.org/web/20170530040432/https://www.amazon.com/p/feature/x4et6o3v69rc8rd

Amrita Vishwa Vidyapeetham. (2018). *Amrita Kripa.* Retrieved November 21, 2018, from https://play.google.com: https://play.google.com/store/apps/details?id=edu.amrita.awna.floodevac&hl=en_IN

Anacleto, R., Figueiredo, L., Almeida, A., & Novais, P. (2014). Mobile application to provide personalized sightseeing tours. *Journal of Network and Computer Applications, 41,* 56–64. doi:10.1016/j.jnca.2013.10.005

Anderson, T. A., Hwang, W. Y., & Hsieh, C. H. (2008). A study of a mobile collaborative learning system for Chinese language learning. *Proceedings of International Conference on Computers in Education,* 217-222.

Andersson, A., & Grönlund, Å. (2009). A conceptual framework for eLearning in developing countries: A critical review of research challenges. *The Electronic Journal on Information Systems in Developing Countries, 38*(1), 1–16. doi:10.1002/j.1681-4835.2009.tb00271.x

Ansari, M. A., & Pandey, N. (2011). *Assessing the potential and use of mobile phones by the farmers in Uttarakhand (India): A special project report.* Pantnagar, India: G.B. Pant University of Agriculture and Technology.

Appflyer. (2018). *State of App Marketing in India Report.* Author.

Arhippainen, L., & Tähti, M. (2003, December). Empirical evaluation of user experience in two adaptive mobile application prototypes. In *MUM 2003. Proceedings of the 2nd International Conference on Mobile and Ubiquitous Multimedia* (No. 011, pp. 27-34). Linköping University Electronic Press.

Armstrong, T. (2016). *The Evolution of Mobile Translation.* Retrieved May 3, 2018, from https://blog.stepes.com/the-evolution-of-mobile-translation/

Arnáiz-Uzquiza, V., & Álvarez-Álvarez, S. (2016). El uso de dispositivos y aplicaciones móviles en el aula de traducción: Perspectiva de los estudiantes. *Revista Tradumàtica: tecnologies de la traducció, 14*, 100-111.

Ashton, K. (2009). That "Internet of Things" Thing. *RFID Journal, 22*(7), 97–114.

Attané, I., & Gu, B. (2014). China's demography in a changing society: Old problems and new challenges. In I. Attane & B. Gu (Eds.), *Analysing China's population: Social change in a new demographic era.* (p.1). Dordrecht: Springer Netherlands. Retrieved from https://journals.library.ualberta.ca/csp/index.php/csp/article/viewFile/28849/21140

Azevedo, A. &Mesquita, A. (Eds.). (2018). *International Conference on Gender Research.* Porto, Portugal: Academic Conferences and Publishing International Limited. Available from: https://books.google.com.tr/books?id=CcFWDwAAQBAJ&pg=PR6&lpg=PR6&dq=International+Conference+on+Gender+Research.+April+12-13+2018&source=bl&ots=7tbPObsaPt&sig=s8ALhMDV2vs-qB0SET2VCVHJGpE&hl=tr&sa=X&ved=0ahUKEwjc9OLChajbAhWGWSwKHe8pA-8Q6AEIbDAJ#v=onepage&q=International%20Conference%20on%20Gender%20Research.%20April%2012-13%202018&f=false

Bader, A., Baldauf, M., Leinert, S., Fleck, M., & Liebrich, A. (2012). Mobile Tourism Services and Technology Acceptance in a Mature Domestic Tourism Market: The Case of Switzerland. In M. Fuchs, F. Ricci, & L. Cantoni (Eds.), *Information and Communication Technologies in Tourism 2012* (pp. 296–307). Vienna: Springer. doi:10.1007/978-3-7091-1142-0_26

Bae, S. M. (2015). The relationships between perceived parenting style, learning motivation, friendship satisfaction, and the addictive use of smartphones with elementary school students of South Korea: Using multivariate latent growth modeling. *School Psychology International, 36*(5), 513–531. doi:10.1177/0143034315604017

Bahri, H., & Mahadi, T. S. T. (2016). The Application of Mobile Devices in the Translation Classroom. *Advances in Language and Literary Studies, 7*(6), 237–242. doi:10.7575/aiac.alls.v.7n.6p.237

Bairagi, A., Roy, T., & Polin, A. (2011). *Socio-Economic Impacts of Mobile Phone in Rural Bangladesh: A case Study in Batiaghata Thana, Khulna District.* Retrieved at www.ijcit.org/ijcit_papers/vol2no1/IJCIT-110738.pdf

Baker, A., Dede, C., & Evans, J. (2014). *The 8 Essentials for Mobile Learning Success in Education.* Retrieved 2 May 2018 from https://www.qualcomm.com/media/documents/files/the-8-essentials-for-mobile-learning-success-in-education.pdf

Baker, L. R., & Oswald, D. L. (2010). Shyness and online social networking services. *Journal of Social and Personal Relationships, 27*(7), 873–889. doi:10.1177/0265407510375261

Bakhshi, S., Shamma, D. A., & Gilbert, E. (2014, April). Faces engage us: Photos with faces attract more likes and comments on Instagram. In *Proceedings of the 32nd annual ACM conference on Human factors in computing systems* (pp. 965-974). ACM. 10.1145/2556288.2557403

Bakia, M., Shear, L., Toyama, Y., & Lasseter, A. (2012). *Understanding the Implications of Online Learning for Educational Productivity.* Washington, DC: U.S. Department of Education, Office of Educational Technology. Retrieved 8 January 2018 from https://www.sri.com/work/publications/understanding-implications-online-learning-educational-productivity

Balakrishnan, V., & Loo, H.-L. (2012). Mobile Phone and Short Message Service Appropriation, Usage and Behavioral Issues among University Students. *Journal of Social Sciences, 8*(3), 364–371. doi:10.3844/jssp.2012.364.371

Bandura, A. (1977). *Social Learning Theory.* Englewood Cliffs, NJ: Prentice Hall.

Banerjee, S., Mandal, K. S., & Dey, P. (2014, April). A Study on the Permeation and Scope of ICT Intervention at the Indian Rural Primary School Level. *CSEDU*, (2), 363-370.

Banik, S. (2017). *A Study on Financial Analysis of Rural Artisans in India: Issues and Challenges.* Academic Press.

Banks, K., & Burge, R. (2004). *Mobile Phones: An Appropriate Tool for Conservation and Development.* Cambridge, UK: Fauna & Flora International.

Bardin, L. (2008). *Análise de conteúdo.* Edições 70.

Barki, E., & Parente, J. (2010). Consumer Behaviour of the Base of the Pyramid Market in Brazil. *Greener Management International,* (56).

Baron, N. S., & Ling, R. (2007). *Emerging Patterns of American Mobile Phone Use: Electronically-mediated Communication in Transition.* Available from: http://doczine.com/bigdata/2/1366989405_23054102ae/emerging-patterns-ofamerican-mobile-phone-use-3.pdf

Baron, R. M., & Kenny, D. A. (1986). The moderator-mediator variable distinction in social psychological research: Conceptual, strategic, and statistical considerations. *Journal of Personality and Social Psychology, 51*(6), 1173–1182. doi:10.1037/0022-3514.51.6.1173 PMID:3806354

Barreneche, C. (2012). Governing the geocoded world: Environmentality and the politics of location platforms. *Convergence (London), 18*(3), 331–351. doi:10.1177/1354856512442764

Battard, N., & Mangematin, V. (2013). Idiosyncratic distances: Impact of mobile technology practices on role segmentation and integration. *Technological Forecasting and Social Change, 80*(2), 231–242. doi:10.1016/j.techfore.2011.11.007

Baumrind, D. (1966). Effects of authoritative parental control on child behaviour. *Child Development, 37*(4), 887–907. doi:10.2307/1126611

Bayes, A., von Braun, J., & Akhter, R. (1999). *Village pay phones and poverty reduction: Insights from a Grameen Bank initiative in Bangladesh* (ZEF Discussion Papers on Development Policy No. 8). TeleCommons Development Group. Retrieved at http://www.telecommons.com/villagephone/ Bayes99.pdf

Beck, U., & Beck-Gernsheim, E. (2002). *Individualization: Institutionalized individualism and its social and political consequences.* London: Sage.

Belair-Gagnon, V. (2015). Book review: Mette Mortensen, Journalism and Eyewitness Images: Digital Media, Participation, and Conflict, Routledge: London, 2015, 182. Media, War & Conflict, 8(3), 384–386.

Bennett, D. (2013). Exploring the impact of an evolving war and terror blogosphere on traditional media coverage of conflict. *Media. War & Conflict, 6*(1), 37–53. doi:10.1177/1750635212469907

Ben-Ze'ev, A. (2004). *Love online: Emotions on the Internet.* Cambridge, UK: Cambridge University Press. doi:10.1017/CBO9780511489785

Ben-Zeev, D., Schueller, S. M., Begale, M., Duffecy, J., Kane, J. M., & Mohr, D. C. (2015). Strategies for mHealth research: Lessons from 3 mobile intervention studies. *Administration and Policy in Mental Health, 42*(2), 157–167. doi:10.100710488-014-0556-2 PMID:24824311

Berger, G. (2011). Empowering the youth as citizen journalists: A South African experience. *Journalism, 12*(6), 708–726. doi:10.1177/1464884911405466

Bethell, P. (2010). Journalism students' experience of mobile phone technology: implications for journalism education. *Asia Pacific Media Educator,* (20), 103.

Bhavnani, A., Chiu, R. W. W., Janakiram, S., Silarszky, P., & Bhatia, D. (2008). *The role of mobile phones in sustainable rural poverty reduction.* Academic Press.

Bhuasiri, W., Xaymoungkhoun, O., Zo, H., Rho, J. J., & Ciganek, A. P. (2012). Critical success factors for e-learning in developing countries: A comparative analysis between ICT experts and faculty. *Computers & Education, 58*(2), 843–855. doi:10.1016/j.compedu.2011.10.010

Bianchi, A., & Phillips, J. G. (2005). Psychological predictors of problem mobile phone use. *Cyberpsychology & Behavior, 8*(1), 39–51. doi:10.1089/cpb.2005.8.39 PMID:15738692

Bianchi, A., & Phillips, J. G. (2005). Psychological predictors of problem mobile phone use. *CyberPsychology and Behavior: The Impact of the Internet. Multimedia and Virtual Reality on Behavior and Society, 8*(1), 39–51.

Bidarra, J., Figueiredo, M., & Natálio, C. (2015). Interactive design and gamification of ebooks for mobile and contextual learning. *International Journal of Interactive Mobile Technologies*, 24-32.

Bijari, B., Javadinia, S. A., Erfanian, M., Abedini, M., & Abassi, A. (2013). The Impact of Virtual Social Networks on Students' Academic Achievement in Birjand University of Medical Sciences in East Iran. *Procedia: Social and Behavioral Sciences, 83*, 103–106. doi:10.1016/j.sbspro.2013.06.020

Billieux, J., Linden, M. V. D., & Rochat, L. (2008). The role of impulsivity in actual and problematic use of the mobile phone. *Applied Cognitive Psychology, 22*(9), 1195–1210. doi:10.1002/acp.1429

Bivens, R. K. (2008). The internet, mobile phones and blogging. *Journalism Practice, 2*(1), 113–129. doi:10.1080/17512780701768568

Bjørnestad, S., Tessem, B., & Nyre, L. (2011, February). Design and evaluation of a location-based mobile news reader. In *New Technologies, Mobility and Security (NTMS), 2011 4th IFIP International Conference on* (pp. 1-4). IEEE. 10.1109/NTMS.2011.5720634

Bolle, C. (2014). *Who is a Smartphone addict? The impact of personal factors and type of usage on Smartphone addiction in a Dutch population* (Unpublished Master's Thesis). University of Twente, Enschede, The Netherlands.

Boniel-Nissim, M., Tabak, I., Mazur, J., Borraccino, A., Brooks, F., Gommans, R., ... Finne, E. (2015). Supportive communication with parents moderates the negative effects of electronic media use on life satisfaction during adolescence. *International Journal of Public Health, 60*(2), 189–198. doi:10.100700038-014-0636-9 PMID:25549611

Bonsignore, E., Quinn, A. J., Druin, A., & Bederson, B. B. (2013). Sharing stories "in the wild": A mobile storytelling case study using StoryKit. *ACM Transactions on Computer-Human Interaction, 20*(3), 18. doi:10.1145/2491500.2491506

Boston Consulting Group. (2011). *The Socio-Economic Impact of Mobile Financial Services: Analysis of Pakistan, India, Bangladesh, Serbia and Malaysia.* Retrieved at https://www.telenor.com/wp-content/uploads/2012/03/The-Socio-Economic-Impact-of-Mobile-Financial-Services-BCG-Telenor-Group-2011.pdf

Boswijk, A., Thijssen, T., & Peelen, E. (2006). *A new perspective on the experience economy. Bilthovenm The Netherlands.* The European Centre for the Experience Economy.

Bothun, D., & Lieberman, M. (2017). *Smart home, seamless life: Unlocking a culture of convenience.* Academic Press.

Botnen, E. O., Bendixen, M., Grøntvedt, T. V., & Kennair, L. E. O. (2018). Individual differences in sociosexuality predict picture-based mobile dating app use. *Personality and Individual Differences, 131*, 67–73. doi:10.1016/j.paid.2018.04.021

Bourdieu, P. (2005). *The social structures of the economy.* London: Polity Press.

Bowlby, J. (1969). *Attachment and loss* (Vol. 1). Attachment.

Bracken, C. C., Jeffres, L. W., Neuendorf, K. A., Kopfman, J., & Moulla, F. (2005). How cosmopolites react to messages: America under attack. *Communication Research Reports, 22*(1), 47–58. doi:10.1080/0882409052000343516

Brandenburg, C., Worrall, L., Rodriguez, A. D., & Copland, D. (2013). Mobile computing technology and aphasia: An integrated review of accessibility and potential uses. *Aphasiology*, *27*(4), 444–461. doi:10.1080/02687038.2013.772293

Brashers, D. E. (2001). Communication and uncertainty management. *Journal of Communication*, *51*(3), 477–497. doi:10.1111/j.1460-2466.2001.tb02892.x

Brashers, D. E., & Hogan, T. P. (2013). The appraisal and management of uncertainty: Implications for information-retrieval systems. *Information Processing & Management*, *49*(6), 1241–1249. doi:10.1016/j.ipm.2013.06.002

Brashers, D. E., Neidig, J. L., Haas, S. M., Dobbs, L. K., Cardillo, L. W., & Russell, J. A. (2000). Communication in the management of uncertainty: The case of persons living with HIV or AIDS. *Communication Monographs*, *67*(1), 63–84. doi:10.1080/03637750009376495

British Broadcasting Corporation. (2011). *What is Ashura?* Retrieved at www.bbc.com/news/world-middle-east-16047713

Brodie, M., Flournoy, R. E., Altman, D. E., Blendon, R. J., Benson, J. M., & Rosenbaum, M. D. (2000). Health information, the Internet, and the digital divide. *Health Affairs*, *19*(6), 255–265. doi:10.1377/hlthaff.19.6.255 PMID:11192412

Bronfenbrenner, U. (1979). *The ecology of human development: experiments by nature and design*. Cambridge, MA: Harvard University Press.

Bröns, P., Greifeneder, E., & Støvring, S. (2013). How Facebook Promotes Students' Academic Life. *Zeitschrift für Bibliothekskultur*, *3*, 116-126.

Brown, B., & Chalmers, M. (2003). Tourism and mobile technology. In *Proceedings of the eighth conference on European Conference on Computer Supported Cooperative Work* (pp 335-354). Helsinki: Kluwer Academic Publishers.

Brubaker, J. R., Ananny, M., & Crawford, K. (2016). Departing glances: A sociotechnical account of 'leaving' Grindr. *New Media & Society*, *18*(3), 373–390. doi:10.1177/1461444814542311

Brundage, M., Avin, S., Clark, J., Toner, H., Eckersley, P., Garfinkel, B., … Amodei, D. (2018). *The Malicious Use of Artificial Intelligence : Forecasting, Prevention, and Mitigation*. Retrieved June 15, 2018, from https://maliciousaireport.com

Buhalis, D., & Law, R. (2008). Progress in information technology and tourism management: 20 years on and 10 years after the internet- The state of eTourism research. *Tourism Management*, *29*(4), 609–623. doi:10.1016/j.tourman.2008.01.005

Buijink, A., Visser, B. J., & Marshall, L. (2013). Medical apps for smartphones: Lack of evidence undermines quality and safety. *Evidence-Based Medicine*, *18*(3), 90–92. doi:10.1136/eb-2012-100885 PMID:22923708

Burt, R. S. (1995). *Structural Holes: The Social Structure of Competition*. Cambridge, MA: Harvard University Press.

Buss, A. H. (1980). Shyness and sociability. *Journal of Personality and Social Psychology*, *41*, 330–339.

Buss, D. M. (1998). Sexual strategies theory: Historical origins and current status. *Journal of Sex Research*, *35*(1), 19–31. doi:10.1080/00224499809551914

Buss, D. M., & Schmitt, D. P. (1993). Sexual strategies theory: An evolutionary perspective on human mating. *Psychological Review*, *100*(2), 204–232. doi:10.1037/0033-295X.100.2.204 PMID:8483982

Buss, D. M., & Schmitt, D. P. (2016). Sexual strategies theory. In T. Shackelford & V. Weekes-Shackelford (Eds.), *Encyclopedia of evolutionary psychological science*. Cham, Switzerland: Springer. doi:10.1007/978-3-319-16999-6_1861-1

Cáceres, R. B., & Fernández-Ardèvol, M. (2012). Mobile phone use among market traders at fairs in rural Peru. *Information Technologies & International Development*, *8*(3), 35.

Campbell. (2015). *The impact of the mobile phone on young people's social life*. Available from: https://www.research-gate.net/publication/27465354_The_impact_of_the_mobile_phone_on_young_people's_social_life

Campbell, W. K., & Foster, J. D. (2007). The narcissistic self: Background, an extended agency model, and ongoing controversies. *Self*, 115–138.

Carr, D. C., & Komp, K. (2011). *Gerontology in the era of the third age: Implications and next steps*. New York, NY: Springer Publishing Company.

Caruth, G. D. (2013). Demystifying Mixed Methods Research Design: A Review of the Literature. *Mevlana International Journal of Education*, *3*(2), 112–122. doi:10.13054/mije.13.35.3.2

Casey, B. M. (2012). *Linking Psychological Attributes to Smartphone Addiction, Face-to-Face Communication, Present Absence and Social Capital* (Unpublished Master's thesis). The Chinese University of Hong Kong, Hong Kong, China.

Cecchini, S., & Scott, C. (2003). Can information and communications technology applications contribute to poverty reduction? Lessons from rural India. *Information Technology for Development*, *10*(2), 73–84. doi:10.1002/itdj.1590100203

Cha, S. S., & Seo, B. K. (2018). Smartphone use and smartphone addiction in middle school students in Korea: Prevalence, social networking service, and game use. *Health Psychology Open, 5*(1).

Chai, C. S., Wong, L. H., & King, R. B. (2016). Surveying and Modeling Students' Motivation and Learning Strategies for Mobile-Assisted Seamless Chinese Language Learning. *Journal of Educational Technology & Society*, *19*(3), 170–180.

Chan, K., Wong, A., Tam, E., & Tse, M. (2014). The use of smart phones and their mobile applications among older adults in Hong Kong: an exploratory study. *GSTF Journal of Nursing and Health Care*. doi:10.5176/2345-718X_1.2.45

Chang, C.-K., & Hsu, C.-K. (2011). A mobile-assisted synchronously collaborative translation–annotation system for English as a foreign language (EFL) reading comprehension. *Computer Assisted Language Learning*, *24*(2), 155–180. doi:10.1080/09588221.2010.536952

Chang, D. T. S., Thyer, I. A., Hayne, D., & Katz, D. J. (2014). Using mobile technology to overcome language barriers in medicine. *Annals of the Royal College of Surgeons of England*, *96*(6), e23–e25. doi:10.1308/003588414X13946184903685 PMID:25198966

Chang, F. C., Chiu, C. H., Miao, N. F., Chen, P. H., Lee, C. M., Chiang, J. T., & Pan, Y.-C. (2015). The relationship between parental mediation and Internet addiction among adolescents, and the association with cyberbullying and depression. *Comprehensive Psychiatry*, *57*, 21–28. doi:10.1016/j.comppsych.2014.11.013 PMID:25487108

Chang, H. Y., Wu, H. K., & Hsu, Y. S. (2013). Integrating a mobile augmented reality activity to contextualize student learning of a socioscientific issue. *British Journal of Educational Technology*, *44*(3), E95–E99. doi:10.1111/j.1467-8535.2012.01379.x

Chan, L. S. (2018). Ambivalence in networked intimacy: Observations from gay men using mobile dating apps. *New Media & Society*, *20*(7), 2566–2581. doi:10.1177/1461444817727156

Chan, M. (2011). Shyness, sociability, and the role of media synchronicity in the use of computer-mediated communication for interpersonal communication. *Asian Journal of Social Psychology*, *14*(1), 84–90.

Chan, M. (2015). Examining the influences of news use patterns, motivations, and age cohort on mobile news use: The case of Hong Kong. *Mobile Media & Communication*, *3*(2), 179–195. doi:10.1177/2050157914550663

Chan, M. (2015). Mobile phones and the good life: Examining the relationships among mobile use, social capital and subjective well-being. *New Media & Society*, *17*(1), 96–113. doi:10.1177/1461444813516836

Cheawjindakarn, B., Suwannatthachote, P., & Theeraroungchaisri, A. (2012). Critical Success Factors for Online Distance Learning in Higher Education: A Review of the Literature. *Creative Education, 03*(08), 61–66. doi:10.4236/ce.2012.38B014

Chen, C. H., & Chou, H. W. (2007). Location-aware technology in Chinese language learning. *IADIS International Conference on Mobile Learning.*

Chen, C., & Leung, L. (2015). Are you addicted to Candy Crush Saga? An exploratory study linking psychological factors to mobile social game addiction. *Telematics and Informatics.*

Cheng, J. F. (2014). *Research on business models of mobile social media* (Master Thesis). Northwestern University.

Chen, G. M. (2011). Tweet this: A uses and gratifications perspective on how active Twitter use gratifies a need to connect with others. *Computers in Human Behavior, 27*(2), 755–762. doi:10.1016/j.chb.2010.10.023

Cheon, J., Lee, S., Crooks, S. M., & Song, J. (2012). An investigation of mobile learning readiness in higher education based on the theory of planned behavior. *Computers & Education, 59*(3), 1054–1064. doi:10.1016/j.compedu.2012.04.015

Chhachhar, R., Chen, C., & Jin, J. (2017). Performance and Efforts Regarding Usage of Mobile Phones among Farmers for Agricultural Knowledge. *Asian Social Science, 13*(8), 1–11. doi:10.5539/ass.v13n8p1

China Internet Network Information Center. (2018). Statistical Report on Internet Development in China of 2018 (42). Retrieved from http://cac.gov.cn/wxb_pdf/CNNIC42.pdf

Chipere, N. (2016). A framework for developing sustainable e-learning programmes. *Open Learning: The Journal of Open. Distance and E-Learning, 32*(1), 36–55. doi:10.1080/02680513.2016.1270198

Choi, M. K., Chan, K. B., & Chan, K. (2013). *Online dating as a strategic game: why and how men in Hong Kong use QQ to chase women in Mainland China.* Retrieved from http://ebookcentral.proquest.com

Chóliz, M. (2012). Mobile-phone addiction in adolescence: The Test of Mobile Phone Dependence (TMD). *Progress in Health Sciences, 2*(1), 33–44.

Chong, A. Y. L., Chan, F. T., & Ooi, K. B. (2012). Predicting consumer decisions to adopt mobile commerce: Cross country empirical examination between China and Malaysia. *Decision Support Systems, 53*(1), 34–43. doi:10.1016/j.dss.2011.12.001

Chowdhury, M. (2015). Socio-economic penetration of mobile penetration in SAARC countries with special emphasis on Bangladesh. *Asian Business Review, 5*(2), 66–71. doi:10.18034/abr.v5i2.56

Christensen, T. P., Flanagan, M., & Schjoldager, A. (2017). Mapping Translation Technology Research in Translation Studies. An Introduction to the Thematic Section. *HERMES-Journal of Language and Communication in Business,* (56), 7-20.

Chua, W. Y., Wan, M. P. H., Chang, K., & Yi, W. (2014). *Improving Mobile Applications Usage Experience of Novice Users through User-Acclimatized Interaction: A Case Study.* Academic Press.

Chuma, W. (2014). The social meanings of mobile phones among South Africa's 'digital natives': A case study. *Media Culture & Society, 36*(3), 398–408. doi:10.1177/0163443713517482

Chumskey, K., & Hjorth, L. (2013). *Mobile media practices, presence and politics. The challenge of being seamlessly mobile.* New York: Routledge.

Ciampa, K. (2014). Learning in a mobile age: An investigation of student motivation. *Journal of Computer Assisted Learning, 30*(1), 82–96. doi:10.1111/jcal.12036

Clark, J. (2015). Mobile dating apps could be driving HIV epidemic among adolescents in Asia Pacific, report says. *BMJ: British Medical Journal, 351*. https://doi-org.gate.lib.buffalo.edu/10.1136/bmj.h6493

Clarke, J., & Bromley, M. (Eds.). (2012). *International news in the digital age: East-west perceptions of a new world order* (Vol. 4). Routledge. doi:10.4324/9780203804674

CNNIC. (2018). *China Social Media Overview' released by CIC*. Available at: http://www.cac.gov.cn/2018zt/cnnic41/index.htm

Cochrane, T. D. (2012). Critical success factors for transforming pedagogy with mobile Web 2.0. *British Journal of Educational Technology, 45*(1), 65-82. doi:10.1111/j.14678535.2012.01384.x

Cohen, N. (2001). *What Works: Grameen Telecom's Village Phones?* A Digital dividend Study by The World Resources Institute. Retrieved at http://www.digitaldividend.org/pdf/grameen.pdf

Cohen, A., & Ezra, O. (2018). Development of a contextualised MALL research framework based on L2 Chinese empirical study. *Computer Assisted Language Learning*, 1–26. doi:10.1080/09588221.2018.1527359

Coleman, J. S. (1988). Supplement: Organizations and Institutions: Sociological and Economic Approaches to the Analysis of Social Structure. *American Journal of Sociology, 94*, S95–S120. doi:10.1086/228943

Collins, R. (2017, February). A Glimpse into the Future of Work. *Huffpost*. Retrieved from https://www.huffingtonpost.com/entry/a-glimpse-into-the-future-of-work_us_5893effee4b061551b3dfd33

Comas-Quinn, A., & Mardomingo, R. (2012). Language learning on the move: a review of mobile blogging tasks and their potential. Innovation and Leadership in English Language Teaching, 6(6), 47–65.

Confucius Institute. (2017). *Confucius Institute Annual Development Report*. Retrieved from http://www.hanban.edu.cn/report/2017.pdf

Constantinides, M. (2015, April). Apps with habits: Adaptive Interfaces for News Apps. In *Proceedings of the 33rd Annual ACM Conference Extended Abstracts on Human Factors in Computing Systems* (pp. 191-194). ACM. 10.1145/2702613.2702622

Constantinides, M., Dowell, J., Johnson, D., & Malacria, S. (2015). *Exploring mobile news reading interactions for news app personalization. 17th International Conference on Human-Computer Interaction with Mobile Devices and Services*, Copenhagen, Denmark.

Corriero, E. F., & Tong, S. T. (2016). Managing uncertainty in mobile dating applications: Goals, concerns of use, and information seeking in Grindr. *Mobile Media & Communication, 4*(1), 121–141. doi:10.1177/2050157915614872

Couch, D., & Liamputtong, P. (2007). Online dating and mating: Perceptions of risk and health among online users. *Health Risk & Society, 9*(3), 275–294. doi:10.1080/13698570701488936

Cox, A. M., & Blake, M. K. (2011, March). Information and food blogging as serious leisure. In P. Willett (Ed.), ASLIB proceedings (Vol. 63, No. 2/3, pp. 204-220). Emerald Group Publishing Limited. doi:10.1108/00012531111135664

Coyne, S. M., Padilla-Walker, L. M., & Holmgren, H. G. (2018). A Six-Year Longitudinal Study of Texting Trajectories During Adolescence. *Child Development, 89*(1), 58–65. doi:10.1111/cdev.12823 PMID:28478654

Craig, R. T. (2007). Issue Forum Introduction: Mobile Media and Communication: What are the Important Questions? *Communication Monographs, 74*(3), 386–388. doi:10.1080/03637750701543501

Creech, B. (2015). Disciplines of truth: The 'Arab Spring', American journalistic practice, and the production of public knowledge. *Journalism, 16*(8), 1010–1026. doi:10.1177/1464884914550971

Crompton, H., Burke, D., & Gregory, K. H. (2017). The use of mobile learning in PK-12 education: A systematic review. *Computers & Education, 110*, 51–63. doi:10.1016/j.compedu.2017.03.013

Crone, E. A., & Konijn, E. A. (2018). Media use and brain development during adolescence. *Nature Communications, 9*(1), 588. doi:10.103841467-018-03126-x PMID:29467362

Cronholm, S., & Hjalmarsson, A. (2011). Experiences from sequential use of mixed methods. *Electronic Journal of Business Research Methods, 9*(2), 87–95.

Cronin, M. (2010). The Translation Crowd. *Revista Tradumàtica: tecnologies de la traducció, 8,* 1–7. Retrieved from http://revistes.uab.cat/tradumatica/article/view/100/pdf_15

Cronin, M. (2013). *Translation in the Digital Age*. Abingdon, UK: Routledge.

Cronin, M. (2017). Response by Cronin to Translation and the materialities of communication. *Translation Studies, 10*(1), 92–96. doi:10.1080/14781700.2016.1243287

Cross, J., & Street, A. (2009). Anthropology at the bottom of the pyramid. *Anthropology Today, 25*(4), 4–9. doi:10.1111/j.1467-8322.2009.00675.x

Czaja, S. J., & Lee, C. C. (2006). The impact of aging on access to technology. *Universal Access in the Information Society, 5*(4), 341–349. doi:10.100710209-006-0060-x

Daffalla, A., & Dimetry, D. A. (2014). The Impact of Facebook and Others Social Networks Usage on Academic Performance and Social Life among Medical Students at Khartoum. *International Journal of Scientific & Technology Research, 3*(5), 3–8.

Dafoe, A., & Lyall, J. (2015). From cell phones to conflict? Reflections on the emerging ICT–political conflict research agenda. *Journal of Peace Research, 52*(3), 401–413. doi:10.1177/0022343314563653

Dale Storie, M. L. I. S. (2014). Mobile devices in medicine: A survey of how medical students, residents, and faculty use smartphones and other mobile devices to find information*. *Journal of the Medical Library Association: JMLA, 102*(1), 22–30. doi:10.3163/1536-5050.102.1.006 PMID:24415916

Darmanto, H. Y., & Hermawan, B. (2016). Mobile learning application to support Mandarin language learning for high school student. *Imperial Journal of Interdisciplinary Research, 2*(4), 402–407.

Datta, D., & Mitra, S. (2010). *M-learning: mobile-enabled educational technology*. Innovating.

Davenport, S. W., Bergman, S. M., Bergman, J. Z., & Fearrington, M. E. (2014). Twitter versus Facebook: Exploring the role of narcissism in the motives and usage of different social media platforms. *Computers in Human Behavior, 32*, 212–220. doi:10.1016/j.chb.2013.12.011

Davis, F. D., Jr. (1986). *A technology acceptance model for empirically testing new end-user information systems: Theory and results* (Doctoral dissertation). Massachusetts Institute of Technology.

Davis, F. D. (1989). Perceived usefulness, perceived ease of use, and user acceptance of information technology. *Management Information Systems Quarterly, 13*(3), 319–340. doi:10.2307/249008

De Silva, H., & Ratnadiwakara, D. (2008). *Using ICT to reduce transaction costs in agriculture through better communication: A case-study from Sri Lanka*. LIRNEasia, Colombo, Sri Lanka.

De Silva, H., & Ratnadiwakara, D. (2008). *Using ICT to reduce transaction costs in agriculture through better communication: A case-study from Sri Lanka*. Retrieved at http://lirneasia.net/ wp-content/uploads/2008/11/transactioncosts.pdf

De Silva, H., Ratnadiwakara, D., & Zainudeen, A. (2009). Social influence in mobile phone adoption: Evidence from the bottom of pyramid in emerging. *Asia*. doi:10.2139srn.1564091

Dean, S., & Illowsky, B. (2010). *Sampling and Data: Frequency, Relative Frequency and Cumulative Frequency*. Retrieved at https://www.saylor.org/site/wp-content/uploads/2011/06/MA121-1.1.3-3rd.pdf

Deloitte. (2017). *The Place of Mobile Technologies in Our Digitalized Life, Deloitte Global Mobile User Survey* [Dijitalleşen Hayatımızda Mobil Teknolojilerin Yeri, Deloitte Global Mobil Kullanıcı Anketi]. Available from: https://www2.deloitte.com/content/dam/Deloitte/tr/Documents/technology-mediatelecommunications/deloitte_gmcs_2017.pdf

Demir, N., & Cakır, F. (2014). *Research on determining university students' smartphone purchasing preferences* [Üniversite öğrencilerinin akıllı telefon satın alma tercihlerini belirlemeye yönelik bir araştırma]. Available from: https://www.researchgate.net/publication/304624501_Universite_Ogrencilerinin_Akilli_Telefon_Satin_Alma_Tercihlerini_Belirlemeye_Yonelik_Bir_Arastirma

Demirci, K., Orhan, H., Demirdas, A., Akpinar, A., & Sert, H. (2014). Validity and reliability of the Turkish Version of the Smartphone Addiction Scale in a younger population. *Bulletin of Clinical Psychopharmacology*, *24*(3), 226–235. doi:10.5455/bcp.20140710040824

Deng, Z., Mo, X., & Liu, S. (2014). Comparison of the middle-aged and older users' adoption of mobile health services in China. *International Journal of Medical Informatics*, *83*(3), 210–224. doi:10.1016/j.ijmedinf.2013.12.002 PMID:24388129

Dhingra, V., Mudgal, R. K., & Dhingra, M. (2017). Safe and Healthy Work Environment: A Study of Artisans of Indian Metalware Handicraft Industry. *Management and Labour Studies*, *42*(2), 152–166. doi:10.1177/0258042X17714071

Dhir, A., Pallesen, S., Torsheim, T., & Andreassen, C. S. (2016). Do age and gender differences exist in selfie-related behaviours? *Computers in Human Behavior*, *63*, 549–555. doi:10.1016/j.chb.2016.05.053

Dhir, A., & Tsai, C. C. (2017). Understanding the relationship between intensity and gratifications of Facebook use among adolescents and young adults. *Telematics and Informatics*, *34*(4), 350–364. doi:10.1016/j.tele.2016.08.017

Diary, O. (2018). *Youth prefers mobile dating apps over matrimonial sites/ads*. Retrieved from http://link.galegroup.com.gate.lib.buffalo.edu/apps/doc/A527756418/STND?u=sunybuff_main&sid=STND&xid=da2f1060

Dickinson, J. E., Ghali, K., Cherrett, T., Speed, C., Davies, N., & Norgate, S. (2014). Tourism and the smartphone app: Capabilities, emerging practice and scope in the travel domain. *Current Issues in Tourism*, *17*(1), 84–101. doi:10.1080/13683500.2012.718323

Dickinson, J. E., Hibbert, J. F., & Filimonau, V. (2016). Mobile technology and the tourist experience: (Dis) connection at the campsite. *Tourism Management*, *57*, 193–201. doi:10.1016/j.tourman.2016.06.005

Digital Education Action Plan. (2018). *Education and training*. European Commission. Retrieved from https://ec.europa.eu/education/initiatives/european-education-area/digital-education-action-plan_en

Dinh, H. T., Lee, C., Niyato, D., & Wang, P. (2013). A survey of mobile cloud computing: Architecture, applications, and approaches. *Wireless Communications and Mobile Computing*, *13*(18), 1587–1611. doi:10.1002/wcm.1203

Division for Sustainable Development Goals. (2015). *17 Sustainable Development Goals*. Retrieved from https://sustainabledevelopment.un.org/sdgs

Djamasbi, S., McAuliffe, D., Gomez, W., Kardzhaliyski, G., Liu, W., & Oglesby, F. (2014, June). Designing for success: Creating business value with mobile user experience (UX). In *International conference on HCI in Business* (pp. 299-306). Springer. 10.1007/978-3-319-07293-7_29

Dong, X. M., & Li, F. Y. (2011). Research on rural private lending on structural hole theory. *South China Finance, 8*, 40–43.

Donnar, J. (2006). The use of mobile phones by microentrepreneurs in Kigali, Rwanda: Changes to social and business networks. *MIT Information Technologies and International Development, 3*(2), 3–19. doi:10.1162/itid.2007.3.2.3

Donner, J. (2008). Research approaches to mobile use in the developing world: a review of the literature. *The Information Society, 24*(3), 140–159. doi:10.1080/01972240802019970

Dredge, S. (2015, May 7). Research says 30% of Tinder users are married. *The Guardian.*

Drew, B., & Waters, J. (1986). Video games: Utilization of a novel strategy to improve perceptual motor skills and cognitive functioning in the non-institutionalized elderly. *Cognitive Rehabilitation, 4*(2), 26–31.

Dugoua, E., & Urpelainen, J. (2014). Relative deprivation and energy poverty: When does unequal access to electricity cause dissatisfaction? *International Journal of Energy Research, 38*(13), 1727–1740. doi:10.1002/er.3200

Duman, G., Orhon, G., & Gedik, N. (2015). Research trends in mobile assisted language learning from 2000 to 2012. *ReCALL, 27*(02), 197–216. doi:10.1017/S0958344014000287

Edge, D., Searle, E., Chiu, K., Zhao, J., & Landay, J. A. (2011). MicroMandarin: mobile language learning in context. *Proceedings of the SIGCHI Conference on Human Factors in Computing Systems*, 3169-3178.

ELearning Market Trends and Forecast 2017-2021. (2016). *Docebo.* Retrieved 1 May 2018 from https://eclass.teicrete.gr/modules/document/file.php/TP271/Additional material/docebo-elearning-trends-report-2017.pdf

Elhai, J. D., & Contractor, A. A. (2018). Examining latent classes of smartphone users: Relations with psychopathology and problematic smartphone use. *Computers in Human Behavior, 82*, 159–166. doi:10.1016/j.chb.2018.01.010

Ellison, N. B., Hancock, J. T., & Toma, C. L. (2012). Profile as promise: A framework for con- ceptualizing veracity in online dating self-presentations. *New Media & Society, 14*(1), 45–62. doi:10.1177/1461444811410395

Ellison, N., Heino, R., & Gibbs, J. (2006). Managing impressions online: Self presentation processes in the online dating environment. *Journal of Computer-Mediated Communication, 11*(2), 415–441. doi:10.1111/j.1083-6101.2006.00020.x

Ellul, J. (1964). *The Technological Society.* New York: Vintage Books.

Emily, M. (2018, Aug. 8). Tinder Sends Match Earnings Blazing Past Estimates. *Bloomberg News.*

Eng, D. S., & Lee, J. M. (2013). The promise and peril of mobile health applications for diabetes and endocrinology. *Pediatric Diabetes, 14*(4), 231–238. doi:10.1111/pedi.12034 PMID:23627878

Eriksson, N., & Strandvik, P. (2009). Possible determinants affecting the use of mobile tourism services. In *International Conference on E-Business and Telecommunications* (pp. 61-73). Springer. 10.1007/978-3-642-05197-5_4

Ernst & Young LLP. (2018). Re-imagining India's M&E Sector. Kolkata: Ernst & Young LLP.

Espada, J. P., Díaz, V. G., Crespo, R. G., Martínez, O. S., G-Bustelo, B. C. P., & Lovelle, J. M. C. (2015). Using extended web technologies to develop Bluetooth multi-platform mobile applications for interact with smart things. *Information Fusion, 21*, 30–41. doi:10.1016/j.inffus.2013.04.008

Esser, F., & Hanitzsch, T. (2012). *The Handbook of Comparative Communication Research.* New York: Routledge.

Eubanks, J., Yeh, H., & Tseng, H. (2018). Learning Chinese through a twenty-first century writing workshop with the integration of mobile technology in a language immersion elementary school. *Computer Assisted Language Learning, 31*(4), 346–366. doi:10.1080/09588221.2017.1399911

Eune, J., & Lee, K. P. 2009. Analysis on Intercultural Differences through User Experiences of Mobile Phone for Glocalization. *Proceedings of International Association of Societies*, 1215–26.

European Commission. (2014). *The 2015 Ageing Report: Underlying Assumptions and Projection Methodologies*. Retrieved from http://ec.europa.eu/economy_finance/publications/european_economy/2014/pdf/ee8_en.pdf

Ezra & Cohen. (2018). Contextualised MALL: L2 Chinese students in target and non-target country. *Computers & Education*, *125*, 158-174.

Farnden, J., Martini, B., & Choo, K. K. R. (2015). *Privacy risks in mobile dating apps*. arXiv preprint arXiv:1505.02906

Fei, X. T. (2006). *Local China*. Shanghai People's Publishing House.

Felson, M., & Spaeth, J. L. (1978). Community Structure and Collaborative Consumption: A Routine Activity Approach. *The American Behavioral Scientist*, *21*(4), 614–624. doi:10.1177/000276427802100411

Feng, L. (2004). Social scene: Psychological field of the communication subject. *China Communication Forum*.

Ferdiana, R., & Hantono, B. S. (2014). *Mobile tourism services model: A contextual tourism experience using mobile services*. 2014 6th International Conference on Information Technology and Electrical Engineering (ICITEE), Yogyakarta.

Fidalgo, A. (2009). PUSHED NEWS: When the news comes to the cellphone. *Brazilian Journalism Research*, *5*(2), 113–124. doi:10.25200/BJR.v5n2.2009.214

Finkel, E. J., Eastwick, P. W., Karney, B. R., Reis, H. T., & Sprecher, S. (2012). Online dating: A critical analysis from the perspective of psychological science. *Psychological Science in the Public Interest*, *13*(1), 3–66. doi:10.1177/1529100612436522 PMID:26173279

Fiordelli, M., Diviani, N., & Schulz, P. J. (2013). Mapping mHealth research: A decade of evolution. *Journal of Medical Internet Research*, *15*(5), e95. doi:10.2196/jmir.2430 PMID:23697600

Fishbein, M., & Ajzen, I. (1975). *Belief, attitude, intention, and behavior: An introduction to theory and research*. Reading, MA: Addison-Wesley.

Flick, U. (2000). Triangulation in Qualitative Research. In U. E. Flick, E. V. Kardoff, & I. Steinke (Eds.), *A Companion to Qualitative Research* (pp. 178–183). London: Sage Publications.

Flora, H. K., Wang, X., & Chande, S. V. (2014). An investigation on the characteristics of mobile applications: A survey study. *International Journal of Information Technology and Computer Science*, *6*(11), 21–27. doi:10.5815/ijitcs.2014.11.03

Forgays, D., Hyman, I., & Schreiber, J. (2013). Texting everywhere for everything: Gender and age differences in cell phone etiquette and use. *Computers in Human Behavior*, *31*, 314–321. doi:10.1016/j.chb.2013.10.053

Forlizzi, J., & Ford, S. 2000. Building Blocks of Experience: An Early Framework for Interaction Designers. In *Proceedings of Designing Interactive Systems Conference (DIS 2000)* (pp. 419–23). Brooklyn, NY: Association for Computing Machinery. 10.1145/347642.347800

Fortunati, L., & Taipale, S. (2014). The advanced use of mobile phones in five European countries. *The British Journal of Sociology*, *65*(2), 317–337. doi:10.1111/1468-4446.12075 PMID:24697752

Foster, C., & Heeks, R. (2014). Nurturing user–producer interaction: Inclusive innovation flows in a low-income mobile phone market. *Innovation and Development*, *4*(2), 221–237. doi:10.1080/2157930X.2014.921353

Foucault, M. (2012). *Discipline and Punishment: the birth of prison*. Life, Reading and New Knowledge SanLian Bookstore Press.

Foulkes, L., & Blakemore, S. J. (2018). Studying individual differences in human adolescent brain development. *Nature Neuroscience*, 1. PMID:29403031

Fox, S., & Duggan, M. (2012). *Mobile health 2012*. Washington, DC: Pew Internet & American Life Project.

Fredriksen-Goldsen, K. I., Kim, H.-J., Shiu, C., Goldsen, J., & Emlet, C. A. (2014). Successful aging among LGBT older adults: Physical and mental health-related quality of life by age group. *The Gerontologist*, *55*(1), 154–168. doi:10.1093/geront/gnu081 PMID:25213483

Frempong, G., Essegbey, G. O., & Tetteh, E. O. (2007). *Survey on the use of mobile telephones for micro and small business development: The case of Ghana*. Retrieved at https://www.idrc.ca/en/project/survey-use-mobile-telephone-micro-and-small-business-development-ghana

Frost & Sullivan. (2006). *Social impact of mobile telephony in Latin America*. Retrieved at http://www.gsmlaa.org/ªles/content/0/94/Social%20Impact%20of%20Mobile%20Telephony%20in%20Latin%20America.pdf

Fukuda, K., Asai, H., & Nagami, K. (2015, October). Tracking the evolution and diversity in network usage of smartphones. In *Proceedings of the 2015 Internet Measurement Conference* (pp. 253-266). ACM. doi:10.1145/2815675.2815697

Fullwood, C., Quinn, S., Kaye, L. K., & Redding, C. (2017). My virtual friend: A qualitative analysis of the attitudes and experiences of Smartphone users: Implications for Smartphone attachment. *Computers in Human Behavior*, *75*, 347–355. doi:10.1016/j.chb.2017.05.029

Futures without Violence. (2009). *"That's not cool" initiative background and development research*. Retrieved from http://www.thatsnotcool.com/tools/index.asp?L1.1

Galperin, H., & Mariscal, J. (2007). *Poverty and mobile telephony in Latin America and the Caribbean. Dialogo Regional sobre Sociedad de la Informacion (DIRSI)*. IDRC.

Gao, J., Bai, X., Tsai, W. T., & Uehara, T. (2014). Mobile application testing: A tutorial. *Computer*, *47*(2), 46–55. doi:10.1109/MC.2013.445

Garai, A. (2011). *Role of mHealth in rural health in India and opportunities for collaboration*. Indira Gandhi National Open University.

Gascon, H., Uellenbeck, S., Wolf, C., & Rieck, K. (2014, March). *Continuous Authentication on Mobile Devices by Analysis of Typing Motion Behavior*. Sicherheit.

Gergen, M. M., & Gergen, K. J. (2001). Positive aging: New images for a new age. *Ageing International*, *27*(1), 3–23. doi:10.100712126-001-1013-6

Gerlich, R. N., Drumheller, K., & Babb, J. (2015). App Consumption: An Exploratory Analysis of the Uses & Gratifications of Mobile Apps. *Academy of Marketing Studies Journal*, *19*(1), 69.

Gerrish, K., Chau, R., Sobowale, A., & Birks, E. (2004). Bridging the language barrier: The use of interpreters in primary care nursing. *Health & Social Care in the Community*, *12*(5), 407–413. doi:10.1111/j.1365-2524.2004.00510.x PMID:15373819

Gibbs, J. L., Ellison, N. B., & Lai, C. H. (2010). First comes love, then comes Google: An investigation of uncertainty reduction strategies and self-disclosure in online dating. *Communication Research*, *38*(1), 70–100. doi:10.1177/0093650210377091

Giddens, A., & Griffiths, S. (2006). *Sociology* (5th ed.). Cambridge, UK: Polity Press.

Gilleard, C., & Higgs, P. (2002). The third age: Class, cohort or generation? *Ageing and Society*, *22*(3), 369–382. doi:10.1017/S0144686X0200870X

Gilleard, C., & Higgs, P. (2008). The third age and the baby boomers: Two approaches to the social structuring of later life. *International Journal of Ageing and Later Life*, *2*(2), 13–30. doi:10.3384/ijal.1652-8670.072213

Gilleard, C., Higgs, P., Hyde, M., Wiggins, R., & Blane, D. (2005). Class, cohort, and consumption: The British experience of the third age. *The Journals of Gerontology. Series B, Psychological Sciences and Social Sciences*, *60*(6), S305–S310. doi:10.1093/geronb/60.6.S305 PMID:16260712

Glackin, B. C., Rodenhiser, R. W., & Herzog, B. (2014). A library and the disciplines: A collaborative project assessing the impact of eBooks and mobile devices on student learning. *Journal of Academic Librarianship*, *40*(3), 299–306. doi:10.1016/j.acalib.2014.04.007

Global E-Learning Market. (2017). *Orbis Research*. Retrieved from https://www.reuters.com/brandfeatures/venture-capital/article?id=11353

Godwin-Jones, R. (2011). Mobile apps for language learning. *Language Learning & Technology*, *15*(2), 2–11.

Goffman, E. (1959). *The presentation of self in everyday life*. New York, NY: Doubleday.

Goffman, E. (2009). The presentation of self in everyday life. *Threepenny Review*, *21*(116), 14–15.

Goggin, G. (2006). *Cell Phone Culture: Mobile Technology in Everyday Life*. London: Routledge.

Goggin, G. (2012). *Cell phone culture: Mobile technology in everyday life*. Oxfordshire, UK: Routledge.

Goggin, G., & Hjorth, L. (2014). *The Routledge companion to mobile media*. New York: Routledge. doi:10.4324/9780203434833

Goggin, G., Martin, F., & Dwyer, T. (2015). Locative news: Mobile media, place informatics, and digital news. *Journalism Studies*, *16*(1), 41–59. doi:10.1080/1461670X.2014.890329

Goh, D. H., Ang, R. P., & Lee, C. K. (2010). Determining Services for the Mobile Tourist. *Journal of Computer Information Systems*, *51*(1), 31–40.

Gohil, A., Modi, H., & Patel, S. K. (2013, March). 5G technology of mobile communication: A survey. In *Intelligent Systems and Signal Processing (ISSP), 2013 International Conference on* (pp. 288-292). IEEE.

Gokaliler, E., Aybar, A., & Gulay, G. (2011). The perception of Iphone branded smart phone as a status consumption symbol. *Selcuk Iletisim, 7*(1), 36-48. Available from: http://dergipark.gov.tr/josc/issue/19023/200589

Gökçearslan, Ş., Mumcu, F. K., Haşlaman, T., & Çevik, Y. D. (2016). Modelling smartphone addiction: The role of smartphone usage, self-regulation, general self-efficacy and cyberloafing in university students. *Computers in Human Behavior*, *63*, 639–649. doi:10.1016/j.chb.2016.05.091

Goldsmith, B. (2014). The smartphone app economy and app ecosystems. In G. Goggin & L. Hjorth (Eds.), *The Routledge Companion to Mobile Media* (pp. 171–180). New York, NY: Routledge.

Gómez-Barroso, J. L., Compañó, R., Feijóo, C., Bacigalupo, M., Westlund, O., Ramos, S., … Concepción García-Jiménez, M. (2010). *Prospects of Mobile Search*. European Commission, JRC, Institute for Prospective Technological Studies (IPTS), EUR 24148 EN 2010.

Goodman, J. (2005). *Linking mobile phone ownership and use to social capital in rural South Africa and Tanzania* (Vodafone Policy Paper Series, 2, pp. 56-65). Retrieved at https://www.vodafone.com/content/dam/vodafone/about/public_policy/policy_papers/public_policy_series_2.pdf

Goodman, D. (2008). *The new rich in China: Future rulers, present lives*. London: Routledge. doi:10.4324/9780203931172

Goyal, S., Sergi, B. S., & Kapoor, A. (2014). Understanding the key characteristics of an embedded business model for the base of the pyramid markets. *Economia e Sociologia (Evora, Portugal)*, *7*(4), 26.

Granovetter, M. S. (1973). The Strength of Weak Ties. *American Journal of Sociology*, *78*(6), 1360–1380. doi:10.1086/225469

Gray, A. (2009). Population aging and health care expenditure. *China Labor Economics*, *1*(10). Retrieved from http://en.cnki.com.cn/Article_en/CJFDTOTAL-ZLDJ200901010.htm

Gray, P. S., Williamson, J. B., Karp, D. A., & Dalphin, J. R. (2007). *The research imagination: An Introduction to qualitative and quantitative methods*. New York, NY: Cambridge University Press. doi:10.1017/CBO9780511819391

Green, N. (2002). On the move: Technology, mobility, and the mediation of social time and space. *The Information Society*, *18*(4), 281–292. doi:10.1080/01972240290075129

Grellhesl, M., & Narissra, M. (2012). Using the Uses and Gratifications Theory to Understand Gratifications Sought through Text Messaging Practices of Male and Female Undergraduate Students. *Computers in Human Behavior*, *28*(6), 2175–2181. doi:10.1016/j.chb.2012.06.024

Gretzel, U., Fesenmaier, D. R., & O'Leary, J. T. (2006). The transformation of consumer behaviour. In D. Buhalis & C. Costa (Eds.), *Tourism business frontiers: Consumers, products and industry* (pp. 9–18). Oxford, UK: Elsevier. doi:10.1016/B978-0-7506-6377-9.50009-2

Gretzel, U., & Jamal, T. (2009). Conceptualizing the creative tourist class: Technology, mobility and tourism experiences. *Tourism Analysis*, *14*(4), 471–482. doi:10.3727/108354209X12596287114219

Grindr. (2018). Retrieved from https://www.grindr.com/privacy-policy/(accessed Nov. 2018)

Grindrod, K. A., Li, M., & Gates, A. (2014). Evaluating user perceptions of mobile medication management applications with older adults: A usability study. *JMIR mHealth and uHealth*, *2*(1), e11. doi:10.2196/mhealth.3048 PMID:25099993

Groot Kormelink, T., & Costera Meijer, I. (2014). Tailor-Made News: Meeting the demands of news users on mobile and social media. *Journalism Studies*, *15*(5), 632–641. doi:10.1080/1461670X.2014.894367

Groves, M., & Mundt, K. (2015). Friend or foe? Google Translate in language for academic purposes. *English for Specific Purposes*, *37*, 112–121. doi:10.1016/j.esp.2014.09.001

Grün, C., Werthner, H., Pröll, B., Retschitzegger, W., & Schwinger, W. (2008). Assisting tourists on the move-an evaluation of mobile tourist guides. *7th International Conference on Mobile Business*, 171–180. 10.1109/ICMB.2008.28

GSMA Intelligence. (2018). *Global Data*. Retrieved January 15, 2018, from https://www.gsmaintelligence.com/

GSMA Mobile for Development. (n.d.). *Overview*. Retrieved from https://www.gsma.com/mobilefordevelopment/

GSMA. (2015). *Bridging the gender gap 2015: Mobile access and usage in low and middle-income countries 2015*. Available from: http://www.altaiconsulting.com/wp-content/uploads/2016/03/GSM0001_02252015_GSMAReport_FINAL-WEB-spreads.pdf

GSMA. (2017). *GSMA Input into report on the digital gender divide*. Available from: http://www.ohchr.org/Documents/Issues/Women/WRGS/GenderDigital/GSMA.pdf

GSMA. (2018). *Mobile Gender Gap Report 2018*. Available from: https://www.gsma.com/mobilefordevelopment/wp-content/uploads/2018/02/GSMA_The_Mobile_Gender_Gap_Report_2018_Final_210218.pdf

GSMA. (n.d.). *About Us*. Retrieved from https://www.gsma.com/aboutus/

Güler, Ç., Kılıç, E., & Çavuş, H. (2014). A comparison of difficulties in instructional design processes: Mobile vs. desktop. *Computers in Human Behavior*, *39*, 128–135. doi:10.1016/j.chb.2014.07.008

Guo, B., Yu, Z., Zhou, X., & Zhang, D. (2014, March). From participatory sensing to mobile crowd sensing. In *Pervasive Computing and Communications Workshops (PERCOM Workshops), 2014 IEEE International Conference on* (pp. 593-598). IEEE. 10.1109/PerComW.2014.6815273

Gupta, S. (2013). The mobile banking and payment revolution. *European Finance Review*, *2*, 3–6.

Gynnild, A. (2014). The robot eye witness. *Digital Journalism*, *2*(3), 334–343. doi:10.1080/21670811.2014.883184

Haddon, L., & Vincent, J. (2015). *UK children's experience of smartphones and tablets: perspectives from children, parents and teachers. LSE*. London: Net Children Go Mobile.

Halder, I., Halder, S., & Guha, A. (2015). *Undergraduate students use of mobile phones: Exploring use of advanced technological aids for educational purpose*. Available from: http://www.academicjournals.org/article/article1427128975_Halder%20et%20al.pdf

Hampton, K., Goulet, L. S., Ja Her, E., & Rainie, L. (2009). *Social Isolation and New Technology*. Retrieved at http://www.pewinternet.org/2009/11/04/social-isolation-and-new-technology/

Hancock, J. T., & Toma, C. L. (2009). Putting your best face forward: The accuracy of online dating photographs. *Journal of Communication, 59*(2), 367-386. doi:.1460-2466.2009.01420.x doi:10.1111/j

Hancock, J. T., Toma, C., & Ellison, N. (2007). The truth about lying in online dating pro-files. In *Proceedings of the SIGCHI Conference on Human Factors in Computing Systems* (pp. 449–452). New York, NY: ACM. 10.1145/1240624.1240697

Hannam, K., Butler, G., & Paris, C. (2014). Developments and Key Issues in Tourism Mobilities. *Annals of Tourism Research*, *44*, 171–185. doi:10.1016/j.annals.2013.09.010

Han, S. H., Yun, M. H., Kwahk, J., & Hong, S. W. (2001). Usability of Consumer Electronic Products. *International Journal of Industrial Ergonomics*, *28*(3-4), 143–151. doi:10.1016/S0169-8141(01)00025-7

Hanson, S. (2010). Gender and mobility: new approaches for informing sustainability. *Gender, Place & Culture, 17*(1), 5–23. Available from: https://www.tandfonline.com/doi/abs/10.1080/09663690903498225

Harkin, J. (2013). *Good media, bad politics? New media and the Syrian conflict. Reuters Institute Fellowship Paper*. University of Oxford.

Harley, D., Winn, S., Pemberton, S., & Wilcox, P. (2007). Using texting to support students' transition to university. *Innovations in Education and Teaching International*, *44*(3), 229–241. doi:10.1080/14703290701486506

Harrison, V., Proudfoot, J., Wee, P. P., Parker, G., Pavlovic, D. H., & Manicavasagar, V. (2011). Mobile mental health: Review of the emerging field and proof of concept study. *Journal of Mental Health (Abingdon, England)*, *20*(6), 509–524. doi:10.3109/09638237.2011.608746 PMID:21988230

Harwit, E. (2017). WeChat:social and poitical development of China's dominant messaging app. *Chinese Journal of Communication*, *10*(3), 312–327. doi:10.1080/17544750.2016.1213757

Hashim, K. F., Tan, F. B., & Rashid, A. (2015). Adult learners' intention to adopt mobile learning: A motivational perspective. *British Journal of Educational Technology*, *46*(2), 381–390. doi:10.1111/bjet.12148

Hassenzahl, M. (2003). The thing and I: Understanding the relationship between user and product. In M. A. Blythe, A. F. Monk, K. Overbeeke, & P. C. Wright (Eds.), *Funology: From usability to enjoyment* (pp. 31–42). Kluwer Academic Publisher. doi:10.1007/1-4020-2967-5_4

Head, M., & Ziolkowski, N. (2012). Understanding student attitudes of mobile phone features: Rethinking adoption through conjoint, cluster and SEM analyses. *Computers in Human Behavior*, *28*(6), 2331–2339. doi:10.1016/j.chb.2012.07.003

Hedegaard, M. (2009). Children's development form a cultural-historical approach: Children's activity in everyday local settings as foundation for their development. *Mind, Culture, and Activity*, *16*(1), 64–82. doi:10.1080/10749030802477374

Heil, C. R., Wu, J. S., Lee, J. J., & Schmidt, T. (2016). A Review of Mobile Language Learning Applications: Trends, Challenges, and Opportunities. *The EuroCALL Review*, *24*(2), 32–50. doi:10.4995/eurocall.2016.6402

Helbostad, J. L., & Vereijken, B. (2016). *Activity app for an aging population*. Retrieved from http://www.preventit.eu/index.php/news_events/activity-app-for-an-ageing-population/

Helbostad, J. L., Vereijken, B., Becker, C., Todd, C., Taraldsen, K., Pijnappels, M., ... Mellone, S. (2017). Mobile Health Applications to Promote Active and Healthy Ageing. *Sensors*, *17*(3), 622. doi:10.339017030622 PMID:28335475

Henan Daily. (2017). The release of Internet development report of Henan 2016. *Henan Daily*. Retrieved from http://www.gov.cn/xinwen/2017-05/17/content_5194606.htm

Hendin, H. M., & Cheek, J. M. (1997). Assessing Hypersensitive Narcissism: A Re-examination of Murray's Narcissism Scale. *Journal of Research in Personality*, *31*(4), 588–599. doi:10.1006/jrpe.1997.2204

Henze, N., Rukzio, E., & Boll, S. (2012, May). Observational and experimental investigation of typing behaviour using virtual keyboards for mobile devices. In *2012 ACM annual conference on Human Factors in Computing Systems (CHI '12)* (pp. 2659-2668). New York: ACM. 10.1145/2207676.2208658

Heryadi, Y., & Muliamin, K. (2016). Gamification of M-learning Mandarin as second language. *Game, Game Art, and Gamification (ICGGAG), 2016 1st International Conference on*, 1-4. 10.1109/ICGGAG.2016.8052645

Hesketh, T., Lu, L., & Xing, Z. W. (2005). The effect of China's one-child family policy after 25 years. Mass Medical Soc. Retrived from https://www.nejm.org/doi/full/10.1056/nejmhpr051833

Hilbert, M. (2011). Digital gender divide or technologically empowered women in developing countries? A typical case of lies, damned lies, and statistics. *Women's Studies International Forum*, *34*(6), 479-489. doi:10.1016/j.wsif.2011.07.001

Holmes, J. S. (1994). The Name and Nature of Translation Studies. In *Translated! Papers on Literary Translation and Translation Studies* (2nd ed.). Amsterdam: Rodopi.

Hong, F., Chiu, S., & Huang, D. (2012). A model of the relationship between psychological characteristics, mobile phone addiction and use of mobile phones by Taiwanese university female students. *Computers in Human Behavior*, *28*(6), 2152–2159. doi:10.1016/j.chb.2012.06.020

Hong, S. J., Thong, J. Y., Moon, J. Y., & Tam, K. Y. (2008). Understanding the behavior of mobile data services consumers. *Information Systems Frontiers*, *10*(4), 431–445. doi:10.100710796-008-9096-1

Hooper, V., & Zhou, Y. (2007, June). *Addictive, dependent, compulsive? A research of mobile phone usage*. Paper presented at the 20th Bled e-Conference e-Mergence: Merging and Emerging Technologies, Processes and Institutions, Bled, Slovenia.

Ho, S. Y. (2012). The effects of location personalization on individuals' intention to use mobile services. *Decision Support Systems*, *53*(4), 802–812. doi:10.1016/j.dss.2012.05.012

Howe, J. (2006). *Crowdsourcing: A Definition*. Retrieved May 10, 2018, from http://crowdsourcing.typepad.com/cs/2006/06/crowdsourcing_a.html

Hui-Lien, C., Chien, C., & Chao-Hsiu, C. (2016). The moderating effects of parenting styles on the relation between the internet attitudes and internet behaviors of high-school students in Taiwan. *Computers & Education, 94*, 204–214. doi:10.1016/j.compedu.2015.11.017

Humphreys, L., Von Pape, T., & Karnowski, V. (2013). Evolving mobile media: Uses and conceptualizations of the mobile internet. *Journal of Computer-Mediated Communication, 18*(4), 491–507. doi:10.1111/jcc4.12019

Hu, X., Chu, T. H., Leung, V., Ngai, E. C. H., Kruchten, P., & Chan, H. C. (2014). A survey on mobile social networks: Applications, platforms, system architectures, and future research directions. *IEEE Communications Surveys and Tutorials, 17*(3), 1557–1581. doi:10.1109/COMST.2014.2371813

Hu, Y., Manikonda, L., & Kambhampati, S. (2014, June). What We Instagram: A First Analysis of Instagram Photo Content and User Types. ICWSM.

Hwang, G. J., & Wu, P. H. (2014). Applications, impacts and trends of mobile technology-enhanced learning: A review of 2008-2012 publications in selected SSCI journals. *International Journal of Mobile Learning and Organisation, 8*(2), 83. doi:10.1504/IJMLO.2014.062346

Hwang, G.-J., & Tsai, C.-C. (2011). Research trends in mobile and ubiquitous learning: A review of publications in selected journals from 2001 to 2010. *British Journal of Educational Technology, 42*(4), 65–70. doi:10.1111/j.1467-8535.2011.01183.x

Hyun, M. Y., Lee, S., & Hu, C. (2009). Mobile-mediated virtual experience in tourism: Concept, typology and applications. *Journal of Vacation Marketing, 15*(2), 149–164. doi:10.1177/1356766708100904

IBIS World. (2017). Retrieved from https://www.ibisworld.com/industry-trends/market-research-reports/other-services-except-public-administration/personal-laundry/dating-services.html

Ibrahim, N., Kamaruddin, S., & Ling, T. (2017). Interactive educational Android mobile app for students learning Chinese characters writing. *Computer and Drone Applications (IConDA), 2017 International Conference on*, 96-101. 10.1109/ICONDA.2017.8270407

Ickin, S., Wac, K., Fiedler, M., Janowski, L., Hong, J. H., & Dey, A. K. (2012). Factors influencing quality of experience of commonly used mobile applications. *IEEE Communications Magazine, 50*(4), 48–56.

İHH. (2015). Technology usage and dependence photo of Turkey's youth [Teknoloji kullanımı ve bağımlılığı açısından Türkiye gençliğinin fotoğrafı]. *İnsani Sosyal Araştırmalar Merkezi*. Available from: http://insamer.com/rsm/files/Teknoloji%20kullanimi%20ve%20bagimliligi.pdf

Ilahiane, H., & Sherry, J. W. (2012). The problematics of the "Bottom of the Pyramid" approach to international development: The case of micro-entrepreneurs' use of mobile phones in Morocco. *Information Technologies and International Development, 8*(1), 13.

Intel. (2018). *A guide to the internet of things: Infographic*. Retrieved from https://www.intel.com/content/www/us/en/internet-of-things/infographics/guide-to-iot.html

Islam, M. S., & Gronlund, A. (2007). Agriculture market information e-service in Bangladesh: A stakeholder-oriented case study. In M. A. Wimmer, H. J. Scholl, & A. Gronlund (Eds.), *Electronic Government* (pp. 167–178). Berlin: Springer-Verlag Berlin Heidelberg. doi:10.1007/978-3-540-74444-3_15

Jabeur, N., Zeadally, S., & Sayed, B. (2013). Mobile social networking applications. *Communications of the ACM, 56*(3), 71–79. doi:10.1145/2428556.2428573

Jack, W., & Suri, T. (2011). *Mobile money: The economics of M-PESA (No. w16721).* National Bureau of Economic Research. doi:10.3386/w16721

Jacobsen, W., & Forste, R. (2011). The wired generation: Academic and social outcomes of electronic media use among university students. *Cyberpsychology, Behavior, and Social Networking, 14*(5), 5. doi:10.1089/cyber.2010.0135 PMID:20961220

Jagun, A., Heeks, R., & Whalley, J. (2007). *Mobile telephony and developing country micro-enterprise: A Nigerian case study.* Retrieved at http://itidjournal.org/itid/article/view/310

Jain, A. K., & Shanbhag, D. (2012). Addressing security and privacy risks in mobile applications. *IT Professional, 14*(5), 28–33. doi:10.1109/MITP.2012.72

Jambulingam, M. (2013). Behavioural intention to adopt mobile technology among tertiary students. *World Applied Sciences Journal, 22*(9), 1262–1271.

Jambulingam, M., & Sorooshian, S. (2013). Usage of mobile features among undergraduates and mobile learning. *Current Research Journal of Social Sciences, 5*(4), 130–133.

Jansson, A. (2007). A sense of tourism: New media and the dialectic of encapsulation/ decapsulation. *Tourist Studies, 7*(1), 5–24. doi:10.1177/1468797607079799

Jayson, S. (2014, Jan. 26). Latest trend in digital dating: Live video chat dates. *USA Today.* Retrieved from https://www.usatoday.com/story/tech/personal/2014/01/26/dating-mobile-phone-video/4674651/(accessed Oct. 3rd 2018)

Jennings, L., Ong'ech, J., Simiyu, R., Sirengo, M., & Kassaye, S. (2013). Exploring the use of mobile phone technology for the enhancement of the prevention of mother-to-child transmission of HIV program in Nyanza, Kenya: A qualitative study. *BMC Public Health, 13*(1), 1. doi:10.1186/1471-2458-13-1131 PMID:24308409

Jensen, R. (2007). The digital provide: Information (technology), market performance, and welfare in the South Indian fisheries sector. *The Quarterly Journal of Economics, 122*(3), 879–924. doi:10.1162/qjec.122.3.879

Jeong, S. H., Kim, H., Yum, J. Y., & Hwang, Y. (2016). What type of content are smartphone users addicted to?: SNS vs. games. *Computers in Human Behavior, 54*, 10–17. doi:10.1016/j.chb.2015.07.035

Jiang, W., & Li, W. (2018). Linking up learners of Chinese with native speakers through WeChat in an Australian tertiary CFL curriculum. *Asian-Pacific Journal of Second and Foreign Language Education, 3*(1), 1–16. doi:10.118640862-018-0056-0

Jimenez-Crespo, M. A. (2016). Mobile apps and translation crowdsourcing: The next frontier in the evolution of translation. *Revista Tradumàtica: tecnologies de la traducció, 14*, 75–84. doi:10.5565/rev/tradumatica.167

Jin, L. (2018). Digital affordances on WeChat: Learning Chinese as a second language. *Computer Assisted Language Learning, 31*(1-2), 27–52. doi:10.1080/09588221.2017.1376687

Johnson, B. K., & Ranzini, G. (2018). Click here to look clever: Self-presentation via selective sharing of music and film on social media. *Computers in Human Behavior, 82*, 148–158. doi:10.1016/j.chb.2018.01.008

Johnson, K. J., & Mutchler, J. E. (2014). The emergence of a positive gerontology: from disengagement to social involvement. *The Gerontologist, 54*(1), 93–100. doi:10.1093/geront/gnt099 PMID:24009172

Johnson, S. C., & Thakur, D. (2015). Mobile phone ecosystems and the informal sector in developing countries–cases from Jamaica. *The Electronic Journal on Information Systems in Developing Countries, 66*(1), 1–22. doi:10.1002/j.1681-4835.2015.tb00476.x

Joiner, R., Stewart, C., & Beaney, C. (2015). *Gender digital divide exist and what are the explanations. The Wiley Handbook of Psychology, Technology, and Society.* Oxford, UK: John Wiley & Sons, Ltd. Available from https://books.google.com.tr/books?hl=tr&lr=&id=Zb0oBwAAQBAJ&oi=fnd&pg=PA74&dq=gender+gap+and+smartphone&ots=QtV6nCtENr&sig=61o5H9gYWCF6mWsTMs1oglCQf34&redir_esc=y#v=onepage&q=gender%20gap%20and%20smartphone&f=false

Jokela, T., Väätäjä, H., & Koponen, T. (2009, September). Mobile Journalist Toolkit: a field study on producing news articles with a mobile device. In *Proceedings of the 13th International MindTrek Conference: Everyday Life in the Ubiquitous Era* (pp. 45-52). ACM. 10.1145/1621841.1621851

Jones, A. C., Scanlon, E., & Clough, G. (2013). Mobile learning: Two case studies of supporting inquiry learning in informal and semiformal settings. *Computers & Education, 61,* 21–32. doi:10.1016/j.compedu.2012.08.008

Joo, J., & Sang, Y. (2013). Exploring Koreans' smartphone usage: An integrated model of the technology acceptance model and uses and gratifications theory. *Computers in Human Behavior, 29*(6), 2512–2518. doi:10.1016/j.chb.2013.06.002

Jordan, P. W. (2000). *Designing Pleasurable Products.* New York: Taylor and Francis. doi:10.4324/9780203305683

Joshi, A. (2009, October). Mobile phones and economic sustainability: perspectives from India. In *Proceedings of the First international conference on Expressive Interactions for Sustainability and Empowerment* (pp. 2-2). British Computer Society.

Kafyulilo, A. (2014). Access, use and perceptions of teachers and students towards mobile phones as a tool for teaching and learning in Tanzania. *Education and Information Technologies, 19*(1), 115–127. doi:10.100710639-012-9207-y

Kalam, A. P. (2004, June 30). *Address at the Technology Day Award Function, Pragati Maidan, New Delhi.* Retrieved November 21, 2018, from http://abdulkalam.nic.in: http://abdulkalam.nic.in/sp300604.html

Karliner, L. S., Jacobs, E. A., Chen, A. H., & Mutha, S. (2007). Do professional interpreters improve clinical care for patients with limited English proficiency? A systematic review of the literature. *Health Services Research, 42*(2), 727–754. doi:10.1111/j.1475-6773.2006.00629.x PMID:17362215

Karpinski, A. C., Kirschner, P. A., Ozer, I., Mellott, J. A., & Ochwo, P. (2013). An exploration of social networking site use, multitasking, and academic performance among United States and European university students. *Computers in Human Behavior, 29,* 1182-1192. doi:10.1016/j.chb.2012.10.011

Kathuria, R., Uppal, M., & Mamta. (2009). *An econometric analysis of the impact of mobile* (Vodafone Policy Paper Series, 9, pp. 5-20). Retrieved at http://www.icrier.org/pdf/public_policy19jan09.pdf

Katz, E. (1959). Mass communications research and the study of popular culture: An editorial note on a possible future for this journal. *Departmental Papers (ASC), 165.*

Katz, J., & Aakhus, M. (Eds.). (2002). *Perpetual Contact: Mobile Communication, Private Talk, Public Performance.* London: Cambridge University Press. Available from: https://books.google.com.tr/books?hl=tr&lr=&id=Wt5AsHEgUh0C&oi=fnd&pg=PR9&dq=mobile+usage+social+interaction&ots=YU_y_cNqkL&sig=RunBsoZK7CiToCIkEj36EvzgZ2M&redir_esc=y#v=onepage&q&f=false

Katz, E., Blumler, J. G., & Gurevitch, M. (1973). Uses and gratifications research. *Public Opinion Quarterly, 37*(4), 509–523. doi:10.1086/268109

Kaur, H., Lechman, E., & Marszk, A. (Eds.). (2017). *Catalyzing development through ICT adoption: the developing world experience.* Springer. doi:10.1007/978-3-319-56523-1

Kazai, G., Yusof, I., & Clarke, D. (2016, July). Personalised News and Blog Recommendations based on User Location, Facebook and Twitter User Profiling. In *Proceedings of the 39th International ACM SIGIR Conference on Research and Development in Information Retrieval* (pp. 1129-1132). ACM. 10.1145/2911451.2911464

Kearney, M., & Maher, D. (2013). Mobile learning in math teacher education: Using iPads to support pre-service teachers' professional development. *Australian Educational Computing*, *27*(3), 76–84.

Kearney, M., Schuck, S., Burden, K., & Aubusson, P. (2012). Viewing mobile learning from a pedagogical perspective. *Research in Learning Technology*, 20.

Keegan, V. (2007, Dec. 6). Dating moves from the PC to the mobile. *The Guardian*. Retrieved from https://www.theguardian.com/technology/2007/dec/06/digitalvideo.mobilephones(accessed Oct. 3rd 2018)

Ke, F., & Hsu, Y. C. (2015). Mobile augmented-reality artifact creation as a component of mobile computer-supported collaborative learning. *The Internet and Higher Education*, *26*, 33–41. doi:10.1016/j.iheduc.2015.04.003

Kendig, H. (2004). The social sciences and successful aging: Issues for Asia–Oceania. *Geriatrics & Gerontology International*, *4*(s1), S6–S11. doi:10.1111/j.1447-0594.2004.00136.x

Kharrazi, H., Chisholm, R., VanNasdale, D., & Thompson, B. (2012). Mobile personal health records: An evaluation of features and functionality. *International Journal of Medical Informatics*, *81*(9), 579–593. doi:10.1016/j.ijmedinf.2012.04.007 PMID:22809779

Kibona, L., & Mgaya, G. (2015). Smartphones' effects on academic performance of higher learning students. *Journal of Multidisciplinary Engineering Science and Technology*, *2*(4), 777–784. Available from https://pdfs.semanticscholar.org/1203/16b911f8e69ec4b79efdc5b6bda9fbf23ec6.pdf

Kim, T.-Y., & Shin, D.-H. (2013). The Usage and the Gratifications About Smartphone Models and Applications. *International Telecommunications Policy Review, 20*(4). Retrieved on March 2, 2013 from: SSRN: http://ssrn.com/abstract=2373428

Kim, C., Park, T., Lim, H., & Kim, H. (2013). On-site construction management using mobile computing technology. *Automation in Construction*, *35*, 415–423. doi:10.1016/j.autcon.2013.05.027

Kim, D., Park, J., & Morrison, A. M. (2008). A Model of Traveler Acceptance of Mobile Technology. *International Journal of Tourism Research*, *10*(5), 393–407. doi:10.1002/jtr.669

Kim, E., Lee, J. A., Sung, Y., & Choi, S. M. (2016). Predicting selfie-posting behavior on social networking sites: An extension of theory of planned behavior. *Computers in Human Behavior*, *62*, 116–123. doi:10.1016/j.chb.2016.03.078

Kim, E., Lin, J. S., & Sung, Y. (2013). To app or not to app: Engaging consumers via branded mobile apps. *Journal of Interactive Advertising*, *13*(1), 53–65. doi:10.1080/15252019.2013.782780

Kim, H., & Law, R. (2015). Smartphones in Tourism and Hospitality Marketing: A Literature Review. *Journal of Travel & Tourism Marketing*, *32*(6), 692–711. doi:10.1080/10548408.2014.943458

Kim, J., Ahn, K., & Chung, N. (2013). Examining the factors affecting perceived enjoyment and usage intention of ubiquitous tour information services: A service quality perspective. *Asia Pacific Journal of Tourism Research*, *18*(6), 598–617. doi:10.1080/10941665.2012.695282

Kim, J., & Hahn, K. H. Y. (2015). The effects of self-monitoring tendency on young adult consumers' mobile dependency. *Computers in Human Behavior*, *50*, 169–176. doi:10.1016/j.chb.2015.04.009

Kim, M. J., Kim, W. G., Kim, J. M., & Kim, C. (2016). Does knowledge matter to seniors' usage of mobile devices? Focusing on motivation and attachment. *International Journal of Contemporary Hospitality Management, 28*(8), 1702–1727. doi:10.1108/IJCHM-01-2015-0031

Kim, Y. H., Kim, D. J., & Wachter, K. (2013). A study of mobile user engagement (MoEN): Engagement motivations, perceived value, satisfaction, and continued engagement intention. *Decision Support Systems, 56*, 361–370. doi:10.1016/j.dss.2013.07.002

Kirakowski, J., & Corbett, M. (1993). SUMI: The Software Usability Measurement Inventory. *British Journal of Educational Technology, 24*(3), 210–212. doi:10.1111/j.1467-8535.1993.tb00076.x

Kircaburun, K., Alhabash, S., Tosuntaş, Ş. B., & Griffiths, M. D. (2018). Uses and Gratifications of Problematic Social Media Use Among University Students: A Simultaneous Examination of the Big Five of Personality Traits, Social Media Platforms, and Social Media Use Motives. *International Journal of Mental Health and Addiction*, 1–23.

Kiritoshi, K., & Ma, Q. (2015). A Diversity-Seeking Mobile News App Based on Difference Analysis of News Articles. In *Database and Expert Systems Applications 26th International Conference, DEXA 2015* (pp. 73-81). Academic Press. 10.1007/978-3-319-22852-5_7

Klettke, B., Hallford, D. J., & Mellor, D. J. (2014). Sexting prevalence and correlates: A systematic literature review. *Clinical Psychology Review, 34*(1), 44–53. doi:10.1016/j.cpr.2013.10.007 PMID:24370714

Kolawole, S. O. (2012). Is every bilingual a translator? *Translation Journal, 16*(2). Retrieved from http://translation-journal.net/journal/60bilingual

Koponen, T., & Väätäjä, H. (2009, September). Early adopters' experiences of using mobile multimedia phones in news journalism. In *European Conference on Cognitive Ergonomics: Designing beyond the Product---Understanding Activity and User Experience in Ubiquitous Environments* (p. 2). VTT Technical Research Centre of Finland.

Krajewska-Kułak, E., Kułak, W., Stryzhak, A., Szpakow, A., Prokopowicz, W., & Marcinkowski, J. T. (2012). Problematic mobile phone using among the Polish and Belarusian University students: A comparative research. *Progress in Health Sciences, 2*(1), 45–50.

Krcmar, M., & Cingel, D. P. (2016). Examining two theoretical models predicting American and Dutch parents' mediation of adolescent social media use. *Journal of Family Communication, 16*(3), 247–262. doi:10.1080/15267431.2016.1181632

Kristensen, N. N., & Mortensen, M. (2013). Amateur sources breaking the news, metasources authorizing the news of Gaddif's death. *Digital Journalism, 1*(3), 352–367. doi:10.1080/21670811.2013.790610

Krone, M., Dannenberg, P., & Nduru, G. (2016). The use of modern information and communication technologies in smallholder agriculture: Examples from Kenya and Tanzania. *Information Development, 32*(5), 1503–1512. doi:10.1177/0266666915611195

Krontiris, I., Langheinrich, M., & Shilton, K. (2014). Trust and privacy in mobile experience sharing: Future challenges and avenues for research. *Communications Magazine, IEEE, 52*(8), 50–55. doi:10.1109/MCOM.2014.6871669

Kujala, S., & Miron-Shatz, T. (2013, April). Emotions, experiences and usability in real-life mobile phone use. In *Proceedings of the SIGCHI Conference on Human Factors in Computing Systems* (pp. 1061-1070). ACM. 10.1145/2470654.2466135

Kuntsman, A. (2010). Online Memories, Digital Conflicts and the Cybertouch of War, Digital Icons: Studies in Russian. *Eurasian and Central European New Media, 4*, 1-12.

Kuntsman, A., & Stein, R. L. (2010). *Another War Zone Social Media in the Israeli-Palestinian Conflict*. Retrieved from http://www.merip.org/mero/interventions/another-war-zone

Kuo, J. H., Huang, C. M., Liao, W. H., & Huang, C. C. (2011). HuayuNavi: a mobile Chinese learning application based on intelligent character recognition. In *International Conference on Technologies for E-Learning and Digital Entertainment* (pp. 346-354). Springer. 10.1007/978-3-642-23456-9_63

Kushchu, I. (2007). *Positive Contributions of Mobile Phones to Society*. Publication of the Mobile Government Consortium International UK. Retrieved at http://www.kiwanja.net/database/document/report_positive_impact.pdf

Kuyucu, M. (2017). Use of smart phone and problematic of smart phone addiction in young people: "smart phone (colic)" university youth [Gençlerde akıllı telefon kullanımı ve akıllı telefon bağımlılığı sorunsalı: "akıllı telefon(kolik)" üniversite gençliği]. *Global Media Journal TR Edition, 7*(14). Available from: http://globalmediajournaltr.yeditepe.edu.tr/sites/default/files/mihalis_kuyucu_-_genclerde_akilli_telefon_kullanimi_ve_akilli_telefon_bagimliligi_sorunsali_akilli_telefonkolik_universite_gencligi.pdf

Kuznekoff, J., & Titsworth, S. (2013). The Impact of Mobile Phone Usage on Student Learning. *Communication Education, 62*(3), 233–252. doi:10.1080/03634523.2013.767917

Kwaku Kyem, P. A., & LeMaire, P. K. (2006). *Transforming recent gains in the digital divide into digital opportunities: Africa and the boom in mobile phone subscription*. Retrieved at http://www.ejisdc.org/ojs2/index.php/ejisdc/ article/viewFile/343/189

Kwon, M., Lee, J.-Y., Won, W.-Y., Park, J.-W., Min, J.-A., Hahn, C., ... Kim, D.-J. (2013). Development and Validation of a Smartphone Addiction Scale (SAS). *PLoS One, 8*(2), 1–7. doi:10.1371/journal.pone.0056936 PMID:23468893

La Polla, M., Martinelli, F., & Sgandurra, D. (2013). A survey on security for mobile devices. *IEEE Communications Surveys and Tutorials, 15*(1), 446–471. doi:10.1109/SURV.2012.013012.00028

Lamsfus, C., Wang, D., Alzua-Sorzabal, A., & Xiang, Z. (2015). Going mobile: Defining context for on-the-go travelers. *Journal of Travel Research, 54*(6), 691–701. doi:10.1177/0047287514538839

Lane, N. D., Miluzzo, E., Lu, H., Peebles, D., Choudhury, T., & Campbell, A. T. (2010). A survey of mobile phone sensing. *IEEE Communications Magazine, 48*(9), 140–150. doi:10.1109/MCOM.2010.5560598

Lan, Y.-J., & Lin, Y.-T. (2016). Mobile Seamless Technology Enhanced CSL Oral Communication. *Journal of Educational Technology & Society, 19*(3), 335–350.

Lasica, J. D., & Firestone, C. M. (2008). *Civic Engagement on the Move: How mobile media can serve the public good*. Washington, DC: The Aspen Institute.

Laslett, P. (1991). A fresh map of life: The emergence of the third age. Massachusetts: Harvard University Press.

Laslett, P. (1987). The emergence of the third age. *Ageing and Society, 7*(2), 133–160. doi:10.1017/S0144686X00012538

Laslett, P. (1994). The third age, the fourth age and the future. *Ageing and Society, 14*(3), 436–447. doi:10.1017/S0144686X00001677

Law, R., Chan, I. C. C., & Wang, L. (2018). A comprehensive review of mobile technology use in hospitality and tourism. *Journal of Hospitality Marketing & Management, 27*(6), 626–648. doi:10.1080/19368623.2018.1423251

Lawson, H. M., & Leck, K. (2006). Dynamics of Internet dating. *Social Science Computer Review, 24*(2), 189–208. doi:10.1177/0894439305283402

Lecturer, J. Y., Dominic, G., & Lecturer, E. (2014). The Impact of WhatsApp Messenger Usage on Students Performance in Tertiary Institutions in Ghana. *Journal of Education and Practice, 5*(6), 157–164.

Lee, D., Moon, J., Kim, Y. J., & Mun, Y. Y. (2015). Antecedents and consequences of mobile phone usability: Linking simplicity and interactivity to satisfaction, trust, and brand loyalty. *Information & Management*, *52*(3), 295–304. doi:10.1016/j.im.2014.12.001

Lee, E., Lee, J. A., Moon, J. H., & Sung, Y. (2015). Pictures Speak Louder than Words: Motivations for Using Instagram. *Cyberpsychology, Behavior, and Social Networking*, *18*(9), 552–556. doi:10.1089/cyber.2015.0157 PMID:26348817

Lee, H., Kwak, N., Campbell, S. W., & Ling, R. (2014). Mobile communication and political participation in South Korea: Examining the intersections between informational and relational uses. *Computers in Human Behavior*, *38*, 85–92. doi:10.1016/j.chb.2014.05.017

Lee, J. H., & Kim, J. (2014). Socio-demographic gaps in mobile use, causes, and consequences: A multi-group analysis of the mobile divide model. *Information Communication and Society*, *17*(8), 917–936. doi:10.1080/1369118X.2013.860182

Lee, K. J., & Kim, J. E. (2013). A Mobile-based Learning Tool to Improve Writing Skills of Efl Learners. *Procedia: Social and Behavioral Sciences*, *106*, 112–119. doi:10.1016/j.sbspro.2013.12.014

Lee, Y., Chang, C., Lin, Y., & Cheng, Z. (2014). The dark side of smartphone usage: Psychological traits, compulsive behavior and technostress. *Computers in Human Behavior*, *31*, 373–383. doi:10.1016/j.chb.2013.10.047

Legris, P., Ingham, J., & Collerette, P. (2003). Why do people use information technology? A critical review of the technology acceptance model. *Information & Management*, *40*(3), 191–204. doi:10.1016/S0378-7206(01)00143-4

Lepp, A., Barkley, J., & Karpinski, A. (2013). *Computers in Human Behavior*. Kent, OH: Elsevier. Available from https://www.sciencedirect.com/science/article/pii/S0747563213003993?via%3Dihub

Leung, L. (2009). User-generated content on the internet: An examination of gratifications, civic engagement and psychological empowerment. *New Media & Society*, *11*(8), 1327–1347. doi:10.1177/1461444809341264

Leung, L., & Wei, R. (2000). More than just talk on the move: A use-and-gratification study of the cellular phone. *Journalism & Mass Communication Quarterly*, *77*(2), 308–320. doi:10.1177/107769900007700206

Levenson, J. C., Shensa, A., Sidani, J. E., Colditz, J. B., & Primack, B. A. (2016). The association between social media use and sleep disturbance among young adults. *Preventive Medicine*, *85*, 36–41. doi:10.1016/j.ypmed.2016.01.001 PMID:26791323

Lewis, S., & Westlund, O. (2015). Actors, Actants, Audiences, and Activities in Cross-Media News Work. *Digital Journalism*, *3*(1), 19–37. doi:10.1080/21670811.2014.927986

Liang, S., Schuckert, M., Law, R., & Masiero, L. (2017). The relevance of mobile tourism and information technology: An analysis of recent trends and future research directions. *Journal of Travel & Tourism Marketing*, *34*(6), 732–748. doi:10.1080/10548408.2016.1218403

Li, D. C. S. (2017). *Multilingual Hong Kong: Languages, literacies and identities (Multilingual Education 19)*. Cham: Springer. doi:10.1007/978-3-319-44195-5

Light, B. (2014). *Disconnecting with Social Networking Sites*. Basingstoke, UK: Palgrave Macmillan. doi:10.1057/9781137022479

Li, L., & Chen, T. (2005). Cyberlove and its ethical issues: A clarification. *Studies in Ethics*, *1*(1), 72–74.

Lilley, W., & Hardman, J. (2017). "You focus, I'm talking": A CHAT analysis of mobile dictionary use in an advanced EFL class. *Africa Education Review*, *14*(1), 120–138. doi:10.1080/18146627.2016.1224592

Li, M., Cao, N., Yu, S., & Lou, W. (2011). Findu: Privacy-preserving personal profile matching in mobile social networks. In *INFOCOM, 2011 Proceedings IEEE*. IEEE. doi:10.1109/INFCOM.2011.5935065

Lin, Y.-H., Chang L.-R., Lee, Y.-H., Tseng, H.-W., Kuo, T. B. J., & Chen, S-H. (2014). Development and Validation of the Smartphone Addiction Inventory (SPAI). *PLOS One, 9*(6), 1-5. doi:10.1371/journal.pone.0098312

Lindlof, T. R., & Shatzer, M. J. (1998). Media ethnography in virtual space: Strategies, limits, and possibilities. *Journal of Broadcasting & Electronic Media, 42*(2), 170–189. doi:10.1080/08838159809364442

Ling, R., Lim, S. S., Fortunati, L., & Goggin, G. (Eds.). (2019). Handbook of mobile communication. Oxford, UK: Oxford University Press. (forthcoming)

Ling, R. (2012). *Taken for grantedness: The embedding of mobile communication into society*. Cambridge, MA: MIT Press. doi:10.7551/mitpress/8445.001.0001

Lin, W. Y., Zhang, X., Jung, J. Y., & Kim, Y. C. (2013). From the wired to wireless generation? Investigating teens' Internet use through the mobile phone. *Telecommunications Policy, 37*(8), 651–661. doi:10.1016/j.telpol.2012.09.008

Lippa, R. A. (2009). Sex differences in sex drive, sociosexuality, and height across 53 nations: Testing evolutionary and social structural theories. *Archives of Sexual Behavior, 38*(5), 631–651. doi:10.100710508-007-9242-8 PMID:17975724

Lippmann, W. (2010). Public opinion and the politicians. *National Municipal Review, 15*(1), 5–8. doi:10.1002/ncr.4110150102

Litman, L., Rosen, Z., Spierer, D., Weinberger-Litman, S., Goldschein, A., & Robinson, J. (2015). Mobile exercise apps and increased leisure time exercise activity: A moderated mediation analysis of the role of self-efficacy and barriers. *Journal of Medical Internet Research, 17*(8), e195. doi:10.2196/jmir.4142 PMID:26276227

Littau, K. (2016). Translation's Histories and Digital Futures. *International Journal of Communication, 10*, 907–928. Retrieved from http://ijoc.org/index.php/ijoc/article/view/3508

Liu, C.Z. (2015). Discussion on profit models of mobile social media. *News World*, (8),188-189.

Liu, Y., Li, H., Kostakos, V., Goncalves, J., Hosio, S., & Hu, F. (2014). An empirical investigation of mobile government adoption in rural China: A case study in Zhejiang province. *Government Information Quarterly, 31*(3), 432–442. doi:10.1016/j.giq.2014.02.008

Livingstone, S., Haddon, L., Görzig, A., & Ólafsson, K. (2011). *Risks and safety on the internet: the perspective of European children: full findings and policy implications from the EU Kids Online survey of 9-16 year olds and their parents in 25 countries*. Academic Press.

Livingstone, S., Cagiltay, K., & Ólafsson, K. (2015). EU Kids Online II Dataset: A cross-national study of children's use of the Internet and its associated opportunities and risks. *British Journal of Educational Technology, 46*(5), 988–992. doi:10.1111/bjet.12317

Livingston, S. (2011). The CNN effect reconsidered (again): Problematizing ICT and global governance in the CNN effect research agenda. *Media, War & Conflict, 4*(1), 20–36. doi:10.1177/1750635210396127

Li, Y., & Perkins, A. (2007). The impact of technological developments on the daily life of the elderly. *Technology in Society, 29*(3), 361–368. doi:10.1016/j.techsoc.2007.04.004

Logan, R. J. (1994). Behavioral and Emotional Usability: Thomson Consumer Electronics. In M. E. Wiklund (Ed.), *Usability in Practice* (pp. 59–82). New York: AP Professional.

LohaChoudhury, B. (2001). *Media Performance: The Experience of Calcutta Media* (PhD thesis). Silchar, Assam: Assam University.

Lomas, N. (2015). *Stepes Is A Bet That A Chat App Can Mobilize Crowdsourced Translation.* Retrieved June 12, 2018, from https://techcrunch.com/2015/12/17/stepes-is-a-bet-that-a-chat-app-can-mobilize-crowdsourced-translation/?ncid=rss

Londhe, B. R., Radhakrishnan, S., & Divekar, B. R. (2014). Socio economic impact of mobile phones on the bottom of pyramid population-A pilot study. *Procedia Economics and Finance, 11*, 620–625. doi:10.1016/S2212-5671(14)00227-5

López-García, X., Silva Rodríguez, A., Vizoso, Á., Westlund, O., & Canavilhas, J. (2019). Periodismo móvil: Revisión sistemática de la producción científica/Mobile journalism: Systematic literature review. *Communicar, 59*, 2019–2. (forthcoming)

Lorenzo-Dus, N., & Bryan, A. (2011). Recontextualizing Participatory Journalists' Mobile Media in British Television News: A Case Study of the Live Coverage and Commemorations of the 2005 London Bombings. *Discourse & Communication, 5*(1), 23–40. doi:10.1177/1750481310390164

Lu, C., Chang, M., Huang, E., & Ching-Wen, C. (2014). Context-Aware Mobile Role Playing Game for Learning-A Case of Canada and Taiwan. *Journal of Educational Technology & Society, 17*(2), 101.

Lu, J., Meng, S., & Tam, V. (2014). Learning Chinese characters via mobile technology in a primary school classroom. *Educational Media International, 51*(3), 166–184. doi:10.1080/09523987.2014.968448

Luo, J. D. (2017). *Complex: Connections, Opportunities and Layouts in the Information Age.* CITIC Publishing Group Co., Ltd.

Luthar, S. S., Cicchetti, D., & Becker, B. (2000). The construction of resilience: A critical evaluation and guidelines for future work. *Child Development, 71*(5), 543–562. doi:10.1111/1467-8624.00164 PMID:10953923

Lu, X. H. (2003). *Network reliance: Challenges between virtual and reality.* Shanghai: Southeast University Press.

MacKay, K., & Vogt, C. (2012). Information technology in everyday and vacation contexts. *Annals of Tourism Research, 39*(3), 1380–1401. doi:10.1016/j.annals.2012.02.001

Madden, M., Lenhart, A., Duggan, M., Cortesi, S., & Gasser, U. (2013). *Teens And Technology.* Available from: http://www.pewinternet.org/Reports/2013/Teens-and-Tech.aspx

Madden, G., & Savage, S. (2000). Telecommunications and economic growth. *International Journal of Social Economics, 27*(7-10), 893–906. doi:10.1108/03068290010336397

Mahamad, S., Hipani, N., Basri, S., Hashim, A., Sarlan, A., & Sulaiman, S. (2016). Development of Chinese language application in learning as a second language for Malaysian. *Computer and Information Sciences (ICCOINS), 2016 3rd International Conference on,* 596-599. 10.1109/ICCOINS.2016.7783282

Mahlke, S. (2005). Understanding Users' Experience of Interaction. In *Proceedings of the 2005 Annual Conference on European Association of Cognitive Ergonomics* (pp. 251–4). Chania, Greece: Academic Press.

Mahlke, S. (2005). Understanding Users' Experience of Interaction. *Proceedings of the 2005 Annual Conference on European Association of Cognitive Ergonomics,* 251–4.

Majumder, S., & Sharma, R. P. (2014). Indian ITES Industry Going Rural: The Road Ahead. *Journal of Business and Economic Policy.*

Malik, A., Dhir, A., & Nieminen, M. (2016). Uses and Gratifications of digital photo sharing on Facebook. *Telematics and Informatics, 33*(1), 129–138. doi:10.1016/j.tele.2015.06.009

Malik, S., Chaudhry, I., & Abbass, Q. (2009). Socio-Economic Impact of Cellular Phone Growth in Pakistan: An Empirical Analysis. Pakistan. *Journal of social Sciences*, *29*(1), 23–37.

Malka, V., Ariel, Y., & Avidar, R. (2015). Fighting, worrying and sharing: Operation 'Protective Edge' as the first WhatsApp war. *Media, War & Conflict*, *8*(3), 329–344. doi:10.1177/1750635215611610

Mallenius, S., Rossi, M., & Tuunainen, V. (2007). *Factors affecting the adoption and use of mobile devices and services by elderly people – results from a pilot study*. Available from: https://www.researchgate.net/publication/228632076_Factors_affecting_the_adoption_and_use_of_mobile_devices_and_services_by_elderly_people-results_from_a_pilot_study

Mancini, F. (Ed.). (2013). *New technology and the prevention of violence and conflict. April 2013*. New York: International Peace Institute. Retrieved from https://ssrn.com/abstract=2902494

Maqsood, M. (2015). *Use of mobile technology among rural women in Pakistan for agriculture extension information*. Retrieved at https://d.lib.msu.edu/.../USE_OF_MOBILE_TECHNOLOGY_AMONG_RURAL_WO

Mare, A. (2014). New Media Technologies and Internal Newsroom Creativity in Mozambique. *Digital Journalism*, *2*(1), 12–28. doi:10.1080/21670811.2013.850196

Markowitz, D. M., Hancock, J. T., & Tong, S. (2018). Interpersonal dynamics in online dating: Profiles, matching, and discovery. In Z. Papacharissi (Ed.), A networked self and love (pp. 50–61). New York, NY: Routledge.

Markowitz, D. M., & Hancock, J. T. (2018). Deception in mobile dating conversations. *Journal of Communication*, *68*(3), 547–569. doi:10.1093/joc/jqy019

Marsh, J., Plowman, L., Yamada-Rice, D., Bishop, J. C., Lahmar, J., & Scott, F. (2015). *Exploring Play and Creativity in Pre-Schoolers' Use of Apps: Final Project Report*. Available at: www.techandplay.org

Martyn, P. H. (2009). The Mojo in the third millennium: Is multimedia journalism affecting the news we see? *Journalism Practice*, *3*(2), 196–215. doi:10.1080/17512780802681264

Mascheroni, G., & Ólafsson, K. (2016). The mobile Internet: Access, use, opportunities and divides among European children. *New Media & Society*, *18*(8), 1657–1679. doi:10.1177/1461444814567986

Mas, I. (2011). Why are banks so scarce in developing countries? A regulatory and infrastructure perspective. *Critical Review*, *23*(1-2), 135–145. doi:10.1080/08913811.2011.574476

Matherson, D., & Allan, S. (2009). *Digital war reporting*. Cambridge, UK: Polity Press.

Mayfield, C. H., Ohara, P. T., & O'Sullivan, P. S. (2013). Perceptions of a mobile technology on learning strategies in the anatomy laboratory. *Anatomical Sciences Education*, *6*(2), 81–89. doi:10.1002/ase.1307 PMID:22927203

McCarthy, J., & Wright, P. C. (2005). Putting 'Felt-Life" at the Centre of Human-Computer Interaction (HCI). *Cognition Technology and Work*, *7*(4), 262–271. doi:10.100710111-005-0011-y

McDaniel, B. T., & Radesky, J. S. (2018). Technoference: Parent distraction with technology and associations with child behavior problems. *Child Development*, *89*(1), 100–109. doi:10.1111/cdev.12822 PMID:28493400

McKnight, A. J., & McKnight, A. S. (1993). The effect of cellular phone use upon driver attention. *Accident; Analysis and Prevention*, *25*(3), 259–265. doi:10.1016/0001-4575(93)90020-W PMID:8323660

McNamara, N., & Kirakowski, J. (2005, July). Defining usability: quality of use or quality of experience? In *Professional Communication Conference, 2005. IPCC 2005. Proceedings. International* (pp. 200-204). IEEE. 10.1109/IPCC.2005.1494178

McQuail, D. (2005). *Mass communication theory* (5th ed.). London: Sage.

Means, B., Toyama, Y., Murphy, R., Bakia, M., & Jones, K. (2009). *Evaluation of Evidence-Based Practices in Online Learning: A Meta-Analysis and Review of Online Learning Studies. Project Report.* Centre for Learning Technology.

Meier, P., & Leaning, J. (2009). *Applying Technology to Crisis Mapping and Early Warning in Humanitarian Settings.* Working Paper Series.

Meissen, U., Faust, D., & Fuchs-Kittowski, F. (2013). WIND-A meteorological early warning system and its extensions towards mobile services. In EnviroInfo (pp. 612-621). Academic Press.

Mendoza, R. U., & Thelen, N. (2008). Innovations to make markets more inclusive for the poor. *Development Policy Review*, *26*(4), 427–458. doi:10.1111/j.1467-7679.2008.00417.x

Meyer, C. O., Baden, C. N., & Frère, M. (2017). Navigating the complexities of media roles in conflict: The INFOCORE approach. *Media, War & Conflict*, 1–19.

Meyrowitz, J. (2010). Shifting worlds of strangers: Medium theory and changes in "them" versus "us". *Sociological Inquiry*, *67*(1), 59–71. doi:10.1111/j.1475-682X.1997.tb00429.x

Miller, G. (2012). The Smartphone Psychology Manifesto. *Perspectives on Psychological Science*, *7*(3), 221–237. doi:10.1177/1745691612441215 PMID:26168460

Miller, V. (2011). *Understanding Digital Culture*. London: Sage.

Mills, J., Paul, E., Omer, R., & Heli, V. (2012). MoJo in Action: The Use of Mobiles in Conflict, Community, and Cross-platform Journalism. *Continuum*, *26*(5), 669–683. doi:10.1080/10304312.2012.706457

Minaz, C. B. (2017). *Investigation of university students smartphone addiction levels and usage purposes in terms of different variables.* Available from: http://dergipark.gov.tr/download/article-file/352363

Ministry of Electronics & Information Technology. (2018a). *Introduction to Digital India.* Retrieved November 21, 2018, from http://digitalindia.gov.in: http://digitalindia.gov.in/content/introduction

Ministry of Electronics & Information Technology. (2018b). *Universal Access to Mobile Connectivity.* Retrieved November 21, 2018, from http://digitalindia.gov.in: http://digitalindia.gov.in/content/universal-access-mobile-connectivity

Ministry of Electronics & Information Technology. (2018c). *Vision and Vision Areas.* Retrieved November 21, 2018, from http://digitalindia.gov.in: http://digitalindia.gov.in/content/vision-and-vision-areas

Mirani, L. (2017, December 12). No, Google's Pixel Buds won't change the world. *1843 Magazine.* Retrieved from https://www.1843magazine.com/technology/the-daily/no-googles-pixel-buds-wont-change-the-world

Mishra, A., Maheswarappa, S. S., Maity, M., & Samu, S. (2018). Adolescent's eWOM intentions: An investigation into the roles of peers, the Internet and gender. *Journal of Business Research*, *86*, 394–405. doi:10.1016/j.jbusres.2017.04.005

Misra, R., & Srivastava, S. (2016). M-education in India: An effort to improve educational outcomes with a special emphasis on Ananya Bihar. *On the Horizon*, *24*(2), 153–165.

Mittal, B. (1995). A comparative analysis of four scales of consumer involvement. *Psychology and Marketing*, *12*(7), 663–682. doi:10.1002/mar.4220120708

Mittal, S., Gandhi, S., & Tripathi, G. (2010). *Socio-economic impact of mobile phones on Indian agriculture.* New Delhi: Indian Council for Research on International Economic Relations.

Mizouni, R., Matar, M. A., Al Mahmoud, Z., Alzahmi, S., & Salah, A. (2014). A framework for context-aware self-adaptive mobile applications SPL. *Expert Systems with Applications*, *41*(16), 7549–7564. doi:10.1016/j.eswa.2014.05.049

Moghaddam, G. (2010). Information technology and gender gap: Toward a global view. *The Electronic Library*, *28*(5), 722–733. doi:10.1108/02640471011081997

Mohapatra, S., & Dash, M. (2011). Problems Associated with Artisans in Making of Handicrafts in Orissa, India. *Management Review: An International Journal*, *6*(1), 56–64.

MoMo Inc. (2016). *MoMo announces unaudited financial results for the second quarter 2016* [Press release]. Retrieved from http://media.corporate-ir.net/media_files/IROL/25/ 253834/2016/Momo2016Q2-final.pdf

Moor, K. D., Berte, K., Marez, L. D., Joseph, W., Deryckere, T., & Martens, L. (2010). User-Driven Innovation? Challenges of User Involvement in Future Technology Analysis. *Science & Public Policy*, *37*(1), 51–61. doi:10.3152/030234210X484775

Moorkens, J., O'Brien, S., & Vreeke, J. (2016). Developing and testing Kanjingo: A mobile app for post-editing. *Revista Tradumàtica: tecnologies de la traducció*, *14*, 58–66.

Morris, J. W., & Patterson, E. (2015). Podcasting and its apps: Software, sound, and the interfaces of digital audio. *Journal of Radio & Audio Media*, *22*(2), 220–230. doi:10.1080/19376529.2015.1083374

Mortensen, M. (2011). When citizen photojournalism sets the news agenda: Neda Agha Soltan as a Web 2.0 icon of post-election unrest in Iran. *Global Media and Communication*, *7*(1), 4–16. doi:10.1177/1742766510397936

Mramba, N., Apiola, M., Sutinen, E., Haule, M., Klomsri, T., & Msami, P. (2015, June). Empowering street vendors through technology: An explorative study in Dar es Salaam, Tanzania. In *Engineering, Technology and Innovation/International Technology Management Conference (ICE/ITMC), 2015 IEEE International Conference on* (pp. 1-9). IEEE.

Mudhai, O. F. (2011). Immediacy and openness in a digital Africa: Networked-convergent journalisms in Kenya. *Journalism*, *12*(6), 674–691. doi:10.1177/1464884911405470

Mundt, K., & Groves, M. (2016). A double-edged sword: The merits and the policy implications of Google Translate in higher education. *European Journal of Higher Education*, *6*(4), 387–401. doi:10.1080/21568235.2016.1172248

Muriithi, M. K., & Muriithi, I. W. (2013). Student's motives for utilizing social networking sites in private universities in Dar Es salaam, Tanzania. *Academic Research International*, *4*(4), 74–83.

Murphie, A., & Potts, J. (2003). *Culture and Technology*. London: Palgrave. doi:10.1007/978-1-137-08938-0

Murray, S., & Sapnar, A. M. (2016). Lez takes time: Designing lesbian contact in geosocial networking apps. *Critical Studies in Media Communication*, *33*(1), 53–69. doi:10.1080/15295036.2015.1133921

Naftali, M., & Findlater, L. (2014, October). Accessibility in context: understanding the truly mobile experience of smartphone users with motor impairments. In *Proceedings of the 16th international ACM SIGACCESS conference on Computers & accessibility* (pp. 209–216). ACM. doi:10.1145/2661334.2661372

Nakata, C., & Weidner, K. (2012). Enhancing new product adoption at the base of the pyramid: A contextualized model. *Journal of Product Innovation Management*, *29*(1), 21–32. doi:10.1111/j.1540-5885.2011.00876.x

National Bureau of Statistics of the People's Republic of China. (2011). *2010 The Sixth Nationwide Population Census Report*. Retrieved from http://www.stats.gov.cn/tjsj/pcsj/rkpc/6rp/indexch.htm

National Bureau of Statistics of the People's Republic of China. (2012). *The average life expectancy reaches to 74.83 in China*. Retrieved from http://www.stats.gov.cn/tjsj/tjgb/rkpcgb/qgrkpcgb/201209/t20120921_30330.html

Nayak, B. (2014). *3 Key Essentials to Achieving Mobile Learning Success*. Retrieved 8 May 2018 from https://cdns3.trainingindustry.com/media/17775903/essentials_of_mobile_learning_white_paper.pdf

Negroponte, N. (1997). *Being Digital*. Hainan Press. doi:10.1063/1.4822554

Neuhofer, B. (2012). An analysis of the perceived value of touristic location based services. In M. Fuchs, F. Ricci, & L. Cantoni (Eds.), *Information and Communication Technologies in Tourism 2012*. Vienna: Springer. doi:10.1007/978-3-7091-1142-0_8

Neuhofer, B., Buhalis, D., & Ladkin, A. (2015). Smart technologies for personalized experiences: A case study in the hospitality domain. *Electronic Markets*, *25*(3), 243–254. doi:10.100712525-015-0182-1

Newspaper, C. (2017). Smart phone usage rate in Turkey [İşte Türkiye'de akıllı telefon kullanım oranı]. *Cumhuriyet*. Available from: http://www.cumhuriyet.com.tr/haber/ekonomi/816129/iste_Turkiye_de_akilli_telefon_kullanim_orani.html

Nicholas, J., Larsen, M. E., Proudfoot, J., & Christensen, H. (2015). Mobile apps for bipolar disorder: A systematic review of features and content quality. *Journal of Medical Internet Research*, *17*(8), e198. doi:10.2196/jmir.4581 PMID:26283290

Niehaves, B., & Becker, J. (2008). The Age-divide in e-government–data, interpretations, theory fragments. In Oya, M., Uda, R., Yasunobu, C (Ed.), IFIP International Federation for Information Processing, Volume 286; Towards sustainable society on ubituitous networks. Boston: Springer. doi:10.1007/978-0-387-85691-9_24

Nieroda, M., Keeling, K., & Keeling, D. (2015). Acceptance of Mobile Apps for Health Self-management: Regulatory Fit Perspective. UMAP Workshops.

Norman, D. (1999). *The Invisible Computer*. Cambridge, MA: MIT Press.

Norman, D. (2004). *Emotional Design: Why We Love (or Hate) Everyday Things*. New York: Basic Books. doi:10.1145/985600.966013

Notley, T. (2009). Young people, online networks, and social inclusion. *Journal of Computer-Mediated Communication*, *14*(4), 1208–1227. doi:10.1111/j.1083-6101.2009.01487.x

Noulas, A., Scellato, S., Lathia, N., & Mascolo, C. (2012, December). Mining user mobility features for next place prediction in location-based services. In *Data mining (ICDM), 2012 IEEE 12th international conference on* (pp. 1038-1043). IEEE. 10.1109/ICDM.2012.113

O'bannon, B. W., & Thomas, K. (2014). Teacher perceptions of using mobile phones in the classroom: Age matters! *Computers & Education*, *74*, 15–25. doi:10.1016/j.compedu.2014.01.006

O'Brien, S. (2017). Machine Translation and Cognition. In J. W. Schwieter & A. Ferreira (Eds.), *The Handbook of Translation and Cognition* (pp. 311–331). Hoboken, NJ: John Wiley & Sons Inc.; doi:10.1002/9781119241485.ch17

O'Brien, S., Moorkens, J., & Vreeke, J. (2014). Kanjingo: A Mobile App for Post-Editing. In M. Tadic, P. Koehn, J. Roturier, & A. Way (Eds.), *Proceedings of the 17th Annual Conference of the European Association for Machine Translation (EAMT 2014)* (pp. 137–141). Dubrovnik, Croatia: EAMT.

O'Hagan, M. (2012). The impact of new technologies on translation studies. In C. Millán & F. Bartrina (Eds.), *The Routledge Handbook of Translation Studies* (pp. 503–518). Routledge. doi:10.4324/9780203102893

O'Hagan, M. (2016a). Massively Open Translation: Unpacking the Relationship Between Technology and Translation in the 21st Century. *International Journal of Communication*, *10*, 929–946. Retrieved from http://ijoc.org/index.php/ijoc/article/view/3507/1572

O'Hagan, M. (2016b). Response by O'Hagan to "Translation and the materialities of communication." *Translation Studies*, *9*(3), 322–326. doi:10.1080/14781700.2016.1170628

Och, F. (2006). *Statistical machine translation live*. Retrieved November 8, 2017, from https://research.googleblog.com/2006/04/statistical-machine-translation-live.html

Ofcom. (2017). *Children and Parents: Media Use and Attitudes Report 2017*. Accessed from https://www.ofcom.org.uk/research-and-data/media-literacy-research/childrens/children-parents-2017

Ofcom. (2018). *Adults Media Use and Attitudes Report 2018*. Accessed from https://www.ofcom.org.uk/__data/assets/pdf_file/0011/113222/Adults-Media-Use-and-Attitudes-Report-2018.pdf

Oh, S., Lehto, X. Y., & Park, J. (2009). Travelers' intent to use mobile technologies as a function of effort and performance expectancy. *Journal of Hospitality Marketing & Management*, *18*(8), 765–781. doi:10.1080/19368620903235795

Oladosu, J. B., & Emuoyibofarhe, J. O. (2012). A Yoruba—English Language Translator for Doctor—Patient Mobile Chat Application. *International Journal of Computers and Applications*, *34*(3), 149–156. doi:10.2316/Journal.202.2012.3.202-3079

Oliveira, R. R., Noguez, F. C., Costa, C. A., Barbosa, J. L., & Prado, M. P. (2013). SWTRACK: An intelligent model for cargo tracking based on off-the-shelf mobile devices. *Expert Systems with Applications*, *40*(6), 2023–2031. doi:10.1016/j.eswa.2012.10.021

Olphert, W., & Damodaran, L. (2013). Older people and digital disengagement: A fourth digital divide? *Gerontology*, *59*(6), 564–570. doi:10.1159/000353630 PMID:23969758

Oluwatobi, S., & Olurinola, O. I. (2015). *Mobile learning in Africa: strategy for educating the poor*. Academic Press.

Onedio.com. (2015). *Akilli Telefon Bagimliliginin Insan Iliskilerine Etkisini Gozler Onune Seren Reklam Kampanyasi*. Available from: https://onedio.com/haber/akilli-telefon-bagimliliginin-insan-iliskilerini-nasil-etkiledigini-gozler-onune-seren-fotograflar-550175

Ong, E. Y., Ang, R. P., Ho, J. C., Lim, J. C., Goh, D. H., Lee, C. S., & Chua, A. Y. (2011). Narcissism, extraversion and adolescents' self-presentation on Facebook. *Personality and Individual Differences*, *50*(2), 180–185. doi:10.1016/j.paid.2010.09.022

Oppegaard, B., & Rabby, M. K. (2016). Proximity: Revealing new mobile meanings of a traditional news concept. *Digital Journalism*, *4*(5), 621–638. doi:10.1080/21670811.2015.1063075

Organista-Sandoval, J., & Serrano-Santoyo, A. (2014). Appropriation and educational uses of mobile phones by students and teachers at a public university in Mexico. *Creative Education*.

Otto, F., & Meyer, C. O. (2012). Missing the story? Changes in foreign news reporting and their implications for conflict prevention. *Media, War & Conflict*, *5*(3), 205–221. doi:10.1177/1750635212458621

Ozascilar, M. (2012). Mobile phone usage and personal security of young individuals: University students' usage of mobile for personal security [Genç bireylerin cep telefonu kullanımı ve bireysel güvenlik: üniversite öğrencilerinin cep telefonunu bireysel güvenlik amaçlı kullanımları]. *Journal of Sociological Research.[Online]*, *15*, 1. Available from http://dergipark.ulakbim.gov.tr/sosars/article/view/5000093023

Ozkan, S., & Koseler, R. (2009). Multi-dimensional students' evaluation of e-learning systems in the higher education context: An empirical investigation. *Computers & Education*, *53*(4), 1285–1296. doi:10.1016/j.compedu.2009.06.011

Pagoto, S., Schneider, K., Jojic, M., DeBiasse, M., & Mann, D. (2013). Evidence-based strategies in weight-loss mobile apps. *American Journal of Preventive Medicine*, *45*(5), 576–582. doi:10.1016/j.amepre.2013.04.025 PMID:24139770

Pankratius, V., Lind, F., Coster, A., Erickson, P., & Semeter, J. (2014). Mobile crowd sensing in space weather monitoring: The Mahali project. *Communications Magazine, IEEE, 52*(8), 22–28. doi:10.1109/MCOM.2014.6871665

Pantti, M. (2013). Seeing and not seeing the Syrian crisis: New visibility and the visual framing of the Syrian conflict in seven newspapers and their online editions. *Journal Journalism, Media and Cultural Studies*. Retrieved from cf.ac.uk/jomec/jomecjournal/4-november2013/Pantti_Syria.pdf

Papadopoulos, K. A. (2014). Citizen camera-witnessing: Embodied political dissent in the age of 'mediated mass self-communication'. *New Media & Society, 16*(5), 753–769. doi:10.1177/1461444813489863

Papadopoulos, K. A., & Pantti, M. (2013). Re-imagining crisis reporting: Professional ideology of journalists and citizen eyewitness images. *Journalism, 14*(7), 960–977. doi:10.1177/1464884913479055

Papadopoulos, K. A., & Pantti, M. (2013). The Media Work of Syrian Diaspora Activists: Brokering Between the Protest and Mainstream Media. *International Journal of Communication, 7*, 2185–2206.

Papanikolaou, K., & Mavromoustakos, S. (2006). *Critical success factors for the development of mobile learning applications*. Retrieved 8 January 2018 from https://www.researchgate.net/publication/221655699_Critical_Success_Factors_for_the_Development_of_Mobile_Learning_Applications

Paramboukis, O., Skues, J., & Wise, L. (2016). An Exploratory Study of the Relationships between Narcissism, Self-Esteem and Instagram Use. *Social Networking, 5*(02), 82–92. doi:10.4236n.2016.52009

Park, N. (2014). Nature of Youth Smartphone Addiction in Korea Diverse Dimensions of Smartphone Use and Individual Traits. *Journal of communication research, 51*(1), 100–132.

Park, N., Kim, Y., Young, H., & Shim, H. (2013). Factors influencing Smartphone use and dependency in South Korea. *Computers in Human Behavior, 29*(4), 1763–1770. doi:10.1016/j.chb.2013.02.008

Park, S. (2017). *Digital capital*. London: Palgrave Macmillan UK. doi:10.1057/978-1-137-59332-0

Park, W. K. (2005). Mobile Phone Addiction: Mobile Communications. *Computer Supported Cooperative Work, 31*(3), 253–272.

Patel, A., & Gadit, A. (2008). Karo-kari: A form of honour killing in Pakistan. *Transcultural Psychiatry, 45*(4), 683–694. doi:10.1177/1363461508100790 PMID:19091732

Patro, A., Rayanchu, S., Griepentrog, M., Ma, Y., & Banerjee, S. (2013, December). Capturing mobile experience in the wild: a tale of two apps. In *Proceedings of the ninth ACM conference on Emerging networking experiments and technologies* (pp. 199-210). ACM. 10.1145/2535372.2535391

Patsakis, C., Zigomitros, A., & Solanas, A. (2015). Analysis of privacy and security exposure in mobile dating applications. In *International Conference on Mobile, Secure and Programmable Networking* (pp. 151-162). Springer. 10.1007/978-3-319-25744-0_13

Pavlik, J. V., & Bridges, F. (2013). The emergence of augmented reality (AR) as a storytelling medium in journalism. *Journalism & Communication Monographs, 15*(1), 4–59. doi:10.1177/1522637912470819

Pearce, K. E. (2013). Phoning it in: Theory in mobile media and communication in developing countries. *Mobile Media & Communication, 1*(1), 76–82. doi:10.1177/2050157912459182

Pearson, C., & Hussain, Z. (2017). Smartphone use, addiction, narcissism, and personality: A mixed methods investigation. In Gaming and Technology Addiction: Breakthroughs in Research and Practice (pp. 212-229). IGI Global. doi:10.4018/978-1-5225-0778-9.ch011

Pendit, U. C., Zaibon, S. B., & Bakar, J. A. A. (2014). Mobile Augmented Reality for Enjoyable Informal Learning in Cultural Heritage Site. *International Journal of Computers and Applications*, *92*(14).

Peng, L. (2013). Evolution of ''Connection'': The Basic Clue of the Development of Internet. *Chinese Journal of Journalism & Communication*, (2), 6-19.

Peng, L. (2015). Scene: new elements of media in the mobile age. *Journalism Review*, (3), 21-27.

Pentina, I., & Tarafdar, M. (2014). From "information" to "knowing": Exploring the role of social media in contemporary news consumption. *Computers in Human Behavior*, *35*, 211–223. doi:10.1016/j.chb.2014.02.045

Persaud, A., & Azhar, I. (2012). Innovative mobile marketing via smartphones: Are consumers ready? *Marketing Intelligence & Planning*, *30*(4), 418–443. doi:10.1108/02634501211231883

Persson, V., & Nouri, J. (2018). A systematic review of second language learning with mobile technologies. *International Journal of Emerging Technologies in Learning*, *13*(2), 188–210. doi:10.3991/ijet.v13i02.8094

Pew Research Centre. (2018). *Mobile Fact Sheet*. Accessed from http://www.pewinternet.org/fact-sheet/mobile/

Phua, J., Jin, S. V., & Kim, J. J. (2017). Uses and gratifications of social networking sites for bridging and bonding social capital: A comparison of Facebook, Twitter, Instagram, and Snapchat. *Computers in Human Behavior*, *72*, 115–122. doi:10.1016/j.chb.2017.02.041

Pielot, M., De Oliveira, R., Kwak, H., & Oliver, N. (2014, April). Didn't you see my message?: predicting attentiveness to mobile instant messages. In *Proceedings of the 32nd annual ACM conference on Human factors in computing systems* (pp. 3319-3328). ACM. 10.1145/2556288.2556973

Pierskalla, J. H. & Hollenbach, F. M. (2013). Technology and Collective Action: The Effect of Cell Phone Coverage on Political Violence in Africa. *American Political Science Review*, *107*(2).

Pinder, R. J., Murphy, D., Hatch, S. L., Iversen, A., Dandeker, C., & Wessely, S. (2009). A Mixed Methods Analysis of the Perceptions of the Media by members of the British Forces during the Iraq War. *Armed Forces and Society*, *36*(1), 131–152. doi:10.1177/0095327X08330818

Pine, B. J., & Gilmore, J. H. (2013). The experience economy: past, present and future. *Handbook on the experience economy*, 21-44.

Pine, B. J., & Gilmore, J. H. (1998). Welcome to the experience economy. *Harvard Business Review*, *76*, 97–105. PMID:10181589

Pine, B. J., Pine, J., & Gilmore, J. H. (1999). *The experience economy: work is theatre & every business a stage*. Harvard Business Press.

Plaza, I., Demarzo, M. M. P., Herrera-Mercadal, P., & García-Campayo, J. (2013). Mindfulness-based mobile applications: Literature review and analysis of current features. *JMIR mHealth and uHealth*, *1*(2), e24. doi:10.2196/mhealth.2733 PMID:25099314

Plowman, L., Stevenson, O., Stephen, C., & McPake, J. (2012). Preschool children's learning with technology at home. *Computers & Education*, *59*(1), 30–37. doi:10.1016/j.compedu.2011.11.014

Pokhrel, P., Fagan, P., Herzog, T. A., Laestadius, L., Buente, W., Kawamoto, C. T., ... Unger, J. B. (2018). Social media e-cigarette exposure and e-cigarette expectancies and use among young adults. *Addictive Behaviors*, *78*, 51–58. doi:10.1016/j.addbeh.2017.10.017 PMID:29127784

Poon, D. C. H., & Leung, L. (2011). *Effects of narcissism, leisure boredom, and gratifications sought on user-generated content among net-generation users.* Academic Press.

Powell, J., & Cook, I. (2000). "A Tiger Behind, and Coming up Fast": Governmentality and the Politics of Population Control in China. *Journal of Aging and Identity, 5.*

Powell, J. L., & Cook, I. G. (2009). Global ageing in comparative perspective: A critical discussion. *The International Journal of Sociology and Social Policy, 29*(7/8), 388–400. doi:10.1108/01443330910975696

Power, D. J. (2013). Mobile decision support and business intelligence: An overview. *Journal of Decision Systems, 22*(1), 4–9. doi:10.1080/12460125.2012.760267

Powers, E. (2012). Learning to do it all: When it comes time to fill one of those precious hiring slots, news outlets are looking for journalists with a wide array of skills. *American Journalism Review, 34*(1), 10–14.

Prahalad, C. K., & Hart, S. L. (2002). The fortune at the bottom of the pyramid. *Strategy and Business, 26,* 54–67.

Prasad, R., Natarajan, P., Stallard, D., Saleem, S., Ananthakrishnan, S., Tsakalidis, S., & Challenner, A. (2013). BBN TransTalk: Robust multilingual two-way speech-to-speech translation for mobile platforms. *Computer Speech & Language, 27*(2), 475–491. doi:10.1016/j.csl.2011.10.003

Qi, H., & Gani, A. (2012, May). Research on mobile cloud computing: Review, trend and perspectives. In *Digital Information and Communication Technology and it's Applications (DICTAP), 2012 Second International Conference on* (pp. 195-202). IEEE.

Qiang, C. Z., Kuek, S. C., Dymond, A., & Esselaar, S. (2012). *Mobile applications for agriculture and rural development.* Academic Press.

Qi, Y., & Mei, W. (2016). Examining the Role of WeChat in Advertising. In X. Xiaoge (Ed.), *Handbook of Research on Human Social Interaction in the Age of Mobile Devices* (pp. 386–405). Hershey, PA: IGI Global.

Quinn, S. (2006). War Reporting and the Technologies of Convergence. In R. D. Berenger (Ed.), *Cybermedia go to war: role of converging media during and after the 2003 Iraq War.* Washington, DC: Marquette Books. Retrieved from http://marquettebooks.com/images/CyberMediaNP.pdf

Quinn, S. (2011). *Mojo-mobile journalism in the Asian Region.* KAS.

Raacke, J., & Bonds-Raacke, J. (2008). MySpace and Facebook: Applying the uses and gratifications theory to exploring friend-networking sites. *Cyberpsychology & Behavior, 11*(2), 169–174. doi:10.1089/cpb.2007.0056 PMID:18422409

Rahimi, M., & Miri, S. S. (2014). The Impact of Mobile Dictionary Use on Language Learning. *Procedia: Social and Behavioral Sciences, 98,* 1469–1474. doi:10.1016/j.sbspro.2014.03.567

Rahman, S. A., Taghizadeh, S. K., Ramayah, T., & Alam, M. M. D. (2017). Technology acceptance among micro-entrepreneurs in marginalized social strata: The case of social innovation in Bangladesh. *Technological Forecasting and Social Change, 118,* 236–245. doi:10.1016/j.techfore.2017.01.027

Rahmati, A., Shepard, C., Zhong, L., & Kortum, P. (2015). Practical context awareness: Measuring and utilizing the context dependency of mobile usage. *IEEE Transactions on Mobile Computing, 14*(9), 1932–1946. doi:10.1109/TMC.2014.2365199

Rains, S. A., & Tukachinsky, R. (2015). An examination of the relationships among uncertainty, appraisal, and information-seeking behavior proposed in uncertainty management theory. *Health Communication, 30*(4), 339–349. doi:10.1080/10410236.2013.858285 PMID:24905910

Rakow, L. F. (1992). *Gender on the line: Women, the telephone, and community life.* Urbana, IL: University of Illinois Press.

Ramanathan, N., Swendeman, D., Comulada, W. S., Estrin, D., & Rotheram-Borus, M. J. (2013). Identifying preferences for mobile health applications for self-monitoring and self-management: Focus group findings from HIV-positive persons and young mothers. *International Journal of Medical Informatics, 82*(4), e38–e46. doi:10.1016/j.ijmedinf.2012.05.009 PMID:22704234

Ranjan, R., & Singh, S. (2017). *Energy deprivation of Indian households: evidence from NSSO data.* Academic Press.

Ranzini, G., & Lutz, C. (2017). Love at first swipe? Explaining Tinder self-presentation and motives. *Mobile Media & Communication, 5*(1), 80–101. doi:10.1177/2050157916664559

Rashid, A. T., & Elder, L. (2009). Mobile Phones and Development: An Analysis of IDRC-Supported Projects. *The Electronic Journal on Information Systems in Developing Countries, 36*(1), 1–16. doi:10.1002/j.1681-4835.2009.tb00249.x

Rawendy, Y., Ying, Y., Arifin, Y., & Rosalin, K. (2017). Design and Development Game Chinese Language Learning with Gamification and Using Mnemonic Method. *Procedia Computer Science, 116*, 61–67. doi:10.1016/j.procs.2017.10.009

Razvi, S., Srivastava, R., & Halder, B. (2016). Mobile Phone: A Public Tool (Analysing the Use of Mobile Technology in Civic Participation, Education & Health). New Delhi: Digital Empowerment Foundation & UNICEF.

Reading, A. (2009). Mobile Witnessing: Ethics and the Camera Phone in the 'War on Terror'. *Globalizations, 6*(1), 61–76. doi:10.1080/14747730802692435

Realini, C., & Mehta, K. (2015). *Financial Inclusion at the Bottom of the Pyramid.* FriesenPress.

Reed, L. A., Ward, L. M., Tolman, R. M., Lippman, J. R., & Seabrook, R. C. (2018). The association between stereotypical gender and dating beliefs and digital dating abuse perpetration in adolescent dating relationships. *Journal of Interpersonal Violence.* doi:0886260518801933

Reed, L. A., Tolman, R. M., & Ward, L. M. (2016). Snooping and sexting: Digital media as a context and tool for dating violence among college students. *Violence Against Women, 22*(13), 1556–1576. doi:10.1177/1077801216630143 PMID:26912297

Reid, A. J., & Thomas, C. N. (2017). A Case Study in Smartphone Usage and Gratification in the Age of Narcissism. *International Journal of Technology and Human Interaction, 13*(2), 40–56. doi:10.4018/IJTHI.2017040103

Reis, O., & Youniss, J. (2004). Patterns in identity change and development in relationships with mothers and friends. *Journal of Adolescent Research, 19*(1), 31–44. doi:10.1177/0743558403258115

Report, S. (2018). Retrieved June 12, 2018, from http://www.sootoo.com/content/675436.shtml

Reychav, I., & Wu, D. (2014). Exploring mobile tablet training for road safety: A uses and gratifications perspective. *Computers & Education, 71*, 43–55. doi:10.1016/j.compedu.2013.09.005

Rice, E., Craddock, J., Hemler, M., Rusow, J., Plant, A., Montoya, J., & Kordic, T. (2018). Associations Between Sexting Behaviors and Sexual Behaviors Among Mobile Phone Owning Teens in Los Angeles. *Child Development, 89*(1), 110–117. doi:10.1111/cdev.12837 PMID:28556896

Richardson, A. V. (2017). Bearing Witness While Black. *Digital Journalism, 5*(6), 673–698. doi:10.1080/21670811.2016.1193818

Riikonen, A., Smura, T., Kivi, A., & Töyli, J. (2013). Diffusion of mobile handset features: Analysis of turning points and stages. *Telecommunications Policy, 37*(6), 563–572. doi:10.1016/j.telpol.2012.07.011

Ritchie, J., Lewis, J., Nicholls, C. M., & Ormston, R. (Eds.). (2013). *Qualitative research practice: A guide for social science students and researchers*. London: Sage.

Robertson, T. W., Yan, Z., & Rapoza, K. A. (2018). Is resilience a protective factor of internet addiction? *Computers in Human Behavior*, *78*, 255–260. doi:10.1016/j.chb.2017.09.027

Rosales, R. G. (2013). Citizen participation and the uses of mobile technology in radio broadcasting. *Telematics and Informatics*, *30*(3), 252–257. doi:10.1016/j.tele.2012.04.006

Rosenberg, J., & Egbert, N. (2011). Online impression management: Personality traits and con- cerns for secondary goals as predictors of self-presentation tactics on Facebook. *Journal of Computer-Mediated Communication*, *17*(1), 1–18. doi:10.1111/j.1083-6101.2011.01560.x

Rosen, L., Carrier, L., & Cheever, N. (2013). Facebook and texting made me do it: Media-induced task-switching while studying. *Computers in Human Behavior*, *29*(3), 948–958. doi:10.1016/j.chb.2012.12.001

Rubin, A. M., & Rubin, R. B. (1982). Contextual age and television use. *Human Communication Research*, *8*(3), 228–244. doi:10.1111/j.1468-2958.1982.tb00666.x

Rubin, R. B., Rubin, A. M., Graham, E., Perse, E. M., & Seibold, D. (2010). *Communication research measures II: A sourcebook*. Routledge. doi:10.4324/9780203871539

Ruggiero, T. E. (2000). Uses and gratifications theory in the 21st century. *Mass Communication & Society*, *3*(1), 3–37. doi:10.1207/S15327825MCS0301_02

Rupert, M., & Hawi, N. (2016). Relationships among smartphone addiction, stress, academic performance, and satisfaction with life. *Computer in Human Behavior. [Online]*, *57*, 321–325. doi:10.1016/j.chb.2015.12.045

Rusbridger, A. (2012, February 29). Alan Rusbridger on open journalism at the Guardian:'Journalists are not the only experts in the world'–video. *The Guardian*.

Rutter, M. (2000). Psychosocial influences: Critiques, findings, and research needs. *Development and Psychopathology*, *12*(3), 375–405. doi:10.1017/S0954579400003072 PMID:11014744

Ryu, Y. S., & Smith-Jackson, T. L. (2006). Usability Questionnaire Items for Mobile Products and Content Validity. *Proceedings of the Human Factors and Ergonomics Society 50th Annual Meeting*.

Sadana, R., Blas, E., Budhwani, S., Koller, T., & Paraje, G. (2016). Healthy ageing: Raising awareness of inequalities, determinants, and what could be done to improve health equity. *The Gerontologist*, *56*(Suppl 2), S178–S193. doi:10.1093/geront/gnw034 PMID:26994259

Sage, C., & Burgio, E. (2018). Electromagnetic fields, pulsed radiofrequency radiation, and epigenetics: How wireless technologies may affect childhood development. *Child Development*, *89*(1), 129–136. doi:10.1111/cdev.12824 PMID:28504324

Sager, R. L., Alderson, K. G., & Boyes, M. C. (2016). Hooking-up through the use of mobile applications. *Computer Communication & Collaboration*, *4*(2), 15–41.

Salehan, M., & Negahban, A. (2013). Social networking on smartphones: When mobile phones become addictive. *Computers in Human Behavior*, *29*(6), 2632–2639. doi:10.1016/j.chb.2013.07.003

Santosham, S. (2015). Closing the gender gap in mobile phone access and use. *Better Than Cash*. Available from: https://www.betterthancash.org/news/blogs-stories/closing-the-gender-gap-in-mobile-phone-access-and-use

Santos, I. M., & Ali, N. (2012). Exploring the uses of mobile phones to support informal learning. *Education and Information Technologies*, *17*(2), 187–203. doi:10.100710639-011-9151-2

Santrock, J. W. (1998). *Adolescence: Exploring Peer Relations*. New Delhi: Tata McGraw Hill Publishing Company.

Sanz-Blas, S., Ruiz-Mafé, C., & Martí-Parreño, J. (2015). Message-driven factors influencing opening and forwarding of mobile advertising messages. *International Journal of Mobile Communications*, *13*(4), 339–357. doi:10.1504/IJMC.2015.070058

Sarmento, T., & Patrício, L. (2012). Mobile Service Experience-a quantitative study. *AMAServsig2012*.

Schall, G., Zollmann, S., & Reitmayr, G. (2013). Smart Vidente: Advances in mobile augmented reality for interactive visualization of underground infrastructure. *Personal and Ubiquitous Computing*, *17*(7), 1533–1549. doi:10.100700779-012-0599-x

Schejter, A. M., & Cohen, A. A. (2013). Mobile phone usage as an indicator of solidarity: Israelis at war in 2006 and 2009. *Mobile Media & Communication*, *1*(2), 174–195. doi:10.1177/2050157913476706

Schmitt, D. P. (2005). Sociosexuality from Argentina to Zimbabwe: A 48-nation study of sex, culture, and strategies of human mating. *Behavioral and Brain Sciences*, *28*(02), 247–275. doi:10.1017/S0140525X05000051 PMID:16201459

Schmitt, D. P., Shackelford, T. K., & Buss, D. M. (2001). Are men really more oriented toward short-term mating than women? A critical review of theory and research. *Psychology Evolution & Gender*, *3*(3), 211–239. doi:10.1080/14616660110119331

Schwalbe, C. B. (2016). Changing faces: the first five weeks of the Iraq war. In R. D. Berenger (Ed.), *Cybermedia go to war: role of converging media during and after the 2003 Iraq War*. Washington, DC: Marquette Books. Retrieved from http://marquettebooks.com/images/CyberMediaNP.pdf

Scolari, C. A., Aguado, J. M., & Feijóo, C. (2012). Mobile Media: Towards a Definition and Taxonomy of Contents and Applications. *International Journal of Interactive Mobile Technologies*, *6*(2).

Sendra, S., Granell, E., Lloret, J., & Rodrigues, J. J. (2014). Smart collaborative mobile system for taking care of disabled and elderly people. *Mobile Networks and Applications*, *19*(3), 287–302. doi:10.100711036-013-0445-z

Seo, D. G., Park, Y., Kim, M. K., & Park, J. (2016). Mobile phone dependency and its impacts on adolescents' social and academic behaviors. *Computers in Human Behavior*, *63*, 282–292. doi:10.1016/j.chb.2016.05.026

Seth, A., & Ganguly, K. (2017). Digital technologies transforming Indian agriculture. *The Global Innovation Index*, 105-111.

Shambare, R., Rugimbana, R., & Zhowa, T. (2012). Are mobile phones the 21st century addiction? *African Journal of Business Management*, *6*(2), 573–577.

Shapiro, J., & Weidmann, N. B. (2015). Is the phone mightier than the sword? Cell phones and insurgent violence in Iraq. *International Organization*, *69*(2), 247–274. doi:10.1017/S0020818314000423

Shapiro-Mathews, E., & Barton, A. J. (2013). Using the patient engagement framework to develop an institutional mobile health strategy. *Clinical Nurse Specialist CNS*, *27*(5), 221–223. doi:10.1097/NUR.0b013e3182a0b9e2 PMID:23942098

Sharma, N. C. (2018, September 8). *Mobile app saves 12,000 flood victims in Kerala*. Retrieved November 21, 2018, from https://www.livemint.com: https://www.livemint.com/Politics/56R0AnrHfdsP2lagpTGrpJ/Mobile-app-saves-12000-flood-victims-in-Kerala.html

Sharples, M., Arnedillo-Sánchez, I., Milrad, M., & Vavoula, G. (2009). Mobile Learning. *Technology-Enhanced Learning*, 233-249. doi:10.1007/978-1-4020-9827-7_14

Sharples, M. (2013). Mobile learning: Research, practice and challenges. *Distance Education in China*, *3*(5), 5–11.

Shaukat, R., & Shah, I. (2014). Farmers Inclinations to Adoption of Mobile Phone: Agriculture Information and Trade System in Pakistan. *Journal of Economics and Social Studies*, *4*(2), 191–220. doi:10.14706/JECOSS11428

Shaw, H., Ellis, D. A., & Ziegler, F. V. (2018). The Technology Integration Model (TIM). Predicting the continued use of technology. *Computers in Human Behavior*, *83*, 204–214. doi:10.1016/j.chb.2018.02.001

Shaw, L., & Gant, L. (2002). Users divided?Exploring the gender gap in internet use. *Cyber Psychology & Behavior*, *5*(6), 517–527. doi:10.1089/109493102321018150 PMID:12556114

Shazad, M., Shazad, N., Ahmed, T., Hussain, A., & Riaz, F. (2015). Mobile phones addiction among university students: Evidence from twin cities of Pakistan. *Journal of Social Sciences*, *1*(11), 416–420.

Shehabat, A. (2013). *The social media cyber-war: the unfolding events in the Syrian revolution 2011* (Australian Edition). Global Media Journal.

Sheldon, P. (2013). Voices that cannot be heard: Can shyness explain how we communicate on Facebook versus face-to-face? *Computers in Human Behavior*, *29*(4), 1402–1407. doi:10.1016/j.chb.2013.01.016

Sheldon, P., & Bryant, K. (2016). Instagram: Motives for its use and relationship to narcissism and contextual age. *Computers in Human Behavior*, *58*, 89–97. doi:10.1016/j.chb.2015.12.059

Sherman, L. E., Greenfield, P. M., Hernandez, L. M., & Dapretto, M. (2018). Peer influence via Instagram: Effects on brain and behavior in adolescence and young adulthood. *Child Development*, *89*(1), 37–47. doi:10.1111/cdev.12838 PMID:28612930

Sherman, L. E., Payton, A. A., Hernandez, L. M., Greenfield, P. M., & Dapretto, M. (2016). The power of the like in adolescence: Effects of peer influence on neural and behavioral responses to social media. *Psychological Science*, *27*(7), 1027–1035. doi:10.1177/0956797616645673 PMID:27247125

Shih, B., Chen, C., & Li, C. (2013). The exploration of the mobile Mandarin learning system by the application of TRIZ theory. *Computer Applications in Engineering Education*, *21*(2), 343–348. doi:10.1002/cae.20478

Shimamoto, D., Yamada, H., & Gummert, M. (2015). Mobile phones and market information: Evidence from rural Cambodia. *Food Policy*, *57*, 135–141. doi:10.1016/j.foodpol.2015.10.005

Shyle, I. (2011). Global crisis and its effects in the developed and emergent countries-"The bottom of the pyramid" as an innovation resource. *EMAJ: Emerging Markets Journal*, *1*(2), 48–58. doi:10.5195/EMAJ.2011.12

Siau, K., Lim, E., & Shen, Z. (2001). Mobile commerce: Promises, challenges, and research agenda. *Journal of Database Management*, *12*(3), 4–13. doi:10.4018/jdm.2001070101

Sife, A. S., Kiondo, E., & Lyimo-Macha, J. G. (2010). Contribution of mobile phones to rural livelihoods and poverty reduction in Morogoro region, Tanzania. *The Electronic Journal on Information Systems in Developing Countries*, *42*(1), 1–15. doi:10.1002/j.1681-4835.2010.tb00299.x

Silverstein, M. (2013, September). *Introducing the Mobile Experience Scorecard*. Retrieved from http://www.thesearchagents.com/2013/09/introducing-the-mobile-experience-scorecard/

Silverstein, M. (2014). *Female-Oriented Sites More Optimized for Mobile?* Retrieved from http://www.thesearchagents.com/2014/03/female-oriented-sites-more-optimized-for-mobile/

Simay, A. E. (2009). Mobile Phone Usage and Device Selection of University Students. *Symposium for Young Researchers*, *20-22*, 185-193.

Sims, F., Williams, M. A., & Elliot, S. (2007, July). Understanding the Mobile Experience Economy: A key to richer more effective M-Business Technologies, Models and Strategies. In *Management of Mobile Business, 2007. ICMB 2007. International Conference on the* (pp. 12-12). IEEE.

Singh, V. K., Freeman, L., Lepri, B., & Pentland, A. S. (2013, September). Predicting spending behavior using socio-mobile features. In *Social Computing (SocialCom), 2013 International Conference on* (pp. 174-179). IEEE. 10.1109/SocialCom.2013.33

Sivapragasam, N. (2008). *Hit me with a missed call: The use of missed calls at the bottom of the pyramid*. Academic Press.

Smith, A., & Anderson, M. (2016). 5 facts about online dating. *Pew Research Center*. Retrieved from http://www.pewresearch.org/fact-tank/2016/02/29/5-facts-about-online-dating/(accessed July 3rd 2018)

Smith, A. (2016). *15% of American adults have used online dating sites or mobile dating apps*. Pew Research Center.

Sneps-Sneppe, M., & Namiot, D. (2013). Smart cities software: customized messages for mobile subscribers. In *Wireless Access Flexibility* (pp. 25–36). Springer Berlin Heidelberg. doi:10.1007/978-3-642-39805-6_3

Solis, R. J. C., & Wong, K. Y. J. (2018). To meet or not to meet? Measuring motivations and risks as predictors of outcomes in the use of mobile dating applications in China To meet or not to meet? Measuring motivations and risks as predictors of outcomes in the use of mobile dating applications in China. *Chinese Journal of Communication*, 1–20. doi:10.1080/17544750.2018.1498006

Song, I., Larose, R., Eastin, M. S., & Lin, C. A. (2004). Internet gratifications and internet addiction: On the uses and abuses of new media. *Cyberpsychology & Behavior*, 7(4), 384–394. doi:10.1089/cpb.2004.7.384 PMID:15331025

Sridhar, K. S., & Sridhar, V. (2007). Telecommunications Infrastructure and Economic Growth: Evidence from Developing Countries. Applied Econometrics and International Development. *Euro-American Association of Economic Development*, 7(2), 37–61.

Srivastava, R., & Sen, A. (2016). Mobile Phones for Social and Behaviour Change + *Initiatives in India*. New Delhi: Digital Empowerment Foundation.

Ssonko, G. W. (2010). *The role of mobile money services in enhancing financial inclusion in Uganda*. Bank of Uganda.

Stald, G. B., Green, L., Barbowski, M., Haddon, L., Mascheroni, G., Sagvari, B., ... Tzaliki, L. (2014). *Online on the mobile: internet use on the smartphone and associated risks among youth in Europe*. Academic Press.

Statista. (2018). *Number of smartphone users worldwide from 2014 to 2010 (in billions)*. Retrieved from https://www.statista.com/statistics/330695/number-of-smartphone-users-worldwide/

Statista. (2018). *Statistics and Market Data on Social Media & User-Generated Content*. Accessed at https://www.statista.com/markets/424/topic/540/social-media-user-generated-content/

Statista. (2018a). Retrieved from https://www.statista.com/outlook/372/100/online-dating/worldwide#market-age

Statista. (2018b). Retrieved from https://www.statista.com/statistics/449390/quarterly-revenue-match-group/(accessed May 1st 2018)

Statista. (2018c). Retrieved from https://www.statista.com/statistics/826778/most-popular-dating-apps-by-audience-size-usa/(accessed May 2018)

Statista.com. (2018). *Number of smartphone users worldwide from 2014 to 2020 (in billions)*. Retrieved from: https://www.statista.com/statistics/330695/number-of-smartphone-users-worldwide/

Statistia. (2018). *Percentage of all global web pages served to mobile phones from 2009 to 2018*. Retrieved June 12, 2018, from https://www.statista.com/statistics/241462/global-mobile-phone-website-traffic-share/

Stephen, C., Stevenson, O., & Adey, C. (2013). Young children engaging with technologies at home: The influence of family context. *Journal of Early Childhood Research, 11*(2), 149–164. doi:10.1177/1476718X12466215

Stig, T., & Ni, A. (2010). He is he and I am I: Individual and collective among China's Elderly. In M. Halskov Hansen & R. Svarverud (Eds.), *iChina: The rise of the individual in modern Chinese society*. Copenhagen: NIAS Press.

Stritzke, W. G., Nguyen, A., & Durkin, K. (2004). Shyness and computer-mediated communication: A self-presentational theory perspective. *Media Psychology, 6*(1), 1–22. doi:10.12071532785xmep0601_1

Ström, R., Vendel, M., & Bredican, J. (2014). Mobile marketing: A literature review on its value for consumers and retailers. *Journal of Retailing and Consumer Services, 21*(6), 1001–1012. doi:10.1016/j.jretconser.2013.12.003

Subrahmanyam, K., Reich, S. M., Waechter, N., & Espinoza, G. (2008). Online and offline social networks: Use of social networking sites by emerging adults. *Journal of Applied Developmental Psychology, 29*(6), 420–433. doi:10.1016/j.appdev.2008.07.003

Sumter, S. R., Vandenbosch, L., & Ligtenberg, L. (2017). Love me Tinder: Untangling emerging adults' motivations for using the dating application Tinder. *Telematics and Informatics, 34*(1), 67–78. doi:10.1016/j.tele.2016.04.009

Sun, L.P.(1996). Relations, social network and social structure. *Sociological Studies*, (5), 20-30.

Sundbo, J., & Darmer, P. (Eds.). (2008). *Creating experiences in the experience economy*. Edward Elgar Publishing. doi:10.4337/9781848444003

Sung, Y., Lee, J. A., Kim, E., & Choi, S. M. (2016). Why we post selfies: Understanding motivations for posting pictures of oneself. *Personality and Individual Differences, 97*, 260–265. doi:10.1016/j.paid.2016.03.032

Sun, H., Hou, J., Hu, X., & Al-mekhlafi, K. (2015). A Context-based Support System of Mobile Chinese Learning for Foreigners in China. *Procedia Computer Science, 60*(1), 1396–1405. doi:10.1016/j.procs.2015.08.215

Sun, L. P. (1993). Free flow resources and free activity space --- China's social structure changes during its reform. *Probe*, (1): 64–68.

Suominen, A., Hyrynsalmi, S., & Knuutila, T. (2014). Young mobile users: Radical and individual–Not. *Telematics and Informatics, 31*(2), 266–281. doi:10.1016/j.tele.2013.08.003

Su, Z., Xu, Q., Zhu, H., & Wang, Y. (2015). A novel design for content delivery over software defined mobile social networks. *IEEE Network, 29*(4), 62–67. doi:10.1109/MNET.2015.7166192

Suzman, R., & Beard, J. (2015). *Global Health and Aging. The troublesome concept of "technological affordance"*. Retrieved from https://www.who.int/ageing/publications/global_health.pdf

Syson, M., Estuar, M., & See, K. (2012). ABKD: Multimodal Mobile Language Game for Collaborative Learning of Chinese Hanzi and Japanese Kanji Characters. *2012 IEEE/WIC/ACM International Conferences on Web Intelligence and Intelligent Agent Technology (WI-IAT), 3*, 311-315.

Szpakow, A., Stryzhak, A., & Prokopowicz, W. (2011). Evaluation of threat of mobile phone – addition among Belarusian University students. *Progress in Health Sciences, 1*(2), 96–101.

Takao, M., Takahashi, S., & Kitamura, M. (2009). Addictive personality and problematic mobile phone use. *Cyberpsychology & Behavior, 12*(5), 501–507. doi:10.1089/cpb.2009.0022 PMID:19817562

Takavarasha, S. Jr, & Adams, C. (Eds.). (2018). *Affordability Issues Surrounding the Use of ICT for Development and Poverty Reduction*. IGI Global. doi:10.4018/978-1-5225-3179-1

Takeuchi, A., & Nagao, K. (1993, May). Communicative facial displays as a new conversational modality. In *Proceedings of the INTERACT'93 and CHI'93 Conference on Human Factors in Computing Systems* (pp. 187-193). ACM. 10.1145/169059.169156

Talbot, D. (2008). *Upwardly mobile*. Retrieved at https://www.technologyreview.com/s/411020/upwardly-mobile/

Tam & Yeung. (2010). Learning to write Chinese characters with correct stroke sequences on mobile devices. *Education Technology and Computer, 2010 2nd International Conference on, 4*, 395-399.

Tam, V., & Luo, N. (2012). Exploring Chinese through learning objects and interactive interface on mobile devices. *Proceedings of IEEE International Conference on Teaching, Assessment, and Learning for Engineering*, H3C7-C9. 10.1109/TALE.2012.6360350

Tam, V., & Luo, N. (2014). An Intelligent Mobile Application to Facilitate the Exploratory and Personalized Learning of Chinese on Smartphones. *Advanced Learning Technologies (ICALT), 2014 IEEE 14th International Conference on*, 411-412. 10.1109/ICALT.2014.123

Tan, W.-K. (2016). The relationship between smartphone usage, tourist experience and trip satisfaction in the context of a nature-based destination. *Telematics and Informatics, 34*(2), 614–627. doi:10.1016/j.tele.2016.10.004

Tatlow, D. K. (2013, July 24). *Apps offer Chinese a path to the forbidden*. Retrieved from http://www.nytimes.com/2013/07/25/world/asia/25iht-letter25.html(accessed July 3rd 2018)

Tavakolizadeh, J., Atarodi, A., Ahmadpour, S., & Pourgheisar, A. (2014). The Prevalence of Excessive Mobile Phone Use and its Relation With Mental Health Status and Demographic Factors Among the Students of Gonabad University of Medical Sciences in 2011 – 2012. *Razavi International Journal of Medicine, 2*(1), 1–7. doi:10.5812/rijm.15527

Temperton, J. (2017). Google's Pixel Buds aren't just bad, they're utterly pointless. *Wired*. Retrieved from http://www.wired.co.uk/article/pixel-buds-review-google

Teodoro, R., Ozturk, P., Naaman, M., Mason, W., & Lindqvist, J. (2014, February). The motivations and experiences of the on-demand mobile workforce. In *Proceedings of the 17th ACM conference on Computer supported cooperative work & social computing* (pp. 236-247). ACM. 10.1145/2531602.2531680

Terras, M. M., & Ramsay, J. (2016). Family digital literacy practices and children's mobile phone use. *Frontiers in Psychology, 7*. doi:10.3389/fpsyg.2016.01957

Terras, M. M., Ramsay, J., & Boyle, E. A. (2015). Digital media production and identity: Insights from a psychological perspective. *E-Learning and Digital Media, 12*(2), 128–146. doi:10.1177/2042753014568179

Terras, M. M., Yousaf, F., & Ramsay, J. (2016). The relationship between Parent and Child Digital Technology use. *Proceedings of the British Psychological Society Annual Conference*.

Thomas, F., Haddon, L., Gilligan, R., Heinzmann, P., & de Gournay, C. (2005). *Cultural factors shaping the experience of ICTs: An exploratory review. In International collaborative research. Cross-cultural differences and cultures of research* (pp. 13–50). Brussels: COST.

Thurman, N., & Schifferes, S. (2012). The future of personalization at news websites: Lessons from a longitudinal study. *Journalism Studies, 13*(5-6), 775–790. doi:10.1080/1461670X.2012.664341 timesofoman.com. (2018). *Oman's mobile phone subscriber base crosses 7 million*. Retrieved from: http://timesofoman.com/article/110822/Oman/Oman%27s-mobile-phone-subscriber-base-crosses-7-million

Tinder. (2017). Retrieved from https://www.gotinder.com/privacy

Toma, C. L. (2015). Online dating. In C. Berger & M. Roloff (Eds.), *The international encyclopedia of interpersonal communication* (pp. 1–5). Hoboken, NJ: John Wiley & Sons, Inc. Retrieved from https://onlinelibrary.wiley.com/doi/pdf/10.1002/9781118540190.wbeic118

Toma, C. L., Hancock, J. T., & Ellison, N. B. (2008). Separating fact from fiction: An examination of deceptive self-presentation in online dating profiles. *Personality and Social Psychology Bulletin, 34*(8), 1023–1036. doi:10.1177/0146167208318067 PMID:18593866

Torche, F. (2015). *Gender differences in intergenerational mobility in Mexico*. Available from: http://www.ceey.org.mx/sites/default/files/adjuntos/dt-011-2015_si.pdf

Torres-Hostench, O., Moorkens, J., O'Brien, S., & Vreeke, J. (2017). Testing interaction with a mobile MT postediting app. *Translation and Interpreting, 9*(2), 138–150. doi:10.12807/ti.109202.2017.a09

Traxler, J. (2007). Defining, discussing and evaluating mobile learning. *International Review of Research in Open and Distance Learning, 8*(2). doi:10.19173/irrodl.v8i2.346

Trivers, R. (1972). Parental investment and sexual selection. In B. Campell (Ed.), *Sexual selection and the descent of man: 1871–1971* (pp. 136–179). Chicago: Aldine-Atherton.

Trujillo, C. A., Barrios, A., Camacho, S. M., & Rosa, J. A. (2010). Low socioeconomic class and consumer complexity expectations for new product technology. *Journal of Business Research, 63*(6), 538–547. doi:10.1016/j.jbusres.2009.05.010

Tseng, C. C., Lu, C. H., & Hsu, W. L. (2007). A mobile environment for Chinese language learning. In *Symposium on Human Interface and the Management of Information* (pp. 485–489). Springer. 10.1007/978-3-540-73354-6_53

Turner, R. (2003). *Desire and the Self*. Retrieved from http://robin.bilkent.edu.tr/desire/desire3.pdf

Tussyadiah, I. P., & Wang, D. (2016). Tourists' Attitudes toward Proactive Smartphone Systems. *Journal of Travel Research, 55*(4), 493–508. doi:10.1177/0047287514563168

Tussyadiah, I. P., & Zach, F. J. (2012). The role of geo-based technology in place experiences. *Annals of Tourism Research, 39*(2), 780–800. doi:10.1016/j.annals.2011.10.003

Uhlenberg, P. (Ed.). (2009). *International handbook of population aging* (Vol. 1). Springer Science & Business Media. doi:10.1007/978-1-4020-8356-3

United States Department of Agriculture. (2016). *Food expenditures*. Retrieved from https://www.ers.usda.gov/data-products/food-expenditures/food-expenditures/#Expenditures%20on%20food%20and%20alcoholic%20beverages%20that%20were%20consumed%20at%20home%20by%20selected%20countries

Väänänen-Vainio-Mattila, K., & Väätäjä, H. (2008, September). *Towards a life cycle framework of mobile service user experience*. In *2nd MIUX Workshop at MobileHCI*, Amsterdam, The Netherlands.

Väätäjä, H., & Roto, V. (2010, April). Mobile questionnaires for user experience evaluation. In CHI'10 Extended Abstracts on Human Factors in Computing Systems (pp. 3361-3366). ACM.

Väätäjä, H., Männistö, A., Vainio, T., & Jokela, T. (2009). Understanding user experience to support learning for mobile journalist's work. *The Evolution of Mobile Teaching and Learning*, 177-210.

Väätäjä, H. (2008, September). Factors affecting user experience in mobile systems and services. In *Proceedings of the 10th international conference on Human computer interaction with mobile devices and services* (pp. 551-551). ACM.

Väätäjä, H. (2010, November). User experience evaluation criteria for mobile news making technology: findings from a case study. In *Proceedings of the 22nd Conference of the Computer-Human Interaction Special Interest Group of Australia on Computer-Human Interaction* (pp. 152-159). ACM.

Väätäjä, H. (2010, November). User experience of smart phones in mobile journalism: early findings on influence of professional role. In *Proceedings of the 22nd Conference of the Computer-Human Interaction Special Interest Group of Australia on Computer-Human Interaction* (pp. 1-4). ACM.

Väätäjä, H., & Egglestone, P. (2012, February). Briefing news reporting with mobile assignments: perceptions, needs and challenges. In *Proceedings of the ACM 2012 conference on Computer Supported Cooperative Work* (pp. 485-494). ACM.

Väätäjä, H., Koponen, T., & Roto, V. (2009, September). Developing practical tools for user experience evaluation: a case from mobile news journalism. In *European Conference on Cognitive Ergonomics: Designing beyond the Product--Understanding Activity and User Experience in Ubiquitous Environments* (p. 23). VTT Technical Research Centre of Finland.

Väätäjä, H., & Männistö, A. A. (2010, October). Bottlenecks, usability issues and development needs in creating and delivering news videos with smartphones. In *Proceedings of the 3rd workshop on Mobile video delivery* (pp. 45-50). ACM. 10.1145/1878022.1878034

Vaidya. (2016). *Mobile phone usage among youth*. Available from: https://www.researchgate.net/publication/299540610_Mobile_Phone_Usage_among_Youth

Valderrama, J. A. (2014). *Running head: Problematic Smartphone use scale development and validation of the problematic Smartphone use scale* (Unpublished PhD dissertation). Alliant International University, San Francisco, CA.

van Den Eijnden, R. J., Spijkerman, R., Vermulst, A. A., van Rooij, T. J., & Engels, R. C. (2010). Compulsive Internet use among adolescents: Bidirectional parent–child relationships. *Journal of Abnormal Child Psychology, 38*(1), 77–89. doi:10.100710802-009-9347-8 PMID:19728076

van Velsen, L., Beaujean, D. J., & van Gemert-Pijnen, J. E. (2013). Why mobile health app overload drives us crazy, and how to restore the sanity. *BMC Medical Informatics and Decision Making, 13*(1), 1. doi:10.1186/1472-6947-13-23 PMID:23399513

Vanden Abeele, M. M. (2016). Mobile youth culture: A conceptual development. *Mobile Media & Communication, 4*(1), 85–101. doi:10.1177/2050157915601455

Venkatesh, V., Brown, S. A., & Bala, H. (2013). Bridging the qualitative-quantitative divide: Guidelines for conducting mixed methods research in information systems. *Management Information Systems Quarterly, 37*(1), 21–54. doi:10.25300/MISQ/2013/37.1.02

Ventola, C. L. (2014). Mobile devices and apps for health care professionals: Uses and benefits. *P&T, 39*(5), 356. PubMed

Verclas, K., & Mechael, P. (2008). *A mobile voice: The use of mobile phones in citizen media. MobileActive. Org, Pact.* USAID.

Verma, A. (2011). *A new look for Google Translate for Android*. Retrieved October 25, 2017, from https://googleblog.blogspot.co.uk/2011/01/new-look-for-google-translate-for.html

Viberg, O., & Grönlund, Å. (2013). Cross-cultural analysis of users' attitudes toward the use of mobile devices in second and foreign language learning in higher education: A case from Sweden and China. *Computers & Education, 69*, 169–180. doi:10.1016/j.compedu.2013.07.014

Villalobos, O., Lynch, S., DeBlieck, C., & Summers, L. (2017). Utilization of a Mobile App to Assess Psychiatric Patients With Limited English Proficiency. *Hispanic Journal of Behavioral Sciences, 39*(3), 369–380. doi:10.1177/0739986317707490

Vincent, J. (2006). Emotional attachment and mobile phones. *Knowledge, Technology & Policy, 19*(1), 39–44. doi:10.100712130-006-1013-7

Vincent, J. (2015). *Mobile Opportunities: Exploring Positive Mobile Opportunities for European Children, POLIS*. London: The London School of Economics and Political Science.

Vollmer, S. (2017). Syrian newcomers and their digital literacy practices. *Language Issues: The ESOL Journal, 28*(2), 66–72.

Von Ahn, L. (2011). *Massive-scale online collaboration*. Retrieved May 10, 2018, from https://www.ted.com/talks/luis_von_ahn_massive_scale_online_collaboration#t-529477

Vygotsky, L. S. (1978). *Mind in society: The development of higher psychological processes*. Harvard University Press.

Wahlster, W. (2013). *Verbmobil: foundations of speech-to-speech translation: Springer Science & Business Media*. Berlin: Springer.

Walrave, M., Heirman, W., & Hallam, L. (2014). Under pressure to sext? Applying the theory of planned behaviour to adolescent sexting. *Behaviour & Information Technology, 33*(1), 86–98. doi:10.1080/0144929X.2013.837099

Walsh, S. P., & White, K. M. (2007). Me, my mobile, and I: The role of self- and prototypical identity influences in the prediction of mobile phone behavior. *Journal of Applied Social Psychology, 37*(10), 2405–2434. doi:10.1111/j.1559-1816.2007.00264.x

Walsh, S. P., White, K. M., Cox, S., Young, R., & Mc, D. (2011). Keeping in constant touch: The predictors of young Australians' mobile phone involvement. *Computers in Human Behavior, 27*(1), 333–342. doi:10.1016/j.chb.2010.08.011

Walsh, S. P., White, K. M., & Young, R. M. (2007, July). Young and connected: Psychological influences of mobile phone use amongst Australian youth. In *Proceedings Mobile Media* (pp. 125-134), University of Sydney.

Wang, Y., Ji, Y., Zhang, C., & Sun, L. (2010). An approach and implementation of Chinese character learning based on Mobile Game-Based Learning. *Network Infrastructure and Digital Content, 2010 2nd IEEE International Conference on*, 169-173. 10.1109/ICNIDC.2010.5657845

Wang, D., Park, S., & Fesenmaier, D. R. (2012). The Role of Smartphones in Mediating the Touristic Experience. *Journal of Travel Research, 51*(4), 371–387. doi:10.1177/0047287511426341

Wang, D., Xiang, Z., & Fesenmaier, D. R. (2014). Adapting to the mobile world: A model of smartphone use. *Annals of Tourism Research, 48*, 11–26. doi:10.1016/j.annals.2014.04.008

Wang, D., Xiang, Z., & Fesenmaier, D. R. (2016). Smartphone Use in Everyday Life and Travel. *Journal of Travel Research, 55*(1), 52–63. doi:10.1177/0047287514535847

Wang, J., & Leland, C. H. (2012). Exploring Mobile Technologies for Learning Chinese. *Journal of the National Council of Less Commonly Taught Languages, 12*, 133–159.

Wang, Y. H. (2016). Could a mobile-assisted learning system support flipped classrooms for classical Chinese learning? *Journal of Computer Assisted Learning, 32*(5), 391–415. doi:10.1111/jcal.12141

Watkins, J., Hjorth, L., & Koskinen, I. (2012). Wising up: Revising mobile media in an age of smartphones. *Continuum*, *26*(5), 665–668. doi:10.1080/10304312.2012.706456

Watson, C., McCarthy, J., & Rowley, J. (2013). Consumer attitudes towards mobile marketing in the smart phone era. *International Journal of Information Management*, *33*(5), 840–849. doi:10.1016/j.ijinfomgt.2013.06.004

Way, A. (2013). Traditional and Emerging Use-Cases for Machine Translation. Proceedings of Translating and the Computer, 35.

Wei, P. S., & Lu, H. P. (2014). Why do people play mobile social games? An examination of network externalities and of uses and gratifications. *Internet Research*, *24*(3), 313–331. doi:10.1108/IntR-04-2013-0082

Wei, R., & Lo, V.-H. (2006). Staying connected while on the move: Cell phone use and social connectedness. *New Media & Society*, *8*(1), 53–72. doi:10.1177/1461444806059870

Weiss, R. S., & Bass, S. A. (2002). *Challenges of the third age: Meaning and purpose in later life*. Oxford University Press.

Wenger, D., Owens, L., & Thompson, P. (2014). Help Wanted Mobile Journalism Skills Required by Top US News Companies. *Electronic News*, *8*(2), 138–149. doi:10.1177/1931243114546807

Westlund, O. (2007). The adoption of mobile media by young adults in Sweden. In Mobile Media 2007 (pp. 116-124). The University of Sydney.

Westlund, O. (2014). The production and consumption of news in an age of mobile media. *The Routledge companion to mobile media*, 135-145.

Westlund, O. (2013). Mobile news. *Digital Journalism*, *1*(1), 6–26. doi:10.1080/21670811.2012.740273

Westlund, O. (2015). News Consumption in an Age of Mobile Media: Patterns, People, Place and Participation. *Mobile Media & Communication*, *3*(2), 151–159. doi:10.1177/2050157914563369

Westlund, O., & Quinn, S. (2018). Mobile Journalism and MoJos. In H. Örnebring (Ed.), *Oxford Research Encyclopedia of Communication*. Oxford, UK: Oxford University Press. doi:10.1093/acrefore/9780190228613.013.841

White, N. R., & White, P. B. (2007). Home and away: Tourists in a Connected World. *Annals of Tourism Research*, *34*(1), 88–104. doi:10.1016/j.annals.2006.07.001

Whiting, A., & Williams, D. (2013). Why people use social media: A uses and gratifications approach. *Qualitative Market Research*, *16*(4), 362–369. doi:10.1108/QMR-06-2013-0041

Wigelius, H., & Väätäjä, H. (2009). Dimensions of context affecting user experience in mobile work. In *Human-Computer Interaction–INTERACT 2009* (pp. 604–617). Springer Berlin Heidelberg. doi:10.1007/978-3-642-03658-3_65

Williams, A. A., & Marquez, B. A. (2015). Selfies| The Lonely Selfie King: Selfies and the Conspicuous Prosumption of Gender and Race. *International Journal of Communication*, *9*, 13.

Winner, L. (2004). Technology as Forms of Life. In Readings in the Philosophy of Technology (pp. 103-113). Rowman & Littlefield Publishers, Inc.

Wolf, C., & Hohlfeld, R. (2012). Revolution in Journalism? In Images in Mobile Communication (pp. 81-99). VS Verlag für Sozialwissenschaften. doi:10.1007/978-3-531-93190-6_5

Woll, S. B., & Cozby, C. P. (1987). Video-dating and other alternatives to traditional methods of relationship initiation. In W. H. Jones & D. Perlman (Eds.), Advances in personal relationships (Vol. 1, pp. 69–108). Greenwich, CT: JAI.

Wong, L. H., Boticki, I., Sun, J., & Looi, C. K. (2011). Improving the scaffolds of a mobile-assisted Chinese character forming game via a design-based research cycle. *Computers in Human Behavior, 27*(5), 1783–1793. doi:10.1016/j. chb.2011.03.005

Wong, L. H., Chin, C. K., Tan, C. L., Liu, M., & Gong, C. (2010). Students' meaning making in a mobile assisted Chinese idiom learning environment. In *Proceedings of the 9th International Conference of the Learning Sciences-Volume 1*(pp. 349-356). International Society of the Learning Sciences.

Wong, L. H., Hsu, C. K., Sun, J., & Boticki, I. (2013). How flexible grouping affects the collaborative patterns in a mobile-assisted Chinese character learning game? *Journal of Educational Technology & Society, 16*(2), 174–187.

Wong, L., Chin, C., Tan, C., & Liu, M. (2010). Students' Personal and Social Meaning Making in a Chinese Idiom Mobile Learning Environment. *Journal of Educational Technology & Society, 13*(4), 15–26.

Wong, L., & Looi, C. (2011). What seams do we remove in mobile-assisted seamless learning? A critical review of the literature. *Computers & Education, 57*(4), 2364–2381. doi:10.1016/j.compedu.2011.06.007

World Health Organization. (2018). *Ageing and health.* World Health Organization. Retrieved from http://www.who.int/news-room/fact-sheets/detail/ageing-and-health

World Travel & Tourism Council. (2017). *Travel & Tourism. Global Economic Impact & Issues 2017.* London: WTTC.

Worsley, L. (2015). Dating apps have killed romance, says historian. *Times.* Retrieved from https://www.bbc.com/news/technology-34455738

Wu, H. (2016). Elderly people and the Internet: a demographic reconsideration. In M. Keane (Ed.), Handbook of the cultural and creative industries in China (pp. 431–444). Cheltenham, UK: Edward Elgar Publishing. doi:10.4337/9781 782549864.00041.

Wu, Yuan, Zhou, & Cai. (2013). A Mobile Chinese Calligraphic Training System Using Virtual Reality Technology. *AASRI Procedia, 5,* 200-208.

Wunmi, B., & Rob, M. (2018). *Gartner says worldwide sales of smartphones recorded first ever decline during the fourth quarter of 2017.* Gartner, Inc. Available from: https://www.gartner.com/newsroom/id/3859963

Wu, W., Wu, Y. J., Chen, C., Kao, H., Lin, C., & Huang, S. (2012). Review of trends from mobile learning studies: A meta-analysis. *Computers & Education, 59*(2), 817–827. doi:10.1016/j.compedu.2012.03.016

Xiang, Z., & Gretzel, U. (2010). Role of social media in online travel information search. *Tourism Management, 31*(2), 179–188. doi:10.1016/j.tourman.2009.02.016

Xiang, Z., Gretzel, U., & Fesenmaier, D. R. (2009). Semantic Representation of Tourism on the Internet. *Journal of Travel Research, 47*(4), 440–453. doi:10.1177/0047287508326650

Xu, X. (2016). *Mobile Studies International.* Available at http://msi.wiki

Xu, X. (2017). Comparing Mobile Experience. In Handbook of Human-Computer Interaction. Hoboken, NJ: Wiley. doi:10.1002/9781118976005

Xu, X. (Ed.). (2014). Interdisciplinary mobile media and communications: Social, political and economic implications. Hershey, PA: IGI Global.

Xu, X. (Ed.). (2016). Handbook of research on human social interaction in the age of mobile devices. Hershey, PA: IGI Global.

Xu, Q., & Peng, H. (2017). Investigating mobile-assisted oral feedback in teaching Chinese as a second language. *Computer Assisted Language Learning*, *30*(3-4), 173–182. doi:10.1080/09588221.2017.1297836

Xu, X. (2018). Comparing mobile experience. In K. Norman & K. Kirakowski (Eds.), *Wiley Handbook of Human-Computer Interaction Set* (Vol. 1, pp. 225–238). Wiley.

Yang, K. C. (2018). Understanding How Mexican and US Consumers Decide to Use Mobile Social Media: A Cross-National Qualitative Study. In Multi-Platform Advertising Strategies in the Global Marketplace (pp. 168-198). IGI Global.

Yang, B., & Zhao, X. (2018). TV, Social Media, and College Students' Binge Drinking Intentions: Moderated Mediation Models. *Journal of Health Communication*, *23*(1), 61–71. doi:10.1080/10810730.2017.1411995 PMID:29265924

Yang, C. H., Maher, J. P., & Conroy, D. E. (2015). Implementation of behavior change techniques in mobile applications for physical activity. *American Journal of Preventive Medicine*, *48*(4), 452–455. doi:10.1016/j.amepre.2014.10.010 PMID:25576494

Yang, C., & Xie, Y. (2013). Learning Chinese idioms through iPads. *Language Learning & Technology*, *17*(2), 12–22.

Yang, K., & Kim, H. Y. (2012). Mobile shopping motivation: An application of multiple discriminant analysis. *International Journal of Retail & Distribution Management*, *40*(10), 778–789. doi:10.1108/09590551211263182

Yang, M., Li, Y., Jin, D., Zeng, L., Wu, X., & Vasilakos, A. V. (2015). Software-defined and virtualized future mobile and wireless networks: A survey. *Mobile Networks and Applications*, *20*(1), 4–18. doi:10.100711036-014-0533-8

Ying, Y., Lin, X., & Mursitama, T. N. (2017). Mobile learning based of Mandarin for college students: A case study of international department' sophomores. *Information & Communication Technology and System (ICTS), 2017 11th International Conference on*, 281-286.

Ying, Y., Rawendy, D., & Arifin, Y. (2016). Game education for learning Chinese language with mnemonic method. *Information Management and Technology (ICIMTech), International Conference on*, 171-175. 10.1109/ICIMTech.2016.7930324

Yu, G.M. (2009). Media revolution: From panorama prison to shared-scene prison. *People's Tribune*, *8*(1), 21.

Yu, G.M. (2016). Reconstruction of media influence under the paradigm of relationship empowerment. *News and Writing*, (7), 47-51.

Yu, G.M., & Ma, H. (2016). New power paradigm in digital era: Empowerment based on relation network in social media --- Social relation reorganization and power pattern dynamics. *Chinese Journal of Journalism & Communication*, (10), 6-27.

Yu, G.M., & Ma, H. (2016). Relationship empowerment: a new paradigm of social capital allocation --- The logical change of social governance under network reconstruction of social connection. *Editorial Friend*, (9), 5-8.

Yu, G.M., Zhang, C., Li, S., Bao, L.Y., & Zhang, S.N. (2015). The era of individual activation: Reconstruction of communication ecology under the logic of the Internet. *Modern Communication*, (5), 1-4.

Yu, G. (2016). *Social currency- The road to business monetizing in the era of mobile social networking*. Posts and Telecom Press.

Zapata, P. (2016). Translating On the Go? Investigating the Potential of Multimodal Mobile Devices for Interactive Translation Dictation. *Revista Tradumàtica: tecnologies de la traducció*, (14), 66-74.

Zarbatany, L., Hartmann, D. P., & Rankin, D. B. (1990). The psychological functions of preadolescent peer activities. *Child Development*, *61*(4), 1067–1080. doi:10.2307/1130876 PMID:2209178

Zenith. (2017). *Smartphone penetration to reach 66% in 2018*. Retrieved from https://www.zenithmedia.com/smartphone-penetration-reach-66-2018/

Zhang, J., & Yasseri, T. (2016). *What Happens After You Both Swipe Right: A Statistical Description of Mobile Dating Communications*. arXiv preprint arXiv:1607.03320

Zhang, L., Wang, J., & Li, H. (2011). A Mobile Learning System for learning Mandarin Pronunciation. *Computer Science and Network Technology (ICCSNT), 2011 International Conference on, 1*, 621-624. 10.1109/ICCSNT.2011.6182034

Zhang, L. (2009). *The research of the western attention economy school*. China Social Sciences Press.

Zhao, Z., & Balagué, C. (2015). Designing branded mobile apps: Fundamentals and recommendations. *Business Horizons, 58*(3), 305–315. doi:10.1016/j.bushor.2015.01.004

Zhu, J., Zhang, W., Yu, C., & Bao, Z. (2015). Early adolescent Internet game addiction in context: How parents, school, and peers impact youth. *Computers in Human Behavior, 50*, 159–168. doi:10.1016/j.chb.2015.03.079

# About the Contributors

**Xiaoge Xu**, Ph.D., is the editor-in-chief of *Advances in Wireless Technologies and Telecommunication* Book Series of IGI Global. He is the founder of *Mobile Studies International, Creative Industries International*, and *Research Labs International*. He has been passionately advocating, promoting, and conducting comparative and interdisciplinary studies of the mobile, creative and experience industries in the context of experience economy. He is currently teaching at the School of International Communications, University of Nottingham Ningbo China. Before rejoining the China campus of University of Nottingham in July 2018, he taught at Xiamen University Malaysia, Botswana International University of Science and Technology, University of Nottingham Ningbo China, Nanyang Technological University (Singapore), and China School of Journalism in Beijing. He is a grandpa of two lovely grandchildren, Peggy Xu and Jonny Xu. Largely based in Singapore, he enjoys travelling and training around the world.

\* \* \*

**Hafidha AlBarashdi** (Ph.D.) is currently working as a Research and Statistics Specialist at The Research Council. Her publications include more than (22) research papers and (6) conference papers. As a membership in the Omani Identity and Heritage Preservation Team of the Future Vision of Oman 2040, she was honored twice on the SQU University Day for Excellence in Research. Her research interests include technology, employment, motivation, national identity and others.

**Belem Barbosa** (Ph.D.) is Adjunct Professor at the University of Aveiro and member of GOVCOPP, the research unit on Governance, Competitiveness and Public Policy. She received her PhD in Business and Management Studies – specialisation in Marketing and Strategy from the University of Porto, Portugal. Her research interests lie primarily in the area of internet marketing and consumer behaviour, including word-of-mouth communication, sustainability marketing, tourism and events marketing.

**Abdelmajid Bouazza** (Ph.D.) is a Professor in the Department of Information Studies, Sultan Qaboos University. He received a PhD in Information Science from the University of Pittsburgh, USA, and a Master's in Information Science from the University of Montreal, Canada. Prof. Bouazza Published 61 papers in international and regional refereed journals and Prepared, translated, and contributed to a number of specialized books in Information Science and with articles published in the Encyclopedia of Library and Information Science, Marcel Dekker Foundation in New York and in Encyclopedie Internationale de Bibliologie, Paris. He has presented 68 papers at national, regional and international conferences. His research interests include Knowledge Management, Internet and Smartphones addiction, Open access, Education of information professionals and the job market needs.

**Dayana Pinzón Callejas** is an Economist and Professional in Finance and International Bussines from Universidad Tecnológica de Bolívar (Colombia). She currently works as Project Research Assistant at Universidad Tecnológica de Bolívar (Colombia). Her research interests include war tourism and mobile tourism.

**Biplab Loho Choudhury** (Ph.D.) is Professor in the Centre for Journalism & Mass Communication, Visva-Bharati at Santiniketan, India from 2011. He went to Hungary, Germany, China, Hongkong, Thailand, Malaysia, Indonesia, Bangladesh and UAE during last few years for academic work including Visiting professorship, research meeting, presenting papers, giving key note address, plenary address, chairing sessions and field visit . His singly authored books include Media Organization Management (2008, Unique, Kolkata),Indian Paradigm of Development Man Standard and Communication (2011, Sampark, Kolkata & Delhi) and NaboMadhyamer Ruprekha (An Outline of Newmedia in Bengali,West Bengal State Book Board,2013).His last edited volumes are Demonetisation: A Historical Review (2018, NDP), Media and Communication Practices and Issues (SBE,2014). His experience includes working as ICDS Project Officer (1990-1996), Superintendent of Observation home, journalist in two English dailies, lecturer in Assam University Silchar (1996-2005), and as faculty (since 2005) in Visva-Bharati at Santiniketan. He worked as Principal Investigator and Project Coordinator in Assam University Rural Communication Project (1998-2001), Media Convener in Assam University (2000-2005), and consultant to tea sector and NGOs in Communication and project management. His current research interest spans audience-creator matchmaking, developing community from within, Indian Communication and Research Traditions, and Policy review.

**Beliz Dönmez** was born in 1995 in Samsun. She was graduated from Information Systems and Technologies Department of Yeditepe University. Her thesis was investigating undergraduate students' gender-based habits on mobile use. She completed her research at the end of spring 2018 under supervision of Assist. Prof. Dr. Cagla Seneler. Now, she is working as Associate Business Intelligence Consultant in a privately owned IT Consultancy Company.

**Katie Ellis** is associate professor and senior research fellow in Internet Studies at Curtin University. She holds an Australian Research Council Discovery Early Career Research award for a project on disability and digital televisions and is series editor of Routledge Research in Disability and Media Studies. Her current projects include co-editing The Routledge Companion to Disability and Media (2019) with Gerard Goggin and Beth Haller, and Manifestos for the Future of Critical Disability Studies (Routledge, 2019) with Rosemarie Garland Thomson, Mike Kent and Rachel Robertson.

**Sadia Jamil** (Ph.D.) has completed her PhD in Journalism at the University of Queensland, Australia. She holds postgraduate degrees in the disciplines of Media Management (Scotland) and Mass Communication (Karachi). To date, she is the recipient of a number of awards and scholarships including: The University of Queensland's Centennial Award, UQ's International Postgraduate Research Support Award, the Norwegian UNESCO Commissions' conference grants (2015-2018), Union Insurance's Cairo Air Crash Journalists Victim Memorial Gold Medal and Daily Jang's and The News' Sardar Ali Sabri Memorial Gold Medal. She is affiliated with the International Association of Media and Communication

Research (IAMCR) and currently acting as the Co Vice-Chair of IAMCR's Journalism Research and Education Section. Her research work includes studies into journalism, safety of journalists and impunity, freedom of expression and press freedom. For 2018 and 2019, Dr Jamil's research work focuses on safety of journalists, digital divide and social inequalities, urban studies and impacts of mobile phone and new media technologies (in relation to journalism practice and broader socio-political aspects).

**Michael Keane** (Ph.D.) is Professor of Chinese Media and Communications at Curtin University, Perth. He is Program Leader of the Digital China Lab. within the Centre for Culture and Technology. He has many long-standing research contacts in China and is a co-organiser, with colleagues from the Chinese Academy of Social Sciences, of the bi-annual U40 China-Australia Summer School. Prof Keane's key research interests are digital transformation in China; East Asian cultural and media policy; television in China, and creative industries and cultural export strategies in China and East Asia. Prof Keane is the author or editor of 16 books since 2002, over 150 peer-reviewed articles, and has been recipient of 5 Australian Research Council Discovery grants since 2003. He is an editorial board member of a number of leading scholarly journals. In 2018, he was appointed the editor of a new book series called 'Digital China' with Anthem Press (UK). Prof Keane is well known internationally and is frequently invited to present at conferences. He was an Eastern Scholar with the Shanghai Theatre Academy from 2011- 2013 and a Distinguished Visiting Professor at Hong Kong Open University from 2016 – 2017.

**Carolina Barrios Laborda** (Ph.D.) is an industrial engineer from Universidad de los Andes (Colombia) and Docteur en Gestion de ENSAM de Paris (now ParisTech), France. She currently teaches marketing related courses as well as decision making under risk at Universidad Tecnológica de Bolivar (Colombia).

**Wan Chi Leung** (Ph.D.) is a lecturer in the Department of Media and Communication and the Coordinator of the Master of Strategic Communication programme at University of Canterbury, New Zealand. Her research interests include new media effects, health campaigns and communication, entertainment, and social network analysis.

**Wendi (Wendy) Li** is a PhD student in the School of Culture and Communication at the University of Melbourne. She completed her master's degree in global media communication at the same university in 2017. She also holds an MA in Linguistics from the Chinese University of Hong Kong and a BA in English from the Beijing International Studies University. Her research focuses on civic capacity and identity in everyday political engagement in the field of global public communication and, increasingly, issues related to youth, communication, and global risk politics. Her current research project explores young citizens' deliberation on climate change in Hong Kong.

**Nancy Xiuzhi Liu** (Ph.D.) received her MA in Translation and Interpreting from the University of Newcastle-upon-Tyne in 2003 and her PhD in International Communication from the University of Nottingham in 2015. Since 2008, she has been working at the University of Nottingham Ningbo China as Assistant Professor in translation and interpreting between mandarin and English. She writes and presents widely on issues of media text translation, sociocultural meanings of translation, translation of texts with cultural-specific items and pedagogy in translation and interpretation.

**Danielle McKain** (Ph.D.), earned her Ph.D. in Instructional Management and Leadership from Robert Morris University. She holds a Master of Science degree in Instructional Leadership with a concentration in Mathematics Education and a Bachelor of Science degree in Applied Mathematics. Her research interests include real world math, test preparation, math anxiety, remedial math, and preparing future teachers.

**Natália Menezes** has a MSc in Marketing by the University of Aveiro. Her main research interests are mobile marketing and smart tourism.

**Judith Ramsay** (Ph.D.) is a Chartered Psychologist, Associate Fellow of the British Psychological Society and Senior Lecturer in Psychology at Manchester Metropolitan University, UK. Her research interests lie in the psychology of mobile and distance learning, and more generally in the psychology of human-computer interaction.

**Cagla Seneler** (Ph.D.) is an Assistant Professor in Management Information Systems (MIS) Department at Yeditepe University, Turkey. She completed her doctorate in Computer Science Department at University of York, UK and MA and BS degree in MIS Department at Bogazici University, Turkey. She started her career in a privately owned enterprise resources planning (ERP) company as a software engineer. Then, she has resigned to continue her academic career at Bogazici University and worked as a Research Assistant in MIS Department for four years. During her PhD, she gave lectures as a part-time instructor both in Yeditepe and Bogazici Universities. She has publications that are cited by many academics and one of her publications awarded as Outstanding Paper Award Winner at the Literati Network Awards for Excellence 2011. Her research interests are human-computer interaction, user website experience, learning styles, cultural differences, personalization, user interface characteristics, technology adoption, digital plagiarism and IoT.

**Anurupa B. Singh** (Ph.D.) is Associate Professor and Mentoring head with Amity Business School, Noida. She has more than 20 years of experience in teaching and corporate. She has presented many research papers in national and International conferences and has published many in renowned journals indexed in Scopus database. She has guided many Ph. D. scholars in area of Marketing.

**Lu Sun** (Ph.D. Candidate) is studying in School of Journalism at Communication University of China. She is also a visiting scholar at State University of New York (SUNY) at Buffalo (2017.10-2018.10). Sun conducts research in international communication, emerging media and journalism. She had intern experiences in CCTV(China Central Television), CNR(China National Radio) and CGTN(China Global Television Network). Her publications include more than 18 research papers and 6 conference papers. She is also a member of IAMCR in international communication section.

**Pooja Sehgal Tabeck** is assistant Professor at Amity Business School, Noida. A post graduate in Management with Marketing specialization and NET qualified. Her area of interest is Bottom of Pyramid Marketing, consumer behavior and Retail.

**Melody M. Terras** (Ph.D.) is a Chartered Psychologist, Associate Fellow of the British Psychological Society and Lecturer in Psychology at the University of the West of Scotland, UK. Her research interests include the psychology of mobile and e-learning, and the inclusion of individuals with learning difficulties and disabilities in educational and health contexts.

**Anan Wan** (Ph.D.) is an assistant professor in the Department of Communication at Georgia College & State University. Wan conducts research in advertising, media technology, social media, and media psychology.

**Matthew Watts** (Ph.D. student) received his MA in Translation Studies from the University of Nottingham in 2017 and is currently a PhD student in Translation Studies in the Department of Modern Languages and Cultures at the University of Nottingham. His research focuses on mobile translation applications, exploring how they are affecting Translation Studies and the translation industry, as well as how users are actually engaging with these apps and implications for the language barrier.

**Li Zhenhui** (Ph.D.). Associate professor and vice dean of the Faculty of Economics and Management, Communication University of China. Her research interests are media economics and media education. She has published research articles on media economics and media education in China. Her recent research projects focus on the trends and development of cultural development in the new age, the impact of OTT TV, strategies for creative talents cultivation, and use of digital resources for teaching.

# Index

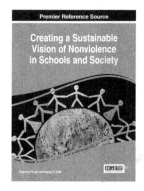

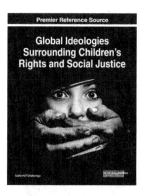
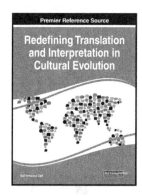

Ensure Quality Research is Introduced to the Academic Community

# Become an IGI Global Reviewer for Authored Book Projects

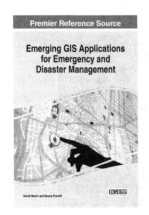

## The overall success of an authored book project is dependent on quality and timely reviews.

In this competitive age of scholarly publishing, constructive and timely feedback significantly expedites the turnaround time of manuscripts from submission to acceptance, allowing the publication and discovery of forward-thinking research at a much more expeditious rate. Several IGI Global authored book projects are currently seeking highly qualified experts in the field to fill vacancies on their respective editorial review boards:

### Applications may be sent to:
development@igi-global.com

Applicants must have a doctorate (or an equivalent degree) as well as publishing and reviewing experience. Reviewers are asked to write reviews in a timely, collegial, and constructive manner. All reviewers will begin their role on an ad-hoc basis for a period of one year, and upon successful completion of this term can be considered for full editorial review board status, with the potential for a subsequent promotion to Associate Editor.

If you have a colleague that may be interested in this opportunity,
we encourage you to share this information with them.

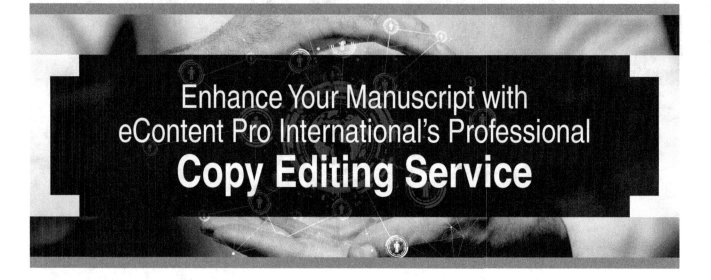

# Are You Ready to Publish Your Research?

**IGI Global**
DISSEMINATOR of KNOWLEDGE

IGI Global offers book authorship and editorship opportunities across 11 subject areas, including business, healthcare, computer science, engineering, and more!

## Benefits of Publishing with IGI Global:

- Free one-to-one editorial and promotional support.

- Expedited publishing timelines that can take your book from start to finish in less than one (1) year.

- Choose from a variety of formats including: Edited and Authored References, Handbooks of Research, Encyclopedias, and Research Insights.

- Utilize IGI Global's eEditorial Discovery® submission system in support of conducting the submission and blind-review process.

- IGI Global maintains a strict adherence to ethical practices due in part to our full membership to the Committee on Publication Ethics (COPE).

- Indexing potential in prestigious indices such as Scopus®, Web of Science™, PsycINFO®, and ERIC – Education Resources Information Center.

- Ability to connect your ORCID iD to your IGI Global publications.

- Earn royalties on your publication as well as receive complimentary copies and exclusive discounts.

Get Started Today by Contacting the Acquisitions Department at:

**acquisition@igi-global.com**

Printed in the United States
By Bookmasters